DYNAMICS OF STELLAR SYSTEMS

DYNAMICS OF STELLAR SYSTEMS

BY

K. F. OGORODNIKOV

TRANSLATED FROM THE RUSSIAN

BY

J. B. SYKES

TRANSLATION EDITED BY

ARTHUR BEER

A Pergamon Press Book

THE MACMILLAN COMPANY

NEW YORK

1965

THE MACMILLAN COMPANY
60 Fifth Avenue
New York 11, N.Y.

This book is distributed by
THE MACMILLAN COMPANY
pursuant to a special arrangement with
PERGAMON PRESS LIMITED
Oxford, England

Library of Congress Catalog Card Number 63-20875

This book is a translation, with author's revision and additions,
of *Динамика звездных систем (Dinamika zvezdnykh sistem)*
published by Fizmatgiz, Moscow

CONTENTS

PREFACE TO THE ENGLISH EDITION

THE English edition of this book is a practically unaltered translation of the original Russian edition of 1958. Although several years have elapsed since the publication of the original, and a great deal of work on the subject has been done in that time, there has not been an opportunity to revise the book, and in consequence the only changes have been the most necessary corrections and some minor alterations.

At the present time the interest of astronomers in the theoretical problems of stellar dynamics has substantially decreased and the number of workers in this field has appreciably diminished. Until a satisfactory theory of the evolution of stellar systems has been worked out, the growth of modern astronomy will always remain unbalanced. The lack of a proper theory has so far prevented the obtaining of a theoretical foundation for a feasible classification of galaxies which should be an adequate substitute for the classic scheme due to Hubble and largely based on the evolutionary ideas of Jeans. The author ventures to hope that the present book may give rise to at least a slight increase in the attention paid to this field of astronomy.

The author's thanks are due to Dr. Arthur Beer of the Cambridge University Observatories for his kindness in initiating the English translation, and to Pergamon Press for undertaking its publication; and most especially to Dr. J. B. Sykes, who not only made an excellent translation of the book but in the course of doing so offered very many helpful suggestions which led to the improvement of a number of obscure or imperfect passages in this edition.

Leningrad University K. F. OGORODNIKOV

PREFACE TO THE RUSSIAN EDITION

THIS monograph is based on lectures given by the author since 1932, first at the University of Moscow and afterwards at the University of Leningrad.

In order to limit the size of the book, the author has made no attempt at a detailed account of the present state of every department of stellar dynamics, nor has he sought to retain the historical sequence of development of the subject, since this would have obstructed the principal purpose of the book. This purpose is to give as complete as possible an exposition of stellar dynamics; that is, to show the general principles on which it is based and how, from these principles, the modern conceptions of motions in stellar systems can be derived. The close relation between stellar dynamics and the other branches of the exact natural sciences, especially hydrodynamics and statistical physics, is explained. The author believes that this interrelation must be the means of bridging the gap still existing between stellar dynamics as a part of astronomy and as a part of theoretical physics. The importance of the methods of statistical physics in stellar dynamics was first pointed out by Charlier and by Jeans. The importance of hydrodynamics, the mechanics of continuous media, in stellar dynamics has still not been fully recognised, despite the fact that the necessary equations were derived by Jeans in 1922. This incorrect estimate of the value of hydrodynamics has been a serious obstacle to the proper development of the theory.

In the course of being written, the book has grown considerably beyond the range of topics originally envisaged. The author realised that it was necessary to give an account of his synthetic method, which combines the statistical and hydrodynamical approaches to the problem. It should be mentioned that all the work on the relevant chapters (IX and X) was done in the winter of 1955 and the spring of 1956, since only then was it possible to arrive at the final form of the solutions to various problems and to elucidate their interrelation. This explains why, despite the time taken by the whole work, the author was compelled for lack of time to omit the treatment of certain mathematical problems which are of importance in the theory. In some cases these are stated but not solved. In consequence, the analysis in these chapters is only qualitative.

Such problems include, in particular, the solution of the differential equations of the level surfaces of star density for various types of galaxy. The author has been obliged to deal only with a first approximation to their solution, and in consequence he has been unable to make a direct comparison between theory and observation, i.e. to ascertain how far the theory predicts even the main structural features of various types of galaxy. Only a qualitative comparison of the theoretical results with photographs of individual galaxies was possible. These photographs appear at the end of the book.

The author has endeavoured to make the book useful even to those who are not astronomers. For this reason smaller type has been used in the discussion of some subjects whose interest is restricted to astronomers, such as the determination of the solar motion, the study of galactic rotation, and the determination of the velocity ellipsoid from observed stellar velocities.

The reader is assumed to have a knowledge of mathematics corresponding to two years' study in the physics and mathematics departments of a university, and also to be acquainted with the elements of statistical physics and mathematical statistics. Hardly any astronomical training is required, except to appreciate the special topics referred to above. The reason is that, at the present stage of its development, the dynamics of stellar systems is able to discuss only the simplest and most general properties of stellar systems. The reader will, of course, benefit from having read Parenago's *Stellar Astronomy* or Smart's *Stellar Dynamics*. For the non-astronomer, a suitable book for general preparation is Hubble's *Realm of the Nebulae*.

A list of astronomical literature cited in the text by numbers in square brackets is given at the end of the book. References to non-astronomical literature (physics, mathematics and mechanics) appear as footnotes to the text.

The author is very grateful to his colleagues who participated in the regular seminar at the Department of Stellar Astronomy and Astrometry, University of Leningrad, where he was able to report the results of his work as it proceeded. Professors A. N. Deĭch and N. N. Pavlov in particular should be mentioned; their contributions to the discussion of the reports and their comments thereon were of great use. The author must also thank T. A. Agekyan, who not only took part in many of the seminar discussions but also undertook to read the whole of the manuscript and, at the author's invitation, wrote § 8 of Chapter IV concerning the interaction of stars with dust clouds.

The manuscript of the book was also read, wholly or in part, by Professors P. P. Parenago and A. I. Lebedinskiĭ and by A. M. Mikisha and F. A. Tsitsin, two members of the staff of the Sternberg State Astronomical Institute. They made many valuable comments, for which the author is sincerely grateful.

Leningrad, 1957

K. F. OGORODNIKOV

INTRODUCTION

§ 1. The Subject of Stellar Dynamics

STELLAR dynamics is the culmination of stellar astronomy, which is one of the great branches of modern astronomical knowledge. The task of stellar astronomy is to study the structure, movements and evolution of stellar systems. It falls naturally into three closely related parts: stellar statistics, stellar kinematics and stellar dynamics.

The first of these, stellar statistics, deals with the theory of star counting and the deduction from star counts of the spatial structure of stellar systems. Little will be said of it in this book; but we shall use the fundamental concepts of stellar statistics to render precise our terminology and to define the basic distribution functions of stellar dynamics. A characteristic of stellar statistics is that it takes no account of stellar motions, that is, it considers each stellar system at a particular time. Thus it gives instantaneous pictures of the stellar systems. By means of such investigations we learn of the various types of stellar system and can undertake their classification. Then, using the results of stellar kinematics and (more especially) stellar dynamics, we may attempt to arrange the known types of stellar system in an evolutionary sequence.

The fundamental stellar system is a galaxy (see Plate I), which is any one of the systems resembling our own Galaxy, of which the Sun is a member. Thus the study of the structure and evolution of galaxies is the fundamental problem of stellar astronomy. However, stellar astronomy is also concerned with smaller systems, the globular and galactic star clusters. Globular clusters (Plate II) are very dense systems, containing hundreds of thousands of stars. They are extremely regular, spherical or slightly ellipsoidal, in shape. The galactic, or open, clusters (Plate III) include only tens to thousands of stars, and are seen as regions of higher star density against the relatively uniform star field of a galaxy. They merge into the stellar associations, first discovered in 1948 by the Soviet astronomers Ambartsumyan and Markaryan; these differ from the galactic clusters in that the relative (partial) density of stars in them is extremely small. The associations are revealed only because they contain stars of unusual types, which are consequently markedly more abundant in regions where there are stellar associations.

Stellar kinematics deals with the various motions of stars within stellar systems, but takes no account of the forces which act to cause these motions. The motions concerned are not those of individual stars, but of large groups of stars. In terms of an analogy with gases, such motions correspond to currents of gas. The most characteristic feature of such motions in stellar systems is rotation.

Finally, stellar dynamics describes the motions of the stars in terms of the forces acting on them. In some cases, general laws concerning the parameters of the orbits of individual stars are considered (for example, the problem of the stability of circular orbits in the equatorial plane of an oblate spheroidal stellar system). Usually, however, stellar dynamics treats the prevailing motions of groups of stars in relation to the distribution of mass in the stellar system, since the gravitational field in a stellar system is self-consistent, that is, it is determined by the stars themselves whose motions are under consideration.

The fundamental problem of stellar dynamics is to ascertain the laws of evolution of stellar systems. The evolutionary development of stellar systems must involve a change in their form, and this amounts essentially to a redistribution of mass. Hence all evolutionary processes are a kind of non-steady motion of stellar systems.

At present, however, it is not yet possible to solve or even to formulate in specific terms the problems of non-steady motion. We are therefore obliged to be content with the solution of problems of steady motion. It is not difficult to show that this is a proper and necessary step towards the study of non-steady problems, since it is improbable that the development of stellar systems consists entirely of a succession of violent changes. The assumption is much more reasonable that the development is slow, at least in successive stages of evolution, and in such stages we may regard a stellar system as being, to a first approximation, in a state of equilibrium. We call such a state quasi-steady.

In a quasi-steady state, the stellar system evolves so slowly that at every instant it takes a form corresponding to a state of dynamic equilibrium. The quasi-steady equilibria are stations along the line of evolution of stellar systems. The various types of stellar system which are observed in great numbers must correspond to these stations, because, the more rapid the process of evolution at any given stage, the fewer the numbers of stellar systems which will be found at that stage. For example, a stellar system must evolve rapidly if its structure departs markedly from uniformity, and this is in accordance with the fact that relatively few such systems are known. The elliptical galaxies, on the other hand, whose mass distribution is extremely uniform, number about one hundred for every galaxy of any other type.

Thus a consideration of quasi-steady systems makes it possible to lead on to a study of the dynamics of the most frequent types of stellar system. Such a procedure also enables us to approach the study of those stellar systems which are at a stage of very rapid evolution. For example, when the needle-shaped equilibrium figures of galaxies are known, the barred galaxies can be regarded as resulting from the emission of a jet of stellar gas as a consequence of the violation of Poincaré's stability condition.

From the above discussion it is evident that the fundamental method of investigation in stellar astronomy, and particularly in stellar dynamics, is the statistical method. Each stellar system is regarded as a statistical ensemble consisting of a large of objects of uniform properties, the stars. This explains the fact that the internal motions of double, triple and multiple stars are not a subject of study in stellar dynamics, although these come within the formal definition of stellar systems. When they are discussed in stellar dynamics it is as particular types of star, that is, each double or multiple star is regarded as an individual member of a statistical ensemble.

§ 2. A Brief Outline of the Historical Development of Stellar Dynamics

Stellar dynamics, like the whole of stellar astronomy, is a relatively recent branch of astronomical science. It originated when the necessity of a detailed study of the stars was first realised. Until the 18th century the stars were used in astronomy only as fixed reference points in the heavens, that is, only for astrometric purposes.

At the beginning of the 18th century, the problem of longitude determination, together with the associated problem of compiling accurate star-catalogues, became of practical importance, and was one of the most urgent questions of the time. In 1718 the proper motions of certain stars were discovered by Halley, and it soon became abundantly clear that the "fixed stars" are by no means fixed. To determine the proper motions of more stars, it was necessary to define a fixed system of astronomical coordinates, which was naturally based on the stars as a whole. To do this, however, both the motions of the stars, giving systematic corrections to star positions, and the spatial distribution of the stars, had to be investigated. Thus the problem of studying the structure of the Galaxy and the internal motions within it arose.

In the 1770's William Herschel began systematic counts of the visible stars in order to determine the shape and size of the Galaxy. He also made the first statistical study of proper motions (based on only 13 stars), and as a result first determined a systematic motion of the stars, which reflects the motion of the Sun among them. This pioneering work was of great importance in the further study of the structure of the Universe.

In 1839 the astronomical observatory at Pulkovo was founded, another important event in the history of stellar astronomy. Its founder and first director, F. G. W. Struve, ensured that the work of the observatory was from the beginning devoted essentially to one broad subject, that of stellar astronomy. By stellar astronomy Struve [126] understood, firstly, the making of the very accurate astronomical observations necessary in geodesy and cartography; secondly, the same observations of star positions as a basis for studying the structure and movement of stellar systems. Thus all the results of stellar astronomy were based on precise astrometry. Conversely, the examination of the positions and motions of the stars was necessary in order to define a coordinate system fixed in space, with respect to which star positions might be measured. In this way the fundamental practical problem of stellar astronomy, and of stellar dynamics, was defined. It remains the fundamental problem today, despite the immensely extended scope of both geodesy and stellar astronomy.

Thus the fundamental problems of stellar dynamics concern the structure and motions of stellar systems, as being part of the universe around us, and the consequent definition of the coordinate system fixed in space. The fundamental constants of precession, nutation, etc., and the reference systems of all astronomical catalogues, are determined in relation to these coordinates.

Stellar dynamics itself originated in the early years of the present century. In 1904 Kapteyn discovered the famous "star streams", and thereby showed that the velocity vectors of the stars, though lying in all directions, are not isotropically distributed as had previously been supposed. In 1907 Schwarzschild showed that Kapteyn's streams may be regarded as a consequence of an ellipsoidal law of distribution of stellar velocities, characterised by a distribution function

$$\varphi(u, v, w) = C\exp(-h^2u^2 - k^2v^2 - l^2w^2),$$

where u, v, w are the components of the star's velocity along three rectangular coordinate axes, and C, h, k, l are coefficients depending on coordinates and time.

By 1917 there had appeared several papers, now classical, by Charlier, Eddington, and Jeans, which laid the foundations of the dynamics of mechanical systems with an ellipsoidal law of particle velocity distribution.†

Until the 1920's, the chief phenomenon to be explained by stellar dynamics was Schwarzschild's law of the residual-velocity distribution of the stars. The attempts by Charlier [19, 20] to explain it by analogy with Maxwell's law in the kinetic theory of gases, in terms of the interactions of stars during close approaches, were unsuccessful. Because of the extremely low density of stars, a time of the order of 10^{14} or 10^{15} years would be needed for star encounters to exercise any noticeable effect on stellar motions. The work of Eddington and of Jeans at this time was therefore concerned with developing a theory of stellar systems which may be called "collision-free stellar dynamics". In this theory, on the basis of the postulate that Schwarzschild's law holds at every point in the system, a great variety of deductions are made in order to obtain as much information as possible concerning the structure of such systems. This line of investigation culminated in the outstanding work of Lindblad and of Chandrasekhar.

An important event in stellar dynamics was Strömberg's discovery, in 1922, of the phenomenon which he called the asymmetry of stellar velocities [125]. This showed that the stellar population of the Galaxy is kinematically by no means homogeneous, but consists of sub-systems of stars with varying residual-velocity dispersions.

In 1926 the rotation of the Galaxy was discovered by Lindblad [65] and by Oort [88, 89].††

B. Lindblad, a pioneer in the study of the spiral structure of the galaxies, developed a dynamical theory on the basis of lengthy investigations, which was entirely confirmed by the observed forms of various types of spiral nebula.†††

In the USSR, problems of stellar dynamics were not studied before the Revolution of 1917.

In the 1920's the first work on the motion of the Sun was carried out by V. G. Fesenkov and his colleagues [10, 32, 33, 53]. He also supervised the initiation of a long series of studies to determine the constant of precession with allowance for the rotation of the Galaxy.

At the beginning of the 1930's there was published the remarkable work of V. A. Ambartsumyan [4], who gave the first physical definition of the quasi-steady state of

† See the historical review by Ohlsson [87].

†† D. Ya. Martynov [69] has shown that the concept of the rotation of the Galaxy was developed mathematically more than forty years earlier by the Russian astronomer M. A. Koval'skiĭ [50], but his discovery passed unnoticed.

††† See the detailed bibliography of Lindblad's work on this subject given by S. Chandrasekhar [17], p. 191.

stellar systems, on which almost the whole of stellar dynamics is now based. Applying to stellar systems certain results (due to Landau [62]), derived for gases with a Coulomb interaction of particles, Ambartsumyan gave a clear and simple picture of the interactions between particles, as a result of which every stellar system tends continually to a state of statistical equilibrium, although this can never be achieved, on account of the absence of any potential barrier at the outside of the system. Consequently, all the stars with positive energies will escape from the system, forming a continual leakage of stars which prevents the establishment of the Maxwell–Boltzmann distribution necessary for statistical equilibrium. Ambartsumyan [5] also obtained important results concerning the evolution of star clusters, and his work has recently been continued by L. Ê. Gurevich [36], who has determined the lifetime of a single cluster and has shown that the dispersal of stars into the surrounding space leads to an increased concentration of stars in the centre of the cluster. These conclusions were confirmed in 1948 by Ambartsumyan and Markaryan's discovery of the stellar associations [6, 7].

Important investigations in galactic stellar dynamics were carried out by P. P. Parenago, who showed that a variety of other theoretical results are applicable to the Galaxy, and determined the values of its dynamical parameters [102, 104, 105]. More recently, he and B. V. Kukarkin developed Lindblad's idea of stellar sub-systems with varying star-velocity dispersions, and applied it to solve cosmogonical problems [52].

The younger generation of Soviet astronomers has also done interesting work on stellar dynamics. T. A. Agekyan [1, 2] has carried out a series of investigations on the interaction between stars and interstellar matter, discussed in Chapter IV, § 8. G. G. Kuzmin has examined the number of independent integrals of the motion in stellar systems of various types [61] and the question of their one-valuedness; the reader's attention is particularly directed to this work.† Kuzmin has also exploited a fruitful idea concerning the importance of quasi-integrals in the dynamics of stellar systems [58, 59]. These problems are briefly considered in Chapter V.

Since the last war, some interesting work has also been published on the gravitational potential of the Galaxy [40, 56, 106, 112] and on the closely related problem of the individual (regular) orbits of stars in the Galaxy. Among Soviet authors, mention should be made of Parenago [104, 106], and also of Idlis [40] and Dzigvashvili [23]. Although their work is somewhat outside the province of the present book, the scientific value of their continuation of classical ideas in stellar dynamics (neglecting encounters) is undisputed, and Chapter VI gives an account of this branch of the subject.

§ 3. The Fundamentals of the Synthetic Method in Stellar Dynamics

This book begins from the postulate that any complete treatment of the dynamics of stellar systems must combine the methods of statistical physics and hydrodynamics.

The first of these methods regards stellar systems as statistical ensembles of gravitating particles, in which, as Charlier and Jeans showed, the effect of encounters

† The latter problem is still insufficiently clear and often leads to misunderstandings; see the discussion in Chapter V, § 8, of Chandrasekhar's theorem on the helical symmetry of galaxies.

between individual stars may be neglected. The stellar systems have been thought of as consisting of isolated points without interaction, moving in a field of force due to the total attraction of all the stars in the system concerned. This regular field causes each star to describe a regular orbit. A serious defect of such a theory is that it leads to no definite law of stellar velocity distribution. Schwarzschild's distribution law is merely postulated to be valid at every point in the stellar system.

The statistical method culminated in the work of Lindblad and of Chandrasekhar already mentioned. These theories, however, despite their mathematical elegance, led to no complete solution of the fundamental problem in the dynamics of stellar systems, namely to give a theoretical interpretation of the observed forms of the various types of galaxy. They are applicable to spirals such as NGC 7217 and NGC 3031, which may be called Crab-type galaxies, having narrow arms and nuclei of dimensions large compared with the widths of the arms, but their applicability to spirals such as NGC 5194 and NGC 598, which have wide arms, is very doubtful. For such galaxies, theory and observation are apparently in conflict, since the arms are found to be trailing, whereas the theory predicts that they should be leading (the galaxy should be "unwinding").

Moreover, the author is quite convinced that these theories really obscure the path to further knowledge of the evolution of stellar systems. One can, of course, formally suppose that all or some of the coefficients in the formulae derived from the ellipsoidal velocity distribution hypothesis are time-dependent. This, however, leads to nothing, and will always do so unless some specific process of physical evolution is considered.

In this respect the hydrodynamic approach to stellar dynamics is promising. It is based on the equations of transfer, which are obtained by averaging the Boltzmann equation over velocities. The similarity of the equations of transfer to the equations of motion of a compressible viscous fluid has long been known, but their application to specific problems of stellar dynamics has been very limited. The reason is that, when the equations of hydrodynamics are obtained from the equations of transfer, new unknown functions appear which render the solution of the problem indeterminate. These functions in the hydrodynamic equations of stellar dynamics correspond to the viscosity coefficients in ordinary hydrodynamics, and likewise represent the interaction between the particles of which the medium is composed. In ordinary hydrodynamics, a phenomenological approach is possible by way of postulating the empirical Hooke's law and its relation to the strain tensor of a liquid particle; in stellar dynamics, this cannot be done.

This difficulty was first overcome for a very simple problem, the dynamics of spherical systems. It was known that the corresponding equation of hydrodynamics (Chapter IX, § 1, eqn. (9.7)) involves, besides the potential U and the star density ν, which are related by Poisson's equation, two further unknown functions: the radial and transverse velocity dispersions, whose presence renders indeterminate the problem of finding ν as a function of the coordinates. Thus, to solve the hydrodynamical problem, it is necessary to know the stellar velocity dispersions, which are essentially statistical quantities, as functions of position.

The synthetic method was, in fact, devised in the search for a general solution of the hydrodynamic equations of stellar dynamics. As applied to spherical systems, it is based on the following procedure. The radial and transverse velocity dispersions which appear in the hydrodynamic equations are found by means of the phase distribution, which is obtained by the methods of statistical physics.

The first main feature of the synthetic method is its wide use of the concept of a macroscopic volume element, which is taken from statistical physics. This signifies a volume which is very small compared with the whole system, but yet so large that it contains many stars. The methods of mathematical statistics can then be applied to the stars within the volume, and in particular their velocity distribution can be found. The synthetic method is applicable only to such systems, and only to those parts of them in which a macroscopic volume element may be defined.

It is easily seen that the possibility of defining a macroscopic volume element is equivalent to that of defining the phase density and thus using the methods of hydrodynamics. Thus a stellar system may be divided into two distinct regions, the inner region of high density, where the system behaves as a continuous medium (a compressible fluid), and the outer region or corona, where the system is essentially discrete and must therefore be regarded as consisting of individual particles (stars).

The concept of the macroscopic volume element, which in [79] I called the ν-volume, makes possible the immediate removal of one considerable fundamental difficulty, which to many workers in the field of stellar dynamics (for example, Kurth and Finlay-Freundlich [34, 55]) has appeared so insuperable that they could see no way of constructing a statistical stellar dynamics. One of the reasons for the occurrence of such difficulties as those mentioned above is the infinite mass which the statistical theory assigns to stellar systems. Since an infinite density is evidently out of the question, the infinite mass implies an infinite radius of the system. In synthetic stellar dynamics this difficulty disappears. For, if we consider models which represent actual galaxies, these must be elliptical, barred or spiral galaxies (since a discussion of the theory of the structure of irregular galaxies may be postponed, and no other types of galaxy are at present known, at least to any extent which would justify an attempt at a theory of their structure). The galaxies of each of the three normal types consist of a central nebula which undoubtedly has "apparent boundaries". The apparent dimensions of a galaxy are a not unreliable indication of its distance, i.e. they have an objective significance. By the use of sensitive photometers, every galaxy can be found to extend beyond its apparent boundaries, but the star density in these outer regions is much less than in the inner part of the galaxy, and moreover the structure of the outer part is entirely different from that of the interior. Shapley [116], Kholopov [49] and others have suggested the name "corona" for the outer regions of stellar systems.

In the spiral galaxies the corona is sharply marked off from the inner region by the fact that the spiral arms do not continue into the corona. In the elliptical galaxies also it is certain that a corona exists beyond the limits of the central part of the system.

Thus the dynamics of stellar systems may be regarded as concerning the theory of the internal structure of galaxies. In this respect there is a close analogy between stellar dynamics and theoretical astrophysics. A single theory would not be expected to cover both the internal structure of a star and also its atmosphere, although the two together form a whole, namely the star itself; for there is a great difference between the laws of the internal structure of stars and those of atmospheric structure. The same could be said of stellar systems. The internal region extends to the point where it becomes impossible to define macroscopic volume elements. Outside this region, the hydrodynamic approximation becomes invalid, and the stellar system must be regarded as consisting of isolated stars moving in individual orbits; but within the region the system may be regarded as a continuous medium. Thus there is a radical difference between the internal part of a stellar system and its corona, and so the problem of infinite mass does not arise.

The question may be asked how the exact position of the outer surface or boundary of a stellar system is to be established. In the author's opinion the situation is like that in the photospheres of the Sun and stars, which are defined as visible surfaces, and the outer surface of a stellar system should be defined in exactly the same way.

As will be shown in Chapter IX, the region where a galaxy is continuous coincides with that of its rigid-body rotation, which in turn is the same as the observed central part of the galaxy, and the outer boundaries are consequently the same. The correctness of our hypothesis, like that of any hypothesis, must be tested by comparing its consequences with the results of observation. The unavoidable slight indeterminacy in this definition is of no importance in practice, since relatively few stars lie outside the surface of a stellar system.

The second main feature of the synthetic method in stellar dynamics is the following. We assume that in every stellar system a statistical mechanism operates to establish within the system a statistical distribution of star coordinates and velocities. This is a basic assumption, which to some extent contradicts established ideas in stellar dynamics, and so it must be discussed in some detail.

The forces of interaction between stars when they pass close to one another will be called irregular forces, as opposed to the regular forces resulting from the gravitational field of the stellar system. The irregular forces are of great importance in the dynamics of the system, since they are responsible for the statistical redistribution of stellar velocities towards equipartition of energy. The rate at which this process occurs is characterised by the relaxation time, defined as the time in which the action of the irregular forces considerably modifies the initial stellar velocity distribution. However, even the first attempts by Charlier in 1916 to define the relaxation time showed that its value is extremely large, of the order of 10^{13} or 10^{14} years, which is much greater than the probable age of the spiral galaxies (not more than 10^{10} or 10^{11} years, according to present ideas).

The results of investigations by Kapteyn, Schwarzschild, Jeans, Eddington and others have shown that the stellar velocities in the Galaxy obey a law which is a natural generalisation of Maxwell's law for gases, namely Schwarzschild's distribution law. This law may be concisely written $f(u, v, w) = e^{-Q}$, where Q is a positive-definite quadratic function of the stellar velocity components u, v, w. Thus a most paradoxical situation arose in modern stellar dynamics: a statistical law of stellar velocity distribution exists, but there is no mechanism whereby it might be established.

For want of any alternative, Schwarzschild's law (or other equivalent laws) appears as an axiom in modern stellar dynamics. On the basis of Jeans' theorem that the phase density can depend only on the integrals of the motion, the hypothesis underlying Schwarzschild's law is often presented in the form of a postulate, namely, that the phase density is a function of a quadratic form in the integrals of the motion of a particle in the regular gravitational field of the stellar system concerned. The laws of physics, however, include no statement that the phase density must necessarily be a quadratic or any other particular function. The hypothesis is therefore a purely mathematical one.

There must consequently exist in stellar systems a statistical mechanism akin to star encounters, which establishes the particular velocity distribution which is statistically most probable for the system in question.

The existence of such a mechanism follows not only from the above-mentioned paradox relating to Schwarzschild's law, but also from the fact that we can point to no other process in stellar systems which could account for their evolution. In the absence of irregular forces, gravitation alone cannot cause evolution: in a steady-state system, every star will move for all time in a fixed regular orbit.

The irregular forces cause a redistribution of energy among the stars. A certain number of stars thereby acquire positive energies, that is, energies exceeding that required to escape from the system. They will then leave the system within a short time, and in general will not return. Thus there is a constant loss or leakage of mass from a stellar system. We can say that the irregular forces tend to dissipate the system. On the other hand, when irregular forces are present, which act like viscosity forces, the regular gravitational forces tend to condense the system. The evolution of stellar systems must be the result of a dynamic equilibrium between these two tendencies.

Of course, we must at present regard the action of the irregular forces as a hypothesis based on various indirect indications. However, the fact that from this hypothesis we can derive a dynamical interpretation of all the principal observed forms of galaxy and, moreover, outline a logically consistent picture of their evolution, seems to support the hypothesis that a statistical mechanism of this type exists in stellar systems. Stellar dynamics which neglects the irregular forces has as yet made no progress whatever in dealing with the evolution of stellar systems. The only such theory, which describes the evolution of the open clusters in the Galaxy, due to V. A. Ambartsumyan [5], is based on a consideration of the interaction of regular and irregular forces.

Thus it is certain that the extremely long relaxation time of galaxies has so far been the chief, though not the only, obstacle to the development of a theory of galactic evolution.

The hypothesis on which the remainder of this book is based is not new. It consists in the assumption that the irregular forces in galaxies are due not to individual stars, but to aggregations such as star clouds or gas-and-dust condensations, whose average mass is 10^5 to 10^6 times the Sun's mass. This concept has been widely used, for instance, in the work of Spitzer and Schwarzschild [124], A. I. Lebedinskiĭ [63], L. É. Gurevich [36] and others. Calculation shows that if, for example, 10 per cent of all stars in the Galaxy are in clouds of 10^6 stars each, then the relaxation time decreases by a factor of 10^5 to a value of 10^8 or 10^9 years, which is comparable with the period of rotation of the Galaxy.

This elementary calculation shows that a solution on this basis is at least possible. The patchy structure of the spiral arms in our Galaxy and in other galaxies indicates the existence of star clouds. Moreover, at the beginning of the evolutionary process this patchiness may well have been even more marked than it is now. Of course, it must be borne in mind that these condensations are not akin to the random fluctuations in star density, since they are genetically related systems. They are much more stable and numerous than the random fluctuations. However, the correctness of this hypothesis must again be finally tested by a comparison of its consequences with the results of observation.

The third feature of the synthetic method is as follows. The hypothesis that the dimensions of stellar systems are finite, and that their relaxation times are relatively short, leads to the method of additive dynamical parameters, which is much used in the later chapters of this book. This method is a direct generalisation of the method of additive integrals well known in statistical physics,† which has been used to derive results concerning the most probable laws of phase distribution for various types of stellar system, as well as the shapes and evolutionary relations of these systems.

Mathematically, the method of additive parameters amounts to an expansion of the logarithm of the phase density in terms of the classical integrals $I_1, I_2, \ldots$: $\log F = \sum k_j f_j(I_1, I_2, \ldots)$, each term of the sum expressing a particular property of the stellar system concerned by means of an equation of the form $f_j(I_1, I_2, \ldots) = \text{constant}$.

We may mention briefly the principal results obtained by the method of additive parameters for rotating stellar systems, which include the majority of galaxies. This method gives the most probable phase distribution, which is found to correspond to rigid-body rotation. The hydrodynamic equations then show the existence of a solution in which the star density is almost constant. On this was based the hypothesis that the forms of galaxies rotating as rigid bodies can differ only slightly from the equilibrium ellipsoids obtained for a homogeneous rotating fluid.

The spiral galaxies do not rotate as rigid bodies. The above analysis is therefore not applicable to them. We know, however, that the star population of the spiral galaxies is not uniform: according to Baade, it consists of two types, Population I being composed of young stars concentrated towards the galactic plane, and Population II forming a kind of spherical corona. It may consequently be supposed that the spiral galaxies are statistically inhomogeneous as regards their star composition. It could then be shown that for spiral galaxies the equilibrium form should be close to the disc-like Maclaurin ellipsoids. The central parts of the galaxies rotate as rigid bodies; outside the ellipsoid, the rotation is not rigid, and this is the region of the corona.

Finally, there is the following evolutionary sequence of galaxies. First, an elongated ellipsoid of the Jacobi type is formed. It rotates as a rigid body and has regions of instability at the ends of its axis. Here stars are detached to form two coiling arms, giving a typical barred spiral. The star population is uniform and of Baade's type I, but thereafter evolves into type II.

The stars arising from the disintegration of the barred spiral form the corona. At this stage another rapid process of star formation begins within the ellipsoid, and

† J.E. Mayer and M.G. Mayer, *Statistical mechanics*, Wiley, New York, 1940.

stars of Population I are formed, with small residual velocities. These occupy uniformly the volume of the new equilibrium figure, a disc-like Maclaurin ellipsoid, corresponding to the prevailing angular velocity of rotation of the galaxy. The remnants of the diffuse matter from the original long, narrow Jacobi ellipsoid are "frozen" into this flat configuration and form spiral arms.

The spiral galaxies cannot be in equilibrium, since the irregular forces must tend to intermingle the two star populations, and so render the galaxies uniform. Analysis shows that in the course of evolution the radius of the disc will increase, and its thickness will decrease, the stars coming to occupy the corona. Ultimately, the spiral structure and disc-like form must entirely disappear, leaving only the nucleus of the original galaxy, consisting of Population II stars only: an elliptical nebula, with no condensations of matter, so that further evolution is slow and the loss of stars occurs in accordance with the Roche model.

Of course, this book does not at all exhaust the possibilities of the method of additive parameters. The more we know of the specific properties of a stellar system, the more independent parameters of the system can be determined, and the more accurately the observed motions of the system can be theoretically reproduced.

Additive integrals are a particular case of additive parameters, distinguished by the fact that they can be derived from the general laws of mechanics for any quasi-steady stellar system. It is therefore useful to begin by ascertaining the additive integrals wherever possible.

§ 4. The Practical Significance of Stellar Dynamics

The fundamental practical problem of stellar astronomy, which led to the development of stellar dynamics, has been mentioned above: it is to define a system of axes fixed in space, from which star coordinates may be measured, and to which all observed motions, celestial and terrestrial, may be referred.

This problem, however, represents only the ancillary position which stellar dynamics holds with respect to astronomy. But stellar dynamics, like other branches of modern astronomy, deals also with the properties of matter in extreme conditions which are inaccessible or nearly so in terrestrial laboratories. From this point of view stellar dynamics occupies a special and unique place among natural sciences: it is the mechanics of the motion of particles in highly rarefied media.

Any medium, rigid or elastic, liquid or gaseous, consists of an immense number of similar particles (molecules or atoms). These particles have disordered motions, which are commonly associated with the concept of heat. In this view any medium is discrete. At the same time, any medium necessarily possesses also the opposite property of continuity. We perceive the continuity of water, of solids, of air, as the property of filling the whole volume which is occupied, without interstices. Both discreteness and continuity are real, objective properties.

The smaller the corpuscular density of a given medium, i.e. the fewer the particles of it per unit volume, the more evident is its discreteness. Conversely, the greater the corpuscular density, the more continuous the medium.

In classical science, fluid mechanics corresponds to the continuous aspect of matter, and the kinetic theory of gases to its discrete aspect. The former entirely ignores the discrete structure of matter, and the latter entirely ignores its continuity. Thus fluid mechanics and statistical physics (the kinetic theory of gases) may be regarded as limiting cases of the theory of matter, which are in reality never completely valid. Any actual medium is intermediate in its properties, in the sense that it is always both continuous and discrete at the same time.

The one-sidedness of hydrodynamics is that it overlooks all phenomena pertaining to diffusion and thermal motion of particles: a classical "particle" of liquid consists of the same particles of matter throughout its motion. These particles move along streamlines in the liquid, like well-drilled soldiers. In hydrodynamics, the phenomena of diffusion of particles from one volume element to another, and their thermal motion, are either completely ignored, or treated purely phenomenologically by means of empirical terms and coefficients.

In statistical physics, on the other hand, thermal motion and diffusion are considered in detail, but the gas itself is assumed to be at rest as a whole.

It can easily be shown that a logically complete statistical physics is impossible without the introduction, perhaps implicit, of a hypothesis concerning the continuity of the gas. For even such fundamental concepts and hypotheses as the volume of the gas, the distinction between ordered and disordered (thermal) motions of particles, and Liouville's theorem, become meaningless if the gas is supposed to be discrete, i.e. not to fill its volume completely.

A measure of the predominance of continuity or discreteness is given by the mean free path of a particle. If this is very small in comparison with the scale of the phenomena under consideration, the hydrodynamic approximation is valid. In the contrary case, the methods of statistical physics must be used. Thus classical science affords no methdo of investigating the motion of media in which the mean free path of the particles is large.

From its beginnings, stellar dynamics has been concerned with such extremely rarefied media as stellar systems, and so it has developed methods whereby the intermediate problem of motion with diffusion may be solved, at least for the simplest cases. For this purpose it was necessary to solve the important and fundamental problem of distinguishing, in a moving medium, two essentially different types of motion: ordered and disordered. In stellar dynamics, this distinction is achieved by means of the concept of the centroid at each point in the medium. The methods of hydrodynamics are thereby extended to rarefied media, where the mean free path of the particles is large.

CHAPTER I

THE FUNDAMENTAL CONCEPTS OF STELLAR STATISTICS

§ 1. Some Properties of Univariate Distribution Functions

THE fundamental method used in stellar astronomy, and in particular in stellar dynamics, is the statistical method. We shall therefore begin this first chapter by discussing some of the basic concepts of mathematical statistics.

One of the primary concepts in mathematical statistics is that of a *statistical variable.* This name is given to any quantity which characterises some definite property of each member of an ensemble.

In order to form a statistical ensemble of any given objects, it is necessary to select one or more specific properties of those objects. If a quantity x characterising some property can take a continuous range of values, we can consider the *distribution function* $f(x)$ for the variable x. Then

$$\mathrm{d}N_x = f(x)\,\mathrm{d}x \tag{1.1}$$

gives the number of members of the ensemble for which the value of this variable is between x and $x + \mathrm{d}x$.

The total number of members is evidently

$$N = \int f(x)\,\mathrm{d}x, \tag{1.2}$$

where the integration is over all possible values of the variable. In practice, the limits of integration are determined by the upper and lower bounds of the values of the variable, but we may, without loss of generality, conventionally take the limits as $\pm\infty$ in all cases, and for brevity omit them. This procedure is permissible because the distribution function may be taken to be zero outside the range of values taken by the variable.

On dividing the distribution function by the total number of members of the ensemble, we obtain the *normalised distribution function*

$$f_0(x) = f(x)/N. \tag{1.3}$$

Clearly

$$\mathrm{d}n_x = f_0(x)\,\mathrm{d}x \tag{1.4}$$

gives the fraction of the total number of members for which the variable has values between x and $x + \mathrm{d}x$. Evidently

$$\int f_0(x)\,\mathrm{d}x = 1. \tag{1.5}$$

This is the normalisation condition.

From probability theory we know that

$$\mathrm{d}n_x = f_0(x)\,\mathrm{d}x \tag{1.6}$$

gives also the *empirical probability* that the value of the variable for a member selected at random from the ensemble lies between x and $x + \mathrm{d}x$.

The quantity

$$\mathscr{E}(x) = \int x f_0(x)\,\mathrm{d}x \tag{1.7}$$

is called the *expectation* of the variable x. It is also the *mean value* $\bar{x}$. The quantity $x - \bar{x}$ is called the *deviation* of the variable from the mean. The expectation of the square of the deviation,

$$\begin{aligned}\sigma_x^2 &= \mathscr{E}[(x - \bar{x})^2] \\ &= \int (x - \bar{x})^2 f_0(x)\,\mathrm{d}x \\ &= (1/N) \int (x - \bar{x})^2 f(x)\,\mathrm{d}x,\end{aligned} \tag{1.8}$$

is called the *variance;* it gives a measure of the *dispersion* of the variable.

From the parameters N, $\bar{x}$ and σ_x for any distribution function $f(x)$ we can form a *Gaussian* or *normal* distribution function

$$\varphi(x) = \frac{N}{\sqrt{(2\pi)}\sigma_x} \exp\,[-\,(x - \bar{x})^2/2\sigma_x^2]. \tag{1.9}$$

It is easily seen by integration that this normal function has the same values of N, $\bar{x}$ and σ_x as the given distribution function $f(x)$. For brevity we shall call it the *corresponding* normal distribution function. Since the definition of the normal function amounts to specifying the values of its three parameters, it can be found for any distribution function $f(x)$, provided that N, $\bar{x}$ and σ_x have finite values.

Instead of the variance σ_x^2, it is often convenient to use another quantity in one-to-one correspondence with it, namely

$$h_x = 1/\sqrt{2} \cdot \sigma_x, \tag{1.10}$$

which is called the *degree of concentration* of the variable.

The point on the x-axis whose abscissa is equal to the mean value $\bar{x}$ of x is called the *centre* of the distribution. Thus

$$\xi = x - \bar{x}, \tag{1.11}$$

which was defined above as the deviation from the mean, can also be called the deviation from the centre. It is easy to see that the expectation or mean value of the deviation from the centre is zero:

$$\bar{\xi} = \mathscr{E}(\xi) = \int \xi f_0(x)\,\mathrm{d}x = 0. \tag{1.12}$$

The expression for the normal or Gaussian distribution function can be written more simply in terms of the degree of concentration and the deviation from the mean:

$$\varphi(\xi) = (Nh/\sqrt{\pi}) \exp(-h^2\xi^2). \tag{1.13}$$

In mathematical statistics the concept of the *moments* of a function frequently occurs. The nth moment of a given distribution function $f(x)$ is defined as

$$m_n = \int x^n f(x)\, \mathrm{d}x \quad (n = 0, 1, 2, \ldots), \tag{1.14}$$

provided, of course, that the integral (1.14) exists. The analogous quantities

$$\mu_n = \int (x - \bar{x})^n f(x)\, \mathrm{d}x = \int \xi^n f(x)\, \mathrm{d}x \tag{1.15}$$

are called the *moments about the centre*. Correspondingly, the moments m_n are called *moments about the origin*.

If the distribution function is normalised, we have the normalised moments (denoted by a superscript zero)

$$m_0^0 = \mu_0^0 = 1, \tag{1.16}$$

and

$$m_1^0 = \bar{x}, \tag{1.17}$$

i.e. the mean value of the variable is equal to the normalised first moment about the origin. (Evidently $\mu_1^0 = 0$.) Finally

$$\mu_2^0 = \sigma_x^2, \tag{1.18}$$

i.e. the variance is equal to the normalised second moment about the centre. It is easy to see that

$$\begin{aligned} m_2^0 &= \mu_2^0 + \bar{x}^2 \\ &= \sigma_x^2 + \bar{x}^2. \end{aligned} \tag{1.19}$$

This equation is often used in the observational determination of σ_x^2.

If the distribution function is not normalised, the corresponding equations are

$$m_0 = \mu_0 = N, \tag{1.16a}$$

$$m_1 = N\bar{x}, \tag{1.17a}$$

$$\mu_2 = N\sigma_x^2, \tag{1.18a}$$

and

$$m_2 = \mu_2 + N\bar{x}^2. \tag{1.19a}$$

Here N is again the total number of members of the ensemble.

If two distribution functions are both normalised and both have their centres at the origin, they differ only in their variances.

The above-mentioned properties of distribution functions depend only on the moments of the second and lower orders, and naturally allow only a rough comparison of distribution functions. For a more exact comparison the moments of higher order must be used. The quantity

$$A = -\mu_3/\sigma_x^3 \tag{1.20}$$

is a measure of the *skewness* of the distribution function, and

$$E = (\mu_4/\sigma_x^4) - 3 \tag{1.21}$$

is called the *excess;* it is a measure of the *peakedness* of the distribution function.

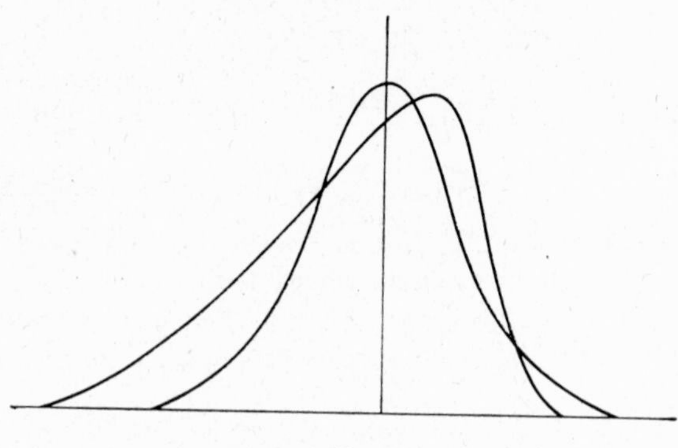

FIG. 1

It is evident that the skewness of any even function of the deviation from the centre is zero. Particular importance attaches to *single-peaked* functions, which have a single maximum. The value of the argument which corresponds to this maximum is called the *mode*. The mode is not usually much different from the mean value of the argument. The normal distribution function is a typical single-peaked symmetrical function. In Fig. 1 one curve shows a single-peaked unsymmetrical function, and the other curve a normal function. A characteristic property of single-peaked unsymmetrical functions is that

$$f(\xi) - f(-\xi) \neq 0. \tag{1.22}$$

If we classify members of an ensemble with an unsymmetrical distribution function in terms of the absolute value of $|\xi|$, and correspondingly form the average

$$\bar{\xi} = [\xi f(\xi) + (-\xi) f(-\xi)]/[f(\xi) + f(-\xi)], \tag{1.23}$$

the result will depend on ξ. For a symmetrical distribution function $f(\xi)$, the value of this $\bar{\xi}$ is zero for all ξ. In other words, for a single-peaked unsymmetrical distribution function, the mean value of the variable for the members of the ensemble having a given absolute magnitude of the variable is a function of that absolute magnitude. This property is the basis for the interpretation of the asymmetry of star velocities in the Galaxy.

The excess (1.21) is easily seen to be zero for a normal distribution function. If the excess is positive, the distribution curve is sharply peaked, and its form relative to the normal curve is shown by the thick line in Fig. 2. Conversely, if the excess is negative,

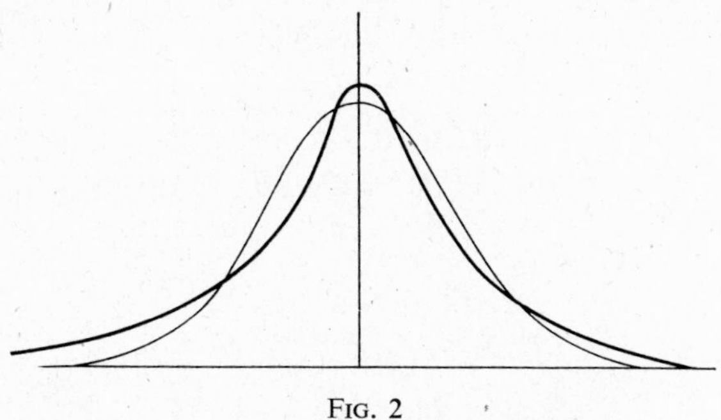

FIG. 2

the distribution curve is flatter than the normal curve (Fig. 3). Thus the excess gives a measure of the extent to which the given distribution function differs from the corresponding normal function. It can be shown [77] that the excess is positive if the ensemble under consideration is statistically inhomogeneous, i.e. consists of a mixture

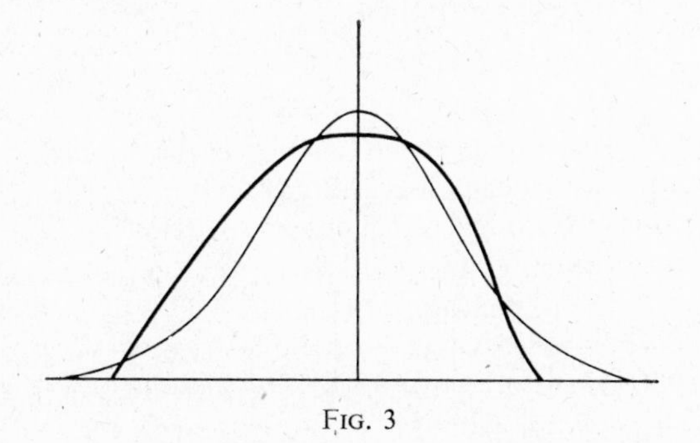

FIG. 3

of several groups each having a normal distribution of the variable. For let the distribution function be composed of a sum of normal functions having varying degrees of concentration but the same centre, which we take to be the origin. Let $f(h)$ be the normalised distribution of h, so that

$$\int_0^\infty f(h)\,\mathrm{d}h = 1.$$

Then the function $\varphi(x)$, after normalisation, is

$$\varphi(x) = \frac{1}{\sqrt{\pi}} \int_0^\infty f(h) h \exp(-h^2 x^2)\,\mathrm{d}h. \tag{1.24}$$

It is evident that

$$\mu_{2k} = \int x^{2k}\, \varphi(x)\, \mathrm{d}x$$

$$= \frac{1}{\sqrt{\pi}} \int_0^\infty f(h) h\, \mathrm{d}h \int_{-\infty}^\infty x^{2k} \exp(-h^2 x^2)\, \mathrm{d}x$$

$$= \frac{(2k-1)!!}{2^k} \int_0^\infty \frac{f(h)}{h^{2k}}\, \mathrm{d}h. \tag{1.25}$$

In particular

$$\mu_2 = \sigma^2 = \frac{1}{2} \int_0^\infty \frac{f(h)}{h^2}\, \mathrm{d}h, \tag{1.26}$$

$$\mu_4 = \frac{3}{4} \int_0^\infty \frac{f(h)}{h^4}\, \mathrm{d}h.$$

Using the auxiliary functions

$$f_1(h) = \sqrt{[\tfrac{1}{2} f(h)]} \quad \text{and} \quad f_2(h) = \sqrt{[\tfrac{1}{2} f(h)]}/h^2, \tag{1.27}$$

we have

$$\int f_1 f_2\, \mathrm{d}h = \mu_2 = \sigma^2, \quad \int f_1^2\, \mathrm{d}h = \tfrac{1}{2}, \quad \int f_2^2\, \mathrm{d}h = \tfrac{2}{3}\mu_4. \tag{1.28}$$

These three integrals obey Schwarz' inequality:

$$\left(\int f_1 f_2\, \mathrm{d}h\right)^2 \leqslant \int f_1^2\, \mathrm{d}h \int f_2^2\, \mathrm{d}h. \tag{1.29}$$

Here we must use the fact that $f_1(h) = h^2 f_2(h)$, and consequently $f_1(h) \neq f_2(h)$. The equality in (1.29) is therefore not possible. After substituting (1.28) in (1.29) we obtain

$$\sigma^4 < \tfrac{1}{3}\mu_4. \tag{1.30}$$

From this it follows at once that the excess (1.21) is positive.

Thus the excess may be regarded as a measure of the non-uniformity of an ensemble. For example, the positive excess observed in the star velocity distribution can be regarded as a result of the non-uniformity of the stellar composition of the Galaxy.

It should be borne in mind that the inclusion or exclusion of a few stars which deviate greatly from the mean affects very considerably the values of the third and fourth moments. Only for very large ensembles can fairly reliable values be obtained for the skewness coefficient and the excess. This also explains the fact that the moments above the fourth are scarcely ever used in applications of the theory.

§ 2. Multivariate Distribution Functions

The preceding discussion can easily be generalised to distribution functions of more than one variable. With a view to subsequent application to stellar velocity distributions, we may briefly discuss some properties of distributions involving three variables, and consider in detail only certain features of the dispersion which occur when there is more than one variable.

Let $u = f(x, y, z)$ be a one-valued distribution function of three independent variables x, y, z. Then the number of members of the ensemble for which the values of the variables are between x and $x + \mathrm{d}x$, y and $y + \mathrm{d}y$, and z and $z + \mathrm{d}z$, is $\mathrm{d}N = f(x, y, z)\,\mathrm{d}x\,\mathrm{d}y\,\mathrm{d}z$. The total number of members is

$$N = \int f(x, y, z)\,\mathrm{d}x\,\mathrm{d}y\,\mathrm{d}z, \tag{1.31}$$

where the integration may be taken to extend over all three-dimensional space. The distribution may be visualised in terms of the level surfaces for the given distribution function, whose equation is $f(x, y, z) = C$. One and only one level surface passes through each point (x, y, z). The intersection of the level surfaces with any plane gives a series of level curves. For example, the intersection with a plane parallel to the yz-plane, $x = x_0 =$ constant, gives the level curves $f(x_0, y, z) = C$.

In applications to stellar dynamics, cases where the level surfaces are of a simple form are of particular importance. In one of the simplest cases, the coordinates appear in the distribution function only as a complete positive-definite quadratic function, i.e.

$$f = f(T), \tag{1.32}$$

where

$$T = ax^2 + by^2 + cz^2 + 2fyz + 2gzx + 2hxy + px + qy + rz + s. \tag{1.33}$$

In this case the level surfaces are concentric coaxial ellipsoids. A distribution function of the form (1.32), (1.33) is said to be *ellipsoidal*. On putting T into a homogeneous form, we have

$$f = f(Q + \sigma), \tag{1.34}$$

where

$$Q = a(x - x_0)^2 + b(y - y_0)^2 + c(z - z_0)^2 + 2f(y - y_0)(z - z_0) + 2g(z - z_0)(x - x_0) + 2h(x - x_0)(y - y_0), \tag{1.35}$$

and σ is a constant term. The quantities x_0, y_0, z_0 are evidently the coordinates of the common centre of the level surfaces. The ellipsoidal distribution function (1.34) has its simplest form when the quadratic form Q is referred to its principal axes. Then

$$Q = H_1^2(x - x_0)^2 + H_2^2(y - y_0)^2 + H_3^2(z - z_0)^2, \tag{1.36}$$

where H_1, H_2, H_3 are evidently inversely proportional to the principal axes of the level ellipsoids.

An important particular case of the ellipsoidal function is *Schwarzschild's function:*

$$f(Q+\sigma) = e^{-Q-\sigma}, \tag{1.37}$$

where Q is a quadratic function of the coordinates given by (1.35) or (1.36).

The quantity σ which appears in Schwarzschild's function (1.37) can be expressed in terms of the total population N. Substituting (1.37) and (1.36) in (1.31) and integrating, we find

$$e^{-\sigma} = H_1 H_2 H_3 N/\pi^{3/2}. \tag{1.38}$$

Hence Schwarzschild's function, when referred to the principal axes of the level ellipsoids, can be written

$$(NH_1H_2H_3/\pi^{3/2}) \exp[-H_1^2(x-x_0)^2 - H_2^2(y-y_0)^2 - H_3^2(z-z_0)^2]. \tag{1.39}$$

To derive an expression for Schwarzschild's function with respect to arbitrary axes, we use the fact that the discriminant of the quadratic form Q, i.e. the determinant of its coefficients

$$A = \begin{vmatrix} a & h & g \\ h & b & f \\ g & f & c \end{vmatrix}, \tag{1.40}$$

is unchanged by any orthogonal transformation of the coordinates. When the principal axes are used, the non-diagonal coefficients are zero. Hence

$$A = \begin{vmatrix} a & h & g \\ h & b & f \\ g & f & c \end{vmatrix} = \begin{vmatrix} H_1^2 & 0 & 0 \\ 0 & H_2^2 & 0 \\ 0 & 0 & H_3^2 \end{vmatrix} = H_1^2 H_2^2 H_3^2,$$

and by (1.38) $e^{-\sigma} = N\sqrt{(A/\pi^3)}$. Substituting this expression in (1.37), we obtain for Schwarzschild's function in arbitrary coordinates

$$Ne^{-Q}\sqrt{(A/\pi^3)}, \tag{1.41}$$

where Q is the quadratic form (1.35) and A is its discriminant (1.40).

The simplest particular case of an ellipsoidal distribution function is that of a *spherical function*, which occurs when $H_1 = H_2 = H_3 = h$. Then we have from (1.34) and (1.36)

$$f = f[(x-x_0)^2 + (y-y_0)^2 + (z-z_0)^2],$$

and therefore the level surfaces are concentric spheres centred at the point (x_0, y_0, z_0). A particular case of a spherical distribution function is the Maxwellian function,

$$(Nh^3/\pi^{3/2}) \exp\{-h^2[(x-x_0)^2 + (y-y_0)^2 + (z-z_0)^2]\}. \tag{1.42}$$

This is evidently also the particular case of Schwarzschild's distribution function (1.39) which corresponds to spherical symmetry in the distribution of the variables.

Moments of various orders are used in the study of multivariate distribution functions. The expression

$$m_{ijk} = \int x^i y^j z^k f(x, y, z)\, d\omega \tag{1.43}$$

is called a moment of order $i + j + k$ of the function f about the origin. Here $\mathrm{d}\omega$ denotes a volume element, and the integration is over all space in the coordinates x, y, z. In particular

$$m_{000} = N = \int f(x, y, z)\, \mathrm{d}\omega$$

is the total population. The normalised distribution function is evidently

$$f_0(x, y, z) = (1/N) f(x, y, z). \tag{1.44}$$

The quantities

$$\left.\begin{aligned} m^0_{100} &= \bar{x} = \int x f_0(x, y, z)\, \mathrm{d}\omega, \\ m^0_{010} &= \bar{y} = \int y f_0(x, y, z)\, \mathrm{d}\omega, \\ m^0_{001} &= \bar{z} = \int z f_0(x, y, z)\, \mathrm{d}\omega \end{aligned}\right\} \tag{1.45}$$

are the mean values or expectations of x, y and z. By means of (1.44), (1.45) and (1.34), (1.36) it is easily seen that, for example, x_0, y_0, z_0 are the mean values of the variables for an ellipsoidal distribution function.

By moving the origin to the centre of the distribution, i.e. to the point $(\bar{x}, \bar{y}, \bar{z})$, we obtain the central moments of the normalised distribution function:

$$\mu^0_{ijk} = \int (x - \bar{x})^i\, (y - \bar{y})^j\, (z - \bar{z})^k\, f_0(x, y, z)\, \mathrm{d}\omega. \tag{1.46}$$

The first-order central moments are obviously zero: $\mu^0_{100} = \mu^0_{010} = \mu^0_{001} = 0$. The second-order central moments give the variances along the coordinate axes:

$$\mu^0_{200} = \mu_{xx}, \quad \mu^0_{020} = \mu_{yy}, \quad \mu^0_{002} = \mu_{zz} \tag{1.47}$$

and the three *mixed moments*

$$\mu^0_{011} = \mu_{yz}, \quad \mu^0_{101} = \mu_{zx}, \quad \mu^0_{110} = \mu_{xy}. \tag{1.48}$$

The order of the suffixes x, y, z is clearly immaterial.

All these moments can be conveniently arranged as a *moment matrix:*

$$\boldsymbol{M} = \begin{bmatrix} \mu_{xx} & \mu_{xy} & \mu_{xz} \\ \mu_{yx} & \mu_{yy} & \mu_{yz} \\ \mu_{zx} & \mu_{zy} & \mu_{zz} \end{bmatrix} = \begin{bmatrix} \mu_{11} & \mu_{12} & \mu_{13} \\ \mu_{21} & \mu_{22} & \mu_{23} \\ \mu_{31} & \mu_{32} & \mu_{33} \end{bmatrix}. \tag{1.49}$$

To simplify the formulae, we shall suppose that the origin is at the centre of the distribution and that the distribution function is normalised. We use x, y, z to denote the deviations from the centre, previously denoted by $x - \bar{x}, y - \bar{y}, z - \bar{z}$. Then the second-order moments are

$$m_{ijk} = \mu_{ijk} = \int x^i y^j z^k f(x, y, z)\, \mathrm{d}\omega,$$

with $i + j + k = 2$.

Let us now determine σ_l^2, the variance of the function in an arbitrary direction l having direction cosines $\alpha_1, \alpha_2, \alpha_3$. We put

$$\sigma_l^2 = \int (\alpha_1 x + \alpha_2 y + \alpha_3 z)^2 f(x, y, z)\, \mathrm{d}\omega \tag{1.50}$$

or, expanding the square,

$$\sigma_l^2 = \sum_{i=1}^{3} \sum_{j=1}^{3} \alpha_i \alpha_j \mu_{ij}. \tag{1.51}$$

This relation is of importance in stellar kinematics. It shows that the variances transform as the components of an affine orthogonal tensor of rank two. The matrix (1.49) can therefore be regarded as the matrix of the variance tensor, whose components are the nine moments μ_{ij}. The moments lying symmetrically about the principal diagonal of the matrix (1.49) are equal, i.e.

$$\mu_{ik} = \mu_{ki}. \tag{1.52}$$

The tensor (1.49) is therefore symmetrical, and has only six different components.

It is known from tensor calculus that any symmetrical tensor has a simple geometrical interpretation in terms of the tensor ellipsoid, whose equation in an arbitrary system of coordinates is

$$\mu_{11} x_1^2 + \mu_{22} x_2^2 + \mu_{33} x_3^2 + 2\mu_{23} x_2 x_3 + 2\mu_{31} x_3 x_1 + 2\mu_{12} x_1 x_2 = 1. \tag{1.53}$$

In the present case the tensor ellipsoid is called the *moment ellipsoid*. The length of the radius vector of the point where the moment ellipsoid intersects the x_1-axis can be found from (1.53) by putting $x_2 = x_3 = 0$. The result is

$$r_1^2 = 1/\mu_{11}. \tag{1.54}$$

Since the direction of the x_1-axis is arbitrary, this equation gives a fundamental geometrical property of every point on the moment ellipsoid, namely that the length of its radius vector is the inverse square root of the variance in the corresponding direction.

On transforming the equation of the moment ellipsoid to its principal axes, we obtain

$$\Sigma_1^2 x_1^2 + \Sigma_2^2 x_2^2 + \Sigma_3^2 x_3^2 = 1. \tag{1.55}$$

The quantities Σ_1^2, Σ_2^2, Σ_3^2 are called the *principal variances*, and the corresponding axes of the ellipsoid are called the *principal axes*. When the directions of these axes and the values of the principal variances are given, the variance in any direction is known.

We have seen in § 1 that the variance of any univariate distribution function is entirely determined by specifying the corresponding normal (Gaussian) function. We shall now show that this result can be generalised to distribution functions of more than one independent variable, and, in particular that for any distribution function of three independent variables the variance in any direction is completely determined by specifying the corresponding Schwarzschild function.

If we know the directions of the principal axes and the magnitudes of the principal variances of the given distribution function, we can normalise it and refer it to its centre, and thus have the following expression for the variance σ_l^2 in an arbitrary direction l whose direction cosines relative to the principal axes are $\alpha_1, \alpha_2, \alpha_3$:

$$\sigma_l^2 = \int (\alpha_1 x_1 + \alpha_2 x_2 + \alpha_3 x_3)^2 f(x, y, z)\, d\omega$$
$$= \alpha_1^2 \Sigma_1^2 + \alpha_2^2 \Sigma_2^2 + \alpha_3^2 \Sigma_3^2 \tag{1.56}$$

since all the mixed moments relative to the principal axes are zero.

By analogy with the case of one variable, we define the *corresponding ellipsoidal function* by the expression

$$\varphi(x_1, x_2, x_3) = \frac{1}{(2\pi)^{3/2}\Sigma_1\Sigma_2\Sigma_3} \exp\left[\frac{1}{2}\left(\frac{x_1^2}{\Sigma_1^2} + \frac{x_2^2}{\Sigma_2^2} + \frac{x_3^2}{\Sigma_3^2}\right)\right]$$
$$= \frac{H_1 H_2 H_3}{\pi^{3/2}} \exp\left[-(H_1^2 x_1^2 + H_2^2 x_2^2 + H_3^2 x_3^2)\right]. \tag{1.57}$$

Here $\Sigma_1^2, \Sigma_2^2, \Sigma_3^2$ denote the principal variances of the given distribution function, and H_1, H_2, H_3 are *degrees of concentration*, analogous to that used in the case of one variable, related to the principal variances by

$$H_1 = 1/\sqrt{2}.\Sigma_1, \quad H_2 = 1/\sqrt{2}.\Sigma_2, \quad H_3 = 1/\sqrt{2}.\Sigma_3. \tag{1.58}$$

The ellipsoidal function thus defined is a Schwarzschild function. It is obviously the product of three normal (Gaussian) functions, each depending on only one argument. The principal variances and the directions of the principal axes of the Schwarzschild function are the same as those of the given distribution function.

Hence it follows immediately that the variance of the Schwarzschild function in any direction l is the same in magnitude as the variance of the given function in that direction. This is seen from the fact that, for the Schwarzschild function, the variance σ_l^2 is given by exactly the same formula as (1.51).

Thus we see that the Schwarzschild function plays the same part with respect to a multivariate distribution as the Gaussian function for a univariate distribution.

§ 3. Statistical Properties of Stars

In stellar statistics no importance attaches to individual stars. When individual stars are considered, it is only as representatives of a wider class of similar objects, the members of a statistical ensemble. In applying the statistical method, we form a statistical ensemble of stars by selecting, out of all the observable stars, those having some one or more particular properties.

The properties most generally considered in stellar astronomy and stellar dynamics may be divided into three groups: geometrical (celestial coordinates, parallax), kinematic (proper motion, radial velocity, spatial velocity) and physical (mass, apparent magnitude, absolute magnitude).† The geometrical properties characterise the disposition (actual and apparent) of the stars in space, the kinematic and geometrical properties together characterise their motion, and the physical properties are determined by the masses, luminosities and temperatures of the stars.

A. Geometrical Properties

(1) *Celestial coordinates.* These are used to define the apparent position of stars in the sky. To set up a system of coordinates on the celestial sphere, we take two diametrically opposite points as the *poles* of the system, and the diameter joining them as its *axis.* The great circle perpendicular to the axis is called the *equator* of the system. For example, the two diametrically opposite points which remain fixed during the diurnal rotation of the stars are the poles of the system of *equatorial coordinates*, which is much used in astronomy, though not in stellar dynamics. These two points are called the *celestial poles*, and the corresponding equator is the *celestial equator.* The latter divides the celestial sphere into two hemispheres, the northern and the southern. Accordingly that pole of any other coordinate system which lies in the northern hemisphere is called the northern pole, and the other the southern pole.

The positive direction of rotation about the axis of the system is taken to be that of counterclockwise rotation when viewed from the northern pole. On the equator of each system†† is its *ascending node*, which is the point where it crosses the celestial equator from south to north in the positive direction. The poles, equator and ascending node are the fundamental elements which define any system of celestial coordinates. In the equatorial system the ascending node is replaced by the *vernal equinox*, which is the ascending node of the *ecliptic*. The latter is the great circle defined by the annual motion of the centre of the Sun. The ascending node is usually denoted by Ω, and the vernal equinox by Υ, a conventional representation of the head of Aries, the Ram.

The great semicircle through any point and the poles of a system is called the *meridian* of that point; the meridian of the ascending node is called the *prime meridian*, and the small circles parallel to the equator of the system are called *circles of latitude*. The meridians and circles of latitude form two families of coordinate curves. The position of any point on the celestial sphere is defined by the two curves which pass through that point. The dihedral angle l between the meridian of a point and the prime meridian, measured in the positive direction from 0 to 360°, is called the *longitude* of the point. The longitude is also given by the arc of the equator between the ascending node and the point where the equator meets the meridian of the point. The angular distance b between the point and the equator is called its *latitude*. This is measured by the arc of the meridian between the equator and the point considered, measured from 0 to $+90°$ towards the northern pole and from 0 to $-90°$ towards the southern pole.

The length Δs of a circle of latitude which corresponds to a longitude difference Δl is evidently

$$\Delta s = \Delta l \cos b. \tag{1.59}$$

† Only those properties which are to be used in what follows are mentioned here.

†† Except, of course, the equatorial system.

In the equatorial coordinate system the longitude is called the *right ascension* and usually denoted by α, and the latitude is called the *declination* and denoted by δ.

Although equatorial coordinates themselves are seldom used in stellar astronomy, they are fundamental because all other systems are defined with respect to them.

The choice of any particular coordinate system is a matter of convenience. In stellar astronomy the most frequently used is that of *galactic coordinates*, defined by the *galactic equator* and the *galactic poles*. The galactic equator is the great circle corresponding to the maximum concentration in the apparent distribution of stars in the sky. By international agreement the values at present accepted for the equatorial coordinates of the northern galactic pole are

$$\alpha = 190°, \quad \delta = +28°.† \tag{1.60}$$

The plane of the galactic equator is parallel to the *galactic plane*, i.e. the plane of symmetry of the Galaxy. This explains the preference given to galactic coordinates. In some problems of stellar dynamics, however, it is more convenient to take an arbitrary system of coordinates, wherein the general properties of stellar motions in the Galaxy are more easily discussed.

Celestial coordinates are used to study the apparent distribution of stars in the sky. By counting the stars per square degree in various parts of the sky we can locate star clusters, which are regions where the star density exceeds the average value. Regions where the star density is below the average are usually due to the presence of "dark nebulae", i.e. masses of dark diffuse matter which lie between the observer and the stars and partly or wholly cut off their light. Finally, by smoothing the star counts and thus eliminating local anomalies, we can find general laws of the apparent distribution of stars, which correspond to features of the structure of the Galaxy.

(2) *Parallax*. The *parallax* of a star is a quantity inversely proportional to its distance from the observer. In astronomy it was originally used as a measure of the distance. The parallax is equal to the angle subtended by the major semiaxis of the ellipse which the star describes on the celestial sphere as a result of the Earth's orbital motion round the Sun. The parallax π, expressed in seconds of arc, and the distance r are related by

$$\pi = d/r \tag{1.61}$$

where d is a constant, which is evidently the distance corresponding to a parallax of 1″. This distance is called a *parsec* (abbreviated ps):††

$$1\,\text{ps} = 3{\cdot}08 \times 10^{18}\,\text{cm}. \tag{1.62}$$

If distances are expressed in parsecs, the formula for the parallax becomes simply

$$\pi = 1/r. \tag{1.63}$$

† According to a decision of the International Astronomical Union (1959) the coordinates of the galactic pole are to be taken as $\alpha = 192°{\cdot}25$, $\delta = +27°{\cdot}4$ (epoch 1950·0), but the difference is of no importance in theoretical discussions.

†† The length of the parsec is the mean distance of the Earth from the Sun ($1{\cdot}495 \times 10^{13}$ cm) multiplied by the number of seconds in one radian ($2{\cdot}063 \times 10^5$).

In stellar dynamics it is distances rather than parallaxes which are chiefly of interest. Wherever necessary we shall therefore eliminate parallaxes by means of formula (1.63). Other units, used in expressing great distances, are the *kiloparsec* (kps) and the *megaparsec* (Mps), equal to one thousand and one million parsecs respectively. The nearest star, Proxima Centauri, is at a distance of 1·3 ps. Only 24 stars, not counting the Sun, lie within 4 ps. Thus even the parsec is a small unit in terms of stellar distances.

It should be noted that a direct trigonometrical determination of stellar parallaxes, even with modern equipment, is possible only up to distances of at most 100 ps, since at greater distances the errors of measurement exceed the quantity to be measured. To determine greater distances numerous indirect methods exist, which we shall discuss below, but they are all based on the direct measurement of trigonometrical parallaxes, whereby indirect measurements of distance can be calibrated.

Together with the celestial latitude and longitude, the distance completely defines the position of any point in space, and forms a system of spherical coordinates centred at the observer. If the celestial coordinates and the parallaxes of the stars are known, therefore, the spatial structure of the stellar system can be investigated.

B. Kinematic Properties

(3) *Proper motion.* The *proper motion* of a star is its apparent angular displacement per year. Unlike the parallax and the aberration, which involve a periodic motion of the star about its mean position, the proper motion is progressive. Thus the proper motion of a star is equal to the annual change in its mean position. In terms of components μ_l in longitude and μ_b in latitude, the proper motion is

$$\mu_l = (\mathrm{d}l/\mathrm{d}t)\cos b, \quad \mu_b = \mathrm{d}b/\mathrm{d}t, \tag{1.64}$$

where the unit of time is one year. The proper motion is usually expressed in seconds of arc per year.†

The quantity

$$\mu = \sqrt{(\mu_l^2 + \mu_b^2)} \tag{1.65}$$

is called the *total proper motion*, and is due to the *transverse motion* of the star, i.e. its spatial motion projected on a plane perpendicular to the line of sight. If the distance and proper motion of a star are known, its *transverse velocity* can be calculated as

$$v_t = k\mu r. \tag{1.66}$$

The coefficient $k = 4{\cdot}738$ is the quotient of the number of kilometres in one astronomical unit by the number of seconds in one year, $3{\cdot}156 \times 10^7$. The quantity v_t in (1.66) is expressed in kilometres per second, as is usual in astronomy.

The proper motions of the stars are small fractions of a second per year. One thousandth of a second per year is a large proper motion. This is mainly due to the great distances of the stars. The transverse velocities of the stars are mostly between 20 and 60 km/s.

† The proper motion in equatorial coordinates α, δ: $\mu_\alpha = (\mathrm{d}\alpha/\mathrm{d}t)\cos\delta$, $\mu_\delta = \mathrm{d}\delta/\mathrm{d}t$, can easily be transformed to any other coordinates. Usually it must be remembered that the catalogues give not μ''_α but $\mu^s_\alpha = \mathrm{d}\alpha/\mathrm{d}t$ with α expressed in seconds of time. Hence $\mu_\alpha = 15\mu^s_\alpha \cos\delta$.

(4) *Radial velocity.* The *radial velocity* of a star is its velocity in the line of sight. It is determined from the Doppler shift of the absorption lines in the star's spectrum. The shift $\Delta\lambda$ of a line at wavelength λ is related to the radial velocity v_r by Doppler's formula $\Delta\lambda/\lambda = v_r/c$, where c is the velocity of light. In practice $\Delta\lambda$ is found from not one but all the lines which can be measured. The value of v_r is found in velocity units (e.g. kilometres per second, the unit commonly used today). In star catalogues, radial velocities are given with respect to the Sun, for which purpose the observed values are corrected for the Earth's orbital motion.

(5) *Spatial velocity.* If the parallax and proper motion of a star are known, the components of the transverse velocity in longitude v_l and in latitude v_b in any coordinate system can be calculated. From formula (1.66) we find

$$v_l = k\mu_l r, \quad v_b = k\mu_b r. \tag{1.67}$$

Together with the radial velocity, v_l and v_b comprise the components of the *spatial velocity* of the star, $\mathbf{V} = (v_l, v_b, v_r)$. In what follows it will be useful to consider the components

$$\mathbf{v}_r = v_r\mathbf{r}^0, \quad \mathbf{v}_l = v_l\mathbf{l}^0, \quad \mathbf{v}_b = v_b\mathbf{b}^0, \tag{1.68}$$

where $\mathbf{r}^0$, $\mathbf{l}^0$, $\mathbf{b}^0$ are unit vectors in the directions of the radius vector, the tangent to the circle of latitude, and the tangent to the meridian respectively (Fig. 4). These form a coordinate trihedral at every point of the celestial sphere, and we shall call it the *moving trihedral.*

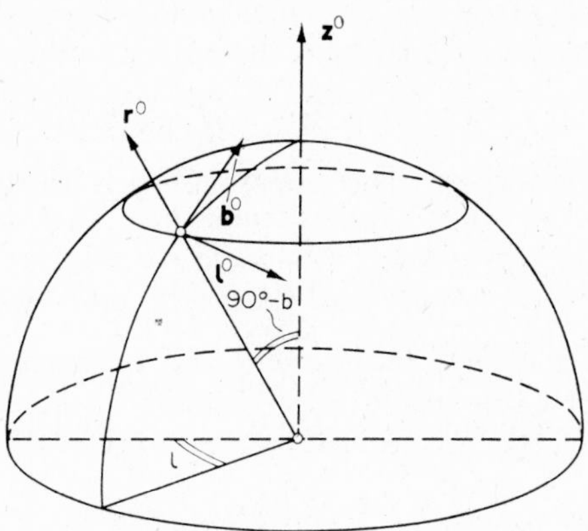

FIG. 4

The unit vectors along the axes of the moving trihedral are in turn determined by the choice of the coordinate system used and by the position of the point considered. Let $\mathbf{z}^0$ be a unit vector in the direction of the northern pole of the coordinate system chosen. If this vector is specified, the position of the coordinate system is evidently determinate. Figure 4 shows that the unit vectors of the moving trihedral are given in terms of $\mathbf{z}^0$, $\mathbf{r}^0$ and the latitude b by

$$\mathbf{l}^0 = \mathbf{z}^0 \times \mathbf{r}^0 \sec b, \quad \mathbf{b}^0 = \mathbf{z}^0 \sec b - \mathbf{r}^0 \tan b. \tag{1.69}$$

Formulae (1.68) and (1.69) are very useful in resolving the spatial velocity into components in a given coordinate system.

C. Physical Properties

(6) *Mass.* The *mass* is a very important property of a star. Unfortunately, our knowledge of the masses of individual stars is as yet scanty. They can be directly determined only for double stars whose orbital motions are completely known. At present such systems number about a hundred. The results have shown that the masses of most stars do not differ greatly from that of the Sun. In general, however, stellar masses can be determined only by indirect methods, which we shall discuss below.

(7) *Apparent magnitude.* The *apparent magnitude* of a star is a measure of the amount of light which reaches us from the star. The concept was introduced by Hipparchus in the 2nd century B.C. as a qualitative classification of the stars. The relation between the magnitudes m_1, m_2 of two stars and their brightnesses i_1, i_2 is given by *Pogson's formula:*

$$m_2 - m_1 = 2{\cdot}5 \log_{10}(i_1/i_2). \tag{1.70}$$

Thus the apparent magnitude gives a measure of the apparent brightness of a star on a logarithmic scale.†

In particular, if $m_2 - m_1 = 5$, formula (1.70) shows that an increase of 5 in the apparent magnitude corresponds to a reduction in brightness by a factor of 100. Putting $m_2 - m_1 = 1$, we find that an increase of 1 in the apparent magnitude corresponds to a reduction in brightness by a factor of $100^{1/5} = 2{\cdot}512$.

The brightest star in the sky, Sirius, has an apparent magnitude $m = -1{\cdot}6$. The faintest stars visible to the unaided eye are of the sixth magnitude. The stars at the limit of visibility by photography with the largest existing telescopes are of the twenty-fourth magnitude, i.e. they are 10^{10} times fainter than a star of magnitude -1.

(8) *Absolute magnitude.* The *absolute magnitude* of a star is the apparent magnitude which it would have at a certain standard distance, and its *luminosity* is the brightness which it would have at that distance.

If the apparent magnitude and brightness of a star are denoted by m and i, and its absolute magnitude and luminosity by M and I, then, using Pogson's formula (1.70), we find

$$M - m = 2{\cdot}5 \log_{10}(i/I). \tag{1.71}$$

Denoting by r_0 the standard distance and by r the actual distance of the star, and assuming that there is no absorption of light in interstellar space, we have

$$i/I = r_0^2/r^2, \tag{1.72}$$

whence

$$M = m + 5 \log_{10}(r_0/r). \tag{1.73}$$

In astrophysics it is convenient and customary to take $r_0 = 10$ ps, but formula (1.73) is simpler if we take $r_0 = 1$ ps. Then the formula for the absolute magnitude as

† The logarithms are to base 10.

generally used in theoretical calculations is

$$M = m - 5 \log_{10} r. \tag{1.74}$$

It should be remembered that the absolute magnitudes given by this formula are five units less than those usually stated.

The apparent magnitude of any object is fairly easily measured. To determine the distance of a star, therefore, we need only know its absolute magnitude and apply formula (1.74). The great majority of indirect methods of measuring distances in modern stellar astronomy are based on estimates of the absolute magnitudes of stars and nebulae obtained from physical considerations. Such are the *spectroscopic parallaxes*, obtained from the relative strength of the absorption lines in stellar spectra; the parallaxes of Cepheids, the parallaxes of the blue giant stars, and so on.

(9) *Spectral class.* The *spectral class* is a very important characteristic of a star, since it is related to the fundamental properties of the star's physical structure. By their spectra we distinguish the different types of star among the members of stellar systems; these different types have different physical structures and also different kinematic properties and different distributions within the systems.

The Harvard classification of stellar spectra, which is now in general use, is based on estimates of the relative strength of absorption lines. Thus it is based on the degree of ionisation of atoms in stellar atmospheres. This in turn depends mainly on two quantities, the temperature and the pressure. The temperatures of the great majority of stellar atmospheres lie between 3000 and 25,000°K, and temperature differences are primarily responsible for the observed properties of stellar spectra. Thus the division of the stars into spectral classes is essentially a classification by temperature. Over 99% of all stars in the present classification are placed in one of the six classes called B, A, F, G, K and M.

If the spectral class of one star precedes that of another star in this list, then the first star is said to be of *earlier* type, and the second of *later* type. The earliest classes are B and A, and the latest are K and M.

Since star temperatures vary continuously, the spectral class must also usually be supposed to take a continuous series of values. The letter of the class is therefore often followed by a figure from 0 to 9, representing a tenth part of the whole spectral class. For example, A9 is a type which is one-tenth of a class earlier than F0.

Within each sub-class the temperature may be taken as almost constant, and so the differences between spectra of a given type arise mainly because of differences in the pressure prevailing in the atmospheres of the stars.

D. Statistical Relations between Properties of Stars

A number of important relations between the various properties of stars have been empirically established, usually by a comparison of the directly observed values of two (or more) quantities for some group of stars. The statistical relation thus obtained is then extrapolated to all stars, or to some larger group. Although such statistical relations are sometimes without the necessary theoretical foundation, they are of great importance in astronomy. In particular, as we shall see, they can be used to determine a number of properties of stars whose direct determination is impossible.

Here we shall discuss only some of these statistical relations.

(1) *The spectrum–luminosity relation: the Hertzsprung–Russell diagram.* In 1913 Russell showed that stars of spectral type G0 and later can be divided fairly definitely into two groups on the basis of their absolute magnitudes. In one group the mean absolute magnitude is constant and equal to 0^m. These are called *giant stars.* The majority of stars, however, belong to the other group, called the *main sequence*.

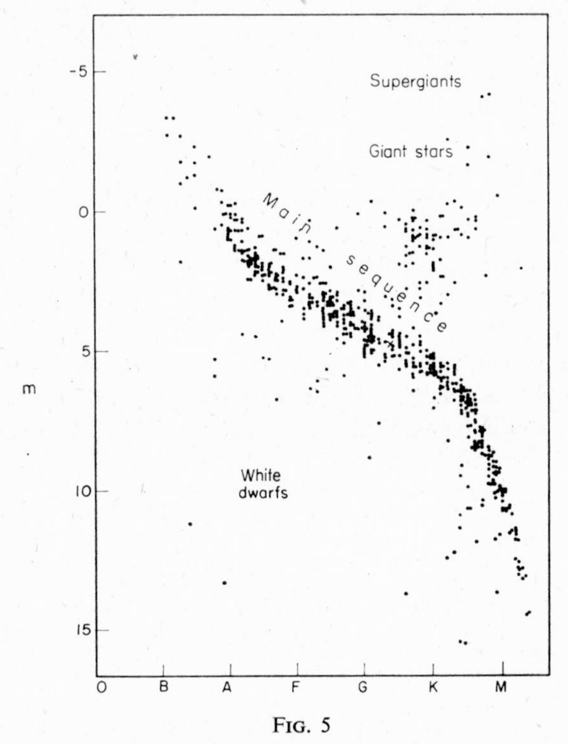

FIG. 5

The main sequence begins with bright stars of the early types, but, as we proceed to later types, the luminosity decreases, i.e. the absolute magnitude increases. The main-sequence stars of type G0 and later are called *dwarfs*. Besides the giants and the main-sequence stars, there are other classes, but these are at present of minor importance in stellar dynamics, either because they are few in number, like the *supergiants*, or because they have not been much studied, like the *white dwarfs*.

The general distribution of the stars among the four classes of main sequence, giants, supergiants and white dwarfs is shown in the Hertzsprung–Russell diagram (Fig. 5). The main sequence forms a diagonal band, and the giants form a horizontal band.

The division of the stars into giants and dwarfs is of importance in stellar dynamics. The giants differ from the dwarfs in having greater masses, smaller velocities, and accordingly a greater galactic concentration, which leads to a considerable difference in the nature of their movement in the Galaxy. The stars are classified with respect to luminosity according to the Yerkes system, whereby the spectral class is followed by a roman numeral giving the *luminosity class:* I signifies a supergiant, II a bright giant, III a giant, IV a subgiant, and V a dwarf.

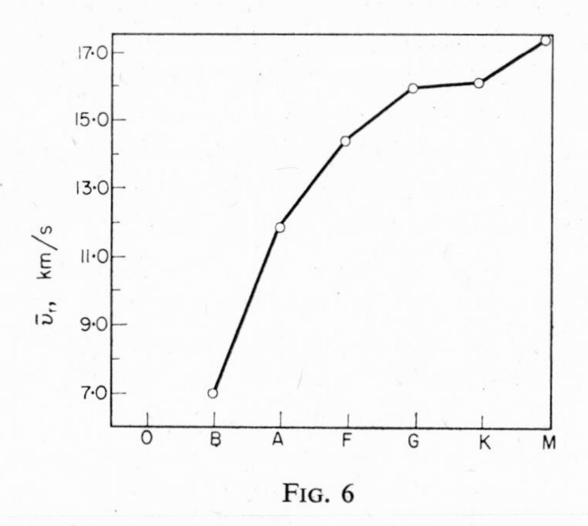

FIG. 6

(2) *The spectrum–velocity relation.* The mean radial velocities $\bar{v}_r$ of the stars exhibit a marked dependence on spectral type, which is illustrated by Fig. 6. The radial velocities of late-type stars are approximately twice those of the early types B and A.

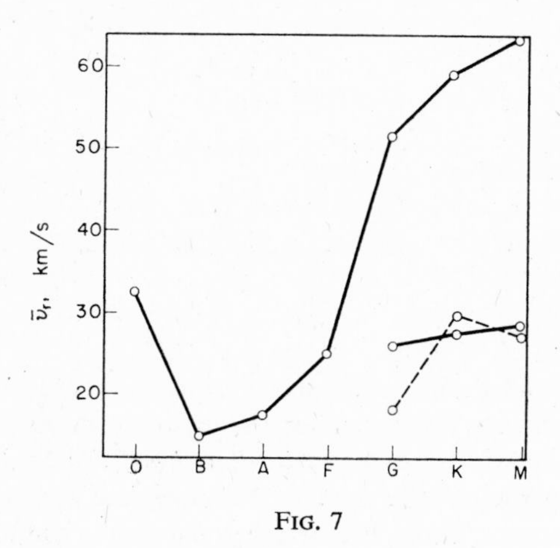

FIG. 7

The difference in radial velocities is, however, merely a consequence of a still greater difference in the spatial velocities of giants and dwarfs, as shown in Fig. 7. The thick line shows the mean spatial velocity of stars of the main sequence, the short continuous line the same quantity for giants, and the broken line the same quantity for supergiants.

From these curves we see that the velocities of the stars depend considerably, not only on spectral type, but also on absolute magnitude: the average velocities of giants are only half those of dwarfs. The above-mentioned difference between the velocities of stars of early and late types is undoubtedly due mainly to the fact that the bright stars of late types which we observe include a considerable number of dwarfs.

The division of the stars into two groups having very different velocities is of great importance in galactic dynamics. It shows, in particular, that the Galaxy is not a homogeneous stellar system, but consists of at least two sub-systems having different dynamical characteristics. This concept, first put forward by Lindblad [65] in 1926, was afterwards confirmed in 1943 by Baade [8], who discovered marked differences in the shapes of spiral nebulae when photographed in light of different colours. Whereas in ordinary photography these nebulae appear flat and lens-shaped, in red light they are almost spherical. Thus the distribution of the late-type red stars in these galaxies is quite different from that of the early-type blue stars. They form two distinct sub-systems, one flat and the other spherical. It is of interest to note that only the blue and white stars have the spiral distribution characteristic of most galaxies. In red light these galaxies exhibit no trace of spiral structure.

The concept of the division of the stars in galaxies into sub-systems was further developed by the Soviet astronomers B.V. Kukarkin and P.P. Parenago and applied with success to the solution of various problems in the structure and evolution of stellar systems.

(3) *The mass–luminosity relation.* By comparing the masses of stars (where these had been directly determined) with their absolute magnitudes, Eddington detected an evident relation between the two quantities. This is called in astrophysics the *mass–luminosity relation*, and is of great importance as regards the internal structure of the stars. We can say that, except for the white dwarfs, the masses of all the stars are approximately given by the linear relation

$$M = 3{\cdot}9 \log_{10} \mathfrak{M} + M_0,$$

which follows from the empirical formula $L = \alpha\mathfrak{M}^{3\cdot9}$. Here M and L are the bolometric absolute magnitude and luminosity, $\mathfrak{M}$ the mass, and α and M_0 constants. Hence it follows that the variations in mass among the stars are much less than the variations in luminosity. If the Sun's mass and luminosity are taken as unity, the luminosities of the stars vary from 1/50,000 to 50,000, but their masses (with few exceptions) vary only from 0·25 to 10.

This result means that, in most problems of stellar dynamics, the differences in mass of the stars may be neglected and a mean value taken. The value usually taken is 10^{33} g, which is half the mass of the Sun.

The statistical properties discussed above form the raw material for theoretical studies in stellar dynamics. The aim of the theory is to describe the structure, motion and evolution of stellar systems. To achieve this, starting from the observed distribution of properties, we must find, firstly, the *star density* (the distribution function for spatial coordinates) and, secondly, the *velocity function* (the distribution function for spatial velocities). The determination of the former function is the main task of stellar statistics, and that of the latter function is the main task of stellar kinematics.

CHAPTER II

THE KINEMATICS OF STELLAR SYSTEMS

§ 1. The Macroscopic Volume Element and the Star Density Function

IN MANY problems of the structure and evolution of stellar systems we can, as a first approximation, ignore the differences in the physical structure of the individual stars and regard them as moving particles. Their masses also may be regarded as equal, since the differences in mass are unimportant and can be significant only in a few particular problems. Thus stellar systems are, as a rule, considered in stellar dynamics as systems of particles of equal mass. In such conditions, as we shall see, there is a distribution function which entirely defines the state of the stellar system and is thus of great importance; it is the *phase density*. If this function is known, all other distribution functions used in stellar astronomy can be mathematically derived from it, and in particular the size, shape and motion of the system at any instant can be found. The most important mathematical problem in stellar astronomy is therefore to determine the phase density function.

In order to find the phase density function we consider a *macroscopic volume element* $d\omega$ in the stellar system; for example, a rectangular parallelepiped with its centre at (x, y, z) and its edges parallel to the coordinate axes. This volume element must satisfy two conditions, which are to some extent contradictory. Firstly, it must be small in comparison with the dimensions of the stellar system itself, since it is to be used primarily to describe the structure and motion of the system in the immediate neighbourhood of the point (x, y, z), i.e. the *local properties* of the system. Secondly, the volume element must contain so many stars that statistical methods may be applied to it. Evidently this means that the volume element must be fairly large. Thus the choice of the most suitable dimensions of the macroscopic volume element must be made for each stellar system separately. The greater the homogeneity of structure of the system at the point considered, and the larger the system, the larger may be the volume element, but within the range which satisfies the above requirements its size and shape may be chosen arbitrarily. For example, in the neighbourhood of the Sun the density of stars is one per 10 ps^3. Hence a cube of side 100 ps will contain about 100,000 stars, yet it will be small compared with the Galaxy, whose diameter is 300 times greater.

Let the number of stars in a macroscopic volume element $d\omega$ be dN. The ratio $dN/d\omega$ gives the number of stars per unit volume at the point considered in the system, i.e. it is the *star density*. By virtue of the conditions imposed on the volume element, the ratio $dN/d\omega$ should be (within certain limits) statistically independent of the shape and size of $d\omega$, depending only on the coordinates of the point. Thus we have

$$dN(x, y, z) = \nu(x, y, z)\, d\omega;$$

the proportionality coefficient is called the *star density function*. It may in general depend on the time as well as on the coordinates.

In what follows we shall sometimes use an abbreviated notation customary in statistical physics, writing $\nu(x, y, z) = \nu(q)$, where q denotes every coordinate. Evidently $\nu(q)$ satisfies the condition $\int \nu(q)\, dq = N$, where N is the total number of stars in the system, and the integral is taken over the whole volume of the system.

Besides $\nu(q)$ we may define the normalised star density function $\nu_0(q)$ by

$$\nu(q) = N\nu_0(q).$$

Then $\nu_0(q)\, dq$ gives the fraction of the total number of stars which is within the volume element dq, or the probability that a star chosen at random will be in that element. Evidently $\int \nu_0(q)\, dq = 1$.

The equation $\nu(q) = C$, where C is any sufficiently small positive constant, clearly defines a level surface of star density in the system. If ν_m denotes the maximum star density in the system, then for values of C between 0 and ν_m we obtain a family of level surfaces which entirely define the size and shape of the system, as well as the distribution of stars within it.

§ 2. The Phase Density Function

Let us now proceed to determine the phase density function. Each star within the chosen volume element has, in a system of coordinates which we suppose fixed, a definite velocity $\mathbf{V}$, say, with components u, v, w.

We now take an arbitrary coordinate system and draw from its origin the velocity vectors of all the stars in the volume element. Then the termini of these vectors form a set of points in *velocity space*, the latter being a three-dimensional space with coordinates u, v, w. We can now apply to this set of points the same procedure as we did in § 1 to the stellar system itself. We take a new macroscopic volume element $d\Omega$, for example a rectangular parallelepiped with its centre at a point (u, v, w) and its edges parallel to the coordinate axes. Let dN be the number of points lying within this volume element. Then we again put this number proportional to the volume $d\Omega$:

$$dN = \varphi(x, y, z, u, v, w)\, d\omega\, d\Omega.$$

It should be mentioned that the conditions which must be imposed on dN are much less stringent than the restrictions on the choice of the volume $d\omega$ in coordinate space. In deriving the star density function it was necessary to ensure the possibility of subsequently establishing the stellar velocity distribution, for which thousands of stars are needed. The volume element $d\Omega$ need contain only tens or hundreds of stars.

Evidently the above result could also be achieved by starting with a six-dimensional Euclidean space and regarding each star as a point with six coordinates, namely three space coordinates x, y, z and three velocity coordinates u, v, w. Such spaces are much used in statistical physics, and are called *phase spaces*. By analogy with the definition of the star density in the coordinate space (x, y, z), we may define the *phase density function* as the proportionality coefficient in the relation between the number of points in a macroscopic volume element in phase space and the volume $\mathrm{d}\omega\,\mathrm{d}\Omega$ of the element.

In the abbreviated notation of statistical physics, we write p for every component of a star's velocity, $\mathrm{d}p$ for the volume element in velocity space, and $\varphi(q, p)$ for the phase density. Then the above equation becomes $\mathrm{d}N = \varphi(q, p)\,\mathrm{d}q\,\mathrm{d}p$.

It is evident that the phase density function may be defined as that function of coordinates and velocities which, multiplied by a macroscopic volume element $\mathrm{d}q\,\mathrm{d}p$, gives the number of stars lying within a macroscopic volume element $\mathrm{d}q$ in coordinate space and having velocity vectors whose termini lie within a macroscopic volume element $\mathrm{d}p$ in velocity space. In general, the phase density of a stellar system depends, not only on the six phase coordinates q and p, but also on the time t. The system is then said to be *non-steady*. It should be borne in mind that the time dependence of the phase density of a non-steady stellar system is different in nature from its dependence on the phase coordinates. With respect to the coordinates, the phase density is a distribution function, but the time can appear in it only as a parameter.

There is an evident relation between the phase density and the star density:

$$\nu(q) = \int \varphi(q, p)\,\mathrm{d}p, \tag{2.1}$$

where the integration is extended over all velocity space. This relation shows that, if the phase density is given, the star density function is determined.

§ 3. The Velocity Distribution Function

If the values of the coordinates in the phase density function are supposed known, we obtain another function of importance in stellar dynamics: the velocity density distribution function or, more briefly, the *velocity function*. This gives the distribution of velocities for a macroscopic volume element whose centre is at the point (x, y, z), these coordinates appearing as parameters. Thus the velocity function gives the distribution, in magnitude and direction, of the velocities of the stars at a particular point in the stellar system.

At each (fixed) point (x, y, z) in coordinate space we can set up a coordinate system (u, v, w) which defines velocity space. In this space we select a macroscopic volume element $\mathrm{d}\Omega$. Then the number of points in this volume is evidently

$$\begin{aligned}\mathrm{d}n &= \varphi(x, y, z, u, v, w)\,\mathrm{d}\Omega \\ &= \varphi(x, y, z, u, v, w)\,\mathrm{d}u\,\mathrm{d}v\,\mathrm{d}w\end{aligned}$$

or, briefly, $\mathrm{d}n = \varphi(q, p)\,\mathrm{d}p$. In actual systems the velocities of the stars are bounded in magnitude, and beyond certain limits the function φ is zero. We must therefore suppose *a priori* that the velocity function is non-negative for all values of its arguments and that, as any velocity component increases *ad infinitum*, the velocity function tends to zero faster than any power of p, i.e.

$$\lim_{p \to \pm\infty} p^m \varphi(q, p) = 0 \tag{2.2}$$

for any $m > 0$. We shall also suppose that all partial derivatives of the velocity function needed in the calculations exist.

The equation $\varphi(q, p) = C$, where C is a positive constant, defines, for given coordinates x, y, z, a level surface of the velocity function. By taking all possible values of C, we obtain a family of level surfaces of the velocity function, which gives the shape and size of the assembly of points in velocity space in the same way as the level surfaces of the star density function give the shape and size of the stellar system itself.

In the simplest case, the level surfaces of velocity are concentric spheres or ellipsoids. The velocity distribution is then said to be *spherical* or *ellipsoidal.* In general, an ellipsoidal law is given by

$$\varphi(q, p) = \Psi(T),$$

where Ψ is an arbitrary function of its argument, and T is the most general positive-definite quadratic function of the velocities:

$$T = au^2 + bv^2 + cw^2 + 2fvw + 2gwu + 2huv + pu + qv + rw + s.$$

The coefficients $a, b, c, \ldots, s$ are functions of the coordinates x, y, z and in general also of the time t.

If the quadratic form T is rendered homogeneous, we have

$$\varphi(q, p) = \Psi(Q + \sigma), \tag{2.3}$$

where

$$\begin{aligned} Q = {} & a(u - u_0)^2 + b(v - v_0)^2 + c(w - w_0)^2 \\ & + 2f(v - v_0)(w - w_0) + 2g(w - w_0)(u - u_0) + 2h(u - u_0)(v - v_0) \end{aligned} \tag{2.4}$$

and σ is some function of the coordinates. The functions u_0, v_0, w_0 of the coordinates are the mean components of the total velocity of a star along the coordinate axes, i.e. the components of the *mean velocity*.† Referring the quadratic form Q to its principal axes, we have

$$Q = h^2(u - u_0)^2 + k^2(v - v_0)^2 + l^2(w - w_0)^2. \tag{2.5}$$

The most important particular ellipsoidal law is Schwarzschild's law

$$\varphi(q, p) = \mathrm{e}^{-(Q + \sigma)}. \tag{2.6}$$

† The mechanical significance of these quantities will be explained in § 4.

From (2.3) and (2.1) it follows that the "total population" for the velocity function is represented by the product $dN = \nu(q)\,dq$. Hence, to determine σ, we can use the results of Chapter I, § 2. In particular, by analogy with (1.39) we obtain for Schwarzschild's function, referred to principal axes, the expression

$$\varphi(q,p) = \nu \frac{hkl}{\pi^{3/2}} \exp\{-[h^2(u-u_0)^2 + k^2(v-v_0)^2 + l^2(w-w_0)^2]\}, \quad (2.7)$$

and by analogy with (1.41), for arbitrary axes,

$$\varphi(q,p) = \nu \sqrt{\frac{D}{\pi^3}} \cdot e^{-Q}, \quad (2.8)$$

where Q is the quadratic form (2.4) and D is its discriminant:

$$D = \begin{vmatrix} a & h & g \\ h & b & f \\ g & f & c \end{vmatrix}. \quad (2.9)$$

When $h = k = l$ the ellipsoidal law (2.5), (2.7) becomes a spherical one:

$$\varphi(q,p) = \Psi\{h^2[(u-u_0)^2 + (v-v_0)^2 + (w-w_0)^2]\}. \quad (2.10)$$

The most frequent particular case of a spherical law is *Maxwell's law*:

$$\varphi(q,p) = \frac{\nu h^3}{\pi^{3/2}} \exp\{-h^2[(u-u_0)^2 + (v-v_0)^2 + (w-w_0)^2]\}, \quad (2.11)$$

which is widely used in statistical physics and also in some problems of stellar dynamics. It is evidently a particular case of Schwarzschild's law.

§ 4. Centroids

The concept of the *centroid* is of importance in describing the motion of stellar systems. This is a point which is at rest relative to the macroscopic volume element to which it belongs. In other words, the centroid is a point which, at a given instant, moves with the mean velocity of the stars in a given macroscopic volume element.

If φ is the velocity function for the volume element considered, then the equations

$$\begin{gathered} \overline{\mathbf{V}} = \frac{1}{\nu} \int \mathbf{V} \varphi \, d\Omega, \\ \bar{u} = \frac{1}{\nu} \int u\varphi \, d\Omega, \quad \bar{v} = \frac{1}{\nu} \int v\varphi \, d\Omega, \quad \bar{w} = \frac{1}{\nu} \int w\varphi \, d\Omega \end{gathered} \quad (2.12)$$

give the velocity of the centroid and its components along the coordinate axes. The centroid coincides with the statistical centre of the velocity distribution in the volume element, i.e. at a given point in the stellar system. For example, in the case of an ellipsoidal law (in particular, Maxwell's or Schwarzschild's law) the velocity components of the centroid are u_0, v_0, w_0 as defined in § 3.

The significance of centroids in stellar dynamics is that they make possible a transition from a stellar system regarded as a discrete medium (consisting of a large number of homogeneous stars) to one which is continuous throughout some volume of space. Without the use of centroids, the interpretation of internal motions in stellar systems is not possible. As an example, we may consider rotation about the axis of the Galaxy. The circular movement about the axis pertains to the centroids, not to the individual stars. The orbits of the stars may be very complex and quite unrelated to the rotation of the system as a whole. Some stars may be moving in the direction opposite to the rotation of the Galaxy. This happens because each star, besides participating in the galactic rotation, has an intrinsic motion of its own.

In order to exhibit more clearly the kinematics of highly rarefied media, and of stellar systems in particular, it is useful to make a comparison with the motion of a fluid. In fluid dynamics the volume element is represented by a "particle" of the fluid, which is assumed homogeneous and continuous. It is known from physics that all fluids actually consist of molecules, which participate in the motion of the fluid particle as a whole and also have a random thermal motion. If we consider a cross-section of a current of fluid, say water flowing in a channel, then at every point in the cross-section there exists a hydrodynamic velocity vector. This velocity is a vector function of the coordinates of the point to which it pertains. The irregular thermal motions of the molecules cause the individual physical particles of the fluid, i.e. the molecules, to move in various directions at any given instant, and some of them to move in a direction opposite to that of the current. Consequently, the rate of flow is given by the velocity of the centroid in a fluid just as in a very rarefied medium such as a stellar system. There is, however, a fundamental difference between an ordinary dense medium and a highly rarefied one. In the former, the mean free path is much less than the diameter of the macroscopic volume element, whereas in the latter medium (for example, a stellar system) the mean free path is much greater than the dimensions of the volume element. For instance, in the neighbourhood of the Sun the mean free path of a star is of the order of 10^8 ps, which is 10,000 times the radius of the Galaxy and a million times the most suitable dimensions of the macroscopic volume element. In a fluid, therefore, a "particle" contains the same molecules as it moves about, whereas in a rarefied medium there is a continual *diffusion* through the boundaries of the volume element to and from the surrounding space. We shall discuss this in more detail in Chapter VIII.

The difference between the velocity vector of a star and that of the centroid may be called the *residual velocity* of the star. Thus

$$\mathbf{V} = \mathbf{V}_0 + \mathbf{V}',$$
$$u = u_0 + u', \quad v = v_0 + v', \quad w = w_0 + w', \tag{2.13}$$

where $\mathbf{V}$ and $\mathbf{V}_0$ are respectively the velocity of the star and that of the centroid, $\mathbf{V}'$ the residual velocity, and u, v, w; u_0, v_0, w_0; u', v', w' their respective components.

These relations are of importance in stellar dynamics. They make possible the separation of the velocity of each star into two qualitatively different parts, the velocity of the centroid and the residual velocity. The velocity of the centroid is the same for all stars in a given volume element, and is a one-valued vector function of the coordinates, which represents the regular or ordered motion of the stellar system. The residual velocity, on the other hand, is different for each star and is irregular. This is seen from the fact that, within a given volume element, the expectation or mean value of the residual velocity is zero, by (2.12) and (2.13):

$$\sum \mathbf{V}' = 0, \qquad \sum u' = 0, \quad \sum v' = 0, \quad \sum w' = 0, \tag{2.14}$$

and the residual velocities themselves are statistically distributed. Thus they represent the irregular or disordered motion of the system.

We shall see later that this twofold nature of the motions of stellar systems has the result that two essentially different methods of investigation must be used in their study: the *hydrodynamic* method, corresponding to the property of continuity, and the *statistical* method, corresponding to that of discontinuity.

It follows from the above discussion that the velocity function must give the distribution of residual velocities, not that of the total velocities of the stars. This may be illustrated in terms of the ellipsoidal (or spherical) velocity distribution. The velocity function $\varphi(T)$, in the form (2.3) and (2.4), or (2.3) and (2.5), involves the components $u - u_0$, $v - v_0$, $w - w_0$ of the residual velocity, not those of the total velocity (u, v, w). Using the notation u', v', w' for the residual velocity components, we can write (2.3) and (2.5) as

$$\varphi(q, p) = \Psi(Q + \sigma), \qquad Q = h^2 u'^2 + k^2 v'^2 + l^2 w'^2. \tag{2.15}$$

In particular, for the spherical distribution (2.10) we have

$$\varphi(q, p) = \Psi\{h^2(u'^2 + v'^2 + w'^2)\} \quad \text{or} \qquad \varphi(q, p) = \Psi(h^2 V'^2), \tag{2.16}$$

where V' is the magnitude of the residual velocity.

Hence we see that, for a spherical distribution, the velocity function depends only on the magnitude of the residual velocity, and not on its direction. In particular, for Maxwell's law (2.11) we have

$$\varphi(q, p) = \frac{\nu h^3}{\pi^{3/2}} \exp(-h^2 V'^2). \tag{2.17}$$

§ 5. The Local Motion of the Sun in the Galaxy

In what follows we shall naturally devote particular attention to the dynamics of our own Galaxy, as being that representative of the class of galaxies (the most important stellar systems) which has received the closest study. In so doing, we encounter the fact that the Sun, being a star, has its own residual velocity. In the classical literature of astronomy, the Sun's residual velocity is referred to as the *solar motion.* It is usually defined by the *elements of the solar motion*, i.e. the components of the Sun's residual velocity vector, or equivalently by the coordinates of the *apex* together with the magnitude of the residual velocity vector. In galactic coordinates the accepted values of the elements are

$$l = 24°, \quad b = +22°, \quad V_\odot = 20 \text{ km/s}. \tag{2.18}$$

The velocity components corresponding to these elements, along the axes of a rectangular system of galactic coordinates, are

$$X_\odot = 16{\cdot}9 \text{ km/s}, \quad Y_\odot = 7{\cdot}6 \text{ km/s}, \quad Z_\odot = 7{\cdot}5 \text{ km/s}. \tag{2.19}$$

The Sun's velocity is evidently just the velocity of an observer relative to the local centroid. It is therefore unimportant in stellar dynamics, with rare exceptions, and is to be regarded as a systematic error in the observed stellar velocities, which must be eliminated as far as possible by applying an appropriate correction. In what follows we shall usually assume that all stellar velocities have been corrected for the Sun's motion.

The solar motion is sometimes used in modern stellar astronomy as a criterion for selecting independent sub-systems among the stars in the Galaxy. Since they participate in the rotation of the Galaxy, different sub-systems have different centroid velocities. This has the result that the Sun's velocity is different relative to different sub-systems. The difference in the solar velocity vectors is equal to the difference in the centroid velocity vectors.

To determine the local motion of the Sun we choose stars at distances not exceeding a few hundred parsecs, so as to be able to neglect any difference in the corresponding centroid velocities. Then the observed velocity $\mathbf{V}_{obs}$ of a star consists of its residual velocity $\mathbf{V}'$ and that of the Sun $\mathbf{V}_\odot$ reversed, i.e. $\mathbf{V}_{obs} = \mathbf{V}' - \mathbf{V}_\odot$, whence

$$\mathbf{V}_\odot = -\mathbf{V}_{obs} + \mathbf{V}'. \tag{2.20}$$

This relation gives the residual (local) velocity of the Sun from the observed velocities of the stars.

The elements of the solar motion can be determined from either radial velocities, total velocities or proper motions of stars. When proper motions are used, however, the magnitude of the Sun's velocity vector remains unknown.

(a) *The Sun's Local Velocity from Radial Velocities*

We take a rectangular system of galactic coordinates X, Y, Z, with the origin at the observer. Let $X_\odot$, $Y_\odot$, $Z_\odot$ be the components of the Sun's velocity, and α, β, γ the direction cosines of a star. Then the component of eqn. (2.20) in the direction of the star gives

$$\alpha X_\odot + \beta Y_\odot + \gamma Z_\odot = -v_r + v_r', \tag{2.21}$$

where v_r is the observed radial velocity of the star and v_r' its residual radial velocity. The quantities v_r' are regarded as accidental errors in the eqns. (2.21). Then the components $X_\odot, Y_\odot, Z_\odot$ of the solar velocity are found from the equations of condition

$$\alpha X_\odot + \beta Y_\odot + \gamma Z_\odot = -v_r, \tag{2.22}$$

using the method of least squares.

In practice, to reduce the effect of the residual velocities and to curtail the amount of work needed, the equations of condition are set up not for single stars but for groups of stars. For this purpose the entire sky is divided into areas, preferably of equal size. Each group includes stars in a given area. The individual radial velocities are replaced by their mean, and the direction cosines used are those of the centre of the area. In this way the individual stars are replaced by mean positions.

The equations of condition (2.22) are often augmented by an unknown constant term called the *K term*, whose mechanical significance is that the group of stars considered is either expanding or contracting as a whole. A detailed discussion of the K term in radial velocities will be given in Chapter III, but here we may note that, as regards the kinematics of the Galaxy, the K term signifies that the divergence of the centroid velocity vector is not zero. Observation shows that there is a positive K term in the neighbourhood of the Sun for the bright stars of the very early classes B0 to B3, and also for the even hotter stars of class O, which are very rare. The mean value of the K term from recent determinations is $+4{\cdot}3$ km/s for the bright O and B stars; for O and B stars fainter than $m=8$, the K term is zero.

(b) *The Elements of the Solar Motion from Proper Motions*

We take the components of eqn. (2.20) along directions tangential to the circle of latitude and to the meridian (the longitude and latitude directions in the moving trihedral). This gives

$$\mathbf{l}^0 \cdot \mathbf{V}_\odot = -v_l + v_l', \quad \mathbf{b}^0 \cdot \mathbf{V}_\odot = -v_b + v_b', \tag{2.23}$$

where $\mathbf{l}^0$ and $\mathbf{b}^0$ are unit vectors in the moving trihedral, v_l and v_b the corresponding components of the observed transverse velocity, and v_l' and v_b' those of the residual transverse velocity.

To convert the transverse velocities into proper motions they must be divided by $4{\cdot}74r\cos b$ and $4{\cdot}74r$ respectively (where r is expressed in parsecs; see Chapter I, § 3). Thus formulae (2.23) become

$$\frac{\mathbf{l}^0 \cdot \mathbf{V}_\odot}{4{\cdot}74r} = (-\mu_l + \mu_l')\cos b, \quad \frac{\mathbf{b}^0 \cdot \mathbf{V}_\odot}{4{\cdot}74r} = -\mu_b + \mu_b',$$

where μ_l, μ_b and μ_l', μ_b' are the observed and residual proper motions respectively. The quantities on the left-hand sides of these equations are called the *parallactic motions* of the star in longitude and in latitude. They are the parts of the respective proper motions which reflect the Sun's motion. Regarding the residual proper motions μ', μ_b' as accidental errors, we obtain the equations of condition:

$$\frac{\mathbf{l}^0 \cdot \mathbf{V}_\odot}{4{\cdot}74r} = -\mu_l\cos b, \quad \frac{\mathbf{b}^0 \cdot \mathbf{V}_\odot}{4{\cdot}74r} = -\mu_b. \tag{2.24}$$

In order to write these equations in a more explicit form we must derive expressions for the unit vectors $\mathbf{l}^0$ and $\mathbf{b}^0$. Taking the vector $\mathbf{z}^0$ in the direction of the galactic pole and $\mathbf{r}^0$ towards the star, we have

$$\mathbf{z}^0 = \mathbf{k}, \quad \mathbf{r}^0 = \frac{x}{r}\mathbf{i} + \frac{y}{r}\mathbf{j} + \frac{z}{r}\mathbf{k} = \mathbf{i}\cos b\cos l + \mathbf{j}\cos b\sin l + \mathbf{k}\sin b.$$

Substitution in (1.68) gives

$$\begin{aligned} \mathbf{l}^0 &= -\mathbf{i}\sin l + \mathbf{j}\cos l, \\ \mathbf{b}^0 &= -\mathbf{i}\sin b\cos l - \mathbf{j}\sin b\sin l + \mathbf{k}\cos b. \end{aligned} \tag{2.25}$$

The solar velocity vector in rectangular galactic coordinates has the form

$$\mathbf{V}_\odot = X_\odot \mathbf{i} + Y_\odot \mathbf{j} + Z_\odot \mathbf{k}. \tag{2.26}$$

Substituting (2.25) and (2.26) in (2.24), we have finally

$$\begin{aligned} \frac{X_\odot}{4{\cdot}74r}\sin l - \frac{Y_\odot}{4{\cdot}74r}\cos l &= \mu_l \cos b, \\ \frac{X_\odot}{4{\cdot}74r}\sin b \cos l + \frac{Y_\odot}{4{\cdot}74r}\sin b \sin l - \frac{Z_\odot}{4{\cdot}74r}\cos b &= \mu_b. \end{aligned} \tag{2.27}$$

These equations have been formally derived for one star, but in practice, to reduce the necessary labour, they are written for mean positions. The distance r is taken as the mean distance of the group of stars considered, i.e. it is assumed to be the same for each star. Equations (2.27) are called *Koval'skiĭ and Airy's equations.* When the mean distance of the stars is unknown, it remains indeterminate, and the solution of eqns. (2.27) by the method of least squares gives the quantities $\xi_\odot = X_\odot/4{\cdot}74r$, $\eta_\odot = Y_\odot/4{\cdot}74r$, $\zeta_\odot = Z_\odot/4{\cdot}74r$. From these we can find the apex of the Sun's way from the formulae

$$\tan l = \frac{\eta_\odot}{\xi_\odot} = \frac{Y_\odot}{X_\odot}; \quad \tan b = \frac{\zeta_\odot}{\sqrt{(\xi_\odot^2 + \eta_\odot^2)}} = \frac{Z_\odot}{\sqrt{(X_\odot^2 + Y_\odot^2)}}.$$

The numerical value of the mean distance r can be found if we assume a value for the Sun's velocity (obtained by other methods, e.g. from radial velocities): r is given by

$$\begin{aligned} \sqrt{(\xi_\odot^2 + \eta_\odot^2 + \zeta_\odot^2)} &= \sqrt{(X_\odot^2 + Y_\odot^2 + Z_\odot^2)}/4{\cdot}74r \\ &= V_\odot/4{\cdot}74r. \end{aligned}$$

(c) *The Sun's Local Velocity from Spatial Velocities*

The determination of the solar motion from the spatial velocities of stars is comparatively rare, because, firstly, the radial velocity, the proper motion and the distance of each star must all be known, which is true of only a relatively small number of stars; secondly, the results of determining the solar motion from radial velocities or proper motions alone are considerably more accurate than those obtained from total velocities, since the distances of the stars are usually affected by large errors, up to 30%. Moreover, the radial velocities and proper motions contain systematic errors which are difficult to elucidate in such calculations, since the methods of determining radial velocities and proper motions are essentially different.

In theoretical discussions, however, we may ignore these purely practical difficulties, and therefore regard the determination of the solar motion from total velocities as the basic method, since it is theoretically the simplest.

The formulae are very easily obtained. To find the solar motion, we need only calculate the components of the total velocity of each star along any three fixed axes. Then the components of the Sun's velocity along these axes are simply minus the means of the corresponding stellar velocity components:

$$X_\odot = -\sum u/N, \quad Y_\odot = -\sum v/N, \quad Z_\odot = -\sum w/N. \tag{2.28}$$

§ 6. The Velocity Ellipsoid and its Determination from Observations

In this section we shall consider the ideas underlying the determination of the velocity function from observations. Although this function can, in principle, be determined at every point in the space occupied by a stellar system, it has as yet been found only for the neighbourhood of the Sun. According to the present practice in stellar astronomy, only the *velocity ellipsoid* is usually determined. There are two essentially different methods of doing so: Schwarzschild's, which is historically the earlier and almost obsolete, and Charlier's, which is theoretically much superior. In Schwarzschild's method it is assumed *a priori* that an ellipsoidal velocity distribution law is valid in the Sun's neighbourhood, and moreover that it has the particular form given by Schwarzschild's law (2.6).

In § 3 we remarked that, for an ellipsoidal velocity distribution, given by formulae (2.3) and (2.6), the level surfaces are ellipsoids, since the equation of a level surface

$$\varphi(Q + \sigma) = \text{constant} \tag{2.29}$$

is equivalent to

$$Q + \sigma = C, \tag{2.30}$$

where C is a constant. If eqn. (2.30) represents an ellipsoid, it is necessary that Q should be a sign-definite quadratic form in the velocity differences $u - u_0$, $v - v_0$, $w - w_0$. The coefficients in the quadratic form Q and the quantities u_0, v_0, w_0, σ are functions of the coordinates only. Hence, if we consider the level surfaces of an ellipsoidal velocity function for any one point, they form a family of similar and similarly situated ellipsoids. This is seen by transforming eqn. (2.30) to the principal axes of the quadratic form and giving C various values. Substituting in (2.30) the canonical expression (2.5) for Q, we obtain

$$h^2(u - u_0)^2 + k^2(v - v_0)^2 + l^2(w - w_0)^2 + \sigma = C; \tag{2.31}$$

this may also be written

$$\frac{(u - u_0)^2}{a^2} + \frac{(v - v_0)^2}{b^2} + \frac{(w - w_0)^2}{c^2} = 1, \tag{2.32}$$

where

$$a^2 = (C - \sigma)/h^2, \quad b^2 = (C - \sigma)/k^2, \quad c^2 = (C - \sigma)/l^2. \tag{2.33}$$

Equation (2.32) represents a family of similar and similarly situated ellipsoids, depending on the parameter C. From (2.32) and (2.33) we see the geometrical significance of the quantity σ. It has the same value for all the ellipsoids and varies only from point to point in the stellar system.

Evidently the shape of the ellipsoids and the directions of their axes are entirely determined if one ellipsoid is known, for example that where $C - \sigma = \frac{1}{2}$, whose semiaxes are given by $a^2 = 1/2h^2$, $b^2 = 1/2k^2$, $c^2 = 1/2l^2$. This is called the *velocity ellipsoid* or *Schwarzschild ellipsoid*. Studies of the residual velocity distribution in the Sun's neighbourhood are usually confined to the determination of this ellipsoid. From the above discussion it is seen that the velocity ellipsoid is a typical representative of the family of ellipsoidal level surfaces at a given point in a stellar system. The reader is referred elsewhere (e.g. [122], pp. 141–188) for a detailed account of the application of Schwarzschild's method, but we may note here that every form of this method presupposes a velocity distribution obeying Schwarzschild's law. From this assumption the theoretical distribution of radial velocities, proper motions and spatial velocities is derived, depending on the parameters of the velocity ellipsoid. A comparison with the observed distribution then gives, by the method of least squares, the most probable values of these parameters.

In contrast to Schwarzschild's method, that of Charlier [21] makes no *a priori* assumptions concerning the nature of the velocity distribution, and the velocity ellipsoid is determined as the ellipsoid whose principal axes are in the direction of the principal axes of the velocity dispersion, while the squares of its semiaxes are equal to the corresponding variances: $a^2 = \Sigma_1^2$, $b^2 = \Sigma_2^2$, $c^2 = \Sigma_3^2$. Thus the velocity ellipsoid as determined by Charlier's method is of the form

$$\frac{x^2}{\Sigma_1^2} + \frac{y^2}{\Sigma_2^2} + \frac{z^2}{\Sigma_3^2} = 1.$$

The theoretical significance of the velocity ellipsoid as determined by Charlier's method is considerably greater than that of the one found by Schwarzschild's method. The reasons for this statement are as follows.

If the velocity distribution is exactly given by Schwarzschild's law, the two methods lead to the same result; if the distribution is ellipsoidal but not a Schwarzschild distribution, then the velocity ellipsoid given by Schwarzschild's method has no physical meaning, whereas that given by Charlier's method retains its physical significance whatever the velocity distribution law.

This fact renders Charlier's method so advantageous in stellar dynamics that we shall henceforward mean by the velocity ellipsoid the one determined by this method.

It should be borne in mind that, besides the velocity ellipsoid, we must consider the associated *velocity moment ellipsoid*. The equation of this ellipsoid, referred to the same coordinate axes as are used in eqn. (2.32), is

$$\Sigma_1^2 \xi^2 + \Sigma_2^2 \eta^2 + \Sigma_3^2 \zeta^2 = 1. \tag{2.34}$$

Thus the directions of the axes of the two ellipsoids are the same, but the lengths of their axes are reciprocal. In order to avoid any confusion arising from the similarity of the names of the two ellipsoids, in the present section we shall call the velocity ellipsoid the *Schwarzschild ellipsoid*. Evidently, if one of these two ellipsoids is given, the other is entirely determined, and a knowledge of one is equivalent to that of the other. The determination of the Schwarzschild ellipsoid rather than the other is a matter of convenience.

(a) *The Schwarzschild Ellipsoid from Total Velocities*

The derivation of the Schwarzschild ellipsoid (the velocity dispersion ellipsoid) from the total velocities of stars is, in principle, extremely simple. For each star we know its total velocity vector $\mathbf{V}_i \equiv (u_i, v_i, w_i)$ in some fixed coordinate system, for example galactic coordinates. By finding the mean values of the velocity components, we obtain the velocity components of the observer's centroid, which are also (with reversed sign) the components of the Sun's local velocity:

$$u_0 = \frac{1}{N}\sum_i u_i = -u_\odot,$$

$$v_0 = \frac{1}{N}\sum_i v_i = -v_\odot,$$

$$w_0 = \frac{1}{N}\sum_i w_i = -w_\odot.$$

Next, calculating the second-order moments $\overline{u^2}$, $\overline{v^2}$, $\overline{w^2}$, $\overline{vw}$, $\overline{wu}$, $\overline{uv}$, we can use the well-known formulae to convert these to the central moments:

$$\left.\begin{aligned} \overline{u'^2} &= \overline{u^2} - u_\odot^2, \\ \overline{v'^2} &= \overline{v^2} - v_\odot^2, \\ \overline{w'^2} &= \overline{w^2} - w_\odot^2, \\ \overline{v'w'} &= \overline{vw} - v_\odot w_\odot, \quad \text{etc.,} \end{aligned}\right\} \tag{2.35}$$

i.e. the second-order moments of the residual velocities. Evidently these moments, with respect to the coordinate axes, are the coefficients in the equation of the velocity moment ellipsoid:

$$\overline{u'^2}x^2 + \overline{v'^2}y^2 + \overline{w'^2}z^2 + 2\overline{v'w'}yz + 2\overline{w'u'}zx + 2\overline{u'v'}xy = 1. \tag{2.36}$$

To convert eqn. (2.36) to principal axes we form the secular equation of its coefficients:

$$\Delta(\lambda) = \begin{vmatrix} \overline{u'^2} - \lambda & \overline{u'v'} & \overline{u'w'} \\ \overline{v'u'} & \overline{v'^2} - \lambda & \overline{v'w'} \\ \overline{w'u'} & \overline{w'v'} & \overline{w'^2} - \lambda \end{vmatrix} = 0, \tag{2.37}$$

whose roots $\lambda_1, \lambda_2, \lambda_3$ are the inverse squares of the semiaxes of the velocity moment ellipsoid. They are therefore the squares of the semiaxes of the Schwarzschild ellipsoid:

$$\lambda_1 = \Sigma_1^2, \quad \lambda_2 = \Sigma_2^2, \quad \lambda_3 = \Sigma_3^2. \tag{2.38}$$

We know from the general theory of quadric surfaces that the direction cosines l, m, n of the principal axes of the ellipsoid can be found from the relations

$$\frac{l}{G_{11}} = \frac{m}{G_{12}} = \frac{n}{G_{13}} = \frac{1}{\sqrt{(G_{11}^2 + G_{12}^2 + G_{13}^2)}}, \tag{2.39}$$

where G_{11}, G_{12}, G_{13} are the minors of the elements in the first row of the determinant which appears in the secular equation, with $\lambda_1, \lambda_2, \lambda_3$ successively substituted for λ. As stated above, these directions are also those of the principal axes of the Schwarzschild ellipsoid.

It should be pointed out that, for the same reasons as were given in discussing the solar motion, the determination of the velocity ellipsoid from the total velocities of stars is in practice unusual, despite its theoretical simplicity.

(b) *The Schwarzschild Ellipsoid from Radial Velocities*

The method of determining the velocity ellipsoid from radial velocities and proper motions of stars was developed in detail during the 1920's by the Swedish mathematician and astronomer Charlier [21]. Since our main purpose is to discuss the determination of the ellipsoid from radial velocities, we shall first consider the purely mathematical problem of the transformation of the moments of an arbitrary distribution function when the coordinate axes are rotated.

Let x_1, x_2, x_3 denote the coordinates of points in a three-dimensional Euclidean space, and $f_0(x_1, x_2, x_3)$ a normalised distribution function giving the density of these points in space. Then a moment of order $i+j+k$ of this function is the expectation of a product of powers of the coordinates:

$$m_{ijk} = \int x_1^i \, x_2^j \, x_3^k \, f(x_1, x_2, x_3) \, \mathrm{d}\omega. \tag{2.40}$$

Let us now effect a rotation of the coordinate axes, the transformation being given by the formulae

$$\left.\begin{aligned} x_1' &= \alpha_{11}x_1 + \alpha_{12}x_2 + \alpha_{13}x_3, \\ x_2' &= \alpha_{21}x_1 + \alpha_{22}x_2 + \alpha_{23}x_3, \\ x_3' &= \alpha_{31}x_1 + \alpha_{32}x_2 + \alpha_{33}x_3, \end{aligned}\right\} \tag{2.41}$$

or, in abbreviated form,

$$x_i' = \sum_j \alpha_{ij} \, x_j \quad (i, j = 1, 2, 3). \tag{2.42}$$

Then the moments of a given order relative to the two sets of axes are evidently related by the linear equations

$$m_{pqr}' = \sum_{i,j,k} A_{ijk} \, m_{ijk} \quad (i+j+k = p+q+r). \tag{2.43}$$

In particular, we have for the three first-order moments and the six second-order moments

$$\left.\begin{aligned} m_{100}' &= \alpha_{11}m_{100} + \alpha_{12}m_{010} + \alpha_{13}m_{001}, \\ m_{010}' &= \alpha_{21}m_{100} + \alpha_{22}m_{010} + \alpha_{23}m_{001}, \\ m_{001}' &= \alpha_{31}m_{100} + \alpha_{32}m_{010} + \alpha_{33}m_{001}; \end{aligned}\right\} \tag{2.44}$$

$$\left.\begin{aligned} m_{200}' &= \alpha_{11}^2 m_{200} + \alpha_{12}^2 m_{020} + \alpha_{13}^2 m_{002} + 2\alpha_{12}\alpha_{13} m_{011} \\ &\quad + 2\alpha_{13}\alpha_{11} m_{101} + 2\alpha_{11}\alpha_{12} m_{110}, \\ m_{0\ 0}' &= \alpha_{21}^2 m_{200} + \alpha_{22}^2 m_{020} + \alpha_{23}^2 m_{002} + 2\alpha_{22}\alpha_{23} m_{011} \\ &\quad + 2\alpha_{23}\alpha_{21} m_{101} + 2\alpha_{21}\alpha_{22} m_{110}, \\ m_{002}' &= \alpha_{31}^2 m_{200} + \alpha_{32}^2 m_{020} + \alpha_{33}^2 m_{002} + 2\alpha_{32}\alpha_{33} m_{011} \\ &\quad + 2\alpha_{33}\alpha_{31} m_{101} + 2\alpha_{31}\alpha_{32} m_{110}, \\ m_{011}' &= \alpha_{21}\alpha_{31} m_{200} + \alpha_{22}\alpha_{32} m_{020} + \alpha_{23}\alpha_{33} m_{002} \\ &\quad + (\alpha_{22}\alpha_{33} + \alpha_{23}\alpha_{32}) m_{011} + (\alpha_{23}\alpha_{31} + \alpha_{21}\alpha_{33}) m_{101} \\ &\quad + (\alpha_{21}\alpha_{32} + \alpha_{22}\alpha_{31}) m_{110}, \\ m_{101}' &= \alpha_{31}\alpha_{11} m_{200} + \alpha_{32}\alpha_{12} m_{020} + \alpha_{33}\alpha_{13} m_{002} \\ &\quad + (\alpha_{32}\alpha_{13} + \alpha_{33}\alpha_{12}) m_{011} + (\alpha_{33}\alpha_{11} + \alpha_{31}\alpha_{13}) m_{101} \\ &\quad + (\alpha_{31}\alpha_{12} + \alpha_{32}\alpha_{11}) m_{110}, \\ m_{110}' &= \alpha_{11}\alpha_{21} m_{200} + \alpha_{12}\alpha_{22} m_{020} + \alpha_{13}\alpha_{23} m_{002} \\ &\quad + (\alpha_{12}\alpha_{23} + \alpha_{13}\alpha_{22}) m_{011} + (\alpha_{13}\alpha_{21} + \alpha_{11}\alpha_{23}) m_{101} \\ &\quad + (\alpha_{11}\alpha_{22} + \alpha_{12}\alpha_{21}) m_{110}. \end{aligned}\right\} \tag{2.45}$$

The relations (2.43), (2.44) and (2.45) give the transformation of moments. We have considered the transformation of moments about the origin, but these formulae are evidently valid for moments about any point, and in particular for moments about the centre of the distribution, whose coordinates are the expectations of the corresponding coordinate values.

Let us now apply the moment transformation formulae to determine the velocity dispersion ellipsoid from the radial velocities of stars. Let u_i, v_i, w_i be the components of the velocity of the ith star along the axes of a rectangular system of galactic coordinates. For simplicity we shall suppose that all stellar velocities are corrected for the motion of the local centroid, and are therefore residual velocities. As we know already, the components of the stellar velocities can be regarded as the coordinates of points in a Euclidean velocity space, the velocity function being also the density distribution function of these points.

We now select a small area $d\omega$ of the sky. In order to find the dispersion of radial velocities, we shall suppose that all the stars in this area are on a line from the origin to the centre of the area. Then the velocity vectors of all these stars can be resolved into components along the axes of the moving trihedral whose origin is at the centre of the area. We denote by

$$v_r, v_l, v_b \tag{2.46}$$

the velocity components along the line of sight, the tangent to the circle of galactic latitude, and the tangent to the galactic meridian respectively. The first of these components is evidently the residual radial velocity, and the other two are the components of the residual transverse velocity in galactic longitude and latitude respectively. Let σ_r^2 be the dispersion of the radial velocities of the stars considered, and α, β, γ the direction cosines of the centre of the area. Taking the axes of the moving trihedral as new coordinate axes, we have $m'_{200} = \sigma_r^2$, $\alpha_{11} = \alpha$, $\alpha_{12} = \beta$, $\alpha_{13} = \gamma$. Hence, rewriting the first formula (2.45) in the present notation, we obtain

$$\sigma_r^2 = \alpha^2 \overline{u'^2} + \beta^2 \overline{v'^2} + \gamma^2 \overline{w'^2} + 2\beta\gamma \overline{v'w'} + 2\gamma\alpha \overline{w'u'} + 2\alpha\beta \overline{u'v'}. \tag{2.47}$$

Let us now suppose the whole sky divided into areas of equal size. Then a formula like (2.47) applies to each of them. In these formulae we can regard the components of the velocity dispersion tensor $\overline{u'^2}$, $\overline{v'^2}$, $\overline{w'^2}$, $\overline{v'w'}$, $\overline{w'u'}$, $\overline{u'v'}$ as unknowns, and the formulae such as (2.47) as forming an oversufficient set of equations of condition, which may be solved by the method of least squares. Thus all the components of the dispersion tensor are found. It then remains to set up and solve the secular equation (2.37), since the problem is now entirely analogous to that of determining the Schwarzschild ellipsoid from total velocities.

A very convenient procedure for practical calculations is due to Charlier. The whole sky is divided once for all into 48 equal areas arranged symmetrically about the galactic equator (if the calculations are done in galactic coordinates). On each side of the equator is a strip 30° wide containing 12 areas. Beyond each of these strips is another containing 10 areas. A simple calculation using the fact that the areas are equal in size shows that the two latter strips extend to latitude $b = 66°26'36''\cdot8$. The remaining two regions at the poles form two areas each. It is easily verified that these areas are equal to the others. Although an infinite variety of such systems of standard areas could be devised, the Charlier system just described has been thoroughly tested in practice and has usually given results which are entirely satisfactory. The choice of the standard areas is ultimately determined by the population involved, i.e. the number of stars used in a particular investigation.

If every area is given the same weight, regardless of the number of stars in it, the solution of the equations of condition becomes very simple. On account of the symmetrical arrangement of the areas, many of the coeficients in the normal equations are zero, and consequently we have a case well known in the theory of the least-squares method, where the normal equations "separate" and we need only calculate the six quantities

$$\begin{aligned} a &= S\alpha^2\sigma_r^2, \quad b = S\beta^2\sigma_r^2, \quad c = S\gamma^2\sigma_r^2, \\ f &= S\beta\gamma\sigma_r^2, \quad g = S\gamma\alpha\sigma_r^2, \quad h = S\alpha\beta\sigma_r^2, \end{aligned} \tag{2.48}$$

where S denotes a summation over all the areas. When these quantities are known, the required components of the tensor are linear combinations of them, with fixed coefficients:

$$\left.\begin{aligned}\overline{u'^2} &= 0{\cdot}123113a - 0{\cdot}029719b - 0{\cdot}033260c,\\ \overline{v'w'} &= 0{\cdot}15056f,\\ \overline{v'^2} &= -0{\cdot}029719a + 0{\cdot}124299b - 0{\cdot}035725c,\\ \overline{w'u'} &= 0{\cdot}15609g,\\ \overline{w'^2} &= -0{\cdot}033260a - 0{\cdot}035725b + 0{\cdot}137821c,\\ \overline{u'v'} &= 0{\cdot}15349h.\end{aligned}\right\} \tag{2.49}$$

(c) *The Schwarzschild Ellipsoid from Proper Motions*

The difficulty in determining the velocity ellipsoid from proper motions arises because we do not know the parallaxes of the individual stars, and consequently cannot convert the proper motions into transverse velocities. Hence all calculations in this case must be made in the units of measurement of proper motion, i.e. seconds of arc per year. Only at the end can we change to linear units by using certain results from the reduction of radial velocities, i.e. extraneous to the problem.

Let μ_l and μ_b be the proper motions of some star in galactic longitude and latitude, and π its parallax, which is the reciprocal of the distance. Then

$$k\mu_l = \pi v_l, \quad k\mu_b = \pi v_b, \tag{2.50}$$

where k is a constant factor which depends on the choice of units. Forming the second-order moments of the proper motions, and temporarily denoting by u_i, v_i the components of the transverse velocity in galactic longitude and latitude, we have

$$k^2 m_{ll} = \frac{k^2}{N}\sum \mu_l^2 = \frac{1}{N}\sum_i u_i^2\pi_i^2,$$

$$k^2 m_{lb} = \frac{k^2}{N}\sum \mu_l\mu_b = \frac{1}{N}\sum_i u_i v_i \pi_i^2,$$

$$k^2 m_{bb} = \frac{k^2}{N}\sum \mu_b^2 = \frac{1}{N}\sum_i v_i^2\pi_i^2.$$

These give by an identical transformation

$$\left.\begin{aligned}k^2 m_{ll} &= \frac{1}{N}\sum_i [(u_i^2 - \overline{u'^2}) + \overline{u'^2}][(\pi_i^2 - \overline{\pi^2}) + \overline{\pi^2}]\\ &= \overline{u'^2}\,\overline{\pi^2} + \frac{1}{N}\sum_i (u_i^2 - \overline{u'^2})(\pi_i^2 - \overline{\pi^2})\\ &= \overline{u'^2}\,\overline{\pi^2} + R_1,\\ k^2 m_{lb} &= \overline{u'v'}\,\overline{\pi^2} + \frac{1}{N}\sum_i (u_i v_i - \overline{u'v'})(\pi_i^2 - \overline{\pi^2})\\ &= \overline{u'v'}\,\overline{\pi^2} + R_2,\\ k^2 m_{bb} &= \overline{v'^2}\,\overline{\pi^2} + \frac{1}{N}\sum_i (v_i^2 - \overline{v'^2})(\pi_i^2 - \overline{\pi^2})\\ &= \overline{v'^2}\,\overline{\pi^2} + R_3,\end{aligned}\right\} \tag{2.51}$$

where R_1, R_2, R_3 are proportional to the correlation coefficients between the random quantities u_i^2, $u_i v_i$, v_i^2 on the one hand and π_i^2 on the other. They are zero if there is no correlation between the three quantities and π_i^2, and are the greater, the more marked the statistical correlation between them. It is impossible to prove the absence of any such correlation for the stars in the Galaxy, but it cannot be marked. It may have some effect, for example, in the Charlier areas away from the galactic equator, since the number of stars of high luminosity (giants and supergiants), whose velocities are small compared with those of other stars, decreases away from the galactic plane. This relation has not yet been investigated in detail. The only possible procedure, therefore, is to assume that the correlation coefficients in R_1, R_2, R_3 are zero. Then eqns. (2.51) become

$$k^2 \overline{\mu_l^2} = \overline{v_l^2}\,\overline{\pi^2}, \quad k^2 \overline{\mu_b^2} = \overline{v_b^2}\,\overline{\pi^2}, \quad k^2 \overline{\mu_l \mu_b} = \overline{v_l v_b}\,\overline{\pi^2}. \tag{2.52}$$

Using these relations, we can rewrite formulae (2.51) as

$$\left.\begin{aligned} k^2 \overline{\mu_l^2} &= [\alpha_{21}^2 \overline{u'^2} + \alpha_{22}^2 \overline{v'^2} + \alpha_{23}^2 \overline{w'^2} + 2\alpha_{22}\alpha_{23} \overline{v'w'} \\ &\quad + 2\alpha_{23}\alpha_{21} \overline{w'u'} + 2\alpha_{21}\alpha_{22} \overline{u'v'}]\overline{\pi^2}, \\ k^2 \overline{\mu_b^2} &= [\alpha_{31}^2 \overline{u'^2} + \alpha_{32}^2 \overline{v'^2} + \alpha_{33}^2 \overline{w'^2} + 2\alpha_{32}\alpha_{33} \overline{v'w'} \\ &\quad + 2\alpha_{33}\alpha_{31} \overline{w'u'} + 2\alpha_{31}\alpha_{32} \overline{u'v'}]\overline{\pi^2}, \\ k^2 \overline{\mu_l \mu_b} &= [\alpha_{21}\alpha_{31} \overline{u'^2} + \alpha_{22}\alpha_{32} \overline{v'^2} + \alpha_{23}\alpha_{33} \overline{w'^2} + (\alpha_{22}\alpha_{33} + \alpha_{23}\alpha_{32}) \overline{v'w'} \\ &\quad + (\alpha_{23}\alpha_{31} + \alpha_{21}\alpha_{33}) \overline{w'u'} + (\alpha_{21}\alpha_{32} + \alpha_{22}\alpha_{31}) \overline{u'v'}]\overline{\pi^2}. \end{aligned}\right\} \tag{2.53}$$

These equations can be written for each Charlier area and regarded as a set of equations of condition from which we find, by the method of least squares, the quantities $\overline{u'^2}\,\overline{\pi^2}$, $\overline{v'^2}\,\overline{\pi^2}$, $\overline{w'^2}\,\overline{\pi^2}$, $\overline{v'w'}\,\overline{\pi^2}$, $\overline{w'u'}\,\overline{\pi^2}$, $\overline{u'v'}\,\overline{\pi^2}$, and these are proportional to the components of the dispersion tensor. The solution of the normal equations is much simplified because, as in the case of radial velocities, the symmetry of the areas causes a separation of these equations. A further simplification can be achieved by expressing the direction cosines α_{11}, α_{22}, α_{33}, α_{23}, α_{31}, α_{12} in terms of the direction cosines α, β, γ of the centre of the area, i.e. those which are used in determining the velocity ellipsoid from radial velocities. To derive these expressions we note that the vectors $\mathbf{z}^0$ and $\mathbf{r}^0$ which give the direction of the galactic pole and of the star are $\mathbf{z}^0 = (0, 0, 1)$; $\mathbf{r}^0 = (\alpha, \beta, \gamma)$. Hence eqns. (1.68) may be written

$$\mathbf{l}^0 = (-\beta \mathbf{i} + \alpha \mathbf{j}) \sec b,$$

$$\mathbf{b}^0 = -\mathbf{i}\alpha \tan b - \mathbf{j}\beta \tan b - \mathbf{k}(\gamma \tan b - \sec b).$$

Also $\sin b = \gamma$, $\cos b = \sqrt{(1 - \gamma^2)} = \sqrt{(\alpha^2 + \beta^2)}$, and so

$$\mathbf{l}^0 = (-\beta \mathbf{i} + \alpha \mathbf{j})/\sqrt{(\alpha^2 + \beta^2)}$$

$$\mathbf{b}^0 = [-\alpha\gamma \mathbf{i} - \beta\gamma \mathbf{j} + (\alpha^2 + \beta^2)\mathbf{k}]/\sqrt{(\alpha^2 + \beta^2)}.$$

Hence the direction cosines to be used in changing from the fixed coordinates to the moving trihedral are

$$\left.\begin{array}{cccc} & x_1 & x_2 & x_3 \\ r & \alpha & \beta & \gamma \\ l & \dfrac{-\beta}{\sqrt{(\alpha^2 + \beta^2)}} & \dfrac{\alpha}{\sqrt{(\alpha^2 + \beta^2)}} & 0 \\ b & \dfrac{-\alpha\gamma}{\sqrt{(\alpha^2 + \beta^2)}} & \dfrac{-\beta\gamma}{\sqrt{(\alpha^2 + \beta^2)}} & \sqrt{(\alpha^2 + \beta^2)} \end{array}\right\}. \tag{2.54}$$

When these two simplifications are used, the problem of finding the coefficients in the equation of the velocity moment ellipsoid (i.e. the components of the dispersion tensor) reduces to the calculation of the six quantities

$$\left.\begin{aligned} a' &= S(\alpha_{11}^2\,\overline{\mu_l^2} + \alpha_{21}^2\,\overline{\mu_b^2} + 2\alpha_{11}\alpha_{21}\,\overline{\mu_l\mu_b}),\\ b' &= S(\alpha_{12}^2\,\overline{\mu_l^2} + \alpha_{22}^2\,\overline{\mu_b^2} + 2\alpha_{12}\alpha_{22}\,\overline{\mu_l\mu_b}),\\ c' &= S\alpha_{23}^2\,\overline{\mu_b^2},\\ f' &= S(\alpha_{22}\alpha_{23}\,\overline{\mu_b^2} + \alpha_{12}\alpha_{23}\,\overline{\mu_l\mu_b}),\\ g' &= S(\alpha_{23}\alpha_{21}\,\overline{\mu_b^2} + \alpha_{11}\alpha_{23}\,\overline{\mu_l\mu_b}),\\ h' &= S[\alpha_{21}\alpha_{22}\,\overline{\mu_b^2} + (\alpha_{11}\alpha_{22} + \alpha_{12}\alpha_{21})\,\overline{\mu_l\mu_b} + \alpha_{11}\alpha_{12}\,\overline{\mu_l^2}]. \end{aligned}\right\} \qquad (2.55)$$

The desired unknown quantities multiplied by the mean square parallax $\overline{\pi^2}$ are then given by

$$\left.\begin{aligned} \overline{\pi^2}\,\overline{u'^2} &= 0{\cdot}040677a' - 0{\cdot}004703b' - 0{\cdot}004400c',\\ \overline{\pi^2}\,\overline{v'^2} &= -0{\cdot}004703a' + 0{\cdot}041117b' - 0{\cdot}004660c',\\ \overline{\pi^2}\,\overline{w'^2} &= -0{\cdot}004400a' - 0{\cdot}004660b' + 0{\cdot}039503c',\\ \overline{\pi^2}\,\overline{v'w'} &= 0{\cdot}437204f',\\ \overline{\pi^2}\,\overline{w'u'} &= 0{\cdot}439381g',\\ \overline{\pi^2}\,\overline{u'v'} &= 0{\cdot}456019h'. \end{aligned}\right\} \qquad (2.56)$$

These equations show that the proper motions make possible a determination of the directions and ratios of the axes of the velocity ellipsoid, but not of the lengths of these axes. In order to find the lengths of the axes, the quantity $\overline{\pi^2}$ must be eliminated in some way. In practice this is most simply done by comparing the results with the dispersions $\overline{u'^2}$, $\overline{v'^2}$, $\overline{u'v'}$ etc. found from radial velocities.

§ 7. Results of Velocity Ellipsoid Determinations and some Inferences therefrom

We may briefly discuss some general conclusions which can be drawn from the numerous determinations of the velocity ellipsoid, both from proper motions and from radial velocities, of stars taken as a whole or grouped by spectral type or absolute magnitude. The pioneers in this work were K. Schwarzschild, S. I. Belyavskiĭ and others in 1907. Table I gives a summary of the results of determining the velocity ellipsoid from radial velocities of stars of the various spectral classes brighter than apparent magnitude 6^m, based on the extensive work of Nordström [76], which is still unsurpassed in completeness.

To show the consequences of calculating kinematic data from spatial velocities, we may give the results of an extensive study made in 1949 by P. P. Parenago [103]. To determine the solar motion and the elements of the velocity ellipsoid, Parenago calculated the velocity components of 12,000 stars. In order to reject components whose values are rendered meaningless by a large error in their determination, he laid down error limits of 7, 15 and 45 km/s for stars in the plane, intermediate and spherical sub-systems respectively. Although these limits are about 50% of the mean velocities in the respective sub-systems, 75% of the stars had to be rejected, and even then the results were less accurate than those of Nordström. A precise comparison of the results is rendered more difficult by the fact that the two authors used different groupings of the stars, but if we take into account the fact that the total numbers of stars were about the same (Nordström 3238, Parenago 3000) and use the results of the summary tables, where Parenago's stars are divided among eight groups and Nordström's among six, then the root-mean-square error in the coordinates of the vertices of the velocity ellipsoid axes is 13° in Parenago's work but only 8° in Nordström's.

Here it should also be borne in mind that Parenago's grouping of the stars was certainly an improvement over that of Nordström, who used only a formal classification by spectrum and apparent magnitude. Thus Parenago's result should be *ceteris paribus* more accurate, and consequently his attempt to use total velocities must be considered unsuccessful.

TABLE I. VELOCITY ELLIPSOID ELEMENTS FOR STARS BRIGHTER THAN APPARENT MAGNITUDE 6^m

Class	σ_1 (km/s)	l_1°	b_1°	σ_2 (km/s)	l_2°	b_2°	σ_3 (km/s)	l_3°	b_3°	$\bar{\sigma}$ (km/s)
B	12·4	235·7	+34·4	10·7	317·9	−11·2	5·1	32·4	+53·3	9·9
A	17·1	357·9	− 6·4	12·4	268·1	− 3·2	8·5	331·8	+83·0	13·1
F	23·5	339·7	+ 4·2	15·6	248·2	+18·9	12·7	82·1	+70·7	17·9
G	27·5	333·4	+21·4	15·8	260·7	−26·2	17·9	214·8	+55·5	21·0
K	24·2	342·2	− 5·0	18·1	252·0	− 3·8	20·9	303·7	+83·8	21·2
M	27·7	333·5	+ 8·7	19·5	245·1	− 9·9	18·3	202·8	+76·9	22·2
All	20·5	340·1	− 2·4	14·9	249·5	−11·9	17·5	261·5	+77·8	17·8

In Table I σ_1^2, σ_2^2, σ_3^2 denote the principal variances of the velocity along the axes of the ellipsoid, and the directions of these axes in galactic coordinates are given in the two subsequent columns for each σ_i. The last column gives the root-mean-square value of σ over all directions. The first column gives the spectral class of the stars in each group. The last line gives the results for all the stars taken together, numbering 3238. The numbers in the various classes were as follows: 554 B, 697 A, 471 F, 353 G, 984 K, 179 M.

From Table I it is seen that, if the B stars are excluded, the following conclusions may be drawn.

(1) The major axis of the velocity ellipsoid, except for the B and G stars, lies approximately in the galactic plane, its mean direction being $l = 340°$ for stars brighter than apparent magnitude 6^m. The great majority of these stars are giants.

The direction of the major axis of the velocity ellipsoid is of importance in problems of stellar dynamics, and the point on the celestial sphere lying in this direction is called the *vertex*. The direction of the centre of the Galaxy has galactic longitude $l = 325°$.

As we shall see, the theory of a steady-state galaxy shows that the direction of the vertex should coincide with that of the centre. The angle actually existing between these two directions therefore gives some measure of the extent to which the Galaxy is not in a steady state. The longitude difference Δl between the vertex and the galactic centre is called the *deviation* of the vertex. Thus for bright stars the deviation of the vertex is about $+15^\circ$. For fainter stars down to 15^m apparent magnitude, Blaauw [11] has shown that the deviation of the vertex is zero. This is apparently a property of dwarf stars, which form the majority of these faint stars. The same conclusion is given by the work of Wilson and Raymond and of Veldt [134], and also by that of Nordström [76].

(2) The minor axis of the velocity ellipsoid is almost perpendicular to the galactic plane (again except for the B stars).

(3) The ratio of axes of the ellipsoid is roughly 8:5:4.

It should be mentioned that there is a method of determining velocity ellipsoids, which has been widely used in stellar astronomy, based on the hypothesis that the distribution of residual stellar velocities in the neighbourhood of the Sun is *exactly* given by Schwarzschild's exponential law; see, for instance, the work of Schwarzschild, Lense, Fleming and others. In consequence, the results of different determinations are difficult to compare, since they are based on fundamentally different methods. We shall see later that the assumption of a Schwarzschild distribution of residual velocities imposes stringent limitations on the dynamical properties of the Galaxy, which can hardly be fulfilled in reality.

The velocity dispersion tensor, on the other hand, always has a definite physical significance, independent of any hypothesis about the actual distribution of the residual velocities. In Chapter VIII it will be shown that the velocity dispersion tensor plays exactly the same part in stellar dynamics as the internal stress tensor in the dynamics of viscous fluids. The determination of the velocity ellipsoid by Charlier's method is considerably simpler than by Schwarzschild's method, and consequently preference should be decidedly given to the former.

Finally, mention should be made of the *two-stream theory* put forward in 1904 by Kapteyn, before the ellipsoidal theory of Schwarzschild. According to this theory, the stars in the Galaxy are divided into two great "streams" which move in diametrically opposite directions and penetrate each other. In 1906 Eddington [24] developed a method of determining the parameters of Kapteyn's streams from observations of proper motions. By this method the observed distribution of proper motions can be represented with almost the same accuracy as by Schwarzschild's method, which is not surprising, because two eccentric spherical distributions can always represent with fair accuracy any given ellipsoidal distribution. On the other hand, whereas the ellipsoidal theory has some dynamical foundation, Kapteyn's two-stream theory has little dynamical significance.

In 1922 Kapteyn [47] attempted to devise a theoretical explanation of the streams. According to this theory the two streams represent two independent galaxies which are at present interlocked in the same region of space and rotate about the same axis but in opposite directions. In this form the stream theory was not successful, because at that time no process leading to such a behaviour of the Galaxy could be conceived.

For this reason Kapteyn's stream theory has been of historical importance only. It was retained, however, in a modified form, in Lindblad's theory of the rotation of the Galaxy [68]. Lindblad regarded the Galaxy as a system consisting of numerous sub-systems, each in turn consisting of stars having identical masses, spectra, etc., and a certain dispersion of residual velocities. The greater this dispersion, the more slowly the sub-system rotates about the axis of the Galaxy and the more it departs from the plane form, i.e. the smaller the degree of concentration of its stars towards the galactic plane. Thus, although the existence of sub-systems rotating in opposite directions is not assumed, that of sub-systems rotating at different angular velocities is assumed. We shall return to this topic in Chapter X.

§ 8. The Determination of the Actual Form of the Ellipsoidal Velocity Function

In the preceding sections we have discussed the problem of determining the velocity ellipsoid from observations of stellar velocities. This determination gives the shape and spatial orientation of the family of concentric ellipsoids which are the level surfaces of the velocity function when Schwarzschild's law applies. For a general ellipsoidal law, however, a knowledge of the velocity ellipsoid is insufficient to determine the velocity function uniquely. We have seen in § 3 that the general ellipsoidal function is of the form

$$\varphi(q, p) = \Psi(Q + \sigma),$$

where Ψ is an arbitrary function and Q a sign-definite homogeneous quadratic function of the residual velocities:

$$Q = a(u - u_0)^2 + b(v - v_0)^2 + c(w - w_0)^2$$
$$+ 2f(v - v_0)(w - w_0) + 2g(w - w_0)(u - u_0) + 2h(u - u_0)(v - v_0).$$

The coefficients a, b, c, f, g, h, the quantity σ, and the centroid velocity components u_0, v_0, w_0 depend only on the coordinates. At any given point in a stellar system they are constants. The equation of the level surfaces for the general ellipsoidal law, $\Psi(Q + \sigma) =$ constant, is therefore equivalent, as in the case of Schwarzschild's law, to $Q = C$; the arbitrary positive constant C can be regarded as a parameter whose value determines the various level surfaces of the velocity function. Hence it follows that the level surfaces for the general ellipsoidal law are again a family of concentric, similar and similarly situated ellipsoids. This family is the same whatever the function Ψ. If Ψ is the exponential function, we have Schwarzschild's law $\varphi(q, p) = e^{-(Q+\sigma)}$.

By merely determining the velocity ellipsoid we ignore any possible difference in the form of the function Ψ, whereas this form is by no means immaterial in stellar dynamics, since the law of the velocity distribution in a stellar system must in general be determined by its dynamical properties (the mass distribution, energy, angular momentum, etc.). The effect of a difference in the form of Ψ on the family of level ellipsoids may be illustrated as follows. Let us take a sequence of decreasing values of $\varphi: \varphi_1, \varphi_2, \ldots, \varphi_k, \ldots$. Then, to each equation

$$\Psi(Q + \sigma) = \varphi_k \tag{2.57}$$

there corresponds an ellipsoid

$$Q = n_k, \tag{2.58}$$

where the constant n_k is a root of the equation $\Psi(n_k + \sigma) = \varphi_k$. It is seen from this equation that, if the sequence of values φ_k is supposed given, the corresponding sequence of n_k will depend on the particular function Ψ. For example, if Ψ is a rapidly decreasing function of its argument, the numbers n_k will be relatively small, and the family of level ellipsoids will be concentrated near the point where Ψ has its maximum value. In the contrary case this will not occur. Figure 8 shows the distribution of level surfaces as a section by a common diametral plane for functions Ψ which decrease (a) rapidly and (b) slowly. Here we take account of the fact that, as eqn. (2.58) shows, the dimensions of the ellipsoids are proportional to $\sqrt{n_k}$. The sequence of numbers φ_k is the same in each case, but the function Ψ is different. Hence the values n_k corresponding to a given φ_k are also different. Accordingly the ellipses are of different sizes.

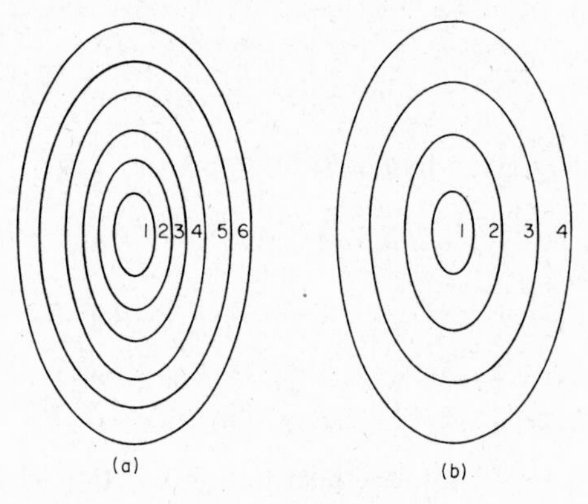

FIG. 8

We shall show how the actual form of the velocity function may be determined on the basis of the observed distribution of residual radial velocities (corrected for the solar motion). The general theory of the determination of spatial velocity functions from the distribution of residual radial velocities was given by Ambartsumyan [3] in 1936. In his theory no assumption is made concerning the form of the required function. To use Ambartsumyan's method we need only have a group of stars whose radial velocities are known and which cover the whole sky fairly uniformly.

Unfortunately, the considerable fluctuations in the distribution of radial velocities have made it impossible to obtain reliable results by Ambartsumyan's method. We shall therefore impose some restrictions on the problem in order to simplify the solution of it [80], and suppose that the residual velocities are distributed according to an ellipsoidal law $\varphi(q, p) = \Psi(Q + \sigma)$, where Q is a sign-definite homogeneous quadratic form in the residual velocities, and σ a function of the coordinates only.

If the coordinate axes are the principal axes of the ellipsoid Q = constant and the origin is at its centre, we can write

$$Q = h^2 u'^2 + k^2 v'^2 + l^2 w'^2, \tag{2.59}$$

where u', v', w' are the residual velocity components.

The problem is to derive, from the observed distribution of radial velocities in a small region of the sky, the actual form of the function $\varphi = \Psi(Q + \sigma)$. If, for example, Schwarzschild's law holds in the Galaxy, our result should be $\varphi = \mathrm{e}^{-Q-\sigma}$; if not, then our result should be something different.

Let $\mathbf{L}_1$ be a unit vector in the direction of the centre of the region considered. We take also two other unit vectors $\mathbf{L}_2$, $\mathbf{L}_3$ to form with $\mathbf{L}_1$ a rectangular coordinate trihedral. These vectors may be defined by their direction cosines with respect to the principal axes of the ellipsoid: $\mathbf{L}_1(\alpha_1, \beta_1, \gamma_1)$, $\mathbf{L}_2(\alpha_2, \beta_2, \gamma_2)$, $\mathbf{L}_3(\alpha_3, \beta_3, \gamma_3)$. Finally let v_1, v_2, v_3 be the components, along these new axes, of the residual velocity of a star. Clearly v_1 is the radial velocity, while v_2, v_3 are transverse components. Then

$$\begin{aligned} u' &= \alpha_1 v_1 + \alpha_2 v_2 + \alpha_3 v_3, \\ v' &= \beta_1 v_1 + \beta_2 v_2 + \beta_3 v_3, \\ w' &= \gamma_1 v_1 + \gamma_2 v_2 + \gamma_3 v_3, \end{aligned}$$

and the quadratic form Q given by (2.59) becomes

$$Q = av_1^2 + bv_2^2 + cv_3^2 + 2dv_2v_3 + 2ev_3v_1 + 2fv_1v_2, \tag{2.60}$$

where

$$\begin{aligned} a &= \alpha_1^2 h^2 + \beta_1^2 k^2 + \gamma_1^2 l^2, & d &= \alpha_2\alpha_3 h^2 + \beta_2\beta_3 k^2 + \gamma_2\gamma_3 l^2, \\ b &= \alpha_2^2 h^2 + \beta_2^2 k^2 + \gamma_2^2 l^2, & e &= \alpha_3\alpha_1 h^2 + \beta_3\beta_1 k^2 + \gamma_3\gamma_1 l^2, \\ c &= \alpha_3^2 h^2 + \beta_3^2 k^2 + \gamma_3^2 l^2, & f &= \alpha_1\alpha_2 h^2 + \beta_1\beta_2 k^2 + \gamma_1\gamma_2 l^2. \end{aligned}$$

Now let $f(v_1)$ be the radial velocity distribution function observed for the region of the sky considered. Evidently

$$f(v_1) = \int\int \Psi(Q + \sigma)\, \mathrm{d}v_2\, \mathrm{d}v_3, \tag{2.61}$$

where the integration over v_2 and v_3 is taken from $-\infty$ to $+\infty$. This integral is most conveniently calculated by first bringing the quadratic function Q to canonical form by means of a "triangular transformation". In the quadratic form Q given by formula (2.60) we complete the square, first for v_3 and then for v_2, obtaining

$$Q = c\left(v_3 + \frac{dv_2 + ev_1}{c}\right)^2 + \frac{(bc - d^2)}{c}\left(v_2 + \frac{fc - de}{bc - d^2}\, v_1\right)^2 + \frac{(ac - e^2)(bc - d^2) - (fc - de)^2}{c(bc - d^2)}\, v_1^2.$$

It is easy to see that the numerator of the last term is equal to c times the discriminant of the quadratic form (2.60),

$$D = \begin{vmatrix} a & f & e \\ f & b & d \\ e & d & c \end{vmatrix},$$

and the denominator is c times the minor a_0 of a in this determinant:

$$a_0 = \begin{vmatrix} b & d \\ d & c \end{vmatrix} = bc - d^2.$$

Hence

$$Q = c\left(v_3 + \frac{dv_2 + ev_1}{c}\right)^2 + \frac{a_0}{c}\left(v_2 + \frac{fc - de}{a_0} v_1\right)^2 + \frac{D}{a_0} v_1^2.$$

Substituting this expression in the integral (2.61) and replacing v_2 and v_3 by new variables of integration:

$$\eta = \sqrt{\frac{a_0}{c}} \cdot \left(v_2 + \frac{fc - de}{a_0} v_1\right),$$

$$\zeta = \sqrt{c} \cdot \left(v_3 + \frac{dv_2 + ev_1}{c}\right),$$

we have

$$f(v_1) = \frac{1}{\sqrt{a_0}} \int\int \Psi\left(\eta^2 + \zeta^2 + \frac{D}{a_0} v_1^2 + \sigma\right) \mathrm{d}\eta\, \mathrm{d}\zeta.$$

Replacing also, for symmetry, v_1 by $\xi = \sqrt{(D/a_0)} v_1$, which evidently amounts to a change of scale, we obtain

$$f\left(\sqrt{\frac{a_0}{D}} \cdot \xi\right) = \frac{1}{\sqrt{a_0}} \int\int \Psi(\xi^2 + \eta^2 + \zeta^2 + \sigma)\, \mathrm{d}\eta\, \mathrm{d}\zeta. \qquad (2.62)$$

The integrand is now converted to polar coordinates $\tau = \eta^2 + \zeta^2$, θ. Integration over θ gives

$$f\left(\sqrt{\frac{a_0}{D}} \cdot \xi\right) = \frac{\pi}{\sqrt{a_0}} \int_0^\infty \Psi(\xi^2 + \tau + \sigma)\, \mathrm{d}\tau.$$

Replacing τ as a variable of integration by $\tau + \xi^2 + \sigma$ gives

$$f\left(\sqrt{\frac{a_0}{D}} \cdot \xi\right) = \frac{\pi}{\sqrt{a_0}} \int_{\xi^2+\sigma}^\infty \Psi(\tau)\, \mathrm{d}\tau. \qquad (2.63)$$

This integral is easily transformed by differentiation with respect to ξ, which gives

$$\sqrt{\frac{a_0}{D}}\cdot f'\left(\sqrt{\frac{a_0}{D}}\cdot\xi\right) = -\frac{2\pi}{\sqrt{a_0}}\,\xi\Psi(\xi^2+\sigma).$$

Putting Q in place of ξ^2, we have

$$\Psi(Q+\sigma) = -\frac{a_0}{2\pi\sqrt{D}}\frac{1}{\sqrt{Q}}f'\left(\sqrt{\frac{a_0 Q}{D}}\right). \tag{2.64}$$

This is the solution of the problem, and it shows that, to find the actual form of the function Ψ, we need only derive from observation the radial velocity distribution in any one part of the sky.

In particular, if the direction of $\mathbf{L}_1$ is along one of the principal axes of the ellipsoid, and those of $\mathbf{L}_2$ and $\mathbf{L}_3$ are along the other two axes, formula (2.64) is simplified: we have $D = h^2k^2l^2$, $a_0 = k^2l^2$ or l^2h^2 or h^2k^2. Thus, for the three axes, the results are

$$\left.\begin{aligned}
\Psi(Q+\sigma) &= -\frac{kl}{2\pi h}\frac{1}{\sqrt{Q}}f'\left(\frac{\sqrt{Q}}{h}\right),\\
\Psi(Q+\sigma) &= -\frac{lh}{2\pi k}\frac{1}{\sqrt{Q}}f'\left(\frac{\sqrt{Q}}{k}\right),\\
\Psi(Q+\sigma) &= -\frac{hk}{2\pi l}\frac{1}{\sqrt{Q}}f'\left(\frac{\sqrt{Q}}{l}\right).
\end{aligned}\right\} \tag{2.65}$$

If Ψ does in fact depend only on the argument $Q+\sigma$, in every case the result should be the same, and an examination of the degree of fulfilment of this prediction will indicate the correctness or otherwise of the assumptions made.

It is easy to see that, if Gauss's law holds for the radial velocity distribution, the function Ψ must be that given by Schwarzschild's law. For let us assume that $f(v_1)$ is given by Gauss's law:

$$f(v_1) = c\exp(-H^2v_1^2).$$

Comparing this expression with, for example, the first eqn. (2.65), we see that $\sqrt{Q}/h$ replaces v_1 as the argument of f and so we can put $Q = h^2v_1^2$, which gives

$$\Psi(Q+\sigma) = -\frac{kl}{2\pi h^2}\frac{1}{v_1}f'(v_1).$$

The right-hand side is $(klH^2/\pi h^2)\exp(-H^2v_1^2)$. Now replacing v_1 by $\sqrt{Q}/h$, we obtain

$$\Psi(Q+\sigma) = \frac{kl}{\pi h^2}H^2\exp(-H^2Q/h^2). \tag{2.66}$$

This is Schwarzschild's law apart from the dimensionless factor H/h.

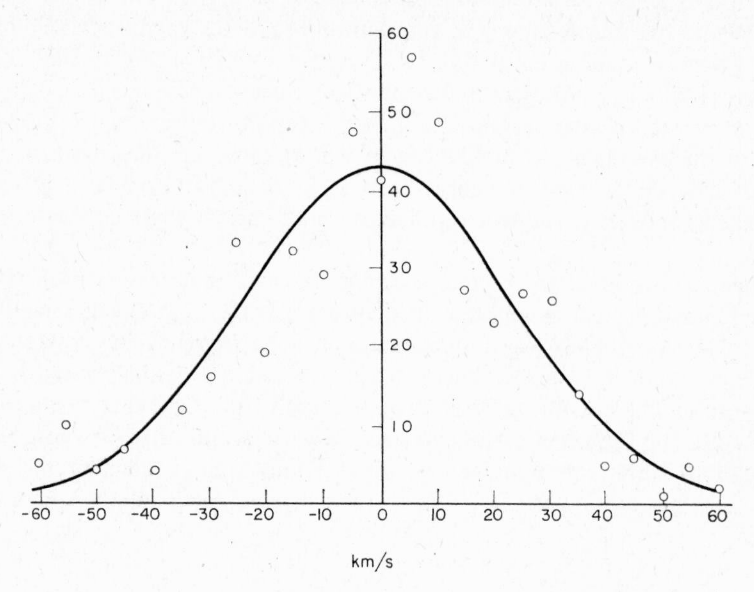
60
50
40
30
20
10
-60
-50
-40
-30
-20
-10
0
10
20
30
40
50
60
km/s

FIG. 9

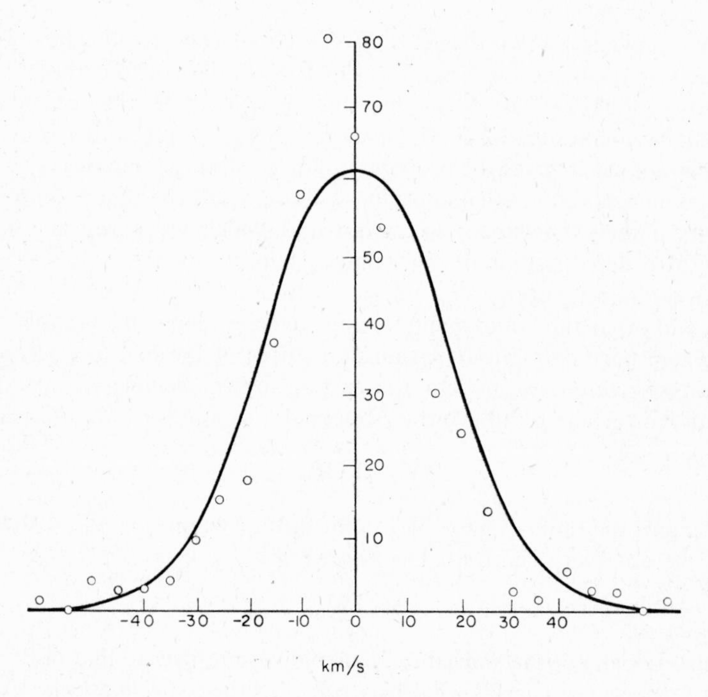
80
70
50
40
30
20
10
-40
-30
-20
-10
0
10
20
30
40
km/s

FIG. 10

Figures 9 and 10 show the observed distribution of radial velocities in the directions of the major and minor axes of the velocity ellipsoid, according to the results of Williamson [138].† The ordinate is the number of stars. The continuous curve is a Gaussian drawn to give the best fit with the observations (represented by circles). We may regard the agreement as satisfactory if we take into account the fact that the deviations of the circles from the curve are systematic and indicate a positive excess (the curve being too low in the centre and at the ends, and too high in the intermediate ranges).

It has been shown in Chapter I that a positive excess is a necessary property of an inhomogeneous statistical ensemble which consists of several groups having different degrees of dispersion. Thus two important conclusions are reached: (1) the observed residual velocities of stars in the Galaxy are distributed approximately in accordance with Schwarzschild's law, (2) the stellar population of the Galaxy appears to be inhomogeneous. The Galaxy must thus consist of various sub-systems whose residual stellar velocities exhibit different degrees of dispersion. As has been mentioned earlier, a number of other arguments in favour of the existence of sub-systems may also be given.

§ 9. The Differential Field of Centroid Velocities. Helmholtz' Theorem for Stellar Systems

It has been shown in § 4 that the velocity vector $\mathbf{V}$ of each star may be resolved into a sum of two vectors, one of which is the velocity $\mathbf{V}_0$ of the centroid at that point, and the other the residual velocity $\mathbf{V}'$ of that star: $\mathbf{V} = \mathbf{V}_0 + \mathbf{V}'$. The consideration of residual stellar velocities leads, as we have seen in §§ 3 and 4, to the concept of the velocity function, which gives the probability law governing the residual velocities.

In the present section we shall consider in more detail the other component of a star's velocity, namely the velocity of the centroid, which arises from the star's participation in the ordered motion of the stellar system.

In general the velocity of the centroid is a continuous function of position. We shall assume *ab initio* that the components of this velocity along the axes of a fixed coordinate system have continuous partial derivatives of the first and second orders. Since no further reference is made in this section to the velocities of individual stars, let $\mathbf{V}$ denote the velocity of an arbitrary centroid:

$$\mathbf{V} = \mathbf{V}(\mathbf{R}), \tag{2.67}$$

where $\mathbf{R}$ denotes the radius vector of a point in the system. Let $\mathbf{V}_0 = \mathbf{V}(\mathbf{R}_0)$ be the velocity of the observer's centroid. Then the vector

$$\mathbf{V} - \mathbf{V}_0 \tag{2.68}$$

is the relative velocity of the "variable" centroid with respect to that of the observer. The vector function (2.67) gives what may be called the *centroid velocity field*. Corre-

† The diagrams themselves are taken from Chandrasekhar's book [17].

spondingly, (2.68) gives the *relative centroid velocity field*. Whereas the residual velocity distribution is a statistical pattern of motions, the centroid velocity field gives a hydrodynamic picture of the motion of the stellar system, since it makes possible an analogy between the motion of stellar systems and that of continuous media.

In stellar dynamics an important problem is that of the distribution of centroid velocities in a *small* region surrounding a given point – the *local* or *differential centroid velocity field*. This importance is due to the fact that, in the development of stellar astronomy and, in particular, of stellar dynamics, the study of the Galaxy began with the investigation of a small region in the neighbourhood of the Sun. Later, after considerable progress had been made in elucidating the structure and motions of the Galaxy as a whole, the neighbourhood of the Sun remained the starting point from which our knowledge of more distant regions of the Galaxy was gradually built up.

In the last fifteen years a beginning has been made in the study of the Metagalaxy. Here the situation is perhaps similar to that concerning the Galaxy itself at the beginning of the 1920's: we must be content with an investigation of a region in the neighbourhood of our Galaxy which is small in comparison with the size of the Metagalaxy.

In the present section we shall discuss the formulae which give the differential (local) field of centroid velocities.

Let C_0 and C be respectively the centroid of the observer and an arbitrary centroid, their radii vectores being $\mathbf{R}_0$ and $\mathbf{R}$. We use a fixed system of rectangular coordinates ξ, η, ζ by means of which the motion of centroids (the macroscopic motion of the stellar system) is to be measured. For example, in the Galaxy this coordinate system could be related to the axis of symmetry, in a globular cluster to the centre of the cluster, and so on.

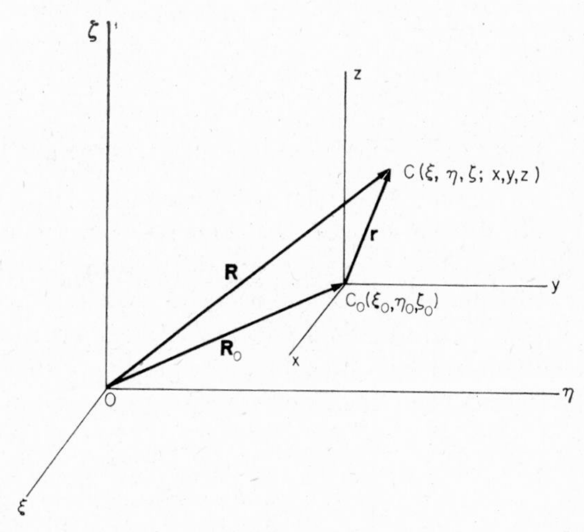

FIG. 11

Let the coordinates of C and C_0 be (ξ, η, ζ) and (ξ_0, η_0, ζ_0). We make the observer's centroid C_0 the origin of a moving system of rectangular coordinates x, y, z whose axes are parallel to the fixed axes (Fig. 11). Let the coordinates of the arbitrary centroid C in the moving system be x, y, z. Then the centroid velocity field (2.67) can be written

$$\mathbf{V} = \mathbf{V}(\mathbf{R}) = \mathbf{V}(\xi, \eta, \zeta) = \mathbf{V}(\xi_0 + x, \eta_0 + y, \zeta_0 + z). \tag{2.69}$$

We shall assume that the topocentric radius vector $\mathbf{r}(x, y, z)$ is small. The vector function $\mathbf{V}(\mathbf{R})$ may be expanded in powers of x, y, z, retaining only the first-order terms. Then

$$\mathbf{V} = \mathbf{V}_0 + \left(\frac{\partial \mathbf{V}}{\partial \xi}\right)_0 x + \left(\frac{\partial \mathbf{V}}{\partial \eta}\right)_0 y + \left(\frac{\partial \mathbf{V}}{\partial \zeta}\right)_0 z,$$

where $\mathbf{V}_0 = \mathbf{V}(\mathbf{R}_0) = \mathbf{V}(\xi_0, \eta_0, \zeta_0)$ is the velocity of the observer's centroid, and the derivatives are taken at the point $C_0(\xi_0, \eta_0, \zeta_0)$.

The quantities ξ_0, η_0, ζ_0 are parameters in eqn. (2.69). Hence $\partial \mathbf{V}/\partial \xi = \partial \mathbf{V}/\partial x$, $\partial \mathbf{V}/\partial \eta = \partial \mathbf{V}/\partial y$, $\partial \mathbf{V}/\partial \zeta = \partial \mathbf{V}/\partial z$, and the above equation becomes

$$\mathbf{V} = \mathbf{V}_0 + \left(\frac{\partial \mathbf{V}}{\partial x}\right)_0 x + \left(\frac{\partial \mathbf{V}}{\partial y}\right)_0 y + \left(\frac{\partial \mathbf{V}}{\partial z}\right)_0 z. \tag{2.70}$$

This is the vector form of the differential centroid velocity field. In the approximation used here, this field is linear in the topocentric coordinates x, y, z. Let u, v, w be the components of the velocity vector of the centroid C, and u_0, v_0, w_0 those of the velocity of C_0, along the fixed axes. Then, writing (2.70) in coordinates, we have

$$\left.\begin{aligned} u &= u_0 + u_x x + u_y y + u_z z, \\ v &= v_0 + v_x x + v_y y + v_z z, \\ w &= w_0 + w_x x + w_y y + w_z z, \end{aligned}\right\} \tag{2.71}$$

where $u_x, u_y, u_z, \ldots, w_z$ denote the *kinematic parameters*, i.e. the partial derivatives of u, v, w at the position of the observer's centroid. Formulae (2.71) give the *differential field of centroid velocities* in rectangular coordinates.

From (2.70) (or (2.71)) it follows at once that there is a complete analogy between the centroid velocity distribution in a small region surrounding any point in a stellar system and the velocity distribution in a "fluid particle" such as is considered in fluid dynamics. Evidently u_0, v_0, w_0 are the components of the translational velocity $\mathbf{V}_0$ of the region, and the other terms are the components of the relative velocity of displacement within that region. Their coefficients may be arranged as a *displacement matrix*

$$\boldsymbol{D} = \begin{bmatrix} u_x & u_y & u_z \\ v_x & v_y & v_z \\ w_x & w_y & w_z \end{bmatrix}. \tag{2.72}$$

We know from fluid mechanics that the elements of this matrix are components of an affine orthogonal tensor $\boldsymbol{D}$ of rank two, called the *displacement tensor*.

Using this tensor in formula (2.70) (or (2.71)), we have

$$\mathbf{V} = \mathbf{V}_0 + \boldsymbol{D} \cdot \mathbf{r}, \tag{2.73}$$

where $\boldsymbol{D} \cdot \mathbf{r}$ denotes the scalar product of the tensor $\boldsymbol{D}$ and the radius vector $\mathbf{r}$. In order to elucidate the mechanical significance of the quantity $\boldsymbol{D} \cdot \mathbf{r}$ in formula (2.73), we may effect a transformation of formula (2.72).† Using the conjugate tensor

$$\boldsymbol{D}_c = \begin{bmatrix} u_x & v_x & w_x \\ u_y & v_y & w_y \\ u_z & v_z & w_z \end{bmatrix}, \tag{2.74}$$

we write

$$\boldsymbol{D} = \boldsymbol{S} + \boldsymbol{A}, \tag{2.75}$$

where

$$\boldsymbol{S} = \tfrac{1}{2}(\boldsymbol{D} + \boldsymbol{D}_c) \tag{2.76}$$

and

$$\boldsymbol{A} = \tfrac{1}{2}(\boldsymbol{D} - \boldsymbol{D}_c). \tag{2.77}$$

From (2.73) and (2.75) we have

$$\mathbf{V} = \mathbf{V}_0 + \boldsymbol{S} \cdot \mathbf{r} + \boldsymbol{A} \cdot \mathbf{r}. \tag{2.78}$$

It is known from tensor calculus that the displacement tensor $\boldsymbol{D}$ and its conjugate $\boldsymbol{D}_c$ may be written

$$\boldsymbol{D} = \mathrm{d}\mathbf{V}/\mathrm{d}\mathbf{r}, \qquad \boldsymbol{D}_c = \operatorname{grad} \mathbf{V}. \tag{2.79}$$

Thus (2.76) and (2.77) give

$$\begin{aligned} \boldsymbol{S} &= \frac{1}{2}\left(\frac{\mathrm{d}\mathbf{V}}{\mathrm{d}\mathbf{r}} + \operatorname{grad} \mathbf{V}\right), \\ \boldsymbol{A} &= \frac{1}{2}\left(\frac{\mathrm{d}\mathbf{V}}{\mathrm{d}\mathbf{r}} - \operatorname{grad} \mathbf{V}\right). \end{aligned} \tag{2.80}$$

Let us now consider the mechanical significance of the quantities $\boldsymbol{S} \cdot \mathbf{r}$ and $\boldsymbol{A} \cdot \mathbf{r}$, whose sum, according to (2.78), gives the relative differential field of centroid velocities. From (2.76), (2.74) and (2.72) we have

$$\boldsymbol{S} = \begin{bmatrix} S_{11} & S_{12} & S_{13} \\ S_{21} & S_{22} & S_{23} \\ S_{31} & S_{32} & S_{33} \end{bmatrix}$$

$$= \begin{bmatrix} u_x & \tfrac{1}{2}(u_y + v_x) & \tfrac{1}{2}(u_z + w_x) \\ \tfrac{1}{2}(v_x + u_y) & v_y & \tfrac{1}{2}(v_z + w_y) \\ \tfrac{1}{2}(w_x + u_z) & \tfrac{1}{2}(w_y + v_z) & w_z \end{bmatrix}. \tag{2.81}$$

† N.E. Kochin, *Vector calculus and the elements of tensor calculus (Vektornoe ischisleniei nachala tenzornogo ischisleniya)*, 5th ed., ONTI, Moscow 1937.

This symmetrical tensor may be called the *local deformation tensor*.

From (2.77), (2.74) and (2.72) we have

$$A = \begin{bmatrix} 0 & -\omega_3 & \omega_2 \\ \omega_3 & 0 & -\omega_1 \\ -\omega_2 & \omega_1 & 0 \end{bmatrix}$$

$$= \begin{bmatrix} 0 & \tfrac{1}{2}(u_y - v_x) & \tfrac{1}{2}(u_z - w_x) \\ \tfrac{1}{2}(v_x - u_y) & 0 & \tfrac{1}{2}(v_z - w_y) \\ \tfrac{1}{2}(w_x - u_z) & \tfrac{1}{2}(w_y - v_z) & 0 \end{bmatrix}. \tag{2.82}$$

The quantities $\omega_1, \omega_2, \omega_3$ which appear in the antisymmetrical tensor A are the components of the vector

$$\boldsymbol{\omega} = \tfrac{1}{2}\,\mathbf{curl}\,\mathbf{V}, \tag{2.83}$$

which is the instantaneous angular velocity vector for rigid rotation of the small region. It is known from the theory of tensors that the scalar product of the tensor A and the radius vector $\mathbf{r}$ is equal to the vector product of $\boldsymbol{\omega}$ and $\mathbf{r}$:†

$$A \cdot \mathbf{r} = \boldsymbol{\omega} \times \mathbf{r}. \tag{2.84}$$

The tensor A may therefore be called the *local rotation tensor*, and formula (2.78) gives *Helmholtz' theorem* for stellar systems: within a small region surrounding any point in a stellar system, the velocities of the centroids consist of (1) the velocity $\mathbf{V}_0$ of rigid translation of the region, (2) the velocity $A \cdot \mathbf{r}$ of rigid rotation of the region with instantaneous angular velocity $\boldsymbol{\omega} = \tfrac{1}{2}\,\mathbf{curl}\,\mathbf{V}$, (3) the velocity $S \cdot \mathbf{r}$ of deformation of the region.

It is easy to see that the components of the vector $S \cdot \mathbf{r}$ can be represented in terms of the partial derivatives of the quadratic function

$$\begin{aligned} F &= \tfrac{1}{2}\mathbf{r} \cdot S \cdot \mathbf{r} \\ &= \tfrac{1}{2}\,[u_x x^2 + v_y y^2 + w_z z^2 + (v_z + w_y)yz + (w_x + u_z)zx + (u_y + v_x)xy]. \end{aligned}$$

This function is therefore the deformation velocity potential of the volume element:

$$\begin{aligned} (S \cdot \mathbf{r})_x &= u_x x + \tfrac{1}{2}(u_y + v_x)y + \tfrac{1}{2}(u_z + w_x)z = \partial F/\partial x, \\ (S \cdot \mathbf{r})_y &= \tfrac{1}{2}(v_x + u_y)x + v_y y + \tfrac{1}{2}(v_z + w_y)z = \partial F/\partial y, \\ (S \cdot \mathbf{r})_z &= \tfrac{1}{2}(w_x + u_z)x + \tfrac{1}{2}(w_y + v_z)y + w_z z = \partial F/\partial z \end{aligned}$$

or, more briefly, $S \cdot \mathbf{r} = \text{grad}\, F$. Thus we have finally Helmholtz' theorem in a form clearer than (2.78):

$$\mathbf{V} = \mathbf{V}_0 + \text{grad}\, F + \boldsymbol{\omega} \times \mathbf{r}. \tag{2.85}$$

† Equation (2.84) is easily verified by using (2.82), (2.83) and the rules of matrix and vector multiplication.

As we mentioned at the beginning of this section, in stellar dynamics the discussion of the centroid velocity field in a small region is of importance, because both in the Galaxy and in the Metagalaxy we find by observation the velocities of only the nearest stars or galaxies. It is therefore necessary to construct a theoretical picture of the motion of the whole system on the basis mainly of data relating to a single volume element. For this reason we shall discuss in detail in § 10 what information concerning the motion of the whole system can be obtained from a study of the centroid velocity distribution in a small region or, more precisely, from a knowledge of the kinematic parameters for any one small region. Since in general this problem is unsolved, we shall consider the converse problem for some of the most important particular cases, i.e. the question of how a given motion of the whole system affects the values of the kinematic coefficients for any one small region.

§ 10. Determination of the Kinematic Parameters of the Differential Field of Centroid Velocities from the Observed Velocities of Stars

(a) *The Parameters from Spatial (Total) Velocities*

The total velocity vector of any star can, as we have seen, be written as a sum: $\mathbf{V} = \overline{\mathbf{V}} + \mathbf{V}'$, where $\overline{\mathbf{V}}$, with components u, v, w, is the velocity of the centroid corresponding to the position of the star, and $\mathbf{V}'$ the residual velocity. Observations of total velocities of stars give, after correction for the velocity of the observer's centroid (which is minus the velocity of the Sun's local motion) the vector

$$\mathbf{V} - \mathbf{V}_{\odot} = (\overline{\mathbf{V}} - \mathbf{V}_{\odot}) + \mathbf{V}', \tag{2.86}$$

which is the velocity of the star relative to the observer's centroid (i.e. the Sun's centroid). If all the stars considered are in the same small region, the components of the relative local velocity $\overline{\mathbf{V}} - \mathbf{V}_{\odot}$ of the centroid along the coordinate axes are, by (2.71), homogeneous linear functions of the star's coordinates. Hence each star whose total velocity is known supplies three equations:

$$\left.\begin{aligned} u - u_{\odot} &= u_x x + u_y y + u_z z + u', \\ v - v_{\odot} &= v_x x + v_y y + v_z z + v', \\ w - w_{\odot} &= w_x x + w_y y + w_z z + w'. \end{aligned}\right\} \tag{2.87}$$

The left-hand sides of these equations are supposed known from observation, and the coordinates x, y, z of the star are given; the unknowns are the kinematic parameters $u_x, u_y, \ldots, w_z$, while the residual velocities u', v', w' are of the nature of accidental errors in the equations of condition for the application of the method of least squares. Hence eqns. (2.87) can be solved by this method to give the kinematic parameters. In practice it is preferable to combine the equations into groups each pertaining to several stars with similar coordinates. A single equation of condition is obtained for each group by averaging the left-hand sides of the equations and the coordinates of the stars. The residual velocities ought to balance out to some extent, and therefore their effect on the errors in the unknowns should be reduced. However, for the same reasons as in previous discussions, the use of total velocities for this purpose is inadvisable, because they depend in an involved manner on the errors in parallax, proper motion and radial velocity.

(b) *The Parameters from Radial Velocities*

The direction cosines of the radius vector **r** of a star are $\alpha = x/r, \beta = y/r, \gamma = z/r$. After multiplying eqns. (2.87) by these direction cosines and summing, we therefore obtain the following expression for the radial velocity v_r of a star relative to the observer's centroid:

$$v_r = r[u_x\alpha^2 + v_y\beta^2 + w_z\gamma^2 + (v_z + w_y)\beta\gamma + (w_x + u_z)\gamma\alpha + (u_y + v_x)\alpha\beta] + v_r'. \tag{2.88}$$

Here v_r' denotes the residual radial velocity of the star.

The radial velocity v_r and the direction cosines α, β, γ may be supposed known from observation. Hence, regarding the residual velocities as accidental errors, we can find from the equations of condition (2.88), by the method of least squares, the quantities

$$u_x, v_y, w_z, \tfrac{1}{2}(v_z + w_y), \quad \tfrac{1}{2}(w_x + u_z), \quad \tfrac{1}{2}(u_y + v_x), \tag{2.89}$$

which are the components of the symmetrical deformation tensor $\boldsymbol{S}$ (2.81). Thus we see that only the deformation tensor of the volume element can be determined from radial velocities alone [82].

The distances of individual stars are usually either entirely unknown, or known so imprecisely that it is undesirable to use them as such. To find the deformation tensor it is preferable to use the same procedure as that which is generally employed in determining the local motion of the Sun (§ 5).

By dividing the sky into standard areas, we can approximately take the direction cosines of the individual stars as equal to those of the centre of the appropriate area. Averaging eqn. (2.88) over all stars in a given area, we obtain a system of equations of condition for the various areas:

$$\bar{v}_r = \bar{r}[u_x\alpha^2 + v_y\beta^2 + w_z\gamma^2 + (v_z + w_y)\beta\gamma + (w_x + u_z)\gamma\alpha + (u_y + v_x)\alpha\beta], \tag{2.90}$$

where the bar denotes the averaging just mentioned. The residual velocities should, at least in theory, disappear on averaging, and so they may be omitted.

The eqns. (2.90) make it possible to determine, by the least-squares method, the quantities

$$\bar{r}u_x, \bar{r}v_y, \bar{r}w_z, \bar{r}(v_z + w_y), \quad \bar{r}(w_x + u_z), \quad \bar{r}(u_y + v_x). \tag{2.91}$$

It now remains to find by some means the mean distance $\bar{r}$ of the group concerned, and then the components of the deformation tensor are known.

(c) *The Parameters from Proper Motions*

In a fairly similar way, we can determine the kinematic parameters from proper motions. In eqns. (2.87) we replace x, y, z by their values in terms of the direction cosines α, β, γ. Multiplying the resulting equations by the respective components of the unit vectors

$$\mathbf{l}^0(\alpha_1, \beta_1, \gamma_1) \quad \text{and} \quad \mathbf{b}^0(\alpha_2, \beta_2, \gamma_2) \tag{2.92}$$

and adding the equations so obtained, we have for each star two equations of condition, of the form

$$\left.\begin{aligned} v_l &= r(u_x\alpha + u_y\beta + u_z\gamma)\alpha_1 + r(v_x\alpha + v_y\beta + v_z\gamma)\beta_1 + r(w_x\alpha + w_y\beta + w_z\gamma)\gamma_1 + v_l', \\ v_b &= r(u_x\alpha + u_y\beta + u_z\gamma)\alpha_2 + r(v_x\alpha + v_y\beta + v_z\gamma)\beta_2 + r(w_x\alpha + w_y\beta + w_z\gamma)\gamma_2 + v_b'. \end{aligned}\right\} \tag{2.93}$$

These equations can be put in a somewhat different form, which makes possible a more convenient comparison with formula (2.90) for radial velocities. To do so, we use the table (2.54) to express the direction cosines (2.92) in terms of those of the radius vector **r**:

$$\alpha_1 = -\beta/\sqrt{(\alpha^2 + \beta^2)}, \quad \beta_1 = \alpha/\sqrt{(\alpha^2 + \beta^2)}, \quad \gamma_1 = 0 \tag{2.94}$$

and

$$\alpha_2 = -\alpha\gamma/\sqrt{(\alpha^2 + \beta^2)}, \quad \beta_2 = -\beta\gamma/\sqrt{(\alpha^2 + \beta^2)}, \quad \gamma_2 = \sqrt{(\alpha^2 + \beta^2)}. \tag{2.95}$$

Then formulae (2.93) become

$$\left.\begin{aligned} v_l &= \frac{r}{\sqrt{(\alpha^2+\beta^2)}}\,[v_x\alpha^2 - u_y\beta^2 - u_z\beta\gamma + v_z\gamma\alpha + (v_y - u_x)\,\alpha\beta] + v_l',\\ v_b &= \frac{-r}{\sqrt{(\alpha^2+\beta^2)}}\,[(u_x - w_z)\,\alpha^2\gamma + (v_y - w_z)\,\beta^2\gamma + (u_z + w_x)\,\alpha\gamma^2\\ &\quad + (v_z + w_y)\,\beta\gamma^2 + (u_y + v_x)\,\alpha\beta\gamma - w_x\alpha - w_y\beta] + v_b'. \end{aligned}\right\} \tag{2.96}$$

We see that these equations do not permit a determination of all the unknowns separately. The number of unknowns is nine, of which three remain indeterminate. These equations make it possible to determine u_y, u_z, v_x, v_z, w_x, w_y individually, but u_x, v_y and w_z only in the combinations $u_x - v_y$, $u_x - w_z$, $v_y - w_z$; the first of these is, apart from a factor $\frac{1}{2}$, the z-component of the vector of rigid rotation of the volume element.

It is again preferable to form the equations of condition for groups of stars, dividing the sky into standard areas and allotting to one group the stars in each area; and the transverse velocities should be converted to proper motions. Then formulae (2.96) become

$$\left.\begin{aligned} k\mu_l &= \frac{1}{\sqrt{(\alpha^2+\beta^2)}}\,[v_x\alpha^2 - u_y\beta^2 - u_z\beta\gamma + v_z\gamma\alpha + (v_y - u_x)\,\alpha\beta] + k\mu_l',\\ k\mu_b &= -\frac{1}{\sqrt{(\alpha^2+\beta^2)}}\,[(u_x - w_z)\,\alpha^2\gamma + (v_y - w_z)\,\beta^2\gamma + (u_z + w_x)\,\alpha\gamma^2\\ &\quad + (v_z + w_y)\,\beta\gamma^2 + (u_y + v_x)\,\alpha\beta\gamma - w_x\alpha - w_y\beta] + k\mu_b'. \end{aligned}\right\} \tag{2.97}$$

Averaging over the stars in a given area, we obtain equations for the quantities

$$u_y,\ u_z,\ v_x,\ v_z,\ w_x,\ w_y,\ u_x - w_z,\ v_y - w_z, \tag{2.98}$$

and also of $u_x - v_y$, which provides a check on the last two. In practice, it is necessary to take account of the fact that the proper motions found by observation are not corrected for the Sun's local motion, i.e. the motion of the observer with respect to the local centroid. If the parallaxes of the individual stars are unknown, we cannot make the necessary corrections to the proper motions. Thus the terms

$$\begin{aligned} L_\odot &= \overline{\frac{1}{r}}\,(\alpha_1 X_\odot + \beta_1 Y_\odot + \gamma_1 Z_\odot),\\ B_\odot &= \overline{\frac{1}{r}}\,(\alpha_2 X_\odot + \beta_2 Y_\odot + \gamma_2 Z_\odot) \end{aligned} \tag{2.99}$$

must be included in (2.97) to allow for the motion of the local centroid. The final equations are

$$\left.\begin{aligned} k\bar{\mu}_l &= L_\odot + \frac{1}{\sqrt{(\alpha^2+\beta^2)}}\,[v_x\alpha^2 - u_y\beta^2 - u_z\beta\gamma + v_z\gamma\alpha + (v_y - u_x)\,\alpha\beta],\\ k\bar{\mu}_b &= B_\odot - \frac{1}{\sqrt{(\alpha^2+\beta^2)}}\,[(u_x - w_z)\,\alpha^2\gamma + (v_y - w_z)\,\beta^2\gamma + (u_z + w_x)\,\alpha\gamma^2\\ &\quad + (v_z + w_y)\,\beta\gamma^2 + (u_y + v_x)\,\alpha\beta\gamma - w_x\alpha - w_y\beta]. \end{aligned}\right\} \tag{2.100}$$

The solution of these equations by the least-squares method gives the elements (2.98) of the local field of centroid velocities.

This concludes our discussion of methods of determining the kinematic parameters of the local field of centroid velocities, and we shall now go on to consider the local fields which are of the greatest astronomical interest.

CHAPTER III

THE ELEMENTARY THEORY OF GALACTIC ROTATION

§ 1. General Formulae of Galactic Rotation

As A simple example of the field of relative centroid velocities, and one which has important applications, let us consider the velocity field of an axial rotation. A typical example of such a motion is the rotation of the Galaxy. In the present section we shall derive formulae which give the relative rotational velocity field of the Galaxy (or of sub-systems of it).

We take the origin of a system of rectangular coordinates at the centre of the Galaxy, with the z-axis along the axis of rotation, and assume for simplicity that the observer's centroid C_0 is in the galactic plane. The symmetry in the observed distribution of stars about the galactic equator shows that the Sun lies very close to the galactic plane. According to recent data, the distance of the Sun from this plane is only 10–15 ps, and we shall neglect it. (According to van Tulder [133], $z_0 = +13{\cdot}5$ ps.) Otherwise, the xy-plane would simply be at a distance z_0 from the galactic plane.

The x-axis is taken to pass through C_0, and the y-axis so as to form a right-handed system (Fig. 12). Let C be an arbitrary centroid, whose position relative to the observer's centroid is given by the radius vector $\mathbf{r} = C_0C$. Let $\mathbf{R}$ and $\mathbf{R}_0$ be the radii vectores of the centroids C and C_0, and $\boldsymbol{\omega}$ and $\boldsymbol{\omega}_0$ the angular velocity vectors at the points C and C_0. Then the linear velocities of the centroids resulting from the rotation

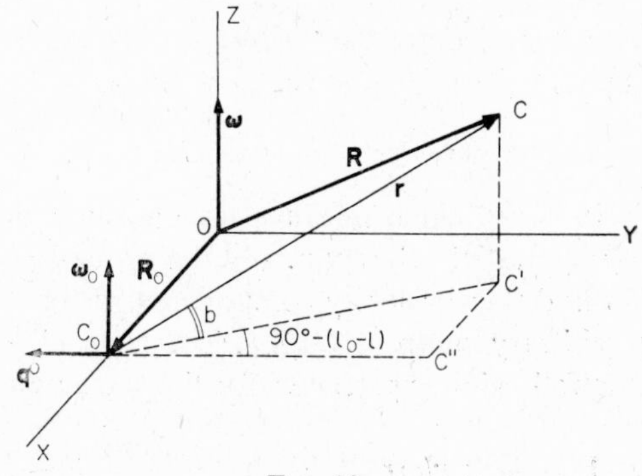

FIG. 12

of the Galaxy are evidently $\mathbf{V}_0 = \boldsymbol{\omega}_0 \times \mathbf{R}_0$, $\mathbf{V} = \boldsymbol{\omega} \times \mathbf{R}$, and the relative velocity of the centroid C with respect to C_0 is $\mathbf{v} = (\boldsymbol{\omega} \times \mathbf{R}) - (\boldsymbol{\omega}_0 \times \mathbf{R}_0)$. Figure 12 shows that $\mathbf{r} = \mathbf{R} - \mathbf{R}_0$. Hence

$$\mathbf{v} = \boldsymbol{\omega} \times \mathbf{r} + (\boldsymbol{\omega} - \boldsymbol{\omega}_0) \times \mathbf{R}_0. \tag{3.1}$$

This equation is the general vector description of the field of relative centroid velocities in the rotation of the Galaxy.

The vectors $\boldsymbol{\omega}$ and $\boldsymbol{\omega}_0$ are evidently parallel to the axis of the Galaxy, and (on the above assumption) $\mathbf{R}_0$ lies in the galactic plane. Hence

$$(\boldsymbol{\omega} - \boldsymbol{\omega}_0) \times \mathbf{R}_0 = -(\omega - \omega_0) R_0 \mathbf{q}^0, \tag{3.2}$$

where $\mathbf{q}^0$ is a unit vector in the galactic plane and perpendicular to $\mathbf{R}_0$ (Figs. 12 and 13; in Fig. 13 O and $\mathbf{R}_0$ have the same significance as in Fig. 12).

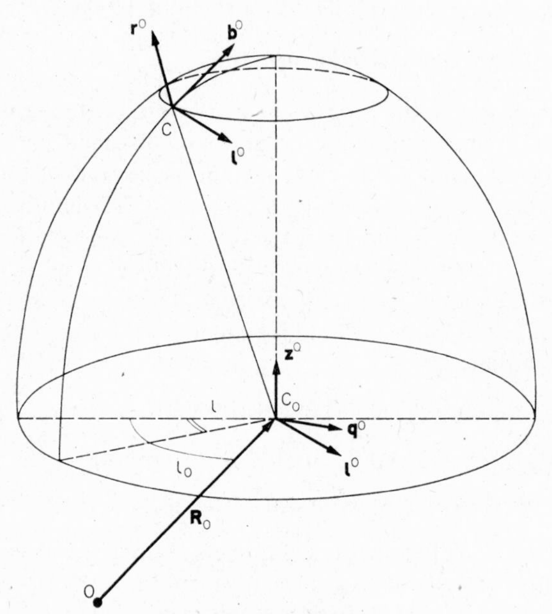

FIG. 13

From (3.1) and (3.2) we find

$$\mathbf{v} = \boldsymbol{\omega} \times \mathbf{r} - (\omega - \omega_0) R_0 \mathbf{q}^0. \tag{3.3}$$

To write this formula in coordinates we must take account of the fact that stellar motions are defined by the radial velocity v_r and the proper motions in longitude μ_l and in latitude μ_b. To find these velocity components, we use the moving trihedral, with its origin at the arbitrary centroid C (Fig. 13). Then $\mathbf{r}^0 = \mathbf{r}/r$, $\mathbf{z}^0 = \boldsymbol{\omega}/\omega$, and formulae (1.69) give the following expressions for the three unit vectors:

$$\mathbf{r}^0 = \mathbf{r}/r, \quad \mathbf{l}^0 = \boldsymbol{\omega} \times \mathbf{r}/\omega r \cos b, \quad \mathbf{b}^0 = \frac{\boldsymbol{\omega}}{\omega \cos b} - \frac{\mathbf{r} \tan b}{r}. \tag{3.4}$$

The scalar products of eqn. (3.3) with the vectors (3.4) give the components of the relative velocity vector (the radial velocity and the two components of the transverse velocity) as

$$v_r = \frac{\boldsymbol{\omega} \times \mathbf{r} \cdot \mathbf{r}}{r} - \frac{(\omega - \omega_0) R_0}{r} \mathbf{r} \cdot \mathbf{q}^0,$$

$$v_l = (\boldsymbol{\omega} \times \mathbf{r})^2 / \omega r \cos b - (\omega - \omega_0) R_0 \mathbf{l}^0 \cdot \mathbf{q}^0,$$

$$v_b = [\boldsymbol{\omega} \times \mathbf{r} - (\omega - \omega_0) R_0 \mathbf{q}^0] \cdot \left(\frac{\boldsymbol{\omega}}{\omega \cos b} - \frac{\mathbf{r}}{r} \tan b \right).$$

These formulae can be simplified, since the triple products involving two equal vectors are zero, and also $\mathbf{q}^0 \cdot \boldsymbol{\omega} = 0$, since the vector $\mathbf{q}^0$ is perpendicular to $\boldsymbol{\omega}$. Thus

$$\left.\begin{aligned} v_r &= -[(\omega - \omega_0) R_0 / r]\, \mathbf{r} \cdot \mathbf{q}^0, \\ v_l &= (\boldsymbol{\omega} \times \mathbf{r})^2 / \omega r \cos b - (\omega - \omega_0) R_0 \mathbf{l}^0 \cdot \mathbf{q}^0, \\ v_b &= [(\omega - \omega_0) R_0 / r]\, \mathbf{r} \cdot \mathbf{q}^0 \tan b. \end{aligned}\right\} \quad (3.5)$$

To find the value of the scalar product $\mathbf{r} \cdot \mathbf{q}^0$, we project the centroid C on the coordinate plane XOY and project the point C' so obtained on the vector $\mathbf{q}^0$ (see Fig. 12). The resulting point C'' is evidently the projection of the terminus of the vector $\mathbf{r}$ on the vector $-\mathbf{q}^0$. Hence $\mathbf{r} \cdot \mathbf{q}^0 = -C_0 C''$. Denoting by l and l_0 the longitudes of the centroid C and the galactic centre, we have from Fig. 12 $C_0 C'' = C_0 C' \sin(l_0 - l) = r \sin(l_0 - l) \cos b$, and therefore $\mathbf{r} \cdot \mathbf{q}^0 = r \sin(l - l_0) \cos b$.

Also, from Figs. 12 and 13, $|\boldsymbol{\omega} \times \mathbf{r}| = \omega r \cos b$, $\mathbf{l}^0 \cdot \mathbf{q}^0 = \cos(l - l_0)$. Substituting these results, we have the following final formulae for the components of the relative field of galactic rotation:

$$\left.\begin{aligned} v_r &= -(\omega - \omega_0) R_0 \cos b \sin(l - l_0), \\ v_l &= -(\omega - \omega_0) R_0 \cos(l - l_0) + \omega r \cos b, \\ v_b &= (\omega - \omega_0) R_0 \sin b \sin(l - l_0) = -v_r \tan b. \end{aligned}\right\} \quad (3.6)$$

In these formulae v_r, v_l, v_b denote the components, along the axes of the moving trihedral, of the relative velocity of an arbitrary centroid C with respect to the observer's centroid C_0, R_0 the distance of C_0 from the axis of the Galaxy, ω and ω_0 the angular velocities of rotation at the points C and C_0, r the distance from C to C_0, l_0 the galactic longitude of the axis of the Galaxy from C_0, and l and b the galactic coordinates of C with respect to C_0 (see Figs. 12 and 13).

Formulae (3.6) hold whatever the dependence of the angular velocity on the distances from the axis and from the galactic plane, and for all relative distances r.

In deriving formulae (3.1)–(3.6) we have supposed that the stellar system rotates as a whole, i.e. that its stellar composition is kinematically homogeneous. Throughout this chapter we shall likewise assume that the stellar system considered (the Galaxy) cannot be divided into kinematically distinct sub-systems with different centroid velocities in each macroscopic volume element. If such sub-systems are present, our conclusions will apply to them individually.

§ 2. The Local Field of Centroid Velocities. Generalisation of Oort's Formulae

Formulae (3.6) have been derived for the relative field of galactic rotation. They may be written, calling the cylindrical radius vector ϱ_0 instead of R_0,

$$\left.\begin{aligned} v_r &= -(\omega - \omega_0)\varrho_0 \cos b \sin(l - l_0), \\ v_l &= -(\omega - \omega_0)\varrho_0 \cos(l - l_0) + \omega r \cos b, \\ v_b &= -v_r \tan b. \end{aligned}\right\} \tag{3.6'}$$

These are valid for any axial rotation, i.e. for any dependence of the angular velocity ω on the distances from the axis of the Galaxy and from the galactic plane, $\omega = \omega(\varrho, z)$.

In the present section we shall assume that the rotation of the Galaxy is *barotropic*, i.e. that the angular velocity ω of the rotation depends only on the distance from the axis:

$$\omega = \omega(\varrho). \tag{3.7}$$

Before proceeding to more general problems, let us consider, as a simple application of the above discussion, the determination of the local field of the centroid velocities in the Galaxy. We first express the cylindrical radius vector ϱ in terms of the quantities r, b, l which characterise the position of an arbitrary centroid C with respect to the observer's centroid. From Fig. 12 we have

$$\varrho^2 = \varrho_0^2 + (C_0 C')^2 - 2\varrho_0 C_0 C' \cos(l_0 - l)$$

or, since $C_0 C' = r \cos b$,

$$\varrho^2 = \varrho_0^2 + r^2 \cos^2 b - 2r\varrho_0 \cos(l - l_0) \cos b. \tag{3.8}$$

For the local field, r is a small quantity. Hence the formulae for the local field of centroid velocities can be obtained by expanding the angular velocity ω as a series of powers of r, using formulae (3.7) and (3.8), and substituting the result in (3.6).

From (3.8) we have approximately

$$\varrho \approx \varrho_0 - r \cos b \cos(l - l_0).$$

Expanding ω in powers of r and taking only the linear terms, we obtain

$$\omega = \omega_0 - \omega_0' r \cos b \cos(l - l_0), \tag{3.9}$$

where

$$\omega_0' = (d\omega/d\varrho)_{\varrho=\varrho_0}.$$

Substitution of (3.9) in (3.6) gives

$$\begin{aligned} v_r &= \omega_0'\varrho_0 r \cos^2 b \sin(l - l_0) \cos(l - l_0), \\ v_l &= \omega_0'\varrho_0 r \cos b \cos^2(l - l_0) + \omega_0 r \cos b, \\ v_b &= -v_r \tan b, \end{aligned}$$

or

$$\left.\begin{aligned} v_r &= rA\sin 2(l-l_0)\cos^2 b, \\ v_l &= r[A\cos 2(l-l_0)+B]\cos b, \\ v_b &= -v_r \tan b, \end{aligned}\right\} \tag{3.10}$$

where

$$A = \tfrac{1}{2}\omega_0'\varrho_0 \quad \text{and} \quad B = \tfrac{1}{2}\omega_0'\varrho_0 + \omega_0 \tag{3.11}$$

are called *Oort's constants*. Formulae (3.10) are *Oort's formulae*, derived by him in 1926. They are valid for values of r not exceeding 1500 ps.

For greater distances, the corresponding formulae as far as any power of r can be obtained by including the appropriate higher terms in the expansion (3.9). This may be done, for example, as follows. Instead of the variable ϱ we can use

$$s = \varrho^2 \tag{3.12}$$

and regard the angular velocity as a function of s: $\omega(\varrho) = \varpi(s)$. The new variable is, according to (3.8), a quadratic in r:

$$s = \varrho_0^2 + \alpha r + \beta r^2, \tag{3.13}$$

where we have put for brevity

$$\begin{aligned} \alpha &= -2\varrho_0 \cos b \cos(l-l_0), \\ \beta &= \cos^2 b. \end{aligned} \tag{3.14}$$

Then it is easy to show that

$$\begin{aligned} \frac{d^n\omega}{dr^n} &= \frac{d^n\varpi}{ds^n}\left(\frac{ds}{dr}\right)^n + \frac{n(n-1)}{2}\,\frac{d^{n-1}\varpi}{ds^{n-1}}\left(\frac{ds}{dr}\right)^{n-2}\left(\frac{d^2s}{dr^2}\right) \\ &+ \frac{n(n-1)(n-2)(n-3)}{2\cdot 4}\,\frac{d^{n-2}\varpi}{ds^{n-2}}\left(\frac{ds}{dr}\right)^{n-4}\left(\frac{d^2s}{dr^2}\right)^2 + \ldots \\ &+ \begin{cases} 1\cdot 3\cdot 5\ldots(n-1)\,\dfrac{d^{n/2}\varpi}{ds^{n/2}}\left(\dfrac{d^2s}{dr^2}\right)^{n/2} & \text{if } n \text{ is even,} \\[2ex] 1\cdot 3\cdot 5\ldots n\,\dfrac{d^{(n+1)/2}\varpi}{ds^{(n+1)/2}}\left(\dfrac{d^2s}{dr^2}\right)^{(n-1)/2}\dfrac{ds}{dr} & \text{if } n \text{ is odd.} \end{cases} \end{aligned} \tag{3.15}$$

Putting

$$\varpi = \varpi_0 + \frac{\varpi_1}{1!}r + \frac{\varpi_2}{2!}r^2 + \cdots + \frac{\varpi_n}{n!}r^n + \cdots \tag{3.16}$$

and using the facts that

$$(ds/dr)_{r=0} = \alpha, \quad (d^2s/dr^2)_{r=0} = 2\cos^2 b,$$

we obtain from (3.15)

$$\varpi_n = \varpi_0^{(n)}\,\alpha^n + \frac{n(n-1)}{1!}\,\varpi_0^{(n-1)}\,\alpha^{n-2}\cos^2 b + \frac{n(n-1)(n-2)(n-3)}{2!}\,\varpi_0^{(n-2)}\,\alpha^{n-4}\cos^4 b + \cdots + \begin{cases} \dfrac{n!}{(\frac{1}{2}n)!}\,\varpi_0^{\frac{1}{2}n}\cos^n b & \text{if } n \text{ is even,} \\[2ex] \dfrac{(n+1)!}{(\frac{1}{2}n+\frac{1}{2})!}\,\varpi_0^{\frac{1}{2}n+\frac{1}{2}}\,\alpha\cos^{n-1} b & \text{if } n \text{ is odd.} \end{cases} \tag{3.17}$$

Here

$$\varpi_0^{(n)} = (\mathrm{d}^n\varpi/\mathrm{d}s^n)_{s=\varrho_0^2};$$

evidently $s = \varrho_0^2$ when $r = 0$.

In practice the expansions (3.16) and (3.17) have so far been used only for $n \leqslant 2$ and for r of the order of 2 to 2·5 kps, the distance of the open galactic clusters and the planetary nebulae.

In order to obtain sufficiently accurate expansions of the galactic rotation terms v_r and v_l, we substitute (3.16) and (3.17) in (3.6) and retain the appropriate terms. To convert to the more usual notation, the derivatives with respect to $s = \varrho^2$ must be converted into those with respect to ϱ. For this we must differentiate with respect to ϱ the equation $\omega(\varrho) = \varpi(s)$ the required number of times and substitute $\varrho = \varrho_0$ in the result. This gives

$$\begin{aligned} \omega_0^{\mathrm{I}} &= \varpi_0^{(1)} \cdot 2\varrho_0, \\ \omega_0^{\mathrm{II}} &= \varpi_0^{(2)} \cdot 4\varrho_0^2 + \varpi_0^{(1)} \cdot 2, \\ \omega_0^{\mathrm{III}} &= \varpi_0^{(3)} \cdot 8\varrho_0^3 + 12\varpi_0^{(2)} \cdot \varrho_0, \\ \omega_0^{\mathrm{IV}} &= \varpi_0^{(4)} \cdot 16\varrho_0^4 + 48\varpi_0^{(3)} \cdot \varrho_0^2 + 12\varpi_0^{(2)} \end{aligned}$$

and in general

$$\omega_0^{[n]} = \varpi_0^{(n)} \cdot (2\varrho_0)^n + \frac{n(n-1)}{1!}\,\varpi_0^{(n-1)}\,(2\varrho_0)^{n-2} + \frac{n(n-1)(n-2)(n-3)}{2!}\,\varpi_0^{(n-2)}\,(2\varrho_0)^{n-4} + \ldots + \begin{cases} \dfrac{n!}{(\frac{1}{2}n)!}\,\varpi_0^{\frac{1}{2}n} & \text{if } n \text{ is even,} \\[2ex] \dfrac{(n+1)!}{(\frac{1}{2}n+\frac{1}{2})!}\,\varpi_0^{\frac{1}{2}n+\frac{1}{2}}\,\varrho_0 & \text{if } n \text{ is odd.} \end{cases} \tag{3.18}$$

From these relations we can successively express $\varpi_0^{(1)}$, $\varpi_0^{(2)}$, ... in terms of ω_0^{I}, ω_0^{II}, ... and substitute the results in (3.6).

As an example, let us calculate the terms of the first and second orders in the expansion (3.16). From (3.18) we have

$$\omega_0^{\mathrm{I}} = 2\varrho_0\bar{\omega}_0^{(1)}, \quad \omega_0^{\mathrm{II}} = 4\varrho_0^2\bar{\omega}_0^{(2)} + 2\bar{\omega}_0^{(1)},$$

whence

$$\bar{\omega}_0^{(1)} = \omega_0^{\mathrm{I}}/2\varrho_0, \quad \bar{\omega}_0^{(2)} = \left(\omega_0^{\mathrm{II}} - \frac{\omega_0^{\mathrm{I}}}{\varrho_0}\right)\Big/ 4\varrho_0^2.$$

Hence eqns. (3.17) give

$$\bar{\omega}_1 = \bar{\omega}_0^{(1)}\alpha = -\omega_0^{\mathrm{I}}\cos b\cos(l - l_0),$$

$$\begin{aligned}\bar{\omega}_2 &= \bar{\omega}_0^{(2)}\alpha^2 + 2\bar{\omega}_0^{(1)}\cos^2 b \\ &= \omega_0^{\mathrm{II}}\cos^2 b\cos^2(l - l_0) + \frac{\omega_0^{\mathrm{I}}}{\varrho_0}\cos^2 b\sin^2(l - l_0).\end{aligned}$$

Thus, from (3.16),

$$\omega = \omega_0 - r\omega_0^{\mathrm{I}}\cos b\cos(l - l_0) + \frac{1}{2}r^2\left[\omega_0^{\mathrm{II}}\cos^2(l - l_0) + \frac{\omega_0^{\mathrm{I}}}{\varrho_0}\sin^2(l - l_0)\right]\cos^2 b,$$

and from (3.6) we obtain the generalised Oort's formulae

$$\left.\begin{aligned} v_r &= rA\cos^2 b\sin 2(l - l_0) - r^2 C\cos^3 b\sin(l - l_0)\cos^2(l - l_0) \\ &\quad - r^2(A/\varrho_0)\cos^3 b\sin^3(l - l_0), \\ v_l &= rA\cos b\cos 2(l - l_0) + rB\cos b - r^2 C\cos^2 b\cos^3(l - l_0) \\ &\quad - 3r^2(A/\varrho_0)\cos^2 b\cos(l - l_0) + r^2(A/\varrho_0)\cos^2 b\cos^3(l - l_0), \\ v_b &= -v_r\tan b. \end{aligned}\right\} \quad (3.19)$$

Here A and B are the first-order Oort's constants of galactic rotation:

$$A = \tfrac{1}{2}\varrho_0\omega_0', \quad B = \omega_0 + \tfrac{1}{2}\varrho_0\omega_0', \quad (3.20)$$

and C denotes the corresponding second-order coefficient

$$C = \tfrac{1}{2}\varrho_0\omega_0''. \quad (3.21)$$

To convert the velocity components v_l and v_b into the corresponding proper motions in longitude and latitude we need only divide them by $4{\cdot}74r\cos b$ and $4{\cdot}74r$ respectively.

A consideration of formulae (3.19)–(3.21) shows that, by adding the second-order terms, we can find without any assumptions the following kinematic parameters of the Galaxy: $\omega_0, \omega_0', \omega_0'', \varrho_0$. Of course, the accuracy with which these are determined will depend on that of the stellar motions used.

The principal result of using the second-order formulae of galactic rotation (the generalised Oort's formulae) is that the theory of galactic rotation is confirmed, at least in general, since the results obtained from these terms agree with those based on the first-order terms.

§ 3. Oort's Formulae for an Arbitrary Plane-parallel Motion

It should be borne in mind that the possibility of representing systematic stellar motions by the first-order formulae in Oort's form (3.10) does not at all prove that the motion of the Galaxy is rotational, since the same formulae can be obtained on much more general assumptions concerning the nature of the motion. Equations (2.71) show that the systematic terms in the relative motions of the stars in a small region ($r \ll \varrho_0$) are such that

$$\Delta u = u_x x + u_y y + u_z z,$$

$$\Delta v = v_x x + v_y y + v_z z,$$

$$\Delta w = w_x x + w_y y + w_z z.$$

If we assume that the motion in a system is parallel to some fixed plane, which we may take as the coordinate plane $z = 0$, then the third equation is $\Delta w \equiv 0$. The terms in z in the other two equations vanish if either (a) the motion is independent of z, i.e. is a plane-parallel motion† or (b) the plane $z = 0$ is a plane of symmetry of the stellar motions. In each case

$$\Delta u = u_x x + u_y y, \qquad \Delta v = v_x x + v_y y. \tag{3.22}$$

The corresponding velocity vector is

$$\Delta \mathbf{V} = (u_x x + u_y y)\,\mathbf{i} + (v_x x + v_y y)\,\mathbf{j}. \tag{3.23}$$

To obtain the radial and two transverse components of the relative velocity of the centroid, we again use the moving trihedral, putting $\mathbf{z}^0 = \mathbf{k}$, a unit vector along the z-axis. Then, by (1.69),

$$\mathbf{r}^0 = \frac{x}{r}\,\mathbf{i} + \frac{y}{r}\,\mathbf{j} + \frac{z}{r}\,\mathbf{k},$$

$$\mathbf{l}^0 = \mathbf{k} \times \mathbf{r}_0 \sec b = \left(-\frac{y}{r}\,\mathbf{i} + \frac{x}{r}\,\mathbf{j}\right) \sec b,$$

$$\mathbf{b}^0 = \mathbf{k} \sec b - \mathbf{r}^0 \tan b.$$

Taking the scalar product of the vector $\Delta\mathbf{V}$ (3.23) with these unit vectors, we have

$$v_r = [u_x x^2 + v_y y^2 + (u_y + v_x)\,xy]/r,$$

$$v_l = [v_x x^2 - u_y y^2 + (v_y - u_x)\,xy] \sec b/r,$$

$$v_b = -\mathbf{r}^0 \cdot \Delta\mathbf{V} \tan b = -v_r \tan b.$$

We now use spherical coordinates

$$x = r \cos b \cos l, \quad y = r \cos b \sin l,$$

† Barotropic rotation is a particular case of this.

obtaining

$$\left.\begin{aligned} v_r &= \cos^2 b\,[rA'\sin 2l + rC'\cos 2l + rK],\\ v_l &= \cos b\,[-rC'\sin 2l + rA'\cos 2l + rB],\\ v_b &= -v_r \tan b, \end{aligned}\right\} \tag{3.24}$$

where

$$\begin{aligned} A' &= \tfrac{1}{2}(u_y + v_x), \quad C' = \tfrac{1}{2}(u_x - v_y),\\ B &= \tfrac{1}{2}(v_x - u_y), \quad K = \tfrac{1}{2}(u_x + v_y). \end{aligned} \tag{3.25}$$

Finally, defining other constants A and l_1 by

$$A' = A\cos 2l_1, \quad C' = -A\sin 2l_1, \tag{3.26}$$

we obtain the following formulae, which, when $K = 0$, are exactly the same as Oort's formulae of galactic rotation:

$$\left.\begin{aligned} v_r &= \cos^2 b\,[Ar\sin 2(l - l_1) + Kr],\\ v_l &= \cos b\,[Ar\cos 2(l - l_1) + Br],\\ v_b &= -\sin b\cos b\,[Ar\sin 2(l - l_1) + Kr]. \end{aligned}\right\} \tag{3.27}$$

To bring the first formula (3.27) into Oort's form (3.10), the K *term* in it (Kr) must be omitted. The coefficient in this term is

$$K = \tfrac{1}{2}(u_x + v_y) = \tfrac{1}{2}\operatorname{div}\mathbf{V},$$

and for the two-dimensional problem it gives the hydrodynamic divergence of the vector field of centroid velocities. Thus formulae (3.27) prove the following theorem: every plane-parallel relative motion of a stellar system can be represented, within a small region surrounding any point, as the sum of an instantaneous Oort rotation and a hydrodynamic divergence (that is, either an expansion or a compression, depending on the sign of the K term).

§ 4. The Physical Basis of the Theory of Galactic Rotation. The Nature of the K Term

The question arises of the way in which the truth of the theory of galactic rotation, which is now accepted, can in fact be demonstrated. We have seen in § 3 that, in the general case of a plane-parallel motion, the kinematic formulae (3.27) for a small region surrounding any given point can be written in Oort's form (3.10) if we put $K = 0$, but the longitude l_1 which appears in (3.27) will not in general be the longitude l_0 of the centre of the Galaxy. If we call l_1 the longitude of the kinematic centre of the local motion, then one of the tests of the correctness of the theory of galactic rotation will be that the directions of the kinematic centre and the galactic centre should coincide (the galactic centre being the centre of the apparent distribution of stars, globular clusters, planetary nebulae, etc.).

Another test is the absence of a K term proportional to the distance in the radial velocities. However, as Chandrasekhar [17] has shown, neither of these tests is sufficient to prove the rotation of the Galaxy, even if it is satisfied at every point and not only in the neighbourhood of one point such as the Sun.

To obtain the necessary and sufficient kinematic conditions for a given plane-parallel motion to be described by Oort's formulae, we change from rectangular coordinates x, y to polar coordinates ϱ, θ in the galactic plane. The velocity of the centroid, which we assume parallel to the galactic plane, is resolved into two perpendicular components: P in the direction of increasing ϱ, and Θ in the direction of increasing θ. We suppose that the Sun lies on the x-axis, and denote by a suffix 0 quantities pertaining to the Sun ($\varrho = \varrho_0$, $\theta = \theta_0 = 0$). For simplicity, longitudes are taken to be measured from the direction of the kinematic centre of the Galaxy. Then $l_1 = 0$, and therefore C' is zero (3.26).

We may express the kinematic parameters u_x, u_y, v_x, v_y in terms of P, Θ and their derivatives. Evidently

$$\begin{aligned} u &= \mathrm{P}\cos\theta - \Theta\sin\theta, \\ v &= \mathrm{P}\sin\theta + \Theta\cos\theta. \end{aligned} \tag{3.28}$$

Differentiating these with respect to ϱ and θ, and using the equations $x = \varrho\cos\theta$ and $y = \varrho\sin\theta$, gives

$$u_x\cos\theta + u_y\sin\theta = \frac{\partial \mathrm{P}}{\partial\varrho}\cos\theta - \frac{\partial\Theta}{\partial\varrho}\sin\theta,$$

$$-u_x\varrho\sin\theta + u_y\varrho\cos\theta = \frac{\partial \mathrm{P}}{\partial\theta}\cos\theta - \frac{\partial\Theta}{\partial\theta}\sin\theta - \mathrm{P}\sin\theta - \Theta\cos\theta,$$

$$v_x\cos\theta + v_y\sin\theta = \frac{\partial \mathrm{P}}{\partial\varrho}\sin\theta + \frac{\partial\Theta}{\partial\varrho}\cos\theta,$$

$$-v_y\varrho\sin\theta + v_y\varrho\cos\theta = \frac{\partial \mathrm{P}}{\partial\theta}\sin\theta + \frac{\partial\Theta}{\partial\theta}\cos\theta + \mathrm{P}\cos\theta - \Theta\sin\theta.$$

Putting here $\varrho = \varrho_0$ and $\theta = 0$, we obtain

$$\begin{aligned} u_x &= \left(\frac{\partial \mathrm{P}}{\partial\varrho}\right)_0, & u_y &= \left(\frac{1}{\varrho}\frac{\partial \mathrm{P}}{\partial\theta} - \frac{\Theta}{\varrho}\right)_0, \\ v_x &= \left(\frac{\partial\Theta}{\partial\varrho}\right)_0, & v_y &= \left(\frac{1}{\varrho}\frac{\partial\Theta}{\partial\theta} + \frac{\mathrm{P}}{\varrho}\right)_0, \end{aligned} \tag{3.29}$$

and hence, by (3.25),

$$\left.\begin{aligned} A &= \frac{1}{2}\left(\frac{\partial \Theta}{\partial \varrho} - \frac{\Theta}{\varrho} + \frac{1}{\varrho}\frac{\partial \mathrm{P}}{\partial \theta}\right)_0, \\ B &= \frac{1}{2}\left(\frac{\partial \Theta}{\partial \varrho} + \frac{\Theta}{\varrho} - \frac{1}{\varrho}\frac{\partial \mathrm{P}}{\partial \theta}\right)_0, \\ C' &= \frac{1}{2}\left(\frac{\partial \mathrm{P}}{\partial \varrho} - \frac{\mathrm{P}}{\varrho} - \frac{1}{\varrho}\frac{\partial \Theta}{\partial \theta}\right)_0, \\ K &= \frac{1}{2}\left(\frac{\partial \mathrm{P}}{\partial \varrho} + \frac{\mathrm{P}}{\varrho} + \frac{1}{\varrho}\frac{\partial \Theta}{\partial \theta}\right)_0. \end{aligned}\right\} \qquad (3.30)$$

Hence the conditions $C' = K = 0$ are equivalent to

$$\left(\frac{\partial \mathrm{P}}{\partial \varrho}\right)_0 = 0, \qquad \left(\mathrm{P} + \frac{\partial \Theta}{\partial \theta}\right)_0 = 0. \qquad (3.31)$$

For a pure Oort rotation we have identically $\mathrm{P} \equiv 0$, $\partial\Theta/\partial\theta \equiv 0$, and eqns. (3.31) are evidently satisfied. These identities are not necessary, however. P can be any function whatever of θ; if Θ is then determined from the second eqn. (3.31), we have $C' = K = 0$ everywhere in the galactic plane.

The generality of the argument is not affected by the fact that the result (3.30) has been derived only for a single point, since this point may be chosen arbitrarily.

Thus neither the presence of a term in twice the longitude in the stellar velocities nor the vanishing of the coefficients C' and K proves that the Galaxy is rotating. The rotation of the Galaxy is, nevertheless, at present the most plausible hypothesis concerning the motion of this system. It is indicated by the following arguments.

Firstly, the rotational symmetry of the Galaxy, noted even by W. Herschel in the eighteenth century, can hardly be reconciled with any other type of motion.†

Secondly, there is the similarity between our Galaxy and the other galaxies. The rotation of the other galaxies has been confirmed in every case where a decision was possible. Where a galaxy of considerable apparent dimension is seen "edge-on", the radial velocities of the points at the opposite ends of its diameter can be found spectroscopically. In some cases we can even determine the manner in which the angular velocity of rotation varies with distance from the axis, and the results in turn confirm the theoretical deductions concerning the variation of the angular velocity of the Galaxy. We shall later return to this subject.

A third argument is the agreement, already mentioned, between the results derived from the investigation of stellar motions by the generalised Oort's formulae and those given by Oort's original formulae.

Finally, a fourth argument in favour of the hypothesis of galactic rotation is the observed asymmetry in the distribution of stellar velocities in the neighbourhood of

† See Chapter VIII, § 1, for a further discussion.

the Sun. This phenomenon, discovered empirically by Strömberg in 1923, consists in the fact that the Sun's velocity, as determined from groups of stars having large velocity dispersions (called *high-velocity* stars), is found to be more and more different from the normal value of 20 km/s as the dispersion increases, and the termini of the solar velocity vectors lie preferentially along the *axis of asymmetry*, a straight line perpendicular to the direction of the galactic centre.

Since the stars with small residual-velocity dispersions form plane sub-systems in the Galaxy, while those with large dispersions form spherical sub-systems, the asymmetry may be described as a motion of the centroids of the spherical sub-systems with respect to those of the plane sub-systems, in the direction of galactic longitude $l = 235°$. The magnitude S of the asymmetry is related to the variance σ^2 of the velocities of the sub-group concerned by

$$S = 0{\cdot}0192\sigma^2 + 10{\cdot}0, \tag{3.32}$$

a relation derived empirically by Strömberg [125].

Since the direction $l = 235°$ is diametrically opposite to the direction of galactic rotation in the neighbourhood of the Sun, the quantity S represents the amount by which the spherical sub-systems lag behind the plane sub-systems in the rotation. We shall see in Chapter VIII that the dispersion of the residual velocities retards the motion of the centroid, and so the asymmetry of stellar velocities is a further indirect proof of the general rotation of the Galaxy.

§ 5. Determination of the Parameters of Galactic Rotation from Observation

The determination of the parameters of galactic rotation from observation is effected in practice by means of formulae (3.24) for radial velocities and proper motions separately.

As in other similar determinations, the sky is divided into equal areas arranged in zones. For radial velocities, galactic zones are generally used. For each area the mean observed radial velocity $\bar{v}_r$ is calculated, and then either the radial component of the Sun's velocity

$$X_\odot\alpha + Y_\odot\beta + Z_\odot\gamma = X_\odot \cos b \cos l + Y_\odot \cos b \sin l + Z_\odot \sin b$$

is subtracted, or $X_\odot$, $Y_\odot$, $Z_\odot$ are regarded as additional unknowns. Then we obtain from (3.24) for each area the equation of condition

$$\bar{v}_r = K\bar{r}\cos^2 b - X_\odot \cos b \cos l - Y_\odot \cos b \sin l - Z_\odot \sin b$$
$$+ A'\bar{r}\cos^2 b \sin 2l + C'\bar{r}\cos^2 b \cos 2l, \tag{3.33}$$

where the bar denotes averaging over all the stars in the area concerned. Solving the equations of condition (3.33) by the method of least squares, we find the quantities K, A' and C'.

If the eqn. (3.33) is derived, as is generally done, for areas in a zone parallel to the galactic equator, for which b = constant, then the resulting equations are equivalent to expansions of the mean radial velocity as a trigonometric polynomial in the galactic longitude. The free term is the K term $K\bar{r}\cos^2 b$; the first-order harmonics correspond to the components of the Sun's local velocity, and the second-order harmonics are the Oort galactic-rotation terms.

In deriving Oort's formulae (3.24) from the formula for the local motion of the centroids, we used hypothesis (a) or (b) in order to remove the terms proportional to z in the expressions for Δu and Δv. If this were not done, terms $ru_z \sin b \cos b \cos l + rv_z \sin b \cos b \sin l$ would appear in the first formula

(3.24) and $\bar{r}u_z \sin b \cos b \cos l + \bar{r}v_z \sin b \cos b \sin l$ on the right of formula (3.33), which, with $b =$ constant, would amount to the introduction of some systematic error in the determination of the solar velocity.

Thus, although in theory hypothesis (a) or (b) of §4 was necessary in deriving Oort's formulae, in practice the same formulae are obtained without either hypothesis, but a systematic error is introduced in the solar motion.

To determine the elements of galactic rotation from proper motions, we use the second formula (3.24), dividing by kr and subtracting the solar-motion terms given by the left-hand side of the first formula (2.27). This gives

$$\mu_l \cos b = \frac{B}{k}\cos b - \frac{X_\odot}{kr}\sin l + \frac{Y_\odot}{kr}\cos l + \frac{A'}{k}\cos b \cos 2l - \frac{C'}{k}\cos b \sin 2l. \tag{3.34}$$

Here, as with the radial velocities, an allowance for the terms in u_z and v_z in reducing proper motions in galactic zones leads to a systematic error in the solar motion. By solving the eqns. (3.34) we can find B, A' and C'.

For the proper motions in galactic latitude the galactic-rotation term is, as shown above, $v_b = -v_r \tan b$. Hence, by analogy with eqn. (3.34), we obtain from (3.33) and (2.27) the equations of condition

$$\begin{aligned}\mu_b = &-\frac{K}{k}\sin b \cos b + \frac{X_\odot}{kr}\sin b \cos l + \frac{Y_\odot}{kr}\sin b \sin l \\ &-\frac{Z_\odot}{kr}\cos b - \frac{A'}{k}\sin b \cos b \sin 2l - \frac{C'}{k}\sin b \cos b \cos 2l,\end{aligned} \tag{3.35}$$

the solution of which gives K, A' and C'. The values of A and l_1 are then found from formulae (3.26).

The determination of the rotation of the Galaxy from proper motions is often combined with that of the constant of lunisolar precession. In this case, however, it is preferable to use the proper motions in equatorial coordinates. The galactic-rotation terms must then be expressed in terms of right ascension and declination, a matter of no great difficulty.

The numerical values of Oort's constants have been derived by many investigators since 1926. Recent values, due to P.P.Parenago [107], are

$$\begin{aligned} A &= +0{\cdot}020 \text{ km/s per ps}, \\ B &= -0{\cdot}013 \text{ km/s per ps}. \end{aligned} \tag{3.36}$$

The units of measurement (km/s per ps) are those used by Oort. In theoretical calculations, however, it is usually more convenient to use units of s^{-1}:

$$\begin{aligned} A &= +6{\cdot}5 \times 10^{-16} \text{ per s}, \\ B &= -4{\cdot}2 \times 10^{-16} \text{ per s}. \end{aligned} \tag{3.37}$$

The longitude of the galactic centre is found to be

$$l_1 = 325°, \tag{3.38}$$

in good agreement with the value l_0 obtained from the spatial distribution of stars.

§ 6. Determination of the Period of Rotation and Angular Velocity of the Galaxy

When Oort's constants A and B are known, two very important kinematic characteristics of the rotation of the Galaxy in the neighbourhood of the Sun can be deduced; they are the rate of rotation ω_0 and the velocity V_0 in a circular orbit.

Formulae (3.11) and (3.37) give at once the angular velocity

$$\begin{aligned} \omega_0 &= B - A = -10{\cdot}7 \times 10^{-16} \text{ per s} \\ &= -0''{\cdot}0070 \text{ per y.} \end{aligned} \tag{3.39}$$

The minus sign signifies that the rotation is in the direction of decreasing longitude. The quantity ω_0 is the angular velocity of the rotation of the Galaxy at the Sun's distance from its centre. The period is

$$\begin{aligned} P_0 &= 2\pi/\omega_0 = 5{\cdot}87 \times 10^{15} \text{ s} \\ &= 1{\cdot}86 \times 10^8 \text{ y.} \end{aligned} \tag{3.40}$$

The circular velocity V_0 is related to the angular velocity ω_0 by

$$\omega_0 = V_0/\varrho_0, \tag{3.41}$$

where ϱ_0 is the distance to the galactic centre. Hence any of the quantities ω_0, V_0, ϱ_0 can be found if the other two are known. Until recently, attempts were made to determine the velocity of the Sun's centroid from observation and thence the distance from the centre of the Galaxy from (3.41). The reason for this procedure was the fear of a large error in determining ϱ_0, because of the difficulty of allowing for the absorption of light.

However, the determination of V_0 also involves considerable difficulties. Firstly, V_0 must be determined from objects which do not participate in the galactic rotation, i.e. which are not physically connected with the Galaxy. Secondly, extragalactic objects such as the other galaxies are not suitable, because the relative motion with respect to them includes the translational velocity of the whole Galaxy with respect to the local centroid in the Metagalaxy. Moreover, the velocities of the galaxies include very large systematic terms, as is shown by the presence of the "red-shift" term, proportional to the distance, in their radial velocities.

Thus the necessity of eliminating the translational motion of the Galaxy makes it necessary to choose objects which form with the Galaxy a single physically connected group having a common translational motion in the Metagalaxy. The system which best meets these two conflicting conditions has been taken to be that of the globular clusters, which is one of the spherical sub-systems. These sub-systems have large residual-velocity dispersions and small centroid velocities in the galactic rotation.

The centroid velocity is not zero, however, and so the use of the globular clusters as a standard of rest involves some error due to the neglect of the centroid velocity.†

For this reason, in accordance with a proposal by Parenago ([107], p. 160), it is now considered better to determine the distance ϱ_0 of the galactic centre from observation and then to calculate V_0 from formula (3.41). The absorption of light in the Galaxy may be regarded as now sufficiently accurately known for this purpose.

Taking

$$\varrho_0 = 7500 \text{ ps}, \tag{3.42}$$

we find from (3.39) and (3.41)

$$V_0 = 250 \text{ km/s}. \tag{3.43}$$

The quantities P_0, ϱ_0 and V_0 are among the fundamental constants of stellar dynamics.††

These quantities have been derived above from the values of Oort's constants A and B for the neighbourhood of the Sun. Omitting the suffix zero in formulae (3.11), we obtain for A and B at any point in the Galaxy

$$A = \tfrac{1}{2}\omega'\varrho, \quad B = \tfrac{1}{2}\omega'\varrho + \omega, \tag{3.44}$$

where the prime denotes differentiation with respect to ϱ. Hence

$$\omega = B - A. \tag{3.45}$$

Replacing ω in (3.44) by V/ϱ, we obtain Oort's constants A and B in terms of the circular velocity V:

$$A = \frac{1}{2}\left(-\frac{V}{\varrho} + \frac{\mathrm{d}V}{\mathrm{d}\varrho}\right), \quad B = \frac{1}{2}\left(\frac{V}{\varrho} + \frac{\mathrm{d}V}{\mathrm{d}\varrho}\right), \tag{3.46}$$

and hence the formulae, which will be useful later,

$$V/\varrho = B - A, \quad \mathrm{d}V/\mathrm{d}\varrho = B + A. \tag{3.47}$$

† The question may be asked, relative to what is the angular velocity of rotation of the Galaxy defined? The answer is that a fixed frame of reference on the celestial sphere is given by two "gyroscopes", the Earth rotating about its axis and the solar system revolving about the Sun. These define the celestial equator and the ecliptic respectively.

†† In Trumpler and Weaver's comprehensive book on stellar astronomy [132], somewhat different values are given for the fundamental constants, namely

$$A = +\,0{\cdot}016 \text{ km/s per ps} = 5{\cdot}20 \times 10^{-16} \text{ per s},$$
$$B = -\,0{\cdot}014 \text{ km/s per ps} = -\,4{\cdot}50 \times 10^{-16} \text{ per s},$$

whence

$$P_0 = 2{\cdot}05 \times 10^8 \text{ y} = 6{\cdot}48 \times 10^{15} \text{ s}.$$

The distance of the Sun from the centre of the Galaxy is taken as

$$\varrho_0 = 9200 \text{ ps} = 2{\cdot}84 \times 10^{22} \text{ cm}.$$

The circular velocity of the local centroid is then

$$V_0 = 275 \text{ km/s}.$$

§ 7. Determination of the Mass of the Galaxy

As we shall find more than once, Oort's constants A and B are of importance in local dynamics, i.e. the dynamics of the Galaxy in the immediate neighbourhood of the Sun. Here we shall show how these constants can be used to make an approximate estimate of the mass of the Galaxy.

Let us consider some group of stars having small residual velocities. The Galaxy contains many such groups: they include, for example, the giant stars of the early spectral classes O, B and A, the long-period Cepheids, etc. All such groups of stars form *plane sub-systems*, i.e. they lie close to the galactic plane. Evidently the velocity of any star in such a group can differ only slightly from the circular-orbit velocity V given by

$$V^2/\varrho = F, \tag{3.48}$$

where F is the force. Replacing V in (3.46) by its value in terms of the force F, we find

$$\begin{aligned} A &= \frac{1}{4}\frac{V}{\varrho}\left(-1 + \frac{\varrho}{F}\frac{dF}{d\varrho}\right), \\ B &= \frac{1}{4}\frac{V}{\varrho}\left(3 + \frac{\varrho}{F}\frac{dF}{d\varrho}\right). \end{aligned} \tag{3.49}$$

Equations (3.46) (or (3.44)) are kinematic formulae, and (3.49) dynamic formulae, for Oort's constants. It should be recalled that (3.49) are valid only for plane sub-systems. The force depends on the distribution of mass in the Galaxy, and this distribution can therefore be studied by means of (3.49).

As a first rough approximation, let us assume that the mass of the Galaxy consists of (a) a central nucleus of spherical or ellipsoidal form, surrounded by (b) a homogeneous and considerably flattened spheroid. Then the attraction at a point in the galactic plane can be written as the sum of (a) a force inversely proportional to the square of the distance ϱ from the centre, which we call the *Newtonian* force, and (b) a force proportional to ϱ, which we call the *quasi-elastic* force. Thus

$$F = F_1 + F_2 = \alpha\varrho^{-2} + \beta\varrho, \tag{3.50}$$

where α and β are constants.†

A physical justification of this formula may be seen in the argument that the actual distribution of mass in the Galaxy must lie between two limiting cases in which the mass is (a) all concentrated at the centre ($\beta = 0$) and (b) all distributed uniformly through a homogeneous spheroid ($\alpha = 0$).

It must be admitted, however, that the force (3.50) does not satisfy the boundary conditions which must be satisfied by the force in any actual galaxy. For, if we consider a point moving to a great distance from the centre, beyond the limits of the Galaxy, the force F must tend asymptotically to the Newtonian form, whereas (3.50)

† The proof that F_2 is proportional to ϱ is given later.

gives the opposite result, the term $\beta\varrho$ being dominant for large ϱ. For small ϱ, on the other hand, when the point considered is inside the central nucleus of constant density,† the force should be given by a term proportional to ϱ, but (3.50) again gives the opposite result. Thus formula (3.50) can be correct only for ϱ lying in a range from the outer edge of the central nucleus to the outer edge of the Galaxy. For an approximate estimate of the mass, however, it may be sufficient to consider such values of ϱ.

We may use the dynamical Oort's formulae (3.49) to find the relation between the forces F_1 and F_2. Using (3.50), we find

$$\varrho\, \mathrm{d}F/\mathrm{d}\varrho = -2F_1 + F_2 = F - 3F_1.$$

Substituting this expression in the first formula (3.49) and using the first formula (3.47), we obtain Oort's first formula:

$$F_1/F = 4A/3(A - B). \tag{3.51}$$

Substitution of the numerical values (3.37) for A and B gives $F_1/F = 0{\cdot}81$, this value pertaining, of course, to $\varrho = \varrho_0$, the distance of the Sun from the centre of the Galaxy.

From (3.50) and (3.51) we derive Oort's second formula:

$$F_2/F = 1 - (F_1/F) = (3B + A)/3(B - A), \tag{3.52}$$

or, numerically, $F_2/F = 0{\cdot}19$.††

To find from Oort's formulae the concentrated and distributed masses in the Galaxy, we must express the forces F_1 and F_2 in terms of these masses. The definition of the Newtonian force gives

$$F_1 = GM_1/\varrho^2, \tag{3.53}$$

where G is the gravitational constant and M_1 the concentrated mass. To determine the force F_2, we use the formulae for the attraction at a point inside a homogeneous spheroid with semiaxes a and c. If the coordinate axes coincide with those of the spheroid, and Oz is the axis of symmetry, then the components of the gravitational acceleration along the axes are

$$X = -G\alpha'\,\delta x, \quad Y = -G\alpha'\,\delta y, \quad Z = -G\beta'\,\delta z,$$

where δ is the density,

$$\alpha' = 2\pi a^2 c \int_0^\infty \frac{\mathrm{d}s}{(a^2 + s)^2 \sqrt{(c^2 + s)}}$$

and

$$\beta' = 2\pi a^2 c \int_0^\infty \frac{\mathrm{d}s}{(a^2 + s)\sqrt{(c^2 + s)^3}}. \tag{3.54}$$

† The constancy of the density is confirmed by measurements of the rotation of the nuclei of other galaxies, which are found to rotate as rigid bodies.

†† For the values of the constants used by Trumpler and Weaver (see the footnote following (3.43)) we have $F_1/F = 0{\cdot}71$, $F_2/F = 0{\cdot}29$.

We may take the x-axis so as to pass through the Sun. Then $x = \varrho, y = z = 0$, and so $X = -F_2$. Hence the force of attraction per unit mass due to the distributed mass in the Galaxy is

$$F_2 = G\delta\alpha'\varrho. \tag{3.55}$$

In order to calculate α', we take $1/a^5$ outside the integral in (3.54) and put $s/a^2 = t$. Then

$$\alpha' = 2\pi \frac{c}{a} \int_0^\infty \frac{\mathrm{d}t}{(1+t)^2 \sqrt{(t + c^2/a^2)}},$$

whence we see that α' depends only on the quantity c/a which gives the degree of flattening of the spheroid.

For the Galaxy we can take c/a as approximately 1/10. To calculate the integral in α', therefore, we can expand in a series of powers of c/a and retain only the first term, neglecting $(c/a)^2$. Thus

$$\alpha' = 2\pi \frac{c}{a} \int_0^\infty \frac{\mathrm{d}t}{(1+t)^2 \sqrt{t}}.$$

Putting $t = \tan^2 u$ and integrating over u from 0 to $\frac{1}{2}\pi$, we have† $\alpha' = \pi^2 c/a$. Substituting this value in (3.55), we have the following expression for the attraction of the distributed mass of the Galaxy:

$$F_2 = \pi^2 G\delta c\varrho/a. \tag{3.56}$$

The density δ can be replaced by means of the evident relation $M_2 = 4\pi a^2 c\delta/3$, giving

$$F_2 = 3\pi G\varrho M_2/4a^3. \tag{3.57}$$

Substituting the forces F_1, F_2 and F given by (3.53), (3.57) and (3.50) in Oort's formulae (3.51) and (3.52), we obtain

$$\left.\begin{aligned} M_1 &= \frac{4A}{3(A-B)} \frac{V^2\varrho}{G}, \\ M_2 &= \frac{3B+A}{3(B-A)} \frac{4}{3\pi} \frac{a^3}{G} \frac{V^2}{\varrho^2}. \end{aligned}\right\} \tag{3.58}$$

Hence

$$\frac{M_2}{M_1} = -\frac{3B+A}{A} \frac{1}{3\pi} \left(\frac{a}{\varrho}\right)^3. \tag{3.59}$$

† A detailed calculation, which we omit for the sake of brevity, shows that by neglecting the square of c/a we commit a relative error of 12·5% in α'.

Substitution of the numerical values of V, ϱ, A and B from (3.42), (3.43), (3.37), and $a = 13$ kps, gives the values of the concentrated and distributed masses of the Galaxy:

$$\left.\begin{aligned} M_1 &= 1{\cdot}75 \times 10^{44}\ \text{g} = 8{\cdot}8 \times 10^{10}\ m_{\odot}, \\ M_2 &= 9{\cdot}2 \times 10^{43}\ \text{g} = 4{\cdot}6 \times 10^{10}\ m_{\odot}, \\ &= 0{\cdot}52\ M_1, \end{aligned}\right\} \quad (3.60)$$

where $m_{\odot} = 2 \times 10^{33}$ g is the Sun's mass. The total mass of the Galaxy is therefore†

$$M = M_1 + M_2 = 1{\cdot}3 \times 10^{11}\ m_{\odot} = 2{\cdot}7 \times 10^{44}\ \text{g}. \quad (3.61)$$

§8. The Velocity of Escape and the Phenomenon of the High-velocity Stars

Using the approximate representation (3.50) of the gravitational acceleration of the Galaxy, let us now determine the *velocity of escape*, i.e. the velocity which a star must acquire in order to escape from the attraction of the system and move away from it to a great distance (theoretically to infinity). This quantity is of some importance in galactic dynamics.

The velocity of escape evidently corresponds to what is called in celestial mechanics the *parabolic velocity*. This name may therefore be applied also, by analogy, to the velocity of escape.

For a particle of unit mass moving in a conservative field of force, the energy integral may be written

$$V^2 + 2W = 2E, \quad (3.62)$$

where V is the velocity of the particle, W the potential energy of the field at a given point in space, and E the total energy of the particle, a constant.

Let us suppose that the velocity of the particle is such that it can move to infinity. Putting in (3.62) $W_\infty = 0$ and denoting by V_∞ the velocity of the particle at infinity, we have

$$V_\infty^2 = 2E. \quad (3.63)$$

Thus, for any particle which can reach infinity, the total energy must be non-negative, i.e. either positive or (if the velocity tends asymptotically to zero at infinity) zero. Hence we find, first of all, a necessary and sufficient condition for a particle to belong to a given dynamical system: its total energy must be negative. The total energy of any particle not belonging to a given system is non-negative with respect to that system.

For the parabolic velocity V_p at any point we have

$$V_p^2 + 2W = 0. \quad (3.64)$$

† For the values of the constants used by Trumpler and Weaver we have:
$M_1 = 1{\cdot}16 \times 10^{11}\ m_{\odot} = 2{\cdot}32 \times 10^{44}$ g, $M_2 = 0{\cdot}77 \times 10^{11}\ m_{\odot} = 1{\cdot}54 \times 10^{44}$ g,
$M = M_1 + M_2 = 1{\cdot}9 \times 10^{11}\ m_{\odot} = 3{\cdot}9 \times 10^{44}$ g.

The *circular velocity* V_c of a particle moving in a circular orbit is given by

$$V_c^2 - \varrho \partial W/\partial \varrho = 0. \tag{3.65}$$

From (3.65) and (3.64) we obtain

$$\frac{V_c^2}{V_p^2} = -\frac{1}{2}\frac{\varrho}{W}\frac{\mathrm{d}W}{\mathrm{d}\varrho}. \tag{3.66}$$

Here ϱ denotes the radius of the circular orbit, which, for a centrally symmetrical system, is the distance from the centre.

If the whole mass of the system is concentrated at the centre, then $W = -GM/\varrho$, and from (3.66) we have the "Newtonian" relation

$$V_p = \sqrt{2}\,.\,V_c = 1{\cdot}4\;V_c. \tag{3.67}$$

More generally, we can assume for the Galaxy, by (3.50),

$$F = F_1 + F_2 = \alpha\varrho^{-2} + \beta\varrho = \mathrm{d}U/\mathrm{d}\varrho.$$

Evidently $U = U_1 + U_2$, where $U_1 = \alpha/\varrho$ and $U_2 = -\frac{1}{2}\beta\varrho^2 + \gamma$ are the gravitational potentials of the concentrated and distributed masses. The term γ is the value of U_2 at the centre of the system.† Substitution of this value in (3.66) gives $V_c^2/V_p^2 = (\alpha\varrho^{-2} + \beta\varrho)/(2\alpha\varrho^{-2} - \beta\varrho + 2\gamma\varrho^{-1})$ or, using (3.50),

$$\frac{V_c^2}{V_p^2} = 1\Big/\left[2 + \frac{F_2}{F}\left(\frac{2\gamma}{\beta}\varrho^{-2} - 3\right)\right]. \tag{3.68}$$

The ratio γ/β can be estimated by means of the formulae for the potential of a homogeneous spheroid. Since, by (3.56) and (3.57), $F_2 = \beta\varrho$, we have $\beta = \pi^2 Gc\delta/a = \frac{3}{4}\pi GM_2/a^3$, where M_2 is the distributed mass of the Galaxy, a its equatorial radius, c the minor semiaxis, and δ the density. The potential at the centre of a spheroid is

$$U(0,0,0) = -\pi a^2 cG\int_0^\infty \frac{\mathrm{d}s}{(a^2+s)\sqrt{(c^2+s)}}.$$

In the above notation this is equal to $-\gamma$. By calculations similar to those used to find α' in § 7, we obtain

$$\gamma = \pi^2 Gac\delta = \tfrac{3}{4}\pi GM_2/a. \tag{3.69}$$

The formulae for β and γ give $2\gamma/\beta\varrho^2 = 2a^2/\varrho^2$. Putting for the Sun $\varrho = \frac{2}{3}a$, we have $2\gamma/\beta\varrho^2 = 9/2$. Then, from (3.68) with $F_2 = 0{\cdot}19F$,

$$V_p = \sqrt{2{\cdot}28} \times V_c = 1{\cdot}5\;V_c. \tag{3.70}$$

A comparison of (3.67) with (3.70) shows that in the Newtonian case the parabolic velocity is 40% greater than the circular velocity, but in the Galaxy near the Sun it is 50% greater. Thus, if we take the circular velocity there to be $V_c = 250$ km/s,

† This term was incorrectly omitted in [79], an error pointed out by G. M. Idlis.

the commonly accepted value, the parabolic velocity $V_p = 375$ km/s. This is of importance in galactic dynamics, and explains the phenomenon of the *high-velocity stars*, first observed and studied by Oort [90].

This phenomenon is that the stars whose velocities exceed 60 km/s show a tendency to be moving predominantly in one of the directions of the axis of symmetry of the velocity ellipsoid, viz. that opposite to the circular velocity (Fig. 14). The greater the root-mean-square residual velocity σ of a group of stars, the more pronounced is this tendency.

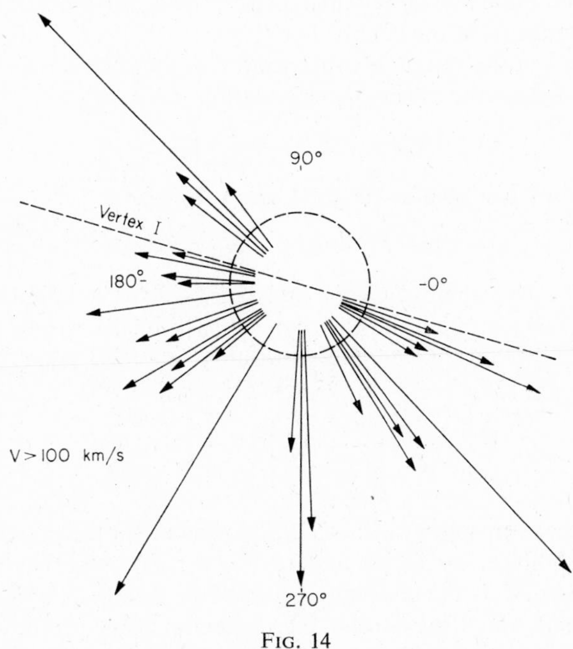

FIG. 14

The phenomenon of the high-velocity stars is explained by the fact that stars whose residual velocity is in the same direction as the circular velocity of the centroid have an absolute velocity exceeding the parabolic value if the residual velocity exceeds $V_p - V_c$. Hence the stars in a group of high-velocity dispersion which are moving in the same direction as the circular velocity of the centroid escape, and only those high-velocity stars remain to be observed which are moving predominantly in the direction opposite to that of the circular velocity.

To determine the fraction of stars which escape from a given sub-system with root-mean-square velocity σ, we proceed as follows. We assume for simplicity that the distribution of residual velocities in the sub-system is Maxwellian. Let $\lambda(v')$ be the fraction of stars having a given residual velocity v' which escape. Then the fraction of stars escaping with any residual velocity and a given dispersion is

$$\varkappa = \int_0^\infty \exp(-h^2 v'^2)\, \lambda(v')\, d\Omega \Big/ \int_0^\infty \exp(-h^2 v'^2)\, d\Omega, \tag{3.71}$$

where $h = 1/\sqrt{2}.\sigma$. As the volume element $d\Omega$ in velocity space we take a spherical shell of radius v' and thickness dv'. Then $d\Omega = 4\pi v'^2\, dv'$, and a calculation of the integral in the denominator of (3.71) gives

$$\varkappa = \frac{4h^3}{\sqrt{\pi}} \int\limits_0^\infty \exp(-h^2 v'^2)\, v'^2 \lambda(v')\, dv'. \tag{3.72}$$

Evidently all stars with residual velocities exceeding $V_p + V_c$ will escape, and none of those with residual velocities less than $V_p - V_c$ will do so. Hence $\lambda(v') = 0$ for $v' < V_p - V_c$ and $\lambda(v') = 1$ for $v' > V_p + V_c$; formula (3.72) becomes

$$\varkappa = \frac{4h^3}{\sqrt{\pi}} \left[\int\limits_{\Delta}^{\Delta_1} \exp(-h^2 v'^2) v'^2 \lambda(v') dv' + \int\limits_{\Delta_1}^{\infty} \exp(-h^2 v'^2) v'^2\, dv' \right], \tag{3.73}$$

where

$$\Delta = V_p - V_c, \quad \Delta_1 = V_p + V_c. \tag{3.74}$$

Let us now find the function $\lambda(v')$.

In Fig. 15, the point O denotes the origin of velocities, and OC the velocity of the centroid of the sub-system. We describe a sphere DD' with centre O and radius equal to the parabolic velocity, and another with centre C and radius equal to some value v' of the residual velocity such that $V_p - V_c \leqslant v' \leqslant V_p + V_c$. The fraction $\lambda(v')$ of stars with the residual velocity v' which escape is clearly equal to the ratio of the area of the part of the sphere of radius v' bounded by DD' to the area of the whole of that

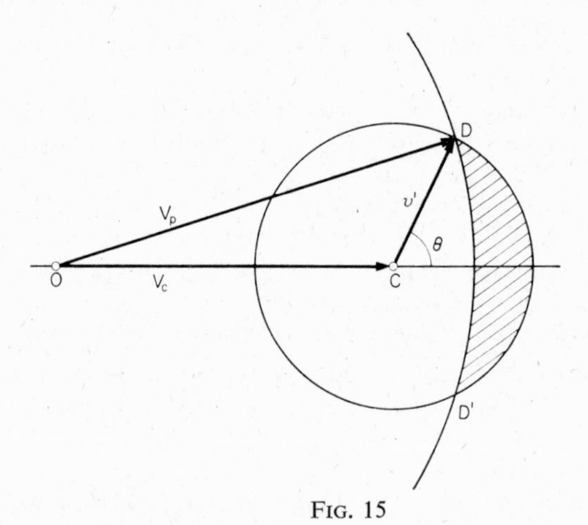

FIG. 15

sphere. It is easy to see that this ratio is $\lambda(v') = \sin^2 \frac{1}{2}\theta$, where θ is half the angle subtended at C by the part of the sphere bounded by DD'. The angle θ can be expressed in terms of v' by means of the triangle OCD, and so

$$\lambda(v') = \sin^2 \tfrac{1}{2}\theta = [(v' + V_c)^2 - V_p^2]/4v' V_c. \tag{3.75}$$

The parabolic velocity in the Galaxy is so large ($V_p = 375$ km/s) that no great error is committed by putting $\Delta_1 = \infty$ in (3.73). Then, substituting $\lambda(v')$, we obtain for the fraction of stars escaping from the sub-system

$$\varkappa = \frac{h^3}{\sqrt{\pi} V_c} \int_{\Delta}^{\infty} \exp(-h^2 v'^2) [(v' + V_c)^2 - V_p^2] v' \, dv', \tag{3.76}$$

where

$$h = 1/\sqrt{2}.\sigma, \quad \Delta = V_p - V_c. \tag{3.77}$$

Integration gives

$$\varkappa = \frac{1}{2\sqrt{\pi} h V_c} \exp(-h^2 \Delta^2) + \tfrac{1}{2}[1 - \operatorname{erf}(h\Delta)], \tag{3.78}$$

where

$$\operatorname{erf} x \equiv \frac{2}{\sqrt{\pi}} \int_0^x \exp(-t^2) \, dt. \tag{3.79}$$

The centroid velocity V_c which appears in these formulae is obtained by subtracting from the velocity of the centroid of the plane sub-systems V_0 the asymmetry S for the corresponding dispersion. The value of S is given by Strömberg's formula (3.32):

$$S = p\sigma^2 + q, \quad p = 0{\cdot}0192, \quad q = 10{\cdot}0, \tag{3.80}$$

where velocities are expressed in km/s. Thus

$$V_c = V_0 - S = 250 - S = 240 - 0{\cdot}0192\sigma^2. \tag{3.81}$$

Calculations using formulae (3.78), (3.77) and (3.81) give the values shown in Table II for the percentage of stars escaping as a function of the variance σ^2 of the

TABLE II

σ (km/s)	0	10	20	30	40	50	60	70	80	90	100	110
V_c	250	238	232	222	209	192	170	146	117	84	48	8
$\varkappa$	0·0	0·0	0·0	0·7	1·7	2·5	3·0	3·1	3·1	3·2	4·1	17·1
Δ/σ	∞	6·2	3·4	2·6	2·3	2·2	2·2	2·2	2·3	2·4	2·5	2·7

residual stellar velocities in the sub-system considered. This table shows that, as the residual-velocity dispersion increases, the centroid velocity rapidly decreases, and in consequence the difference Δ between the parabolic velocity and the centroid velocity remains between two and three times σ, as is shown by the last line of the table.

Since $\varkappa$ does not exceed $\sim 3\%$ for $\sigma < 100$ km/s, the high-velocity stars are not noticeable in any physical sub-system, even in the spherical sub-systems, whose root-mean-square velocities reach 100 km/s. The phenomenon is seen only as a slight truncation of the velocity ellipsoid in the direction of the circular velocity. Projected on the galactic plane, the truncation is bounded by a circle whose centre is at a fixed point and whose radius is the parabolic velocity. This is seen in Figs. 16(a) and (b),

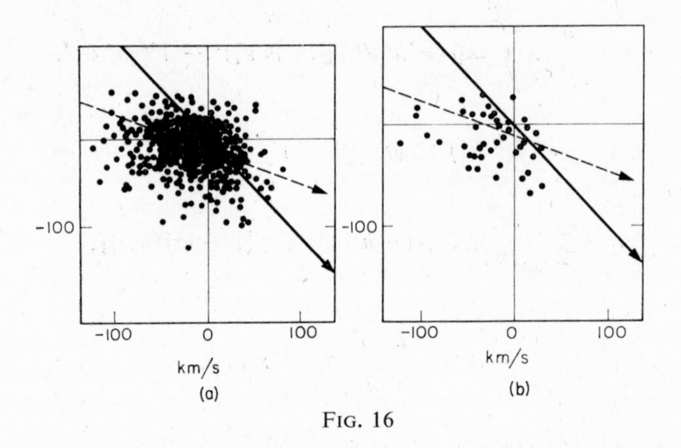

FIG. 16

which reproduce a diagram drawn by Parenago [103] to illustrate the truncation of the velocity ellipsoid. The points show the projections on the galactic plane of the termini of the velocity vectors for two groups of stars. Figure 16(a) is for giants of types G, K and M, and Fig. 16(b) for the similar, but less numerous, giants of types A and F. The origin in each diagram corresponds to the velocity of the centroid. The bottom left-hand corner of each diagram is the fixed point. The continuous arrows point towards the centre of the Galaxy. Both diagrams show clearly the truncation of the ellipsoid, since if there were no truncation the distribution of points would be symmetrical about the broken arrow. Figure 16(b) shows that the truncation is apparent even for a small number of points. Figure 16(a) shows a number of other interesting anomalies in the stellar velocity distribution: it is asymmetrical, as is seen from the fact that the centre of gravity of the points is not at the centroid, and there is a deviation of the vertex, as is seen from the fact the observed line of symmetry (the broken arrow) deviates by 20° from the direction of the centre.

Table II shows that for $\sigma = 110$ km/s the circular velocity V_c is almost zero and $\varkappa$ increases rapidly, reaching the unlikely value of 17% for that value of σ. The reason for this is certainly that Strömberg's empirical formula (3.80) is no longer valid for large σ. We shall therefore assume that $\varkappa = 3\%$ for the spherical sub-systems.

In his study of spatial velocities [105] Parenago points out that the empirical formula (3.80) which we have used has almost no physical significance, since it combines the kinematic properties of plane sub-systems of stars having no genetic connection. In our calculations, however, this objection is not important.

§ 9. Kinematics of Centroids in the Metagalaxy

As a further example of the use of the kinematics of centroids, let us consider the motion of the centroids in the Metagalaxy [82]. The *Metagalaxy* is that part of the universe which is at present accessible to scientific investigation and which lies outside the Galaxy. The population of the Metagalaxy consists of *galaxies*, which are great stellar systems similar to our own Galaxy. The limits of the Metagalaxy are determined by the galaxies which can be observed by present-day methods. Its radius is thus a quantity which increases in the course of time with the progress of astronomical techniques and methods of scientific investigation, and is to some extent arbitrary. At any given time, the radius of the Metagalaxy can be defined as the greatest distance which can be estimated for a single remote galaxy, and at the present time it may be taken as 10^9 light-years, or about 10^{27} cm. Up to this limit we can (at present) determine the distances of individual clusters of galaxies seen by astronomical photography.

We shall make the following assumptions concerning the Metagalaxy.

(1) All galaxies accessible to observation, including our own, form part of one large system, in which the particles are galaxies. This system we call the *Hypergalaxy* or *second-order Galaxy*.

(2) The Hypergalaxy has finite dimensions, but they are large compared with the radius of accessibility, 10^{27} cm. Accordingly we suppose that the Metagalaxy, i.e. the part of the Hypergalaxy accessible to observation, is a small region near the Galaxy.

(3) Finally, we assume that the Hypergalaxy is dynamically united and is not in the course of rapid disintegration. That is, all kinematic parameters will be assumed not explicitly dependent on time.

The first two of these assumptions may be justified as follows. They arise, first of all, from the general Lambert–Charlier cosmology, in which the universe is regarded as an infinite hierarchy of galaxies of various orders, each composed of particles which are galaxies of the next lower order. This cosmology may be regarded as confirmed by the whole development of astronomy during the last two hundred years. It was originally put forward in opposition to the metaphysical idea that space is uniformly occupied by matter. This latter view leads to the insuperable paradoxes of Olbers and Seeliger and, among other things, the conclusion that the universe is finite.

There are, furthermore, at least two facts which indicate that the present radius of accessibility, 10^{27} cm, is only a small part of the second-order Galaxy.

The first of these is that all attempts to determine from observation of numbers of galaxies the space gradient of their distribution density have failed. It is therefore natural to suppose that, as regards the investigation of the Hypergalaxy, we are at present in the position where the astronomy of the 1920's was as regards the investigation of the Galaxy. At that time, in consequence of the limited knowledge of the spatial distribution of stars, it was not possible to detect the variation of star density with distance from the centre of the Galaxy. The apparent uniform decrease in density in all directions away from the Sun was, as we now know, in reality due to the interstellar absorption of light.

The second indication is the observed dependence on distance of the "red shift" of lines in the spectra of the galaxies. The shift of lines in the spectra of celestial objects from their normal positions is usually interpreted on the basis of the Doppler effect, as being the result of a motion of the source of light in the line of sight relative to the observer. In other words, the shift of spectral lines is usually regarded as a measure of radial velocity, and a red shift corresponds to a positive radial velocity, i.e. to recession. The radial velocities of stars are usually a few tens of kilometres per second.

The radial velocities of galaxies have certain peculiarities. Firstly, except for a few of the nearest galaxies, all the radial velocities are positive. Secondly, the radial velocity v_r is, to a first approximation, proportional to the distance r:

$$v_r = Kr. \tag{3.82}$$

If the distance is expressed in megaparsecs, the value of the red-shift coefficient K is [39]†

$$K = 180 \text{ km/s per Mps} = 5{\cdot}84 \times 10^{-18} \text{ per s.} \tag{3.83}$$

Thirdly, the radial velocities of the most distant galaxies are 20 to 40% of the velocity of light, which is most unusual for bodies of macroscopic dimensions.

The consideration of these peculiarities has led to attempts to explain the red shift otherwise than by the Doppler effect, but they have not as yet been successful. The linear dependence of radial velocity on distance (3.82) necessarily occurs within a small region, as we see from (3.27). The same linear dependence will approximately hold for individual galaxies if their residual velocities are small compared with the centroid velocities. This is to be expected in the Metagalaxy if, for example, we make the natural assumption that the residual velocities of the galaxies are, on average, independent of distance. Then, for large r, the centroid velocity will exceed the residual velocity.

Thus the linear increase of v_r may be regarded as an independent indirect confirmation of the hypothesis that the present Metagalaxy is only a small part of the Hypergalaxy, and so the kinematics of centroids of a small region is applicable to the Metagalaxy.

As regards the unusually large radial velocities of the galaxies, we shall now attempt to show that, in view of the scale of the system, these velocities are of the order which might be expected *a priori*. We have seen in § 8 that the velocity of escape is only 40 to 50% greater than the circular velocity. We can therefore suppose that the circular velocity is, in order of magnitude, a measure of the mean velocity of particles in the system concerned.

Let us now consider two systems S_1 and S_2 of different orders, having masses M_1 and M_2, the system S_2 being composed of systems such as S_1. Then the mean particle velocity at a point in S_2 is given by

$$V_2^2 = GM_2/R_2. \tag{3.84}$$

† The recently discovered correction to the zero point of the period – luminosity law for Cepheids, and the consequent doubling of distances in the Metagalaxy, mean that this value of K must be halved. The general conclusions are unchanged (cf., for example, [113a]).

A similar relation can be written for points in S_1, and we have approximately

$$V_1^2 R_1/M_1 = V_2^2 R_2/M_2. \tag{3.85}$$

We take S_1 to be the Galaxy, R_1 the distance of the Sun from its centre; S_2 the Hypergalaxy and R_2 the distance of our Galaxy from its centre. Let the mean density of galaxies in the Hypergalaxy be ν_2. Then

$$M_2 = \frac{4}{3}\pi R_2^3 M_1 \nu_2, \tag{3.86}$$

and eqn. (3.85) gives

$$V_2/R_2 = \sqrt{\left(\frac{4}{3}\pi\nu_2 R_1\right)} V_1. \tag{3.87}$$

The right-hand side of this formula involves only quantities known from observation. If we assume that the velocity in the radial direction is of the order of v_r, then the left-hand side is equal in order of magnitude to the red-shift coefficient K. According to the data of Hubble [38], $\nu_2 = 1$ galaxy per $2{\cdot}5 \times 10^{16}$ or $6{\cdot}6 \times 10^{16}$ ps^3 (depending on the value taken for the mean luminosity of the galaxies). Substituting in (3.87) $R_1 = 7{\cdot}6$ kps, $V_1 = 250$ km/s, and the above values of ν_2, we have $V_2/R_2 = 280$ or 173 km/s per Mps, which in order of magnitude agrees with the observed value of $K = 180$ km/s per Mps [39]. Thus the observed and theoretical values of K are of the same order of magnitude.

From this elementary calculation we see that the large radial velocities of the galaxies are of an order of magnitude entirely commensurate with the probable scale of the Hypergalaxy. The above discussion therefore gives a further independent argument in favour of the applicability of the general kinematics of centroids to our neighbourhood in the Hypergalaxy. We may apply to this "small" region, i.e. to the Metagalaxy, eqns. (2.71), which give the differential field of centroid velocities. Taking the origin at the observer's centroid, we have

$$\left.\begin{aligned} u &= u_x x + u_y y + u_z z, \\ v &= v_x x + v_y y + v_z z, \\ w &= w_x x + w_y y + w_z z. \end{aligned}\right\} \tag{3.88}$$

These equations give the velocities of the *relative* motion of the centroids in the volume element. For the galaxies we know only the radial velocities, since their proper motions are too small to be observed.

Taking the components of (3.88) along the line of sight and proceeding as in Chapter II, § 10, we obtain for each galaxy the equation of condition

$$v_r = r[u_x\alpha^2 + v_y\beta^2 + w_z\gamma^2 + (v_z + w_y)\beta\gamma + (w_x + u_z)\gamma\alpha + (u_y + v_x)\alpha\beta] + v_r'. \tag{3.89}$$

Here v_r and v_r' are the observed and residual radial velocities of the galaxy, r its distance from the observer, and α, β, γ the direction cosines of r. Since $v_r, \alpha, \beta, \gamma$ can be found from observation, we obtain by solving the equations of condition (3.89) the quantities $u_x, v_y, \ldots, u_y + v_x$.

If the distances of the individual galaxies are known, eqn. (3.89) may be divided by r in order to reduce the effects of the residual velocities. If the distances are not known, the mean distance $\bar{r}$ must appear as a factor in each unknown.

A comparison of eqn. (3.89) with (2.71) shows that only the components of the symmetrical tensor $\boldsymbol{S}$ can be found from radial velocities alone. In particular, the angular velocity of rotation ω cannot be found from radial velocities without additional hypotheses. Using (2.81) and omitting the small errors v_r' in (3.89), we obtain the equations of condition for determining the components of the tensor $\boldsymbol{S}$ in the form

$$v_r = r[S_{11}\alpha^2 + S_{22}\beta^2 + S_{33}\gamma^2 + 2S_{23}\beta\gamma + 2S_{31}\gamma\alpha + 2S_{12}\alpha\beta]. \quad (3.90)$$

Taking the scalar product of $\boldsymbol{r}$ and the tensor $\boldsymbol{S}$ found from these equations, we have for the rate of deformation of a small region of the Hypergalaxy the expression

$$\mathbf{V}_g = \boldsymbol{S} \cdot \mathbf{r}. \quad (3.91)$$

The nature of this deformation may be determined by transforming eqn. (2.81) for the tensor $\boldsymbol{S}$. Solving the equation

$$\begin{vmatrix} S_{11} - K & S_{12} & S_{13} \\ S_{21} & S_{22} - K & S_{23} \\ S_{31} & S_{32} & S_{33} - K \end{vmatrix} = 0, \quad (3.92)$$

we find the *principal values* K_1, K_2, K_3 of the tensor $\boldsymbol{S}$. Then this tensor, referred to principal axes, is

$$\boldsymbol{S} = \begin{bmatrix} K_1 & 0 & 0 \\ 0 & K_2 & 0 \\ 0 & 0 & K_3 \end{bmatrix}. \quad (3.93)$$

We may also consider the *planar tensor*

$$\boldsymbol{P} = \boldsymbol{S} - K\boldsymbol{J}, \quad (3.94)$$

where $\boldsymbol{J}$ is the unit tensor

$$\boldsymbol{J} = \begin{bmatrix} 1 & 0 & 0 \\ 0 & 1 & 0 \\ 0 & 0 & 1 \end{bmatrix} \quad (3.95)$$

and K is any one of the principal values.

From (3.94)

$$\boldsymbol{S} = \boldsymbol{P} + K\boldsymbol{J}. \quad (3.96)$$

Substitution of this expression in (3.91), using (3.95), gives for the velocity of local deformation

$$\mathbf{V}_g = K\mathbf{r} + \boldsymbol{P} \cdot \mathbf{r}. \tag{3.97}$$

The first term in this formula is an isotropic expansion, i.e. one which is the same in all directions. To elucidate the mechanical significance of the term $\boldsymbol{P} \cdot \mathbf{r}$ we use the fact that $\boldsymbol{P}$ is a planar tensor. It is a theorem of tensor calculus that any vector $\boldsymbol{P} \cdot \mathbf{r}$ formed from the radius vector $\mathbf{r}$ by scalar multiplication by a planar tensor $\boldsymbol{P}$ is parallel to a certain plane. In the present case this plane is orthogonal to the direction in which all three components of the tensor $\boldsymbol{P}$ are zero. This direction is the axis for which all three components are zero in the tensor obtained by subtracting $\boldsymbol{K} \cdot \mathbf{r}$ from $\boldsymbol{S}$ (3.93).

Thus it follows from formula (3.97) that, for a small region of the Hypergalaxy, the relative motions of the centroids as determined from radial velocities can be resolved into (1) an isotropic expansion (the pure red-shift) and (2) a plane-parallel motion which, as has been shown in § 3, can in turn be resolved into a "plane" expansion, or K effect, and an Oort differential rotation, or (more precisely) deformation.

The application of these theoretical considerations to the radial velocities of 105 galaxies [82] has shown that

(a) the red-shift coefficient is $K = 436$ km/s per Mps, which is 2·4 times the value usually accepted, or 1·2 times when the distance correction is applied, so that the agreement is good;

(b) Oort's constant A for the Hypergalaxy is $A = 72$ km/s per Mps, and the galactic coordinates of the pole of the Hypergalaxy are $l = 110°$, $b = 0°$;

(c) a plane K effect parallel to the hypergalactic plane is also observed, with coefficient $K_1 = 152$ km/s per Mps;

(d) if we assume by analogy with the Galaxy that the Oort term results from an axial rotation, then the most probable direction of the centre of the Hypergalaxy is $l = 20°$, $b = 88°$, corresponding to the great cluster of galaxies in the constellations of Virgo and Coma Berenices;

(e) taking the quantity $\omega_2 = V_2/R_2$ found above from formula (3.87) as an approximate value of the angular velocity, we conclude that the period of rotation of the Hypergalaxy is $P_2 = 2\pi/\omega_2 = 2{\cdot}2 \times 10^{10}$ y or $3{\cdot}7 \times 10^{10}$ y, depending on the value taken for v_2. These are respectively 120 and 200 times the period of rotation of the Galaxy.

Some comments should be made concerning the calculations just given. Firstly, it is evident that they are valid only in order of magnitude. For this reason we could equate V_2/R_2 to the local angular velocity of rotation of the Hypergalaxy. The expansion and the rotation are formally quite independent. In particular, we can easily imagine a system with rotation but no expansion, or vice versa. It may be supposed, however, that in stellar systems these two types of motion should, at least as a rule, coexist and be of the same order. P_2 gives the expected value of the period of the observed angular rotation of the neighbouring galaxies. It becomes of practical value in connection with the considerable task, now being undertaken, of referring star catalogue positions to the distant galaxies. From the above values of P_2 we easily find that the rotation of the system of galaxies may give rise to a systematic proper motion of 3 to 6×10^{-5} seconds per year, i.e. may become perceptible if the errors in the proper motions given in star catalogues become less than $0''{\cdot}0001$ per year.

§ 10. The Problem of Setting up a Fundamental Coordinate System

To conclude this chapter, let us consider the problem of defining a fundamental inertial coordinate system; in theoretical mechanics all motions are considered relative to such a system, which is called an *absolute* system of coordinates.

It has been mentioned in the Introduction that the problem of defining the fundamental system, to which all observed motions (mechanical displacements) of bodies are referred, has always been the main practical problem of stellar astronomy.

To avoid any misunderstanding it should be emphasised immediately that, astronomically, there is no coordinate system at rest with respect to the whole universe. The reason for this is, as we shall see, that the universe is a Lambert–Charlier hierarchy, a conception which is confirmed by the whole history of astronomy.

Let us imagine an observer who, wishing to define a coordinate system at rest relative to the whole universe, describes about himself some small sphere and proceeds to increase its radius *ad infinitum*. When the radius reaches a value of a few parsecs, the sphere includes stars, star clusters, nebulae and other galactic objects. The velocity of any object (a star, for example) determined by the observer will be its *intragalactic* velocity. This velocity will certainly depend on the size of the sphere, since the Galaxy has an internal *regular* motion defined by the kinematics of the centroids, and the mean velocity of the centroids in the volume has no intrinsic significance.

When the radius increases to hundreds of thousands or millions of parsecs, the sphere includes not only individual stars but whole galaxies, i.e. systems of another order, both in mass and in velocity. Whereas the spherical volume previously represented a region within one galaxy, it now represents a region within the Hypergalaxy, and the parameters of mass, velocity, dimensions, etc., change discontinuously from one region to the other.

If we suppose the radius to increase still further, the volume will be successively a region of a galaxy of the second order, the third order, and so on *ad infinitum*. Each transition from one order to the next involves a discontinuous change in the mechanical parameters, whereas within any one galaxy they vary continuously. Evidently such a process cannot converge, and so it cannot lead to the establishment of a fixed coordinate system.

Here a difficulty of procedure arises: we do not know any properties of galaxies above the second order. The preceding discussion shows that any attempt to define the velocity of the observer with respect to the entire universe is equivalent to extending the laws and relations established for cosmic systems of low order to other systems which are qualitatively different.

Thus no absolute frame of reference exists. In mechanics a coordinate system is used instead which is called an *inertial* system. This term denotes any system relative to which a body under no interaction with neighbouring bodies (in astronomical terminology *unperturbed*) moves uniformly in a straight line.

In textbooks of theoretical mechanics the subject of defining the fundamental frame of reference is either omitted, or dealt with by mentioning that such a system is defined in astronomy with reference to the "fixed" stars.†

It need hardly be said that this postulate is a complete anachronism, based on the scientific concepts of more than two centuries ago, when the proper motions of the stars had not yet been discovered.

It has been known since the eighteenth century that all the stars are in motion. The next step, therefore, was to attempt to define an inertial system which should take account of the stellar motions in the Galaxy. This may now be regarded as accomplished. We know from the preceding discussion that the motion of the Sun in the Galaxy†† consists of its motion with respect to the local centroid together with the motion of the centroid about the centre of the Galaxy with angular velocity $\omega = B - A$ and linear velocity $V = \omega \varrho$. Here A and B are Oort's constants and ϱ the distance of the Sun from the centre of the Galaxy. If the unit of acceleration is taken as $g = 978{\cdot}0$ cm/s^2, the acceleration due to gravity at the Earth's equator, then the acceleration due to the Earth's rotation at the equator is 1/288; that due to the Earth's orbital motion round the Sun is 1/1642. The acceleration due to the rotation of the Galaxy is 10^{-8} of that due to the Earth's rotation.

Evidently, in applications of mechanics in terrestrial laboratories, the acceleration due to the Earth's orbital motion is usually of no importance. It must, however, be taken into account when the scale of the motion is comparable with the dimensions of the Earth (Beer's law, the trade winds, etc.). In considering the motions of stars in the Galaxy the acceleration due to the rotation of the Galaxy must be taken into account, and so on.

Hence it follows immediately that there can exist no "universal" inertial coordinate system uniquely fitted for the solution with absolute precision of every problem of mechanics regardless of the linear dimensions involved. Any system of coordinates is valid only within certain limits of accuracy and under certain conditions which depend mainly on the scale of the problem.

The inertial coordinate system at present used by astronomers is the result of historical developments. Until 1718, when Halley discovered the proper motions of the stars, the inertial system was related to the planetary system and its development was a gradual evolution by progressive refinement of the fundamental constants such as the length of the day, the tropical year, the precession, etc.

The eighteenth century marked a turning point in the development of methods of establishing the inertial system, since it was thereafter defined with respect to the galactic system. The problem of investigating the systematic motions of the stars made it necessary to examine the distribution of mass in the Galaxy, i.e. to begin the modern science of stellar astronomy. The pioneers in this subject were William Herschel and, especially, F.G.W.Struve. The latter, throughout his career, continually related the problems of astrometry to those of stellar astronomy.

The second, "galactic", phase of the development of the problem of defining an inertial system lasted until about 1950. It was accompanied by a gradual extension of

† See, for instance, A.Sommerfeld, *Mechanics*, Academic Press, New York, 1952, p. 10; C.J. de la Vallée Poussin, *Leçons de la mécanique analytique*, 2nd ed., Paris, 1938; E.L.Nikolai, *Lectures on theoretical mechanics* (*Lektsii po teoreticheskoĭ mekhanike*), Part 2, 5th ed., ONTI, Moscow, 1934, pp. 5–6; L.G.Loĭtsyanskiĭ and A.L.Lur'e, *Textbook of theoretical mechanics* (*Kurs teoreticheskoĭ mekhaniki*), Part 2, 4th ed., Gostekhizdat, Moscow, 1948, p. 15.

†† The rotation of the Earth and its revolution about the Sun may be supposed known.

knowledge concerning the distribution of mass and the systematic motions of stars in the Galaxy, and a refinement of the values of the kinematic constants A and B.

In the last few years, with the beginning of international work on the positions of the stars with respect to extragalactic objects (the galaxies), the problem has entered the third, "extragalactic", phase of its development, the coordinate system being defined with respect to a system of objects outside the Galaxy. This in turn sets stellar astronomy a new problem: to investigate the distribution of mass and the systematic motions in the Metagalaxy.

The inclusion of each new cosmic system of a higher order makes possible an improvement of the inertial system in the sense of embracing motions on larger and larger scales and increasing the accuracy with which these motions are observed. For example, the inclusion of extragalactic objects not only opens the way to the study of the laws of mechanics, physics, etc., on a much larger scale, but also makes possible a more accurate assessment of the fundamental coordinate system at present used in practice.

CHAPTER IV

IRREGULAR FORCES IN STELLAR SYSTEMS

§ 1. The Fundamental Quantities which Characterise the State of a Stellar System

IT HAS been mentioned in the Introduction that stellar systems must be regarded as very rarefied media. The degree of rarefaction can be characterised by (1) the mean number of stars per unit volume, that is the *star density*, (2) the ratio of the mean distance between neighbouring stars to the mean radius of a star for purposes of stellar dynamics, called the *characteristic ratio*.

Let ν denote the star density. Then $1/\nu$ is the mean volume of space per star, and

$$d = \nu^{-\frac{1}{3}} \tag{4.1}$$

is the mean distance between neighbouring stars.

In the Galaxy in the neighbourhood of the Sun, $\nu = 0{\cdot}1$ star per cubic parsec $= 3{\cdot}41 \times 10^{-57}$ star/cm³. Thus the Galaxy has an extremely low star density. Also

$$d = 10^{\frac{1}{3}} \text{ ps} = 2{\cdot}15 \text{ ps} = 6{\cdot}64 \times 10^{18} \text{ cm}. \tag{4.2}$$

We should expect that, in other galaxies, the situation would be much the same as that observed in our own Galaxy.

In the globular clusters the star density is, on the average, about 10 times the density in the galaxies, and in the central parts of these clusters the density is doubtless still greater. Although the coalescence of the individual star images prevents a direct measurement of this density, the order of magnitude is probably the same.

For the Metagalaxy we have, according to Hubble [38],

$$\nu = 1 \text{ galaxy per } 1{\cdot}4 \times 10^{17} \text{ ps}^3, \tag{4.3}$$

whence

$$d = \nu^{-\frac{1}{3}} = 520{,}000 \text{ ps}. \tag{4.4}$$

Let us now estimate the characteristic ratio. We take as the mean effective radius of a star for purposes of stellar dynamics that of the solar system, which is approximately equal to the radius of a supergiant star:

$$a = 40 \text{ AU} = 6{\cdot}00 \times 10^{14} \text{ cm}. \tag{4.5}$$

Then the characteristic ratio in the Galaxy is

$$d/a = 1{\cdot}11 \times 10^4 \sim 10{,}000. \tag{4.6}$$

This means that, in the Galaxy, the distance between the stars is ten thousand times their effective radius. If the effective radius is replaced by the ordinary astrophysical radius, the ratio d/a is much greater still. For example, its value for the Sun is 10^8. In problems of mechanical interaction of stars, however, the physical radius is not the determining quantity.

The mean radius of a galaxy is approximately

$$a = 10{,}000 \text{ ps}, \tag{4.7}$$

which, with (4.4), gives

$$d/a = 52, \tag{4.8}$$

so that the characteristic ratio for galaxies in the Metagalaxy is 200 times less than that for stars in the Galaxy.

From these estimates we can draw the following conclusion. In stellar systems the dimensions of the stars are so small compared with the distances between them that for purposes of stellar dynamics they can be regarded as particles in the mechanical sense, but in the Metagalaxy the dimensions of the galaxies must be taken into account.

In stellar dynamics we are concerned with systems of various orders, which may be quite different in both dimensions and mass. Already we have in stellar astronomy, on the one hand, globular clusters of diameter only a few tens of parsecs and masses 10^4 to 10^5 times the Sun's mass and, on the other hand, galaxies whose diameters are measured in kiloparsecs and whose masses are 10^9 to 10^{11} times that of the Sun. Systems like the Hypergalaxy must be of still greater size and mass. We have seen in Chapter III, § 9, that the systems of different orders must have internal velocities of different orders.

Thus it is reasonable to suppose that characteristic units of length and time can be defined for systems of each order. The unit of length is most simply taken as the radius of the system, which we denote by R.

The unit of time must be used for measuring the duration of various phenomena in stellar systems, and is most naturally defined on the basis of the dynamical properties of the stellar system. Evidently only a definition in order of magnitude is significant. Let the mass of the system be M and its radius R. For simplicity, we assume the star density distribution to be spherically symmetrical. As the unit of time we shall arbitrarily take the period P of circular motion at the boundary of the system.

If V is the circular velocity and G the gravitational constant, then $V^2/R = GM/R^2$. Since, for motion in a circle, $V = 2\pi R/P$, we have

$$P^2 = 4\pi^2 R^3/GM. \tag{4.9}$$

Substituting, for example, the galactic values $M = 1{\cdot}3 \times 10^{11}\, m_\odot = 2{\cdot}6 \times 10^{44}$ g and $R = 13$ kps $= 4{\cdot}01 \times 10^{22}$ cm, we obtain

$$P_0 = 3{\cdot}83 \times 10^8 \text{ y} = 1{\cdot}21 \times 10^{16} \text{ s}. \tag{4.10}$$

This is only twice the usually accepted value ($1{\cdot}83 \times 10^8$ y) for the period of rotation of the Galaxy. For the globular clusters we have approximately $M = 300{,}000\, m_\odot$, $R = 35$ ps, giving

$$P = 3000 \text{ y}. \tag{4.11}$$

Formula (4.9) becomes especially simple if the mass density δ is assumed constant in the system. Then $M = \frac{4}{3}\pi\delta R^3$, and (4.9) becomes

$$P^2 = 3\pi/G\delta. \tag{4.12}$$

The mass density δ is related to the star density ν by the evident relation

$$\delta = m\nu, \tag{4.13}$$

where m is the mean mass of one star.

Formula (4.12) is particularly convenient because P in it depends only on the density δ. In particular, if δ is taken to have the value

$$\delta = 6{\cdot}0 \times 10^{-24} \text{ g/cm}^3, \tag{4.14}$$

which, according to G.G.Kuzmin [57] and P.P.Parenago [104], corresponds to the neighbourhood of the Sun in the Galaxy, we have immediately

$$P_0 = 1{\cdot}54 \times 10^8 \text{ y}. \tag{4.15}$$

This is only 16% different from the period of rotation of the Galaxy.

Let us now apply formula (4.12) to the Metagalaxy, considering the neighbourhood of the Galaxy, and taking the mean galactic mass as $M = 2{\cdot}6 \times 10^{44}$ g, the mass of the Galaxy, and Hubble's value for ν. Then (4.13) gives $\delta = 6{\cdot}34 \times 10^{-29}$ g/cm^3, and from (4.12) we obtain for the Metagalaxy the unit of time

$$P_1 = 4{\cdot}73 \times 10^{10} \text{ y}. \tag{4.16}$$

The quantity P is of importance as regards the evolution of stellar systems. In the development of such systems, the two opposite tendencies of *dissipation* and *condensation* play a leading part.

The term *condensation* is here used to denote the tendency of stellar systems to take up an equilibrium configuration corresponding to the dynamical conditions, which might also be called "consolidation". We propose to retain the term *condensation* because in many cases, as has been shown by L.É.Gurevich [36], dissipation is in fact accompanied by an increase in the density of the remaining system.

Dissipation of stellar systems is caused by loss of mass resulting from (a) the escape of stars whose velocity exceeds the velocity of escape, (b) rotational acceleration, (c) corpuscular emission, (d) loss of mass from the stars, and so on. During dissipation the mass, energy, angular momentum and other dynamical parameters of the system change.

If the dissipation occurs sufficiently slowly, the stellar system will assume the equilibrium figures corresponding to the various values of the dynamical parameters. Then the shape and mass distribution of the system will remain regular but will vary with time. In these conditions the system will pass through an evolutionary sequence of varying shapes. The globular clusters, and the elliptical and spiral nebulae, are probably stellar systems of this kind.

If, however, the rate of dissipation is high, and the system is not able to assume the equilibrium figure under the action of gravitational forces, it will be irregular in form. Such irregular systems, for example, are the small galaxies like the Magellanic Clouds and other members of the *local group* of galaxies which lie close to our own in the Metagalaxy. However, this example of the Magellanic Clouds shows that in actual galaxies and other stellar systems there is an intermediate case where the process of condensation merely lags somewhat behind that of dissipation.

The dissipation process determines the rate of evolution for a given stellar system. In this sense dissipation is the active side of the evolutionary process, while condensation is the passive side. Both processes must operate if evolution is to take place.

We shall now show that the rate of condensation in stellar systems is also given by a time of the order of P. Let us consider a simplified model, consisting of a non-rotating spherical system of mass M and radius R in dynamic equilibrium, and apply to this system a "perturbation" in the form of an irregular projection from its surface, similar to a solar prominence. We assume the mass of the projection to be small compared with the mass M of the whole system, but its height above the surface is taken to be of the same order as the radius R of the system.

Let us determine the time P' required for the projection to fall to the surface of the system and coalesce with it. Clearly P' is of the order of the time required for a particle with zero initial velocity to fall from height R to the surface. Without affecting the order of magnitude, we can suppose that the acceleration is constant and equal to its value at the midpoint of the path, $GM/(\frac{3}{2}R)^2$. Then $R \approx 2GMP'^2/9R^2$, whence

$$P'^2 \approx 9R^3/2GM. \tag{4.17}$$

A comparison with P (4.9) gives†

$$P' \approx \tfrac{1}{3}P. \tag{4.18}$$

Thus P' is of the same order as the unit of time for stellar systems.

Strictly speaking, this calculation is valid only when the systems concerned have a property analogous to the viscosity of a liquid. Otherwise, instead of being dispersed, the projection might begin to execute oscillations indefinitely about some mean position, or form something resembling a star stream moving *ad infinitum* in some regular orbit.

This is a very instructive example. It shows that, in the absence of viscosity, stellar systems could never acquire the regular form and homogeneous structure of, for example, E galaxies (elliptical nebulae) or the nuclei of the spiral nebulae, since it is difficult to suppose that these bodies had such regular forms from the very beginning, and even then it would follow that evolution of galaxies is impossible, since any evolutionary process involves the conversion of one regular form into another.

Thus we conclude that stellar systems must possess viscosity. In gases, viscosity is due to molecular collisions. In stellar systems, the corresponding events are encounters

† An exact calculation by integration of the differential equation of "free fall", $d^2r/dt^2 = -GM/r^2$, gives $P'^2 = (1 + \frac{1}{2}\pi)^2 R^3/GM$, and $P' = P(\frac{1}{4} + 1/2\pi) \approx 0{\cdot}4P$.

between stars. But stellar encounters are, as we shall see, very inefficient, at least in stellar systems such as galaxies. Yet there is undoubtedly viscosity. This is shown not only by the regular shape of most galaxies but also by various other phenomena, such as the rigid rotation of almost all galaxies (with certain restrictions). The very word "rigid" signifies that the galaxies rotate as if they were rigid bodies, of iron or steel, rather than stellar systems. We can therefore take it as extremely likely that the galaxies possess some efficient mechanism, other than encounters of stars, which causes viscosity in them. The elucidation of this problem will be one of the main purposes of the remainder of this chapter.

§ 2. Star Encounters. Regular and Irregular Forces

Let us consider the mechanical forces which act in stellar systems. We shall discuss mainly galaxies, but the arguments will be largely applicable, with obvious slight changes, to other types of system.

In a stellar system, in consequence of the small star density ν under ordinary conditions, the forces of interaction between individual stars will be imperceptible, and the motion of the stars will take place under the combined attraction of all the stars in the system. The motion of each star will be exactly the same as if it were moving in a non-uniform ideal fluid whose density at every point is equal to the mass density of the system.†

The total force of attraction of the whole system at each point is called the *regular force*. This force forms a continuous gravitational field, uniquely defined at every point in space, both inside and outside the system. Under the action of the regular force, each star describes a path in space, which we call its *regular orbit*. This motion, however, is disturbed from time to time because the star passes close to other individual stars, i.e. has encounters with them.

By an *encounter* of two or more stars we shall mean an approach so close that the acceleration exerted on each star by the others is of the same order of magnitude as the acceleration due to the regular force.††

The forces resulting from star encounters are called *irregular forces*. The simplest type of encounter is one involving two stars. This is called a *simple* or *binary* encounter; one which involves three or more stars is called a *multiple* encounter.

The *encounter radius* δ is the maximum distance between two stars for which an encounter as defined above can take place. We determine this quantity from the condition that the regular and irregular forces should be equal. Since we do not know the way in which the density varies within the system, the encounter radius can be found only in order of magnitude.

As before, we denote by m the mean mass of one star, and by M and R the mass and radius of the system. Then the condition which gives δ at the edge of the system is

$$m/\delta^2 = M/R^2, \tag{4.19}$$

† It may be noted that, as soon as we begin to discuss motions in stellar systems, it is necessary to regard these systems as having the property of continuity.

†† An exact definition of an encounter is given later.

whence

$$\delta = R\sqrt{(m/M)}. \tag{4.20}$$

For points inside the system, except possibly the centre, δ is of the same order.

Using the numerical values given in § 1 we find for the Galaxy

$$\delta = 0{\cdot}036 \text{ ps} = 7440 \text{ AU}. \tag{4.21}$$

This is almost 60 times less than the mean distance d between the stars.

When the ratio δ/d of the encounter radius to the mean distance between the stars is small, the probability of encounters is small. The probability of a k-fold encounter is given by Poisson's formula

$$p_k = (\nu\omega)^k \, e^{-\nu\omega}/k!, \tag{4.22}$$

where $\nu = d^{-3}$ is the star density, and ω the volume of the *encounter sphere:*

$$\omega = \tfrac{4}{3}\pi\delta^3. \tag{4.23}$$

In the Galaxy in the neighbourhood of the Sun,

$$\nu\omega = \tfrac{4}{3}\pi(\delta/d)^3 \approx 2 \times 10^{-5}, \tag{4.24}$$

whence it is clear that the probability of multiple encounters, and therefore the frequency of their occurrence in the Galaxy, is very small in comparison with that of binary encounters. We can therefore neglect the effect of multiple encounters.

In what follows we shall, for convenience, distinguish *encounters* from *passages.* Let two stars be moving in straight lines before an encounter. The *passage distance* is the shortest distance at which the stars would pass if they continued to move in these straight lines, exerting no mutual attraction. A *passage* is a relative motion of two stars at any passage distance. An *encounter* is a passage for which the passage distance p is less than the encounter radius δ.

It must be borne in mind, however, that the encounter radius defined above is purely conventional. Although, as we shall see in § 3, a passage of stars at a distance equal to the encounter radius perturbs the motion of the stars only very slightly, passages at distances exceeding the encounter radius occur continually. Their individual effect is slight, but their number is very large, because, when passages at all possible distances are considered, there is a gradual transition to the regular force field.

We have shown above that multiple encounters are of no significance because of their low probability in comparison with binary collisions. For passages the reverse is true: the greater the distance r between two passing stars compared with the mean distance $d = \nu^{-\frac{1}{3}}$ between stars, the greater will be the probability that other stars than these two will be within the sphere of volume $\frac{4}{3}\pi r^3$, leading to a *multiple passage.* At first sight it may seem that, in consequence of this, the study of star passages would involve great mathematical difficulties. This is not so, however, since the small effects resulting from passages are easily seen to obey the *law of superposition:*

$$f(a) + f(b) = f(a + b), \tag{4.25}$$

where $f(a)$ and $f(b)$ denote respectively the effects of passing star a and star b, and $f(a+b)$ the effect of the multiple passage. This law holds because the effects of passages may be regarded as small perturbations undergone by the star considered as it moves in its regular orbit. Such small perturbations obey the laws of addition of differentials, one of which is the superposition law given above. The significance of this law in the problem of star passages has not yet been fully appreciated, although it is of very great importance in stellar dynamics. In what follows we shall often use this property of star passages.

For the moment we may mention one immediate simplification which results from it: since the effects of star passages are simply superposed, we can base our calculations entirely on binary passages, i.e. consider the dynamical problem of two bodies.

§ 3. The Effect of Irregular Forces: Individual Encounters

In § 2 we saw that binary encounters are the most important. Let us therefore consider the effect of a binary encounter on the residual velocities of the stars.

Let a star S_1, of mass m_1, encounter a star S_2, of mass m_2; let C be their common centre of mass. Since the result of the encounter is independent of the motion of the centre of mass, it may be assumed to be at rest. Let the origin of coordinates be at C; let p be the passage distance, V the initial relative velocity, i.e. the relative velocity of the stars at the beginning of the encounter, and V_1, V_2 the initial velocities of the two stars. The quantities m_1, m_2, V and p are called *parameters of the encounter.* These quantities, as we shall see, entirely determine the result of the encounter.

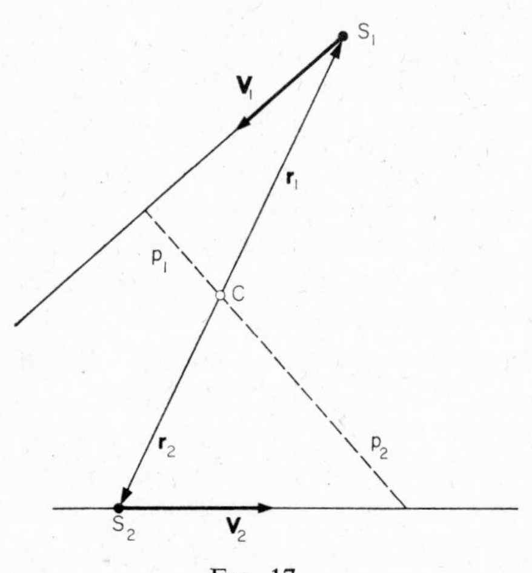

FIG. 17

Let the positions of the stars S_1 and S_2 be given by vectors $\mathbf{r}_1 = CS_1$ and $\mathbf{r}_2 = CS_2$, and that of S_2 relative to S_1 by $\mathbf{r}_{12} = S_1S_2$ (Fig. 17). Then

$$m_1\mathbf{r}_1 + m_2\mathbf{r}_2 = 0, \quad \mathbf{r}_{12} = \mathbf{r}_2 - \mathbf{r}_1. \tag{4.26}$$

Hence, in particular, $m_1 r_1 = m_2 r_2$, $r_{12} = r_1 + r_2$. The motion of the stars takes place under their mutual attraction $F = Gm_1 m_2/r_{12}^2$. If we consider the motion of S_1 relative to S_2, the force F may be conveniently written, by (4.26), as $F = G'(m_1 + m_2)/r_1^2$, where

$$G' = Gm_2^3/(m_1 + m_2)^3. \tag{4.27}$$

Then the differential equations for the motion of S_1 relative to S_2 are

$$\ddot{x}_1 = -G'(m_1 + m_2)x_1/r_1^3,$$

$$\ddot{y}_1 = -G'(m_1 + m_2)y_1/r_1^3,$$

$$\ddot{z}_1 = -G'(m_1 + m_2)z_1/r_1^3.$$

From these equations we find in the usual way the energy integral and the area integral:

$$v_1^2 - V_1^2 = 2G'(m_1 + m_2)/r_1, \tag{4.28}$$

$$v_1 r_1 = l, \tag{4.29}$$

where v_1 and r_1 are the velocity and radius vector of the star S_1 at any instant, and l the constant value of the area integral. The orbit of S_1 relative to C is a hyperbola with C at one focus. Let r_0 be the shortest distance of S_1 from C, and p_1 the distance

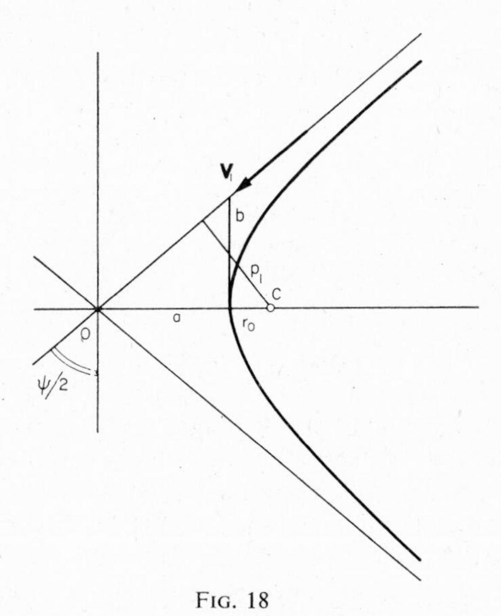

FIG. 18

of C from the asymptotes of the hyperbola (Fig. 18). The integrals (4.28) and (4.29) for the time when S_1 passes through the vertex of the hyperbola (the *pericentre*) give

$$v_0^2 - V_1^2 = 2G'(m_1 + m_2)/r_0, \tag{4.30}$$

$$r_0 v_0 = p_1 V_1, \tag{4.31}$$

where v_0 is the velocity of S_1 at that point. These equations enable us to deduce the effect of the encounter. Eliminating v_0 gives

$$\frac{p_1^2}{r_0^2} - 1 = \frac{2G'(m_1 + m_2)}{V_1^2 r_0}. \tag{4.32}$$

Let ψ be the angle through which the velocity V_1 is turned by the encounter. This angle is evidently equal to the external angle between the asymptotes (i.e. the angle which is bisected by the imaginary axis). We may express the left-hand side of (4.32) in terms of ψ. Let a and b be the real and imaginary semiaxes, and c the distance from focus to centre. Then (see Fig. 18)

$$\tan\tfrac{1}{2}\psi = a/b = a/p_1 = (c - r_0)/p_1 = \sec\tfrac{1}{2}\psi - (r_0/p_1),$$

whence

$$r_0/p_1 = (1 - \sin\tfrac{1}{2}\psi)/\cos\tfrac{1}{2}\psi, \tag{4.33}$$

and so

$$\frac{p_1^2}{r_0^2} - 1 = \frac{2\sin\tfrac{1}{2}\psi}{1 - \sin\tfrac{1}{2}\psi}.$$

Substitution in (4.32) and multiplication of both sides by $\tfrac{1}{2}r_0/p_1$, using (4.33), gives

$$\tan\tfrac{1}{2}\psi = G'(m_1 + m_2)/V_1^2 p_1. \tag{4.34}$$

Finally, we express V_1 and p_1 in terms of the relative velocity V and the passage distance p. From (4.26) we have

$$V_1 = m_2 V/(m_1 + m_2), \quad p_1 = m_2 p/(m_1 + m_2). \tag{4.35}$$

Using also (4.27) to express G' in terms of G, we find

$$\tan\tfrac{1}{2}\psi = G(m_1 + m_2)/V^2 p. \tag{4.36}$$

This formula was first derived by Jeans [45]. It gives the solution of the problem, since it entirely determines the result of the encounter if the values of the parameters are known. It gives the angle through which the direction of the initial relative velocity is turned by the encounter. At first sight it may appear paradoxical that, by formula (4.36), the angle ψ through which the velocity is turned is symmetrical in the masses of the two stars. This is because the more massive star always has a smaller velocity relative to the centre of mass. The absolute velocity of the more massive star changes less than that of the less massive one. The magnitude of the relative velocity is unchanged. If the velocity vector after the encounter is $\mathbf{V}'$, then the modulus ΔV of the vector increment of velocity is obtained from the isosceles triangle with sides $\mathbf{V}$ and $\mathbf{V}'$ and vertical angle ψ (Fig. 19):

$$\Delta V/V' = 2\sin\tfrac{1}{2}\psi. \tag{4.37}$$

The vector $\Delta\mathbf{V}$ is at an angle $90° + \tfrac{1}{2}\psi$ to the initial velocity vector.

We should expect that each individual encounter has only a very slight effect, except for very close encounters, whose probability is extremely small. We call an encounter for which the passage distance p is close to the encounter radius δ a *normal encounter*. In the Galaxy this means that $p \approx \delta = 0{\cdot}036$ ps $= 7500$ AU. The value $a = 40$ AU may be arbitrarily taken as a measure of the maximum passage distance for *close encounters*, which endanger the planetary system (if any) or the star itself if it is a giant. If $p = 4{\cdot}5$ AU, the deviation ψ calculated from (4.36) is 90°. Such an

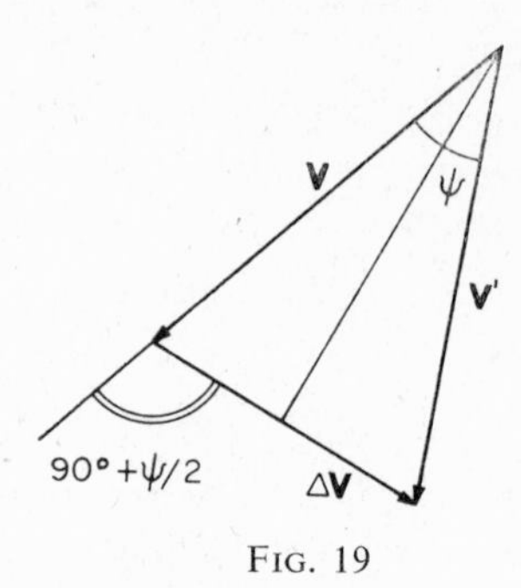

FIG. 19

encounter, which entirely changes the direction of motion of a star, may be called a *very close encounter*, and the distance 4·5 AU may be arbitrarily taken as the corresponding distance. Such an encounter must certainly have a catastrophic effect on the stars involved. For example, if the Sun were one of the stars, the other would pass within the orbit of Jupiter. The probability of such an encounter, however, is practically zero, as we shall see.

Table III gives the calculated values of quantities characterising encounters of these three types; the value of V is taken as equal to $\sigma = 20$ km/s, and $m_1 = m_2 = 2 \times 10^{33}$ g, the mass of the Sun.

TABLE III. COMPARISON OF THE EFFECTS OF STAR ENCOUNTERS

Type of encounter	Very close	Close	Normal
p (AU)	4·5	40	7500
ψ	90^0	$12^0{\cdot}7$	$3^0{\cdot}3$
t_0 (y)	$4{\cdot}9 \times 10^{14}$	$6{\cdot}2 \times 10^{12}$	$1{\cdot}8 \times 10^8$
l_0 (ps)	$7{\cdot}9 \times 10^9$	$1{\cdot}0 \times 10^8$	$2{\cdot}9 \times 10^3$
$E(\psi)$	$0''{\cdot}12$	$1''{\cdot}0$	$1''{\cdot}1$

The first line of the table gives the passage distance; the second line, the angle ψ through which the velocity is turned, is given by formula (4.36); the third line, the mean time t_0 between two successive encounters of the type concerned, is calculated from the following simple arguments. Let a star be moving with velocity V in a homogeneous star field of density ν, the field stars being for simplicity assumed to be at rest. Then the number of encounters with passage distance not exceeding p undergone by this star per unit time is $\pi p^2 \nu V$, and therefore the reciprocal of this is

$$t_0 = 1/\pi p^2 \nu V. \tag{4.38}$$

In calculating t_0 we take $\nu = 0{\cdot}1$ star per ps^3 and $V = \sigma = 20$ km/s.

Finally, the fourth line in the table gives the *mean free path*

$$l_0 = t_0 V \tag{4.39}$$

corresponding to each value of p. Here it must be borne in mind that t_0 and l_0 are calculated for the notion of the star relative to coordinate axes which move with the velocity of the centroid corresponding to the initial position of the star, and therefore l_0 is the path in the *rotating* Galaxy.

From Table III we see that:

(1) Encounters are very rare events in the life of a star. Even "normal" encounters take place, on average, no more than once in each rotation of the Galaxy. Close encounters occur only once in tens of thousands of rotations. Very close encounters are practically impossible. Of the 2×10^{11} stars in the Galaxy, only one pair undergo a very close encounter in 2000 years.

(2) The average effect of individual encounters is very slight. A measure of this effect is given by the expectation of the angle ψ during a time P_0 equal to the period of rotation of the Galaxy. The corresponding values of $E(\psi)$ are given in the last line of Table III. These are calculated as the product of ψ and the corresponding probability of a single passage, given by Poisson's formula $P(1) = (t_0/P_0)\exp(-t_0/P_0)$.

(3) The duration of star encounters is very short. The order of magnitude of this quantity can be determined for any type of encounter as the time that a star moving with velocity $V = \sigma = 20$ km/s takes to cross a sphere of radius $\delta = 1440$ AU. The average time is $1{\cdot}76 \times 10^3$ y. This is five orders of magnitude less than the period P_0 of the rotation of the Galaxy.

(4) During an encounter the coordinates of the stars remain almost unchanged, since the change in the coordinates is of the order of the encounter radius, which is much less than the mean distance between stars and many orders of magnitude less than the dimension of the macroscopic volume element. Thus the stars involved in an encounter remain in the same macroscopic volume element with the same centroid. In its very short duration ("instantaneous" on the galactic time scale) the encounter can change only the residual velocities of the stars. In this respect close encounters bear a marked resemblance to the collisions undergone by molecules in a gas or suspended particles in a colloidal solution.

Summarising the above discussion of the action of irregular forces in encounters, we may say that:

(a) irregular forces are of the impulsive type; they act for a very short time and can change only the relative velocities of the stars involved in each encounter, and do not change either their coordinates or the macroscopic volume element to which they belong;

(b) irregular forces are random in nature, in the sense that the result of an encounter does not depend on the results of previous encounters.

§ 4. The Effect of Irregular Forces: the Cumulative Effect

In §3 we have seen that individual encounters are unable to bring about significant changes in the velocities of the stars. The more distant the passage, the smaller will be its effect, since the angle ψ is inversely proportional to the passage distance. As the passage distance increases, however, the probability of such passages increases rapidly. We may therefore attempt to estimate the cumulative effect.

Before doing so, however, let us derive a formula from which we may calculate the results of a binary passage at a great distance. It will appear below that such a formula could easily be derived from the exact formula (4.36) of §3. We prefer, however, to give an independent derivation, taken from the work of L.D.Landau [62]. This derivation is simple and makes possible an elucidation of certain circumstances of distant passages which will be of importance in the subsequent discussion.

Let a star S, which we shall arbitrarily call the perturbing star, pass at a great distance from another star S_0. Since only a small effect is involved, the perturbation of S by S_0 may be neglected in considering the perturbation of S_0 by S, and we may suppose that S moves in a straight line relative to S_0.

We may determine the change ΔV in the velocity of the star S_0 as a result of the passage. Since this change in independent of the initial velocity, the latter may for simplicity be taken as zero. From the results of § 3 we may also neglect the change in position of S_0 during the passage; that is, we assume that S_0 remains at rest but changes its velocity. Thus the star S_0 will behave as if it received a sudden impulse from the perturbing star S.

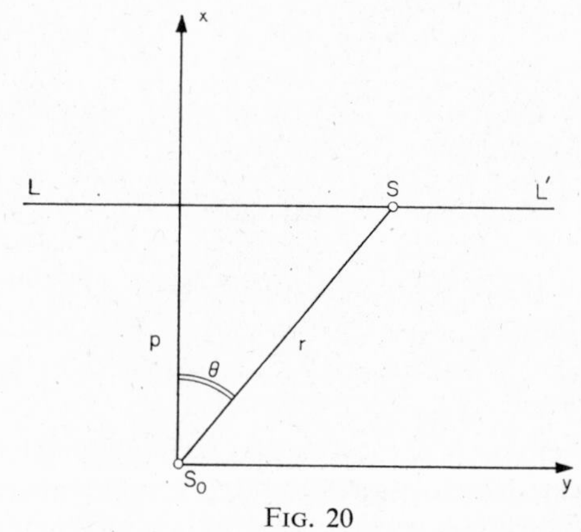

FIG. 20

The length of the perpendicular from S_0 to the line LL' of the relative path of S is the passage distance p (Fig. 20), which, like the initial relative velocity V, is a parameter of the passage. Let S_0 be the origin of a rectangular coordinate system, with the x-axis along the perpendicular and the y-axis parallel to the relative path LL' and in the direction of motion of the star S. Let r be the distance and θ the polar angle of S, measured clockwise from the x-axis (Fig. 20).

At any instant the acceleration of S_0 is along the radius vector $S_0 S$, and its magnitude is Gm/r^2, where m is the mass of the perturbing star S. The components of this acceleration along the coordinate axes are $a_x = (Gm/r^2) \cos\theta$, $a_y = (Gm/r^2) \sin\theta$, and so the velocity increment of S_0 has components

$$\Delta v_x = Gm \int_{-\infty}^{\infty} \frac{\cos\theta}{r^2} \, dt, \quad \Delta v_y = Gm \int_{-\infty}^{\infty} \frac{\sin\theta}{r^2} \, dt. \tag{4.40}$$

Evidently $\Delta v_y = 0$, since to each position of S on the line LL' with negative y there corresponds a symmetrical position with positive y. Hence Δv_x is the whole velocity increment, i.e.

$$\Delta V = Gm \int_{-\infty}^{\infty} \frac{\cos\theta}{r^2} \, dt. \tag{4.41}$$

To calculate the integral, we replace the time as a variable of integration by the polar angle θ, using the area integral

$$r^2 \, d\theta/dt = h. \tag{4.42}$$

By means of this integral for $t = \pm\infty$, we can express the area constant h in terms of the parameters p and V of the encounter: $h = pV$. Substitution in (4.42) gives $dt/r^2 = d\theta/pV$, so that (4.41) gives at once

$$\Delta V = 2Gm/pV. \tag{4.43}$$

It is useful to compare the result (4.43) with the exact solution previously obtained (4.36), (4.37). For this purpose we write (4.43) as

$$\Delta V/V = 2Gm/pV^2. \tag{4.44}$$

For distant encounters, the angle ψ is very small, and so $\tan\frac{1}{2}\psi = \sin\frac{1}{2}\psi = \frac{1}{2}\psi$. Hence (4.36) and (4.37) give

$$\Delta V/V = 2G(m_1 + m_2)/V^2 p. \tag{4.45}$$

A comparison of (4.44) and (4.45) shows that formula (4.43) is obtained from the exact formulae if the mass of the perturbed star S_0 is assumed very small in comparison with the mass of the perturbing star S. The change in velocity given by (4.43) is therefore independent of the mass of the star itself, and depends only on that of the perturbing star.

This result, at first sight unimportant, has a consequence which is of great significance in stellar dynamics. Hitherto we have considered only the results of the individual interaction of passing stars, and have concluded that such effects are small. We now see that the small perturbations are additive: if at any point in the Galaxy there is a local condensation of N stars, it will perturb the motion of neighbouring stars like one star of mass N times the average. For example, a globular cluster consisting of 500,000 stars of equal mass has, for a given initial relative velocity, an effect 500,000 times that of a single star at that point. Star clouds may, as we shall see later, have a still greater effect.

We have seen that the effect of individual passages is negligible. The time necessary for the velocities of the stars to be considerably affected by encounters is much greater than any possible time of evolution of stellar systems, since we can scarcely suppose that the Galaxy might rotate tens of thousands of times without undergoing any significant evolutionary change.

It remains to consider whether there is any cumulative result of the small effects of individual passages. Since these effects are random (see § 3), we must regard them as random quantities in effecting the summation. Let us assume for simplicity that the velocity distribution is spherically symmetrical. Then the relative-velocity distribution function $f(V)$ is independent of direction. Evidently

$$\int_0^\infty f(V)\,\mathrm{d}V = \nu, \tag{4.46}$$

where ν is the star density. During a time t, the star will travel a distance tV, and will undergo an average number

$$2\pi t V p\,\mathrm{d}p \cdot f(V)\,\mathrm{d}V = Q\,\mathrm{d}p\,\mathrm{d}V \tag{4.47}$$

of passages with parameters between p and $p + \mathrm{d}p$ and between V and $V + \mathrm{d}V$ (Fig. 21).

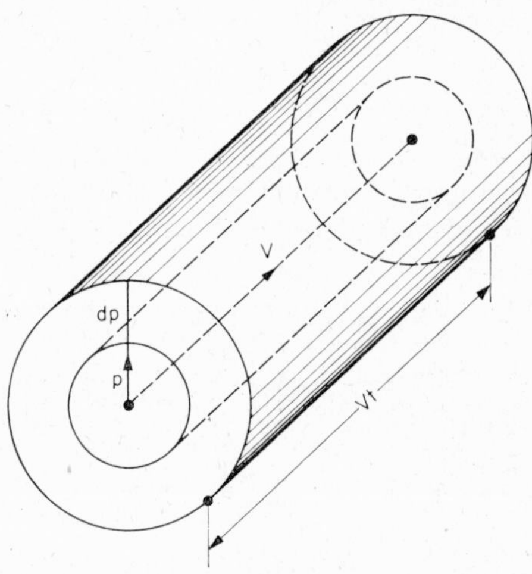

FIG. 21

Let δV be the root-mean-square change in the velocity of the star by the cumulative effect. For simplicity we assume that the masses of all the stars are equal. Then the law of addition of dispersions gives

$$(\delta V)^2 = \int (\Delta V)^2 Q\,\mathrm{d}p\,\mathrm{d}V,$$

where the integration is over all possible values of V and p. The change in the star's velocity as a result of one encounter is given by (4.43). Substituting ΔV from (4.43) and Q from (4.47), we have

$$(\delta V)^2 = 8\pi G^2 m^2 t \int_0^\infty \frac{f(V)\,\mathrm{d}V}{V} \int_{p_1}^{p_2} \frac{\mathrm{d}p}{p} = 8\pi G^2 m^2 t \log(p_2/p_1) \int_0^\infty \frac{f(V)}{V}\,\mathrm{d}V. \tag{4.48}$$

Here p_1 and p_2 denote respectively the minimum and maximum values of the passage distance. In (4.48)

$$\int_0^\infty \frac{f(V)}{V}\,\mathrm{d}V = \nu \overline{V^{-1}}, \tag{4.49}$$

where ν is the star density and $\overline{V^{-1}}$ the mean reciprocal relative velocity. Since we are calculating only in order of magnitude, we can put

$$\overline{V^{-1}} = 1/\hat{v}, \tag{4.50}$$

where $\hat{v}$ is the root-mean-square residual stellar velocity.

From (4.48), (4.49) and (4.50) we have finally

$$(\delta V)^2 = (8\pi G^2 t m^2 \nu/\hat{v}) \log(p_2/p_1). \tag{4.51}$$

In this formula the ratio of the extreme passage distances is as yet undefined.

We shall show that, in order of magnitude,

$$p_2/p_1 = \sqrt{N}, \tag{4.52}$$

where N is the total number of stars in the system. For, from our previous discussion of the law of superposition in star encounters, we can put $p_2 = R$, where R is the radius of the system. As p_1 we take the encounter radius δ. If, as stated above, the masses of all stars in the system are supposed equal, then $M = Nm$, and from (4.20) $p_1 = \delta = R/\sqrt{N}$. Thus p_2/p_1 is given by (4.52).

For the Galaxy we can take $N = 1{\cdot}2 \times 10^{11}$, and then $\log(p_2/p_1) = 12{\cdot}8$. If, instead of $p_1 = \delta$, we take $p_1 = a = 40$ AU, the radius of the solar system, which is 20 times less than δ, then $\log(p_2/p_1)$ would be 15·8, i.e. less than 25% greater. Thus the fact that the ratio p_2/p_1 appears in a logarithm in (4.51) renders quite unimportant the indeterminacy in the value of this ratio. By means of (4.52) we can rewrite formula (4.51) as

$$(\delta V)^2 = (4\pi G^2 m^2 \nu t/\hat{v}) \log N. \tag{4.53}$$

From this formula we see that the expectation of the squared velocity increment from the cumulative effect of passages increases proportionally to the time. To obtain an estimate of the cumulative effect we rewrite (4.53) as

$$\frac{(\delta V)^2}{V^2} = \left(4\pi G^2 \frac{m^2}{\hat{v}} \frac{\nu}{V^2} \log N\right) t, \tag{4.54}$$

where V denotes the initial velocity of a given star. For example, putting for the Galaxy $m = m_{\odot}$, $\nu = 0{\cdot}1$ per ps^3, $V = \hat{v} = 20$ km/s and $N = 1{\cdot}2 \times 10^{11}$, we find

$$\frac{(\delta V)^2}{V^2} = 0{\cdot}38 \times 10^{-13} t. \tag{4.55}$$

Hence we see that, during one rotation of the Galaxy, i.e. for $t = P_0 = 1{\cdot}85 \times 10^8$ y, the velocity of a star changes only by 1/375. For δV to be of the same order as V a time of the order of 10^{13} y must elapse, which is five orders of magnitude greater than the galactic time unit P_0.

Thus we can conclude, from the foregoing analysis, that encounters of stars are of practically no significance in stellar systems. If irregular forces play any part in such systems, they must arise from the passage of stars near masses greatly exceeding those of individual stars.

§ 5. Various Types of Equilibrium of Stellar Systems. The Quasi-steady State

If we neglect, as a first approximation, the effect of irregular forces on stars and consider only their regular motions, we can imagine a state of a stellar system in which all physical parameters, such as shape, size and mass distribution, remain constant, and also the nature of the stellar motions within the system does not vary with time. We call this a state of *dynamic equilibrium*.

The motion of the individual stars does not violate the dynamic equilibrium of the system, and so the problem of finding such states may be treated by the methods of that branch of fluid dynamics which concerns the equilibrium figures of rotating compressible fluids. The fundamentals of these methods as applied to the study of equilibrium figures of rotating stellar systems will be discussed in Chapter X. At present we shall merely remark that this treatment describes the properties of a stellar system as if it were a continuous medium.

The movements of the individual stars must be treated by the methods of statistical mechanics. In dynamic equilibrium without encounters, the stars evidently behave like the molecules of a perfect gas in a given field of the regular forces. It is immaterial whether we consider the motion of the whole system or of a sub-system distinguished by some physical property. The only difference is that in the former case the gravitational field will be *self-consistent*, i.e. it is due to the attraction of the stars whose motion is being considered, but in the latter case this is not so. We shall return to this subject in due course.

Let us suppose that a stellar system is in a state of dynamic equilibrium. In the course of time, the irregular forces will tend to destroy this equilibrium. As a result of encounters in various parts of the system, stars will continually acquire velocities exceeding the velocity of escape. Since encounters are very rare, such stars will practically always leave the system.† Thus the system undergoes a continual gradual loss of mass, and therefore some internal redistribution of mass, so that the previous state

† The stars will practically always have time to leave the system before experiencing another encounter, the only thing which could prevent them from doing so.

of equilibrium is destroyed. If, however, the accumulation of the effects of irregular forces is sufficiently slow, it may be that the stellar system can achieve at every instant a state of dynamic equilibrium corresponding to its dynamical parameters (mass, angular momentum, etc.). Such a state of the system, when at every instant it is in dynamic equilibrium but that equilibrium is continually changing, is called a *quasi-steady state.*

In mechanics a *steady state* of a mechanical system is often considered, i.e. one in which the dynamical parameters of the system do not depend explicitly on the time. From the preceding discussion we see that a strictly steady state cannot be achieved by a stellar system, and in stellar dyramics the quasi-steady state to some extent replaces it in importance.

The definition of the quasi-steady state can be made more precise if we consider also another aspect of the action of the irregular forces from stellar encounters. These, as we have seen, are of the nature of brief random impulses. Hence, besides destroying the dynamic equilibrium, the irregular forces have a smoothing effect and set up a statistical distribution of the residual velocities of the stars.

If this statistical smoothing of the residual velocities in the system occurs sufficiently rapidly, then each instantaneous state of dynamic equilibrium is also a state of *quasi-steady equilibrium*, this being the state in which the residual velocities are smoothly distributed in the manner described.

Thus the irregular forces in stellar systems have two opposite effects. Firstly, they cause individual stars to leave the system, which thereby "disintegrates", and the original state of dynamic equilibrium is destroyed. Secondly, they tend to smooth the residual velocities of the stars and establish a state of statistical equilibrium. The relative rate of these two processes determines the manner of evolution of the system concerned.

In the statistical physics of molecules the smoothing of the residual velocities is related to the concept of *statistical equilibrium*. This is defined as follows: when the system has reached a state of statistical equilibrium, it will remain in that state indefinitely until external agencies cause it to depart therefrom. Such states may or may not exist. If the system is removed from a state of statistical equilibrium by some external agency and then left to itself, it will return to its previous state after some finite time. For example, the state of statistical equilibrium for a gas enclosed in a fixed closed vessel is that in which the molecular velocity distribution is Maxwellian. In the terminology of A. M. Lyapunov's theory of the stability of motion,† we can say that statistical equilibrium is asymptotically stable.

Unlike a confined volume of gas, a stellar system cannot be in a state of statistical equilibrium. This is a formal consequence of the fact that any statistical equilibrium is necessarily a steady state, and we have seen that such a state cannot be achieved by a stellar system. From the practical point of view the explanation is that, unlike the gases of the laboratory, stellar systems are not contained in vessels which prevent stars with very high velocities from escaping. In other words, stellar systems have no potential barrier at their outer boundaries.

† See [64] and J. La Salle and S. Lefschetz, *Stability by Liapunov's direct method*, Academic Press, New York, 1961.

§ 6. The Relaxation Time and the Disintegration Half-life

Any free system for which a state of statistical equilibrium exists will spontaneously tend to assume that state. In statistical physics the *relaxation time* is the name given to the time during which a system left to itself restores a state of statistical equilibrium after this has been destroyed.

Strictly speaking, of course, the relaxation time so defined is to some extent indeterminate, since, firstly, it depends on the magnitude and character of the initial perturbation, even if this perturbation is supposed small (as is usual in physics). Secondly, the restoration of equilibrium may be asymptotic, complete restoration being achieved only at infinite time. Nevertheless, in this case also we can define the relaxation time as that required for the magnitude of the perturbation to decrease to some prescribed fraction of its initial value. It has been found in practice that the indeterminacy does not cause difficulty if the relaxation time is defined only in order of magnitude.

We have seen that a state of statistical equilibrium can never be achieved by stellar systems, and so the relaxation time cannot be defined in the usual way. Following V.A. Ambartsumyan, we shall take the relaxation time of a stellar system to be the time during which the irregular forces significantly affect an original state of dynamic equilibrium.

This definition shows that the relaxation time for quasi-steady stellar systems pertains to the rate at which the continuous variation of the state of dynamic equilibrium occurs.

In order to arrive at specific values, we shall take the relaxation time to be that after which the increments in the stellar velocities by the action of the irregular forces become of the same order as the velocities themselves.

The change in the velocity of a star as a result of the cumulative effect of encounters is given by formula (4.54). The coefficient of t on the right-hand side of this formula evidently has the dimensions of reciprocal time. Putting therefore

$$T = \left(4\pi G^2 \frac{m^2}{\hat{v}} \frac{v}{V^2} \log N\right)^{-1} = \frac{\hat{v} V^2}{4\pi G^2 m^2 v \log N}, \tag{4.56}$$

we can write (4.54) as

$$(\delta V)^2/V^2 = t/T. \tag{4.57}$$

From this it is seen that, when t is of the order of T, the change in the initial velocity of the star becomes comparable with the velocity itself. Hence T may be taken as the relaxation time of the stellar system.

The expression for T includes the two velocities $\hat{v}$ and V, which in general have quite different values. The former is the root-mean-square relative velocity of the stars, and so is always of the order of the residual velocities of the stars in the macroscopic volume element concerned. The quantity V is the initial velocity of the star considered, whose change as a result of encounters is the subject of these calculations.

In a stellar system where the centroid velocity is not zero, therefore, V can represent either the velocity of the absolute motion of the star in its regular orbit or its velocity relative to the local centroid, i.e. the residual velocity.

In the former case the expectation of V^2 is $V_0^2 + \hat{v}^2$, where V_0 is the centroid velocity and $\hat{v}$ the root-mean-square residual velocity. In the latter case $V^2 = \hat{v}^2$. The relaxation time is different in the two cases. Let it be T_1 and T_0 respectively. Then we have from (4.56)

$$(\delta V)^2/(V_0^2 + \hat{v}^2) = t/T_1, \tag{4.58}$$

$$(\delta V)^2/\hat{v}^2 = t/T_0, \tag{4.59}$$

and therefore

$$T_1/T_0 = 1 + (V_0/\hat{v})^2. \tag{4.60}$$

From formula (4.60) it follows that the relaxation time T_1 defined with respect to the absolute stellar velocities is in general greater than the relaxation time T_0 defined with respect to the residual velocities. Accordingly, we call T_1 the *long* relaxation time and T_0 the *short* relaxation time. We shall see that each of these has a definite physical significance.

The *characteristic number* γ is the ratio of the centroid velocity to the root-mean-square residual velocity (more precisely, to the major semiaxis of the velocity ellipsoid):

$$\gamma = V_0/\hat{v}. \tag{4.61}$$

We shall find that the dynamical properties of stellar systems depend largely on the value of γ.

In terms of the characteristic number γ, the relation (4.60) between the long and short relaxation times is

$$T_1 = T_0(1 + \gamma^2). \tag{4.62}$$

Evidently the above distinction of two possible definitions of the relaxation time is important only for systems in which the dispersion of residual velocities is small in comparison with the centroid velocities. For such systems $\gamma \gg 1$. This is true, for example, in the plane sub-systems of the Galaxy, where the centroid velocity is an order of magnitude greater than the residual velocities, and also in plane sub-systems of other galaxies.

If the residual velocities of stars in a system differ only slightly from the absolute velocities, $\gamma < 1$. In such systems the long and short relaxation times are almost equal; they include the spherical sub-systems of galaxies, the globular clusters and, according to present data, the open clusters (on account of the absence of any marked rotation).

No distinction of relaxation times occurs in statistical physics, since only systems with no internal motion are considered (i.e. the centroid velocities are zero).

Let us now ascertain the physical significance of T_1 and T_0, and first consider the long relaxation time T_1. If at the initial instant the absolute velocity of some star in its motion within a stellar system is eaual to the circular velocity V_0, then after a time t its squared velocity V'^2 will be, by the law of addition of random quantities, $V_0^2 + \delta V^2$. If $t = T_1$, then by (4.58) $\delta V \approx V_0$, and†

$$V' = \sqrt{2} . V_0 . \tag{4.63}$$

After the long relaxation time T_1, therefore, the absolute velocity of each star is increased by an average factor of $\sqrt{2}$. We have seen in Chapter III that, for a Newtonian attraction, the velocity of escape is $\sqrt{2} . V_0$. At any instant, and in particular at the initial instant, about half the stars in a given macroscopic volume element will have an absolute velocity less than the circular velocity,†† and about half will have one greater. We can therefore define the long relaxation time as the time during which half the stars which initially belong to a stellar system acquire velocities exceeding the velocity of escape.

Any star whose velocity exceeds the velocity of escape leaves the system "immediately", i.e. in a time small compared with the fundamental time unit of the system, and does not return. We can therefore lastly define the relaxation time as the *disintegration half-life* of the system, and henceforward this definition of T_1 will be used.

It has been shown above that the ordinary state of a stellar system is quasi-steady. Such a state is characterised by a continuous slow variation of the state of dynamic equilibrium, which occurs because of the gradual decrease in the mass of the system by the escape and dissipation of stars into the surrounding space. The destruction of any particular "original" dynamic equilibrium becomes perceptible when a considerable fraction of the stars originally in the system have escaped from it. We have seen that, on average, half the stars escape in a time T_1. Hence T_1 defines the rate of evolution of a stellar system in a quasi-steady state.

It is easy to see that, after a time $t = T_1/\log_{10} 2 = 3{\cdot}3\, T_1$ only one-tenth of the stars remain, i.e. the system has practically ceased to exist. Since, however, the time T_1 itself is defined only in order of magnitude, we can say that T_1 gives the mean statistical lifetime of a rotating stellar system. Here the word "statistical" signifies that we neglect any other possible causes of the evolution of stellar systems (for example, dissipation of mass by radiation).

The disintegration half-life T_1 must not be interpreted literally. It relates to the rate of dissipation of a stellar system in the state in which it is at the instant considered. During the time T_1 the state of the system will itself change considerably.

Thus dissipation determines the rate of evolution of quasi-steady systems.

Let us now consider the short relaxation time T_0, which henceforward will be called simply the *relaxation time*. From (4.59) it follows that at time $t = T_0$ the increment δV of a star's velocity is equal to the residual velocity $\hat{v}$. It has been mentioned earlier that, at any point in a stellar system, the residual velocity distribution is subject to a definite statistical law. For example, observation shows that in the Galaxy, in the neighbourhood of the Sun, the distribution of residual velocities is given, to a certain approximation, by Schwarzschild's law. In other stellar systems some other

† From (4.58) with $t = T_1$ it follows that $\delta V = \sqrt{(V_0^2 + \hat{v}^2)}$. For systems in which T_1 differs considerably from T_0, $\hat{v} < V_0$ and therefore $\delta V \approx V_0$.

†† Here we ignore the phenomenon of asymmetry of stellar velocities and the fact that the circular velocity of the centroid is always somewhat less than that of a free particle.

law may hold, but at every point there must be some definite statistical law of distribution of residual-velocity components. It is natural to assume that, to each dynamic equilibrium state of the system, there corresponds a certain most probable distribution law. However, in order that this law should in fact be established in a stellar system, there must be some mechanism which acts to smooth the residual velocities, which may naturally be taken to be star encounters. For, as we have seen earlier in this chapter, such encounters exhibit a marked analogy with collisions of molecules in gases, which lead to the establishment of the statistical Maxwell–Boltzmann law (the most probable in the conditions concerned). From the above discussion we may assume that the residual-velocity distribution law in stellar systems is set up by the irregular forces.

The rate at which this process occurs can evidently be characterised by the time in which the velocity increment δV becomes equal to the residual velocity, and this time is T_0. Thus the relaxation time T_0 pertains to the rate at which the most probable phase distribution law corresponding to given physical conditions is set up in the stellar system concerned.

For stellar systems with internal motions, i.e. with $\gamma \gg 1$, the relaxation time T_0 is, by (4.62), considerably less than the disintegration half-life T_1 (in the Galaxy, for example, $\hat{v} = 20$ km/s, $V_0 = 250$ km/s, $T_1 \approx 150\, T_0$). Let a state of dynamic equilibrium be set up in some manner in such a system. Then the irregular forces will, firstly, alter the absolute (total) velocities of the stars and, secondly, smooth the residual velocities. Since $T_0 \ll T_1$, however, the latter process will occur much more rapidly than the destruction of the state of dynamic equilibrium. Hence, at every instant, a phase distribution law close to the most probable one will hold in a system with $\gamma \gg 1$. The establishment of the exact most probable law is prevented by the dissipation of stars (the absence of potential barriers).

If, however, a system has no significant internal motions, the relaxation time T_0 will be close to the disintegration half-life T_1. The "statistical lifetime" of such a system must therefore be considerably less than for a system in which there are internal motions.

§ 7. The Fundamental Paradox in the Classical Dynamics of Stellar Systems

So far we have obtained an almost completely natural and clear picture of the statistical processes occurring as a result of stellar encounters. Now, however, we meet with an unexpected internal contradiction, which may be called the paradox of the classical dynamics of stellar systems: the relaxation time for the majority of stellar systems, defined on the basis of encounters between stars, is practically infinite. In other words, although stellar encounters and passages occur in all stellar systems, the star density is so low that their effect is completely imperceptible during any time conceivably representing the age of the systems. For example, in the Galaxy, taking $m = m_\odot$, $\nu = 0{\cdot}1$ per ps^3, $V = \hat{v} = 20$ km/s and $N = 1{\cdot}2 \times 10^{11}$, we find from (4.56)

$$T_0 = 2{\cdot}59 \times 10^{13}\,\mathrm{y} = 8{\cdot}19 \times 10^{20}\,\mathrm{s}. \tag{4.64}$$

This is five orders of magnitude greater than the galactic unit of time. Since $\hat{v} = 20$ km/s and $V_0 = 250$ km/s, we have

$$\gamma = 12 \cdot 5 \tag{4.65}$$

and so

$$T_1 = 4 \cdot 05 \times 10^{15} \text{ y}. \tag{4.66}$$

This is two orders of magnitude greater even than T_0. Thus the irregular forces due to stellar encounters in the Galaxy are of no practical importance.

This remarkable conclusion has had a considerable effect on the development of stellar dynamics, which became a kind of statistical mechanics with no statistical interaction of the particles (the stars). Each star, on this view, describes a regular orbit in the general gravitational field of the stellar system, and continues to do so unimpeded for practically all time. The stellar-velocity distribution is determined by the distribution of elements of the regular orbits or, what is the same thing, of the integrals of the motion. The ellipsoidal distribution of residual velocities (Schwarzschild's law) observed in the Galaxy has usually been taken as an initial postulate in stellar dynamics.

The problem of finding the law of distribution of mass in a stellar system in which the residual-velocity distribution everywhere satisfies Schwarzschild's law (or more generally, the generalised ellipsoidal law) is sometimes called *Chandrasekhar's problem* or the *inverse Jeans' problem*.† Although, in the solution of this problem, a number of important properties of stellar systems have been discovered, they are all *local* properties, i.e. they pertain to the nature of stellar motions in a small neighbourhood of a given point, and this approach has proved quite ineffective in solving the fundamental problem, which is, in a sense, the ultimate aim of every theory of stellar dynamics: to study the structure of stellar systems *as a whole*, and first of all to devise a theory of the equilibrium figures of stellar systems.

The resolution of the above-mentioned paradox in stellar dynamics lies, in the author's opinion, in the fact that the irregular forces should be regarded as much more efficient than is assumed in classical stellar dynamics. However, the most important encounters of stars are not with other single stars, but with condensations of matter whose mass is of the order of 10^5 to 10^6 $m_\odot$. These may be star clouds, which are found in great numbers in the spiral arms of the galaxies, or clouds of diffuse matter.

The idea that our own and other galaxies contain numerous condensations of stellar and diffuse matter, with masses of the order just mentioned, has already been put forward by various authors, including Greenstein [35], Bok [13], Spitzer and Schwarzschild [124], Lebedinskiĭ [63] and Gurevich [36].

It is easy to see that the presence of such condensations greatly reduces the relaxation time. Formula (4.56) for the relaxation time T may be rewritten, using (4.52), as

$$T = A/m^2 \nu, \tag{4.67}$$

where

$$A = \hat{v} V^2 / 8\pi G^2 \log(p_2/p_1) \tag{4.68}$$

† This problem will be discussed in detail in Chapter V.

includes, besides constants, the quantities $\hat{v}$, the root-mean-square relative velocity of the stars; V, the initial velocity of the perturbed star; and p_2/p_1, the ratio of the maximum and minimum passage distances.

Formulae (4.67) and (4.68) can also be used to take account of the effect of encounters of stars with condensations. V is unchanged. In encounters between stars, $\hat{v}^2$ must equal twice the residual-velocity variance, since $\hat{v}^2 = \overline{V_1^2 + V_2^2}$, where V_1 and V_2 are the residual velocities of the perturbed and perturbing stars. Large condensations must have very small residual velocities. In their encounters with stars, therefore, $\hat{v}^2$ will be half its former value, and $\hat{v}$ will be reduced by a factor $\sqrt{2}$. In the factor $\log(p_2/p_1)$ in (4.68), p_2 can again be taken as R, the radius of the galaxy; p_1 must now be taken as the mean radius of the condensation. If we put, for example, $p_1 = 100$ ps, we have (with $p_2 = 30{,}000$ ps) $\log(p_2/p_1) = 8{\cdot}0$, and even with $p_1 = 1000$ ps, $\log(p_2/p_1) = 5{\cdot}7$, so that the change as compared with the previous value 12·7 is only by a factor slightly exceeding 2. Thus the numerator in formula (4.68) is reduced by a factor $\sqrt{2}$, and the denominator by a factor 2, in comparison with the case of encounters between stars. Hence the factor A remains approximately unchanged.

The mass m increases by 5 or 6 orders from encounters with stars to those with condensations. Accordingly, the relaxation time (4.67) is reduced by 10 to 12 orders. This decrease is partly compensated by a reduction in ν, since the number of massive condensations must be considerably less than the number of stars. To estimate the required value of ν, let us assume that, in the Galaxy with a total number $N = 10^{11}$ stars, one-tenth of all stars are members of star clouds each of mass 10^5 to $10^6\, m_\odot$. Then the total number of clouds in the Galaxy is $N_1 = 10^4$ to 10^5. The reduction in the number of perturbing objects therefore decreases ν by 5 or 6 orders, and the relaxation time is correspondingly increased. The net result is that $m^2\nu$ is increased by 5 or 6 orders of magnitude, and the relaxation time is correspondingly decreased, when encounters with condensations, as opposed to encounters with individual stars only, are taken into account.

For encounters between stars we found the relaxation time (4.64) $T_0 = 2{\cdot}59 \times 10^{13}$ y and the disintegration half-life (time of existence) $T_1 = (1 + \gamma^2)\, T_0 = 157\, T_0 \approx 4{\cdot}07 \times 10^{15}$ y. These are now replaced by

$$T_0 \approx 10^7 \quad \text{to} \quad 10^8 \text{ y}, \tag{4.69}$$

$$T_1 \approx 10^9 \quad \text{to} \quad 10^{10} \text{ y}. \tag{4.70}$$

Although these calculations are somewhat tentative the order of magnitude of the values obtained for T_1 and T_0 is promising, since we have taken into account only the possibility of stellar condensations, and neglected the presence of clouds of diffuse matter such as the Orion nebula. To allow for these would increase N_1 and ν_1 and therefore reduce T_1 and T_0. Thus the above calculations are a further indirect argument in favour of the existence in galaxies of many massive condensations of stellar and diffuse matter.

We shall therefore assume henceforward that the irregular forces in galaxies arise from clouds of stellar and diffuse matter, with masses of the order of 10^5 to $10^6\, m_\odot$.

The total number of these clouds in the Galaxy is 10^4 to 10^5. The relaxation time resulting from their effects is 10^7 to 10^8 y, and the disintegration half-life is 10^9 to 10^{10} y.

Thus the statistical distribution of residual velocities of stars in galaxies is established in a time which is about ten times less than the period of rotation. The lifetime of SB and S galaxies is probably a few tens of times the period of rotation.

However, even if we had not reached these conclusions from the above arguments, they would follow from the simple fact that a lifetime of hundreds or thousands of rotations of SB and S galaxies is inconsistent with their form, which indicates the occurrence of violent processes in them.

§ 8. The Interaction of Stars with Dust Clouds

To obtain a complete idea of the nature of the forces acting in stellar systems it is necessary to examine also the interaction of stars with diffuse matter. Work on the origin of stars has shown the great importance of gas and dust in the cosmogony of stellar systems. It has also been found that the mass of the diffuse matter in the Galaxy is comparable with the total mass of the stars. The interaction of the stars with the diffuse matter must therefore be taken into account in problems of stellar dynamics.

The principal distinguishing feature of the interaction between stars and diffuse matter is that not only gravitational forces but also radiation pressure play an important part. Moreover, the stars may be regarded as points but the clouds have a considerable extent. Since the forces of radiation pressure are involved in addition to gravitational forces, the nature of the interaction will depend on the luminosity of the star as well as its mass.

In this section we shall consider the problem of elucidating the effect on stellar dynamics of the interaction between stars and dust clouds. The dynamics of the dust will be of interest only in so far as is necessary for a correct understanding of the dynamics of the stars. Consequently, we shall neglect such phenomena as collisions between dust clouds and processes of magnetic gas dynamics, which may be of considerable importance at great distances from stars of high luminosity.

Following Agekyan [2], let us consider an encounter between a dust particle P of mass μ and a star S of mass m. Let V_0 be the velocity of the star in the inertial frame of reference in which the dust particle is at rest before the encounter. We shall at first assume that the star exerts no radiation pressure on the particle. Then the two bodies will move in hyperbolic orbits under their mutual attraction. As has been shown in § 3, the relative velocity vector $\mathbf{V}$ is unchanged in magnitude by the encounter, but is turned through an angle ψ given by

$$\tan\tfrac{1}{2}\psi = GM/V^2 p, \tag{4.71}$$

where $M = m + \mu$ is the sum of the masses of the star and of the particle and p the passage distance. The length of the relative-velocity increment vector is $2V\sin\frac{1}{2}\psi$, and this vector is at an angle $90° + \frac{1}{2}\psi$ to $\mathbf{V}$. Hence, if the x-axis is antiparallel to $\mathbf{V}$, and the y-axis along the perpendicular from the star to the asymptote of the branch of the hyperbola along which the particle is approached, then the components of the change in the absolute velocity of the particle along these axes are (Fig. 22)

$$(2mV/M)\sin^2\tfrac{1}{2}\psi, \quad -(2mV/M)\sin\tfrac{1}{2}\psi\cos\tfrac{1}{2}\psi. \tag{4.72}$$

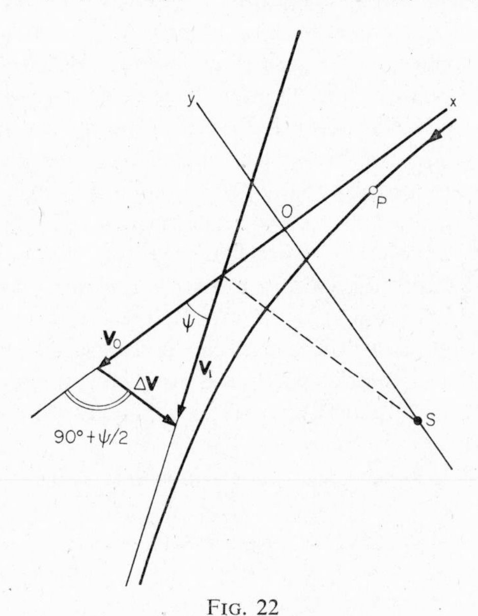

FIG. 22

Eliminating ψ from (4.72) by means of (4.71) and multiplying by μ, we obtain the components of the momentum increment of the particle:

$$2m\mu V/M(1+\varkappa^2), \quad -2m\mu V\varkappa/M(1+\varkappa^2), \tag{4.73}$$

where

$$\varkappa = V^2 p/GM. \tag{4.74}$$

The total momentum of the star and particle must remain constant in the absence of external forces, and therefore the components of the change in the momentum of the star are equal to (4.73) with the signs reversed.

Let us now take into account the radiation pressure of the star, assuming that the resulting force, like that of gravitation, varies inversely as the square of the distance between the star and the particle. We denote by q the dimensionless quantity

$$q = (F - S)/F, \tag{4.75}$$

where F is the gravitational force, and S the force of radiation pressure, exerted by the star on the particle. If there is no radiation pressure, $q = 1$; if the force of radiation pressure is equal to the gravitational force, $q = 0$. When the radiation pressure increases further, q becomes negative and can take large absolute values. The value of q depends on the mass and luminosity of the star and on the size, shape, surface reflectivity and density of the dust particles. If the latter are spherical and homogeneous, q is independent of the position of the particle relative to the star.

We shall first suppose that the force of radiation pressure does not exceed that of gravitation. Then $0 < q < 1$, and the star attracts the particle with a force $F - S = qF$. Hence we need only replace the mass of the star in (4.73) by qm in order to obtain the components of the momentum change of the particle with allowance for radiation pressure. The system of star and cloud can no longer be regarded as closed, however. The total momentum vector of the system changes because the radiation of the star is screened by the dust particles and its intensity is not the same in all directions. The excess radiation in the direction opposite to that of the particle causes a reaction force in the direction of the shadow of the particle. Hence the components of the momentum change of the star can no longer be taken as equal in magnitude to (4.73). They can, however, be easily determined as follows. The particle is subject to both the gravitational force and the radiation pressure from the star, but the star is subject only to the gravitation of the particle (whose radiation pressure on the star is negligible). The quantity q defined by eqn. (4.75) therefore gives the ratio of the forces exerted by the star on the particle and by the particle on the star. Hence it follows that the momentum given to the star by the particle is equal to minus the momentum given to the particle by the star, divided by q. Thus, replacing m by qm in (4.73), dividing by q and changing the signs, we have

$$\begin{aligned} m\Delta V_x &= -2m\mu V/M(1+\varkappa^2), \\ m\Delta V_y &= 2m\mu V\varkappa/M(1+\varkappa^2), \end{aligned} \tag{4.76}$$

where

$$M = \mu + qm. \tag{4.77}$$

Let us now assume that the force of radiation pressure exerted by the star on the particle exceeds the force of gravitation. Then the star will repel the particle with a force $S - F$. In this case also, the star and the particle will move in hyperbolic paths, but each will be at the external, not the internal, focus of the path of the other (Fig. 23).

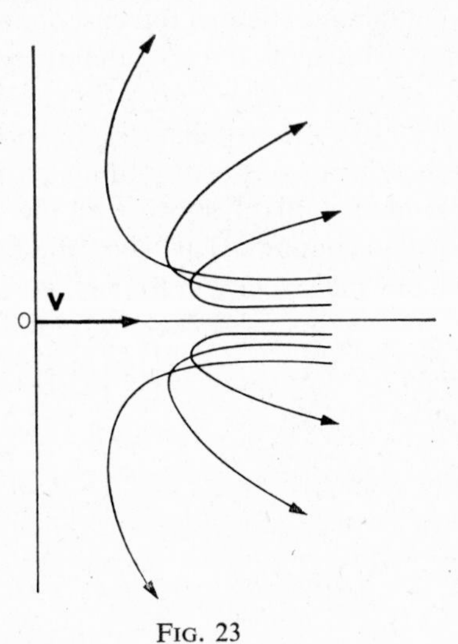

FIG. 23

Nevertheless, in this case also there is an area integral (since the force is central) and an energy integral. By the same arguments as in § 3, therefore, we again obtain (4.36) and the formulae of the present section. Thus formulae (4.76) give the components of the momentum change of the star even when the radiation pressure of the star on the dust particles exceeds the gravitational force.

Formulae (4.76) show that the signs of the components of the star velocity increment depend on the sign of M. If $M < 0$, i.e. $q < -\mu/m$, then ΔV_x is positive, and so has the same sign as V_x. The result of the encounter with the dust particle is therefore to accelerate the star. The sign of ΔV_y is unimportant: since $V_y = 0$, ΔV_y increases the velocity of star whichever sign it has.

Let us determine ΔV, the increment in the velocity of the star as a result of the encounter. Since the vector $\mathbf{V}$ is along Ox, we have

$$\Delta V = \sqrt{[(V + \Delta V_x)^2 + (\Delta V_y)^2]} - V. \qquad (4.78)$$

Substituting in (4.78) ΔV_x and ΔV_y from (4.76), and using the notation (4.77), we find

$$\Delta V = V\left\{\sqrt{\left[1 - \frac{4(\mu/M)(1 - \mu/M)}{1 + \varkappa^2}\right]} - 1\right\}.$$

Since μ/M is small, we can expand the square root in series and keep only the terms of the first and second orders. Then, with (4.77), we have

$$\Delta V = V\left[-\frac{2\mu q m}{M^2(1 + \varkappa^2)} - \frac{\mu^2}{2M^2(1 + \varkappa^2)^2}\right]. \qquad (4.79)$$

On passing through a region filled with dust, a star undergoes accelerations from many particles. If these particles are assumed to be initially at rest and of equal mass, the velocity increment (4.79) which each particle gives to the star depends only on the passage distance p.

Let us determine the cumulative effect of the dust on the velocity of the star, as was done in § 4 for encounters with stars. Let ν be the number of dust particles per unit volume. In a time t, the star undergoes $2\pi V t p\, dp$ encounters at a passage distance between p and $p + dp$. Unlike the cumulative effect for star encounters described in § 4, we here sum the increments of the magnitude of the velocity vector, not those of the vector itself. In calculating this effect we must therefore take the algebraic sum of the results of individual encounters. This gives the following formula for the increment in the magnitude of the velocity in time t:

$$\delta V = \frac{dV}{dt} t = 2\pi \nu V t \int_0^{p_1} \Delta V p\, dp, \qquad (4.80)$$

where p_1 is the maximum passage distance and dV/dt the acceleration.

Substituting (4.79) in (4.80) and integrating, we obtain

$$\frac{dV}{dt} = -\frac{2\pi G^2 \delta m q}{V^2} \log(1 + \varkappa_1^2) - \frac{2\pi G^2 \delta \mu}{V^2} \frac{\varkappa_1^2}{1 + \varkappa_1^2}, \qquad (4.81)$$

where $\delta = \nu\mu$ is the mean density of diffuse matter, and

$$\varkappa_1 = V^2 p_1/GM. \tag{4.82}$$

It is known that the dust is not uniformly distributed, but forms separate clouds of varying shape and density. Two limiting cases may be taken: (1) all the dust is in separate small clouds of high density, (2) the dust is uniformly distributed in a given region of space.

Let us first neglect the finite extent of the dust clouds and imagine them replaced by very dense points, each of mass μ equal to the mass of a cloud. We shall suppose, however, that they remain subject to radiation pressure. Equations (4.79) and (4.81) remain valid when $\mu/M \ll 1$.

The second term on the right-hand side of (4.81) contains not only a factor δ but also a factor μ, which in the present discussion is the mass of one cloud. This term therefore depends on the mass of each dust cloud; it is always negative. Since it arises from the discreteness of the medium with which the star interacts, it may be called the (negative) *statistical acceleration* of the star.

The first term on the right in (4.81) does not depend on the mass of a cloud, but does depend on the mean density, which is a characteristic of a continuous medium. If there is no radiation pressure ($q = 1$), this term causes a negative acceleration. In that case the expression (4.81) is negative, as we should expect, since we have assumed the dust clouds to be at rest and the star to be moving: the interaction and exchange of energy in the closed system gives the clouds some velocity, and the star is retarded. The first term in (4.81) may therefore be called the *acceleration by dynamical friction.*

When the force of radiation pressure is large, and therefore $q < 0$, the first term in (4.81) is positive, and the "dynamical friction" changes sign, the star thereby being accelerated. This term is greater than the second term even for very small magnitudes of q, and the total acceleration of the star is then positive. The dust clouds, which before the encounter were at rest, also begin to move. The increase in the kinetic energy of the star and clouds together is derived from the energy of the star's radiation.

Let us now consider the second limiting case mentioned above, neglecting the cloud formation of dark matter and assuming that the dust particles are uniformly distributed in space. Then, on account of the negligible mass of the particles, the second term in (4.81) is practically zero. This is as we should expect, since in this case it may be assumed that the medium is continuous, and so there is no statistical acceleration.

In reality we have a case intermediate between the two limiting cases considered: the dust forms discrete clouds of considerable extent. Hence the second term in (4.81) should be multiplied by a coefficient $a < 1$. This gives

$$\frac{\mathrm{d}V}{\mathrm{d}t} = -\frac{2\pi G^2 \delta m q}{V^2}\log(1 + \varkappa_1^2) - a\frac{2\pi G^2 \delta\mu}{V^2}\frac{\varkappa_1^2}{1 + \varkappa_1^2}, \tag{4.83}$$

where the quantity a, which characterises the degree of discreteness of the medium, is such that $0 < a < 1$.

The necessity of multiplying the statistical acceleration by a coefficient less than unity shows that, even when μ/M is comparable with unity but the clouds are large, the acceleration by dynamical friction is the principal effect and the expansion (4.79) in powers of μ/M is valid.

Let us now consider what the value of q may be. Since the gravitational force F is proportional to the mass m of the star, while the force of radiation pressure is proportional to the luminosity L of the star, formula (4.75) can be written $q = (m - \alpha L)/m$, where α is a proportionality factor depending on the properties of the dust particles. This factor may be determined from observational data.

It has been found that the absorption of light in the interstellar medium is due mainly to particles of radius $\sim 10^{-5}$ cm. If we assume that the density of these particles is 3, then a calculation shows that the force of radiation pressure from the Sun on such particles is approximately equal to the gravitational force, i.e. $q = 0$. Hence, applying the formula for q just given, we have for the Sun $\alpha = m_{\odot}/L_{\odot}$. If the mass and luminosity of the Sun are taken as unity, then $\alpha = 1$. Hence, assuming that particles of radius about 10^{-5} cm form the main mass of the dust clouds, we can write $q = (m - L)/m$, where m and L are the mass and luminosity of the star, expressed in terms of the mass and luminosity of the Sun. For main-sequence stars, the values of q for various spectral classes are shown in Table IV.

This table is based on mean luminosities of the spectral classes as given by P. P. Parenago [107] and masses from the mass–luminosity diagram given by the same author [96].

From these values of q we may calculate the two components of acceleration on the right-hand side of (4.81) for stars of classes O, B0, B4 and B8. It must be borne in mind that, the greater the luminosity of the star, the greater the radius of the sphere over which it exerts an effect. The value of p_1 for stars of these classes may be taken as 100, 50, 20 and 10 ps respectively, and their masses as 25, 20, 10 and 6 times $m_{\odot}$

TABLE IV

Spectrum	q	Spectrum	q
O	−7000	F0	−2
B0	−2000	F5	−0·6
B2	−600	G0	−0·3
B4	−200	G5	+0·1
B6	−100	K0	+0·5
B8	−50	K5	+0·7
A0	−25	M0	+0·85
A5	−6	M5	+0·98

respectively. Numerous large diffuse clouds are observed in association with concentrations of hot giant stars. We therefore take $\delta = 10^{-23}$ g/cm³ and $\mu = 50\, m_{\odot}$. In each case we have $|\mu/M| = |\mu/(\mu + qm)| \ll 1$, so that formula (4.81) is sufficiently exact. Finally, on account of the accepted cosmogonical fact that the hot giants have a close genetic relationship to the large clouds, we shall assume that their relative velocity is small, putting $V = 2 \times 10^5$ cm/s.

The calculated values of the acceleration by dynamical friction for the classes considered are shown in the second column in Table V.

TABLE V

Spectrum	dV/dt	μ/M	T_0 (y)	$(dV/dt)_{max}$
O	6×10^{-10}	$-1/1750$	1×10^7	1.1×10^{-7}
B0	5×10^{-10}	$-1/400$	1.4×10^7	2.3×10^{-8}
B4	1.3×10^{-10}	$-1/19$	5×10^7	2.3×10^{-9}
B8	3×10^{-11}	$-1/2$	14×10^7	8.3×10^{-10}

The second term, i.e. the statistical acceleration, must be about μ/M times the first term. The third column in Table V shows that this factor is very small for all the classes considered except B8, so that the statistical acceleration can be of importance only in this class.

We can use these accelerations and the formula $\delta V = t\,dV/dt$ to find the time T_0 in which the assumed initial velocity of 2 km/s is doubled. This time in years is shown in the fourth column in Table V. Its value for O and B stars is seen to be of the order of 10^7 to 10^8 y. Thus the relaxation time derived from the interaction of hot stars with diffuse matter confirms the results obtained from the interaction of stars with condensations of stars (see (4.69)). Of course, the calculation of the values of T_0 in Table V cannot be regarded as rigorous, since the effects of diffuse matter and star condensations on the star ought to be taken into account simultaneously, but the order of magnitude must be correct.

Let us find now a theoretical upper limit to the acceleration of stars of various spectral classes, derived from the energy of their radiation as a result of their interaction with dark matter. To do this, we may assume that the star is at rest at the centre of a hemisphere (Fig. 24) which absorbs the radiation of the star and has a mass such that the force of radiation pressure on the hemisphere is just equal to the force of gravitation. Then the relative position of the star and the hemisphere will remain unchanged, but on account of the free emergence of the energy of radiation in one direction the whole system will move in the opposite direction with acceleration

$$\left(\frac{dV}{dt}\right)_{max} = \frac{E}{c(m+\mu)} \cdot \frac{1}{4\pi} \int_0^{\frac{1}{2}\pi} \int_0^{2\pi} \cos\theta \sin\theta \, d\theta \, d\varphi \approx E/4cm, \qquad (4.84)$$

where E is the rate of emission of energy by the star, c the velocity of light, m the mass of the star and μ that of the hemisphere. A straight-forward calculation shows that μ is negligible in comparison with m, and we shall neglect it.

From a comparison of the bolometric luminosities of the Sun and the stars the energies emitted as radiation by the stars concerned are found, and thence the quantities $(dV/dt)_{max}$. The results are shown in the last column of Table V.

Finally, we may note the assumptions which have been made in deriving the fundamental formula (4.81). The first and principal assumption is that the force of radiation pressure on the particle is taken to vary inversely as the square of the distance. This is true only if we may neglect the absorption of light by the dust, or at least by that part of it which is mainly responsible for the acceleration of the star. The available data concerning dust clouds indicate that this assumption is valid, and it is confirmed also by a comparison of the second and fifth columns of Table V. For, if δ in formula (4.83) is increased, the acceleration increases and ultimately exceeds the value given by (4.84).

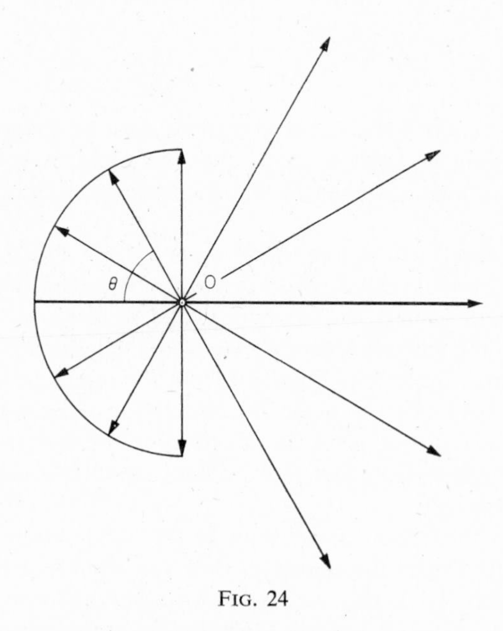

FIG. 24

This impossible result indicates that the density is then so high that the absorption of light is considerable and can not be neglected, so that (4.83) is invalid. The quantities in the second column of Table V, however, are about two orders of magnitude less than those in the last column, and this shows that the absorption of light can still be neglected at the assumed value of the density.

The second assumption is that mutual collisions between dust particles (or, where dust clouds are considered as individual objects, collisions between clouds) may be neglected. The validity of this is confirmed by calculation, which shows that the mean free path of a dust particle is very large. Moreover, a consideration of the intersections of paths of particles demonstrates that collisions between particles cannot change the property of their motion which is of importance in the present discussion, namely that they do not enter the region near the direction of motion behind the star, but fill the region in front of the star, attract it, and so accelerate its motion.

§ 9. The Invariance of Maxwell's and Schwarzschild's Velocity Distribution Laws in Stellar Encounters

An important problem in the dynamics of stellar systems is that of the invariance or otherwise of the stellar velocity distribution as a result of the action of the irregular forces. If these forces do not affect the velocity distribution law, we shall say that it is invariant with respect to irregular forces.

The problem really concerns the invariance of the phase distribution. Let us divide the whole phase volume of a stellar system into a large number of phase cells, $d\gamma = d\omega\, d\Omega$, where $d\omega$ and $d\Omega$ are the macroscopic volume elements in coordinate space and velocity space respectively. Then the phase points will move discontinuously from one phase cell to another as a result of stellar encounters. If the phase distribution is invariant, these transitions will occur in such a way that the number of phase points in each cell remains constant. However, an investigation of the general case of transitions in phase space offers considerable practical difficulties of computation. Hence, to simplify the problem, it is usually assumed that the encounter radius is small compared with the dimension of the macroscopic volume element $d\omega$. In other words, it is assumed that, during an encounter, the coordinates of the stars remain almost unchanged, only their velocities being changed. In this case the problem of the invariance of the phase density reduces to that of the invariance of the velocity distribution law. For this reason the title of the present section has been framed in such terms. It is evident that not every law of velocity distribution has the property of invariance. By analogy with statistical physics we might propose the problem of finding all the distribution laws which do have this property, but that would take us too far into the mathematical complications involved in the solution of the resulting functional equation, and moreover it would be foreign to our method of exposition, which has always in view a particular type of stellar system, to wit the galaxy. Thus only those velocity distribution laws which can occur in actual galaxies should be investigated for invariance. We shall see in Chapter X that there are in practice only two such distribution laws. They are Maxwell's law in the rigid rotation of a stellar system and Schwarzschild's law in quasi-rigid rotation. They will be discussed below.

(a) *Maxwell's Law in a Stellar System at Rest*

As a first step, let us consider Maxwell's law in a stellar system at rest (i.e. not rotating). We shall consider the following simplified model: the system is assumed to consist of two types of star, those of one type having equal masses m_1 and those of the other type having equal masses m_2. We shall take account only of binary (not multiple) encounters involving one star of each type. If we afterwards put $m_1 = m_2$ we obtain a stellar system with a uniform population, but if we put $m_1 \ll m_2$ we have a star field, consisting of individual stars, containing a system of large condensations of mass. This type of problem corresponds to the discussion in Chapter X. The encounters between stars and those between condensations are neglected: the former because of the small masses of the stars and the low density of stars, the latter because of the low residual velocities of the condensations.

Finally, we shall make two further assumptions.

(1) In each macroscopic volume element $d\omega$, the stellar velocity distribution for each type satisfies Maxwell's law, which we write in the form

$$f = \exp(\varkappa_i + \lambda m_i H), \tag{4.85}$$

where m_i is m_1 or m_2, $H = \frac{1}{2}v^2 - U$ is the total energy of a particle of unit mass moving in the field of the regular force, v the velocity of the star, U the potential of the regular force field, and $\varkappa_i$ and λ constants, of which the latter is the same for each type but the former, in general, is not. It is easily seen that the equality of λ for the two types corresponds to equipartition of energy (i.e. equal temperatures of the two "stellar gases"). When this condition holds, the stars of greater mass have the smaller velocities.

The values of $\varkappa_1$ and $\varkappa_2$ are related to the partial densities ν_1 and ν_2 of stars of the two types by relations which follow at once from the usual form of Maxwell's function:

$$\exp \varkappa_1 = \nu_1(-\lambda/\pi)^{3/2} \quad \text{and} \quad \exp \varkappa_2 = \nu_2(-\lambda/\pi)^{3/2},$$

or $\exp(\varkappa_2 - \varkappa_1) = \nu_2/\nu_1$, whence we see *inter alia* that λ must be negative.

(2) We shall consider only binary encounters. When $m_1 = m_2$, this is equivalent to assuming that the star density is small. When $m_1 \ll m_2$, however, it corresponds to the assumption that the mass m_1 is so small that the interaction of stars of the first type is negligible.

Let us therefore consider a binary encounter which converts the initial velocities $\mathbf{v}_1$ and $\mathbf{v}_2$ of the stars m_1 and m_2 into final velocities $\mathbf{v}_3$ and $\mathbf{v}_4$ respectively.† Let $d\Omega_1$, $d\Omega_2$ be macroscopic volume elements in velocity space surrounding the termini of the initial velocity vectors $\mathbf{v}_1$ and $\mathbf{v}_2$, and $d\Omega_3$, $d\Omega_4$ similar elements for $\mathbf{v}_3$ and $\mathbf{v}_4$. To each of these four volumes there correspond phase cell volumes $d\gamma_i = d\omega\, d\Omega_i$ ($i = 1, 2, 3, 4$), where $d\omega$ is a volume element in coordinate space. As a result of the encounter stars from the phase cells $d\gamma_1$ and $d\gamma_2$ are transferred to the cells $d\gamma_3$ and $d\gamma_4$. Let us calculate the number dn_{12} of such stars. In one macroscopic volume element $d\omega$ we can regard the regular force as constant, and so can consider the two stars involved as forming a closed mechanical system. Then their relative motion is along a hyperbola.

Let p be the *passage distance* of the encounter, that is, the length of the perpendicular from one star to the direction of the inital relative velocity of the other (i.e. to the initial asymptote to the relative orbit), and φ the *azimuth* of the encounter, that is, the dihedral angle between the plane of the relative orbit and some fixed plane passing through the initial relative velocity vector $\mathbf{v}_{12}$. Then

$$ds_{12} = p\, dp\, d\varphi \tag{4.86}$$

is called the *differential cross-section* for the encounter.

† See, for instance, E. H. Kennard, *Kinetic theory of gases*, McGraw-Hill, New York, 1938.

We can now find $\mathrm{d}n_{12}$, which is easily seen to be

$$\mathrm{d}n_{12} = (\mathrm{d}\omega)^2 f_1 f_2 v_{12}\, \mathrm{d}s_{12}\, \mathrm{d}\theta\, \mathrm{d}\psi, \tag{4.87}$$

where θ and ψ are some pair of angles defining the direction of the velocity $\mathbf{v}_{12}$, and f_1, f_2 denote for brevity the value of the Maxwellian for velocities $\mathbf{v}_1$ and $\mathbf{v}_2$ respectively. The initial velocities $\mathbf{v}_1$ and $\mathbf{v}_2$ and the cross-section $\mathrm{d}s_{12}$ entirely determine the result of the encounter. Thus, if the initial phase volumes $\mathrm{d}\gamma_1$ and $\mathrm{d}\gamma_2$ are given, the final phase volumes $\mathrm{d}\gamma_3$ and $\mathrm{d}\gamma_4$ are determined. Since the factor $\mathrm{d}\omega$ is assumed unchanged, we can say also that, if the volumes $\mathrm{d}\Omega_1$ and $\mathrm{d}\Omega_2$ are given, and if $\mathbf{v}_1$ and $\mathbf{v}_2$ and $\mathrm{d}s_{12}$ are known, then the volumes $\mathrm{d}\Omega_3$ and $\mathrm{d}\Omega_4$ are determined.

These four volumes are related by the equation

$$\mathrm{d}\Omega_1\, \mathrm{d}\Omega_2 = \mathrm{d}\Omega_3\, \mathrm{d}\Omega_4. \tag{4.88}$$

For the velocities $\mathbf{v}_1$ and $\mathbf{v}_2$ can be expressed in terms of the velocity $\mathbf{v}_0$ of the centre of mass of the stars m_1 and m_2 and their initial relative velocity $\mathbf{v}_{12}$. These velocities are in one-to-one relation with $\mathbf{v}_1$ and $\mathbf{v}_2$: $\mathbf{v}_1 = \mathbf{v}_0 - \mathbf{v}_{12}$, $\mathbf{v}_2 = \mathbf{v}_0 + \mathbf{v}_{12}$. We therefore have identically

$$\mathrm{d}\Omega_1\, \mathrm{d}\Omega_2 \equiv \mathrm{d}\Omega_0\, \mathrm{d}\Omega_{12}, \tag{4.89}$$

where $\mathrm{d}\Omega_0$ and $\mathrm{d}\Omega_{12}$ denote macroscopic volume elements surrounding the termini of the vectors $\mathbf{v}_0$ and $\mathbf{v}_{12}$ respectively. The eqn. (4.89) follows because each side is a measure of the same manifold in velocity space, the only difference being the parameters used to express it.

As a result of the encounter, $\mathbf{v}_0$ remains unchanged and $\mathbf{v}_{12}$ changes sign. After the encounter, therefore, we can write by analogy with (4.89)

$$\mathrm{d}\Omega_3\, \mathrm{d}\Omega_4 \equiv \mathrm{d}\Omega_0\, \mathrm{d}\Omega_{12}. \tag{4.90}$$

A comparison of the left-hand sides of (4.89) and (4.90) shows that (4.88) holds.

The encounters considered transfer stars from phase cells $\mathrm{d}\gamma_1$ and $\mathrm{d}\gamma_2$ to cells $\mathrm{d}\gamma_3$ and $\mathrm{d}\gamma_4$. We shall show that the resulting loss of stars from the two former cells is made up as a result of *inverse* encounters. These are encounters which transfer stars from the cells $\mathrm{d}\gamma_3$ and $\mathrm{d}\gamma_4$ to $\mathrm{d}\gamma_1$ and $\mathrm{d}\gamma_2$. This corresponds to *detailed balancing* in the phase distribution: it is in general possible to imagine a process of compensation of the loss of stars from cells in which the supply of stars to a given cell comes not from inverse encounters but from encounters which take place in some third pair of cells.

We shall first show that inverse encounters are mechanically possible. To do so, we use two properties of motion which are well known in theoretical mechanics. The first of these is the *reversibility of motion*, and follows from the fact that the equation of motion of a particle,

$$\mathrm{d}^2\mathbf{r}/\mathrm{d}t^2 = \mathbf{F}, \tag{4.91}$$

is unchanged if t is replaced by $-t$. Hence, to every solution of (4.91)

$$\mathbf{r} = \mathbf{r}(t) \tag{4.92}$$

which carries a particle from A to B along a given path, there corresponds another solution

$$\mathbf{r}_1 = \mathbf{r}_1(t) \tag{4.93}$$

which carries it along the same path from B to A, and the velocities at any point on the path are equal and opposite: $d\mathbf{r}_1/dt = -d\mathbf{r}/dt$.

The second property is the *symmetry of motion;* it follows from the fact that, if the signs of $\mathbf{r}$ and $\mathbf{F}$ in (4.91) are simultaneously changed, the equation is unaltered. In the case of a central force (as in the present problem, since in the encounter we allow only for the mutual attraction of two stars), we can therefore consider a possible motion which is symmetrical to the original one. At any instant the position of a particle on the symmetrical path is at the same distance in the opposite direction from the centre. We shall express this by saying that the symmetrical motion carries the particle from $-A$ to $-B$.

If we apply the symmetry principle to the reverse motion from B to A, we obtain a motion which carries the particle from $-B$ to $-A$, and the initial and final velocities of this motion are respectively the final and initial velocities of the original motion.

These relations are shown diagrammatically in Fig. 25. O is the centre of mass of the two stars, and we assume it to be at rest. The points A and B denote the initial and final positions of one star; the arrows denote the initial and final velocities. The points $-A$ and $-B$ denote the initial and final positions of the star in the symmetrical motion. Finally, the path $-B$ to $-A$ is that of the reversed symmetrical motion.

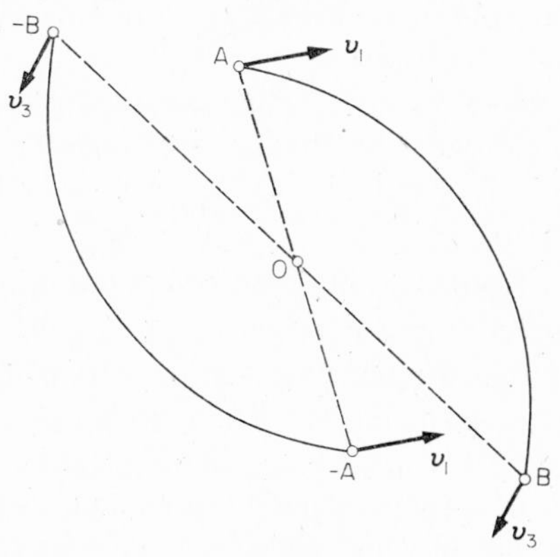

FIG. 25

We see that the initial velocity is in fact the final velocity of the original motion from A to B, and conversely.

It is evident that the reversed symmetrical motion takes the star from the phase cell $d\gamma_3$ back to $d\gamma_1$, i.e. it is the inverse motion. This is because we neglect the change in coordinates of the stars, and so transitions between phase cells are entirely determined by transitions between macroscopic volume elements in velocity space. Thus inverse encounters are possible, since it is evident that, if we consider the reversed symmetrical motions for both the stars, we obtain the inverse motion.

It remains to show that the probabilities of direct and inverse encounters are equal, so that the two types balance. It is clear that the number of inverse encounters is

$$dn_{34} = (d\omega)^2 f_3 f_4 v_{12}\, ds_{12}\, d\Omega_3\, d\Omega_4\, d\theta\, d\psi, \tag{4.94}$$

which is obtained by the same arguments as gave dn_{12} in the form (4.87). Here f_3 and f_4 denote the results of substituting the velocities v_3 and v_4 in the Maxwellian (4.85).

A comparison of (4.94) and (4.87) shows that the difference $dn_{12} - dn_{34}$ is proportional to

$$f_1 f_2 - f_3 f_4 = \exp[(\varkappa_1 + \varkappa_2) + \lambda(m_1 H_1 + m_2 H_2)] - \exp[(\varkappa_1 + \varkappa_2) + \lambda(m_1 H_3 + m_2 H_4)]. \tag{4.95}$$

It is easy to see that $m_1 H_1 + m_2 H_2$ and $m_1 H_3 + m_2 H_4$ are equal to the total energy of the system of two stars. Since this system is regarded as closed during the encounter, its energy remains unchanged, and so

$$m_1 H_1 + m_2 H_2 = m_1 H_3 + m_2 H_4, \tag{4.96}$$

whence it follows that $dn_{12} - dn_{34} = 0$, and therefore the direct and inverse encounters balance.

(b) *Maxwell's Law in Rigid Rotation of a Stellar System*

We shall now show that the detailed arguments just given for the simple case of a non-rotating stellar system can easily be extended to the case of Maxwell's law in a rigidly rotating system. We use ordinary cylindrical coordinates in space (Fig. 26): ϱ is the cylindrical radius vector, θ the azimuthal angle and z the altitude.

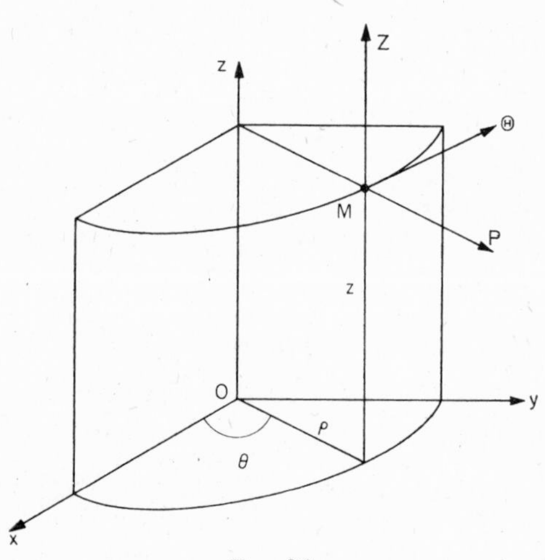

FIG. 26

For the linear velocity components we use Jeans' notation, which is widely employed in stellar dynamics: P is the component in the ϱ-direction, Θ the component perpendicular to the ϱz-plane and Z the component along the z-axis.

Let the z-axis be the axis of rotation, and let the system have rotational symmetry. Then the Maxwellian becomes

$$\exp(\varkappa_i + \lambda m_i H + \mu m_i K), \tag{4.97}$$

where $H = \frac{1}{2}(v^2 - 2U)$ is again the energy of a particle of unit mass, and $K = \varrho\Theta$ is the angular momentum of such a particle about the z-axis.

It is easy to see that the expression (4.97) corresponds to rigid rotation of the system, since the last two terms in the exponent may be transformed as follows:

$$\lambda m_i(\mathrm{P}^2 + \Theta^2 + Z^2 - 2U) + \mu m_i \varrho\Theta = \lambda m_i[\mathrm{P}^2 + (\Theta - \Theta_0)^2 + Z - 2U_1], \tag{4.98}$$

where

$$\Theta_0 = -\mu\varrho/2\lambda \quad \text{and} \quad U_1 = U + \mu^2\varrho^2/4\lambda^2. \tag{4.99}$$

Here Θ_0 is the centroid velocity component in the direction of Θ. Since this direction is perpendicular to the meridian plane at the point considered, and the other two components (in the directions of P and Z) are evidently zero, we have a rotation of the system about the z-axis. Finally, we see that the rotational (circular) velocity of the centroid is proportional to ϱ, i.e. it can be written in the form

$$\Theta_0 = \omega_0\varrho, \tag{4.100}$$

where

$$\omega_0 = -\mu/2\lambda = \text{constant} \tag{4.101}$$

is the angular velocity of the rigid rotation. The quantity U_1 is easily seen to be the gravity potential, i.e. the potential of the resultant of the force of attraction and the centrifugal force:

$$U_1 = U + \tfrac{1}{2}\omega_0^2\varrho^2. \tag{4.102}$$

Instead of (4.95) we have in this case

$$\begin{aligned} f_1 f_2 - f_3 f_4 = {} & \exp[(\varkappa_1 + \varkappa_2) + \lambda(m_1 H_1 + m_2 H_2) + \mu(m_1 K_1 + m_2 K_2)] \\ & - \exp[(\varkappa_1 + \varkappa_2) + \lambda(m_1 H_3 + m_2 H_4) + \mu(m_1 K_3 + m_2 K_4)]. \end{aligned} \tag{4.103}$$

The invariance of Maxwell's law for this case follows from the facts that $m_1 H_1 + m_2 H_2$ is the total energy, i.e. an integral of the motion for a closed system, and $m_1 K_1 + m_2 K_2$ is the angular momentum, which is also an integral of motion and is unaltered by the encounter.

(c) *Schwarzschild's Law in Quasi-rigid Rotation of a Stellar System*

Using the same notation as in (b), we may now consider a stellar system in which Schwarzschild's law holds; we write this law in the form

$$f = \exp[\varkappa_i + \lambda m_i H + \mu m_i K + \nu m_i K^2] \tag{4.104}$$

or

$$f = \exp\{\varkappa_i + \lambda m_i[\mathrm{P}^2 + (1 + \nu\varrho^2/\lambda)(\Theta - \Theta_0)^2 + Z^2 - 2U_1]\}, \tag{4.105}$$

where

$$\Theta_0 = -\mu\varrho/2\lambda(1 + \nu\varrho^2/\lambda)$$

is again the circular velocity of the centroid, but is here not the velocity of rigid rotation. For small ϱ, however, we can put approximately

$$\Theta_0 = \omega_0\varrho, \tag{4.106}$$

where, as before,

$$\omega_0 = -\mu/2\lambda = \text{constant}. \tag{4.107}$$

Thus the rotation is in this case only quasi-rigid, that is, the movement of the inner part of the system is approximately a rigid rotation. U_1 is again the gravity potential.

To discuss the invariance of (4.104) we form the expression

$$f_1 f_2 = \exp[(\varkappa_1 + \varkappa_2) + \lambda(m_1 H_1 + m_2 H_2) + \mu(m_1 K_1 + m_2 K_2) + \nu(m_1 K_1^2 + m_2 K_2^2)]. \tag{4.108}$$

The first three terms in the exponent are unchanged by the encounter. This has been shown in (a) and (b). The last term (in which ν is a constant) can be transformed by putting

$$m_1 K_1^2 + m_2 K_2^2 = \varrho^2(m_1\Theta_1^2 + m_2\Theta_2^2) = \varrho^2[(m_1 + m_2)\Theta_0^2 + (m_1\Theta_1'^2 + m_2\Theta_2'^2)], \tag{4.109}$$

where Θ_0 is the velocity of the centre of mass of the two stars, and Θ_1', Θ_2' their initial velocities relative to the centre of mass.

If the residual velocities of the stars in a system are small in comparison with the circular velocity, then the absolute velocity Θ_0 of the centre of mass will be of the same order as the centroid velocity, while Θ_1' and Θ_2' will be of the order of the residual velocities. The smaller the ratio of the residual-velocity variance σ_θ^2 to the squared circular velocity of the centroid,

$$p = \sigma_\theta^2/\Theta_0^2, \tag{4.110}$$

the smaller will be the error committed by retaining only the first term on the right of (4.109). If we neglect the term $(m_1\Theta_1'^2 + m_2\Theta_2'^2)/(m_1 + m_2)$ in comparison with Θ_0^2, we obtain approximately

$$m_1 K_1^2 + m_2 K_2^2 = (m_1 + m_2)\varrho^2\Theta_0^2; \tag{4.111}$$

this quantity is unaltered by the encounter.

Thus we do not have exact invariance of Schwarzschild's law in quasi-rigid rotation. The invariance is only approximate, but the approximation is the closer, the more rapid the rotation of the stellar system and the smaller the residual velocities in it. Hence we shall say that in this case there is *quasi-invariance*. For example, in the Galaxy the residual velocities of stars in plane sub-systems are of the order of 20 km/s, while the circular velocity of the centroid is 250 km/s. Thus the quantity p in (4.110) is 1/150. For stars of the spherical sub-systems, the residual velocities are of the order of 100 km/s, and $p = 1/6$, so that the invariance condition does not hold. We shall see in Chapter X that this happens because stellar systems such as the Galaxy are not in equilibrium, but the plane component of it has an equilibrium figure, differing only slightly from a flat disc. In the spherical component, considerable deviations from Schwarzschild's law should be observed.

CHAPTER V

STATISTICAL STELLAR DYNAMICS NEGLECTING ENCOUNTERS

§ 1. The Fundamental Differential Equation of Stellar Dynamics

IT HAS been shown in Chapter II that the state of any stellar system is entirely determined by specifying the *phase density*, i.e. the function f of the coordinates x, y, z, the velocities u, v, w and the time t, such that the product

$$\mathrm{d}N = f(q, p, t)\,\mathrm{d}\omega\,\mathrm{d}\Omega = f(x, y, z, u, v, w, t)\,\mathrm{d}\omega\,\mathrm{d}\Omega \tag{5.1}$$

gives the number of stars for which the coordinates lie within a rectangular parallelepiped with centre (x, y, z) and edges $\mathrm{d}x$, $\mathrm{d}y$, $\mathrm{d}z$ parallel to the coordinate axes, and the termini of the velocity vectors lie within a parallelepiped with centre (u, v, w) and edges $\mathrm{d}u$, $\mathrm{d}v$, $\mathrm{d}w$ parallel to the coordinate axes. The volume elements are $\mathrm{d}\omega = \mathrm{d}x\,\mathrm{d}y\,\mathrm{d}z$ and $\mathrm{d}\Omega = \mathrm{d}u\,\mathrm{d}v\,\mathrm{d}w$. Formula (5.1) may be more concisely written by using six-dimensional phase space. Denoting by x_k $(k = 1, 2, \ldots, 6)$ the phase coordinates x, y, z, u, v, w and by $\mathrm{d}\gamma$ the phase volume $\mathrm{d}\omega\,\mathrm{d}\Omega$, we can put $\mathrm{d}N = f(x_k)\,\mathrm{d}\gamma$. We may now derive a partial differential equation which must be satisfied by the phase density in any stellar system.

Let us consider a six-dimensional parallelepiped in phase space with edges parallel to the coordinate axes and volume $\mathrm{d}\gamma = \mathrm{d}x_1\,\mathrm{d}x_2\,\mathrm{d}x_3 \ldots \mathrm{d}x_6$. The equations of motion of stars will be

$$\begin{aligned} \dot{x}_1 &= x_4, & \dot{x}_2 &= x_5, & \dot{x}_3 &= x_6, \\ \dot{x}_4 &= F_1(x_1, x_2, x_3), & \dot{x}_5 &= F_2(x_1, x_2, x_3), & \dot{x}_6 &= F_3(x_1, x_2, x_3), \end{aligned} \tag{5.2}$$

where F_1, F_2, F_3 denote the components of the regular force per unit mass. As the stars move in the system, their "image" points move in phase space.

Let us find the flux of phase points through a volume element $\mathrm{d}\gamma$, assumed fixed in phase space. The number of points entering the volume in time $\mathrm{d}t$ through the face of the parallelepiped $\mathrm{d}\gamma$ having coordinate x_k (perpendicular to the x_k-axis) is $\dot{x}_k f(x_k)\,(\mathrm{d}\gamma/\mathrm{d}x_k)\,\mathrm{d}t$. The number leaving the volume through the opposite face, with coordinate $x_k + \mathrm{d}x_k$, is $\dot{x}_k f(x_k + \mathrm{d}x_k)\,(\mathrm{d}\gamma/\mathrm{d}x_k)\,\mathrm{d}t$. Consequently, the movement through these two faces decreases the number of points in the volume $\mathrm{d}\gamma$ by $\dot{x}_k[f(x_k + \mathrm{d}x_k) - f(x_k)]\,(\mathrm{d}\gamma/\mathrm{d}x_k)\,\mathrm{d}t$, or, to within terms of higher order, $\dot{x}_k(\partial f/\partial x_k)\,\mathrm{d}\gamma\,\mathrm{d}t$.

ss 5a

Summing over $k = 1, 2, \ldots, 6$, we obtain for the loss of phase points from the volume $\mathrm{d}\gamma$ in time $\mathrm{d}t$ by the action of the regular forces

$$\mathrm{d}\gamma\,\mathrm{d}t \sum_{k=1}^{6} \dot{x}_k \frac{\partial f}{\partial x_k}.$$

In general, however, points are lost also by the action of the irregular forces, i.e. by encounters between stars. Let the corresponding loss of phase points per unit phase volume and unit time be $-(\partial f/\partial t)_i$. Then the loss of points from the volume $\mathrm{d}\gamma$ in time $\mathrm{d}t$ is $-(\partial f/\partial t)_i\,\mathrm{d}\gamma\,\mathrm{d}t$.

Adding these two expressions, we have the total loss of points from the volume $\mathrm{d}\gamma$ in time $\mathrm{d}t$ by the action of both regular and irregular forces:

$$\left[\sum_{k=1}^{6} \dot{x}_k \frac{\partial f}{\partial x_k} - \left(\frac{\partial f}{\partial t}\right)_i\right] \mathrm{d}\gamma\,\mathrm{d}t.$$

This loss can evidently be written also as $-(\partial f/\partial t)\,\mathrm{d}\gamma\,\mathrm{d}t$. Equating the two forms gives, after cancelling,

$$\frac{\partial f}{\partial t} + \sum_k \dot{x}_k \frac{\partial f}{\partial x_k} = \left(\frac{\partial f}{\partial t}\right)_i. \tag{5.3}$$

This is the required fundamental equation of stellar dynamics. It is of great importance in all problems of the local structure of stellar systems. With respect to the phase density it is an equation "with a right-hand side", namely the term depending on the irregular forces. This term is usually very complex. To determine its explicit form we must know the actual manner of interaction of stars in encounters. Hence eqn. (5.3) is in general a functional equation.

In stellar dynamics, however, we usually consider only the particular case of eqn. (5.3) where the right-hand side is zero. It then becomes simply

$$\frac{\partial f}{\partial t} + \sum_k \dot{x}_k \frac{\partial f}{\partial x_k} = 0. \tag{5.4}$$

This is a linear homogeneous partial differential equation of the first order. We shall take (5.4) as the basic equation in the following discussion, and call it the *Boltzmann equation*.

It is easy to see that the right-hand side of eqn. (5.3) is zero in the two following cases:

(a) when the effect of the irregular forces is negligible,

(b) when the phase density is invariant with respect to the irregular forces, i.e. when the number of points leaving any phase volume as a result of encounters is exactly (or, more precisely, with a probability bordering on certainty) balanced by those which enter the volume for the same reason.

Case (a) corresponds to the hitherto usual viewpoint in stellar dynamics, and we shall retain it in the present chapter.

Case (b) will not be discussed in detail here; we merely mention that the two cases are not so greatly different as might at first sight appear. For, in the absence of irregular forces, each star describes a certain regular orbit in the system. At each instant the orbit may be regarded as given if the instantaneous values of the six phase coordinates x_k are known. If the star undergoes an encounter and its phase point moves discontinuously to a different phase volume, there will be another star whose phase point moves in the opposite way if the phase density is invariant. Thus the second star acquires the coordinate values x_k which the first star formerly had, and consequently it moves in the same regular orbit as that left by the first star. In other words, if the phase density is invariant, there is detailed balancing of changes from one regular orbit to another. The orbits themselves, like railway lines, remain unchanged. The only difference is that in case (a) the same star moves on each orbit all the time, while in case (b) different stars successively occupy each orbit, jumping from one track to another. Evidently this has no practical effect on the dynamics of the whole system. Thus we can say that a principle of indistinguishability of individual stars operates in stellar systems which are in a steady state: only the distribution of regular orbits, and the distribution of stars among the orbits, are of importance in the dynamics of stellar systems. Which particular star is moving in each orbit at any given instant is of no significance as regards the dynamics of the system.

The principle of indistinguishability can easily be extended to non-steady systems. The only difference is that, in the course of time, both the shapes and the distribution of the regular orbits and the distribution of stars among the orbits undergo changes. It is again of no importance in the dynamics of the system which particular star is moving in each orbit. For non-steady systems, however, it is important that the transfer of a star from one orbit to another should be "instantaneous", i.e. should take place in a time very short compared with that in which the orbits undergo a considerable change.

We shall see in Chapter X that the principle of indistinguishability is of importance in the justification of separating any stellar system into a nucleus and a corona.

Classical stellar dynamics, in which encounters are neglected, is based on the solution of eqn. (5.4), whose right-hand side is zero (the Boltzmann equation). We see that the results of the classical theory hold good not only when encounters are neglected, but also when the phase density is invariant with respect to the irregular forces.

The theory of stellar systems in a steady state is of central importance in classical stellar dynamics based on statistical methods. By a *steady state* we mean a state of a system in which its phase density, and therefore all the macroscopic parameters, do not depend explicitly on time. This is, of course, a formal definition of a steady state. In reality, it can apply only approximately in stellar systems, which may be at most in a quasi-steady state (see Chapter IV, §§ 5–9). In a steady state, $\partial f/\partial t \equiv 0$, and the Boltzmann eqn. (5.4) becomes

$$\sum_k \dot{x}_k \frac{\partial f}{\partial x_k} = 0. \tag{5.5}$$

§ 2. The Boltzmann Equation in Curvilinear Coordinates

It is easy to see that the left-hand side of eqn. (5.4) is just the Stokes derivative (total time derivative) of the phase density in six-dimensional phase space. Hence (5.4) can be very concisely written as

$$\mathrm{D}f/\mathrm{D}t = 0, \tag{5.6}$$

from which the explicit form of the Boltzmann equation in any system of curvilinear coordinates in phase space can be obtained.

We use curvilinear coordinates λ, μ, ν in three-dimensional coordinate space to define the position of a star, and resolve the velocity **V** at a given point in coordinate space into components tangential to the coordinate lines; let these components be Λ, M, N. Then

$$f(q, p, t) = f(\lambda, \mu, \nu, \Lambda, \mathrm{M}, \mathrm{N}, t).$$

Substituting this expression in (5.6), we obtain the Boltzmann equation in curvilinear coordinates:

$$\frac{\partial f}{\partial t} + \frac{\partial f}{\partial \lambda}\dot{\lambda} + \frac{\partial f}{\partial \mu}\dot{\mu} + \frac{\partial f}{\partial \nu}\dot{\nu} + \frac{\partial f}{\partial \Lambda}\dot{\Lambda} + \frac{\partial f}{\partial \mathrm{M}}\dot{\mathrm{M}} + \frac{\partial f}{\partial \mathrm{N}}\dot{\mathrm{N}} = 0. \tag{5.7}$$

We may consider some important particular cases.

In ordinary rectangular (Cartesian) coordinates x, y, z and velocities u, v, w,

$$f(q, p, t) = f(x, y, z, u, v, w, t).$$

Hence (5.7) becomes

$$\frac{\partial f}{\partial t} + \dot{x}\frac{\partial f}{\partial x} + \dot{y}\frac{\partial f}{\partial y} + \dot{z}\frac{\partial f}{\partial z} + \dot{u}\frac{\partial f}{\partial u} + \dot{v}\frac{\partial f}{\partial v} + \dot{w}\frac{\partial f}{\partial w} = 0. \tag{5.8}$$

The differential equations of the motion of a star under the regular forces (5.2) are, in the usual notation,

$$\begin{aligned} &\mathrm{d}x/\mathrm{d}t = u, && \mathrm{d}y/\mathrm{d}t = v, && \mathrm{d}z/\mathrm{d}t = w, \\ &\mathrm{d}u/\mathrm{d}t = \partial U/\partial x, && \mathrm{d}v/\mathrm{d}t = \partial U/\partial y, && \mathrm{d}w/\mathrm{d}t = \partial U/\partial z, \end{aligned} \tag{5.9}$$

where U is the potential. Hence eqn. (5.8) is explicitly

$$\frac{\partial f}{\partial t} + u\frac{\partial f}{\partial x} + v\frac{\partial f}{\partial y} + w\frac{\partial f}{\partial z} + \frac{\partial U}{\partial x}\frac{\partial f}{\partial u} + \frac{\partial U}{\partial y}\frac{\partial f}{\partial v} + \frac{\partial U}{\partial z}\frac{\partial f}{\partial w} = 0. \tag{5.10}$$

In considering stellar systems of certain types, the Boltzmann equation is used in cylindrical and in spherical coordinates. In cylindrical coordinates (see Fig. 26, Chapter IV, § 9)

$$f(q, p, t) = f(\varrho, \theta, z, \mathrm{P}, \Theta, Z, t),$$

and the Boltzmann eqn. (5.7) becomes

$$\frac{\partial f}{\partial t} + \dot{\varrho}\frac{\partial f}{\partial \varrho} + \dot{\theta}\frac{\partial f}{\partial \theta} + \dot{z}\frac{\partial f}{\partial z} + \dot{\mathrm{P}}\frac{\partial f}{\partial \mathrm{P}} + \dot{\Theta}\frac{\partial f}{\partial \Theta} + \dot{Z}\frac{\partial f}{\partial Z} = 0. \tag{5.11}$$

We can express $\dot{\varrho}, \dot{\theta}, \ldots, \dot{Z}$ in terms of the coordinates. Evidently

$$\mathrm{P} = \mathrm{d}\varrho/\mathrm{d}t = \dot{\varrho}, \quad \Theta = \varrho\, \mathrm{d}\theta/\mathrm{d}t = \varrho\dot{\theta}, \quad Z = \mathrm{d}z/\mathrm{d}t = \dot{z}. \tag{5.12}$$

In our notation, the equations of motion of a free particle are

$$\ddot{\varrho} - \varrho\dot{\theta}^2 = \partial U/\partial \varrho, \quad (1/\varrho)\, \mathrm{d}(\varrho^2\dot{\theta})/\mathrm{d}t = (1/\varrho)\, \partial U/\partial \theta, \quad \ddot{z} = \partial U/\partial z, \tag{5.13}$$

where $U(\varrho, \theta, z)$ is the gravitational potential. From (5.12) and (5.13) we easily find

$$\dot{\mathrm{P}} = \partial U/\partial \varrho + \Theta^2/\varrho, \quad \dot{\Theta} = (1/\varrho)\, \partial U/\partial \theta - \mathrm{P}\Theta/\varrho, \quad \dot{Z} = \partial U/\partial z. \tag{5.14}$$

Substituting $\dot{\varrho}, \dot{\theta}, \dot{z}, \dot{\mathrm{P}}, \dot{\Theta}, \dot{Z}$ from (5.12) and (5.14) in (5.11), we have finally

$$\frac{\partial f}{\partial t} + \mathrm{P}\frac{\partial f}{\partial \varrho} + \frac{\Theta}{\varrho}\frac{\partial f}{\partial \theta} + Z\frac{\partial f}{\partial z} + \left(\frac{\partial U}{\partial \varrho} + \frac{\Theta^2}{\varrho}\right)\frac{\partial f}{\partial \mathrm{P}}$$
$$+ \left(\frac{1}{\varrho}\frac{\partial U}{\partial \theta} - \frac{\mathrm{P}\Theta}{\varrho}\right)\frac{\partial f}{\partial \Theta} + \frac{\partial U}{\partial z}\frac{\partial f}{\partial Z} = 0. \tag{5.15}$$

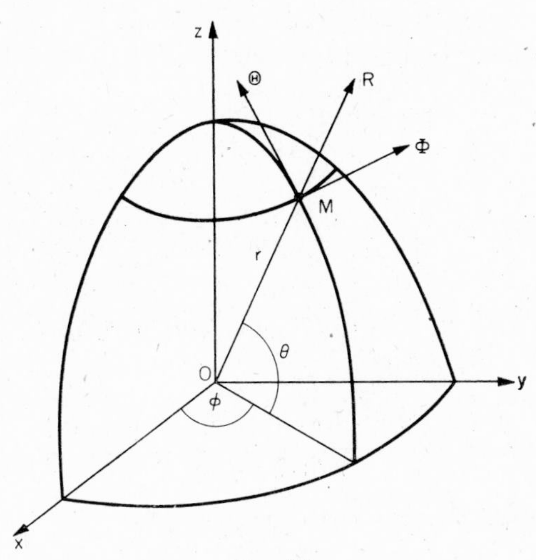

FIG. 27

For spherical coordinates we use the notation (Fig. 27) r = radius vector, θ = polar distance, φ = azimuth, R = velocity component in the direction of the radius vector, Θ = velocity component along the meridian, Φ = velocity component along the circle of latitude. Then

$$f(q, p, t) = f(r, \theta, \varphi, R, \Theta, \Phi, t).$$

The Boltzmann eqn. (5.7) in this case is

$$\frac{\partial f}{\partial t} + \dot{r}\frac{\partial f}{\partial r} + \dot{\theta}\frac{\partial f}{\partial \theta} + \dot{\varphi}\frac{\partial f}{\partial \varphi} + \dot{R}\frac{\partial f}{\partial R} + \dot{\Theta}\frac{\partial f}{\partial \Theta} + \dot{\Phi}\frac{\partial f}{\partial \Phi} = 0. \tag{5.16}$$

We can express the derivatives $\dot{r}, \dot{\theta}, \ldots, \dot{\Phi}$ in terms of the phase coordinates themselves:

$$R = \dot{r}, \quad \Theta = r\dot{\theta}, \quad \Phi = r\dot{\varphi}\sin\theta. \tag{5.17}$$

To eliminate $\dot{R}, \dot{\Theta}, \dot{\Phi}$ we use the Lagrangian equations of motion for a free particle in spherical coordinates:

$$\left.\begin{aligned} &\frac{d\dot{r}}{dt} - (r\dot{\theta}^2 + r\dot{\varphi}^2\sin^2\theta) = \frac{\partial U}{\partial r}, \\ &\frac{d}{dt}(r^2\dot{\theta}) - r^2\dot{\varphi}^2\sin\theta\cos\theta = \frac{\partial U}{\partial \theta}, \\ &\frac{d}{dt}(r^2\dot{\varphi}\sin^2\theta) = \frac{\partial U}{\partial \varphi}. \end{aligned}\right\} \tag{5.18}$$

From (5.17) and (5.18) we find

$$\left.\begin{aligned} \dot{R} &= \frac{\Theta^2 + \Phi^2}{r} + \frac{\partial U}{\partial r}, \\ \dot{\Theta} &= -\frac{R\Theta}{r} + \frac{\Phi^2}{r}\cot\theta + \frac{1}{r}\frac{\partial U}{\partial \theta}, \\ \dot{\Phi} &= -\frac{R\Phi}{r} - \frac{\Theta\Phi}{r}\cot\theta + \frac{1}{r\sin\theta}\frac{\partial U}{\partial \varphi}. \end{aligned}\right\} \tag{5.19}$$

Substituting $\dot{r}, \dot{\theta}, \dot{\varphi}, \dot{R}, \dot{\Theta}, \dot{\Phi}$ from (5.17) and (5.19) in (5.16), we have the Boltzmann equation in spherical coordinates:

$$\begin{aligned} &\frac{\partial f}{\partial t} + R\frac{\partial f}{\partial r} + \frac{\Theta}{r}\frac{\partial f}{\partial \theta} + \frac{\Phi}{r\sin\theta}\frac{\partial f}{\partial \varphi} + \left(\frac{\Theta^2 + \Phi^2}{r} + \frac{\partial U}{\partial r}\right)\frac{\partial f}{\partial R} \\ &\quad + \left(-\frac{R\Theta}{r} + \frac{\Phi^2}{r}\cot\theta + \frac{1}{r}\frac{\partial U}{\partial \theta}\right)\frac{\partial f}{\partial \Theta} \\ &\quad + \left(-\frac{R\Phi}{r} - \frac{\Theta\Phi}{r}\cot\theta + \frac{1}{r\sin\theta}\frac{\partial U}{\partial \varphi}\right)\frac{\partial f}{\partial \Phi} = 0. \end{aligned} \tag{5.20}$$

§ 3. Jeans' Theorem and Liouville's Theorem

Using the fundamental equation of stellar dynamics, we can easily prove two very important theorems. We have seen that the Boltzmann eqn. (5.4) signifies that the total time derivative of the phase density is zero:

$$\mathrm{D}f/\mathrm{D}t = 0. \tag{5.21}$$

Let us consider a point in phase space, arbitrary except that its coordinates shall have actually possible values, i.e. the three space coordinates correspond to a position within the stellar system, and the three velocity coordinates correspond to a velocity which does not differ too greatly from the mean absolute velocity of the stars in the system.

Let us now imagine a star which at time t has the phase coordinates of the point considered. In the course of time the phase coordinates of the star will vary. Any function of these phase coordinates and the time which remains constant during the motion of the star is called an *integral of the motion*.† The characteristic property of all integrals of the motion is that their total time derivatives are zero. The eqn. (5.21) can therefore be expressed in the form of *Jeans' theorem* ([45]; [42], pp. 229ff.): in any motion of a stellar system its phase density is an integral of the motion.

By means of Jeans' theorem we can prove *Liouville's theorem*: in the motion of a stellar system any volume of phase space remains constant. For, let us consider an infinitesimal volume $\mathrm{d}\gamma$ in phase space such that the coordinates of the points in it are "possible" in the sense explained above. Then

$$\mathrm{d}N = f(t, x_k)\,\mathrm{d}\gamma \tag{5.22}$$

gives the number of phase points in the volume $\mathrm{d}\gamma$ which represent actual stars. After some time, at the instant t', these points will have moved in phase space, and their number is given by $\mathrm{d}N' = f(t', x_k')\,\mathrm{d}\gamma'$. But evidently

$$\mathrm{d}N = \mathrm{d}N', \tag{5.23}$$

since only those points which were originally in the moving phase volume $\mathrm{d}\gamma$ will be in it at any time. This is shown as follows. As the volume $\mathrm{d}\gamma$ moves, no other point can enter it, since in that case the point entering would have at some instant the same phase coordinates as a point in the volume. Any point in phase space can represent an actual star. Hence the entry of another point into the phase volume would mean that two stars could have, at a certain instant, the same coordinates and velocities and yet move in different orbits in the system. This is clearly impossible, since by the laws of mechanics the position and velocity of a particle in a given field of force completely and uniquely determine its orbit. In exactly the same manner we can show that no point can leave the phase volume, since it would have to enter some other volume.

† This definition of the integrals of the motion differs from that used in elementary mechanics, where they are constants in the equations.

From the above discussion, the total time derivative of $\mathrm{d}N$ is zero: $\mathrm{D}(\mathrm{d}N)/\mathrm{D}t = 0$, or, by (5.22),

$$\mathrm{d}\gamma \frac{\mathrm{D}f}{\mathrm{D}t} + f\frac{\mathrm{D}(\mathrm{d}\gamma)}{\mathrm{D}t} = 0. \tag{5.24}$$

The first term is zero, by Jeans' theorem. Hence $\mathrm{D}(\mathrm{d}\gamma)/\mathrm{D}t = 0$. This is Liouville's theorem.

It is seen from eqn. (5.24) that the theorems of Jeans and Liouville are complementary, and either may be regarded as a consequence of the other. In statistical physics Liouville's theorem is of considerable importance, and is usually proved first.† In stellar dynamics, on the other hand, Jeans' theorem is the more important, and we shall therefore give another and more direct proof of it, which also leads to an elucidation of some other important matters.

To solve the Boltzmann eqn. (5.10), which is a linear homogeneous equation in f, we form the corresponding Lagrange's equations:

$$\frac{\mathrm{d}x}{u} = \frac{\mathrm{d}v}{y} = \frac{\mathrm{d}z}{w} = \frac{\mathrm{d}u}{\partial U/\partial x} = \frac{\mathrm{d}v}{\partial U/\partial y} = \frac{\mathrm{d}w}{\partial U/\partial z} = \frac{\mathrm{d}t}{1}. \tag{5.25}$$

In the most general case the solution of the partial differential eqn. (5.10) is an arbitrary function of independent integrals of the system of ordinary differential eqns. (5.25). By taking successively each member of (5.25) with the last we obtain six independent equations. These are just the equations of motion of a free particle in the form (5.9). The integrals of (5.25) are therefore the integrals of the motion.

The integration of the system (5.25) gives in general six independent integrals, which may be written

$$I_k(t, x, y, z, u, v, w) = C_k \quad (k = 1, 2, \ldots, 6). \tag{5.26}$$

These equations in principle determine completely the motion of any star if its coordinates x_0, y_0, z_0 and velocities u_0, v_0, w_0 at some initial instant t_0 are given: substitution of these initial values in (5.26) determines the values of the arbitrary constants C_k, and by solving the system (5.26) we then find x, y, z, u, v, w as functions of t and of the six constants C_k.

Thus we have in general

$$f = F(I_1, I_2, \ldots, I_6), \tag{5.27}$$

where F denotes an arbitrary function. Throughout the motion the quantities I_1, $I_2, \ldots, I_6$ are by definition constant. Hence, from (5.27), the phase density also is constant throughout the motion, i.e. f is an integral of the motion. This proves Jeans' theorem. However, this direct proof makes it possible to formulate the theorem in considerably more detail: eqn. (5.27) shows that the coordinates and velocities can appear in the phase density only in combinations which are themselves integrals of the motion. That is, the phase density f is an integral of the motion and must be a function of other integrals of the motion. In this respect the integration of eqns. (5.25) gives

† See, for example, J. E. Mayer and M. G. Mayer, *Statistical mechanics*, Wiley, New York, 1940.

no further information; it merely specifies the nature and number of the independent integrals which are the arguments of the phase density.

We have proved Jeans' theorem (in this last formulation) by using the Boltzmann equation in rectangular coordinates, but it is easy to see that the theorem remains valid independently of the coordinate system used, since the integrals of the motion are invariants.

§ 4. The Integrals of the Motion in Some Typical Cases

The most interesting case with regard to applications is that of a stellar system in a steady state, for which the potential energy U and the phase density f do not depend explicitly on time. Then we have instead of (5.10)

$$u\frac{\partial f}{\partial x}+v\frac{\partial f}{\partial y}+w\frac{\partial f}{\partial z}+\frac{\partial U}{\partial x}\frac{\partial f}{\partial u}+\frac{\partial U}{\partial y}\frac{\partial f}{\partial v}+\frac{\partial U}{\partial z}\frac{\partial f}{\partial w}=0. \quad (5.28)$$

The number of independent integrals of the motion is reduced to five:

$$I_k(x,y,z,u,v,w)=C_k \quad (k=1,2,3,4,5).$$

These integrals together determine the phase path of any star when the arbitrary constants are suitably chosen.

Let us consider the nature of the integrals I_k for various types of stellar system. It is seen from (5.28) that the phase density in a steady state is determined by the distribution of mass in the system, that is, by the nature of the potential function U. We shall first take the three most typical mass distributions.

1. *Arbitrary Distribution of Mass*

In this case U is subject to no restriction, and we can find only one first integral, namely the energy integral. The Lagrange's equations for a steady state are of the same form as (5.25), except that the time t no longer appears explicitly in the derivatives of the potential energy U.

To derive the energy integral I_1, it is sufficient to combine the first and fourth, second and fifth, and third and sixth members of (5.25), obtaining

$$u\,\mathrm{d}u=(\partial U/\partial x)\,\mathrm{d}x,\quad v\,\mathrm{d}v=(\partial U/\partial y)\,\mathrm{d}y,\quad w\,\mathrm{d}w=(\partial U/\partial z)\,\mathrm{d}z.$$

Adding, we obtain $\frac{1}{2}\,\mathrm{d}(u^2+v^2+w^2)=\mathrm{d}U$, and integration gives

$$I_1\equiv u^2+v^2+w^2-2U=C_1, \quad (5.29)$$

or

$$I_1\equiv V^2-2U=C_1, \quad (5.30)$$

where V is the magnitude of the velocity.

Without further specialising the form of the potential function U, i.e. without taking a more definite distribution of mass in the system, we can obtain no integrals other than the energy integral. Hence

$$f=f(I_1).$$

2. *Rotational Symmetry*

A fairly general case, though more special than the preceding one, is that where the mass distribution in a system in a steady state possesses rotational symmetry.† If the axis of symmetry is the z-axis, then

$$U = U(\varrho, z), \tag{5.31}$$

where $\varrho = \sqrt{(x^2 + y^2)}$ is the cylindrical radius.

In this case we have, besides the energy integral I_1, a further independent integral, the area integral or angular momentum integral. For, from (5.31), $x\,\partial U/\partial y = y\,\partial U/\partial x$. Hence the fourth and fifth members of (5.25) give $y\,\mathrm{d}u = x\,\mathrm{d}v$, or $\mathrm{d}(yu - xv) = 0$, since $\mathrm{d}x = u\,\mathrm{d}t$ and $\mathrm{d}y = v\,\mathrm{d}t$. Hence we obtain the integral of angular momentum about the z-axis:

$$I_2 \equiv xv - yu = C_2. \tag{5.32}$$

If no further restriction is made on the mass distribution, no integrals other than I_1 and I_2 can be found. Hence

$$f = f(I_1, I_2).$$

3. *Spherical Symmetry*

The last general case which we shall consider is that of a system in a steady state in which the mass distribution possesses spherical symmetry. Then

$$U = U(r), \tag{5.33}$$

where $r = \sqrt{(x^2 + y^2 + z^2)}$. Since in this case any three mutually perpendicular axes are axes of symmetry, we can write down three integrals of angular momentum about the three coordinate axes:

$$\left.\begin{aligned} I_2 &\equiv yw - zv = C_2, \\ I_3 &\equiv zu - xw = C_3, \\ I_4 &\equiv xv - yu = C_4. \end{aligned}\right\} \tag{5.34}$$

Without further specialising the mass distribution we can find no integrals other than I_1, I_2, I_3 and I_4. For spherical symmetry, therefore, the most general expression for the phase density is

$$f = f(I_1, I_2, I_3, I_4). \tag{5.35}$$

† We draw a distinction between *rotational* symmetry and *axial* symmetry of functions. In rotational symmetry the value of the function depends only on the distance ϱ from the axis of symmetry and on the coordinate z. In axial symmetry there is, for each point (ϱ, θ, z), a point $(\varrho, \theta + \pi, z)$ which lies symmetrically with respect to the axis and at which the function has the same value. In rotational symmetry, but not in axial symmetry, the level surfaces of the function are surfaces of revolution. A spheroid is a surface with rotational symmetry; an ellipsoid is a surface with axial symmetry. Rotational symmetry of the mass distribution is evidently a particular case of axial symmetry. In rotational symmetry $U = U(\varrho, z)$; in axial symmetry $U = U(\varrho, \theta, z) = U(\varrho, \theta + \pi, z)$.

Thus it is not possible to find five integrals in any of the three cases considered. The more precise the assumed mass distribution (and potential), the more definite is the resulting phase density. With an arbitrary potential, the phase density depends only on the energy integral; when the potential is spherically symmetrical, which is the most stringent assumption, the phase density depends on four integrals of the motion.

§ 5. One-valued and Many-valued Integrals. The Ergodic Hypothesis in Stellar Dynamics

In any particular stellar system in a steady state, whether or not it is symmetrical, there are other integrals besides those mentioned above, and the total number of integrals is always five. The question therefore arises of the relation between the "missing" integrals and the phase density. At first sight it might appear that the absence of these integrals from the arguments of the phase density is explained simply by the fact that they are unknown. In other words, it might seem that the phase density for any particular stellar system should always depend on five integrals of the motion, and our inability to determine all five is due simply to our having considered only general types of system, having fairly general properties of symmetry. If this were so, there would be insuperable obstacles to the construction of a complete dynamical theory of stellar systems, since we can practically never start from a potential distribution given *a priori*. Instead, we should have to determine for any given system both the potential U and the phase density, which are related by the fundamental eqn. (5.28) and by Poisson's equation

$$\triangle U = -4\pi G m \int_{-\infty}^{\infty} \int_{-\infty}^{\infty} \int_{-\infty}^{\infty} f(x, y, z, u, v, w)\, \mathrm{d}u\, \mathrm{d}v\, \mathrm{d}w.$$

In reality, however, the situation is not so difficult, because there is a further condition which must be satisfied by integrals of the motion which appear as arguments in the phase density: they must be one-valued. For the phase density, by its physical significance, must be a one-valued function of the six phase coordinates. We know from Jeans' theorem that the phase coordinates appear in the phase density only in the integrals of the motion $I_1, I_2, \ldots$. If these integrals are one-valued, the phase density will be so; if they are not, the phase density may be many-valued, in contradiction to its physical significance. We can therefore formulate Jeans' theorem more precisely thus: the phase density can have as arguments only the one-valued integrals of the motion. Hence the problem reduces to that of determining the number of one-valued integrals for each type of stellar system.

Following von der Pahlen [95] and G.G. Kuzmin [58], let us consider this problem from the point of view of the geometry of phase space. Suppose that we have a stellar system S which is in a steady state. As the stars in the system move, the corresponding phase points will describe phase paths in six-dimensional phase space.

The Boltzmann equation represents the fact that the phase density is constant along the phase paths. If there exists a one-valued integral of the motion

$$I_1(x, y, z, u, v, w) = C_1, \tag{5.36}$$

then in phase space there correspond to this integral a family of hypersurfaces P_1, which are obtained by giving all possible values to the constant of integration C_1. If the phase path of any star has any point on one of the hypersurfaces (5.36), then it lies entirely on that hypersurface. No phase path lying on one of the hypersurfaces (5.36) can intersect another hypersurface in the same family. Since each hypersurface is also a phase space of dimension one less than that of the original six-dimensional phase space, we can say that, if the integral I_1 exists, this six-dimensional phase space consists of a family of non-intersecting five-dimensional sub-spaces, and any continuous manifold of phase paths filling the original space consists of a family of manifolds of paths depending on one parameter (C_1), each of which is everywhere dense in the corresponding five-dimensional sub-space.

Let there be a second one-valued integral,

$$I_2(x, y, z, u, v, w) = C_2. \tag{5.37}$$

Then, by the same arguments, we conclude that the original six-dimensional space consists of a family of non-intersecting four-dimensional sub-spaces depending on two parameters (C_1 and C_2), and any manifold of phase paths consists of a family of manifolds of paths depending on the same two parameters, each of which fills the corresponding four-dimensional sub-space.

By continuing in the same manner to add further integrals, we evidently obtain families of sub-spaces of decreasing numbers of dimensions, and families of manifolds of paths depending on increasing numbers of parameters. The final case is that where there are five independent one-valued integrals,

$$I_k(x, y, z, u, v, w) = C_k \qquad (k = 1, 2, \ldots, 5); \tag{5.38}$$

then the intersection of five hypersurfaces defines a sub-space of one dimension, i.e. a single phase path, and the manifold of paths depends on five parameters C_k ($k = 1, 2, \ldots, 5$).

The result is entirely different if one of the integrals is many-valued. In order to see the effect of many-valued integrals on the distribution of paths in phase space, we may note that, among the possible distributions of mass in stellar systems, two limiting cases are (1) a central point mass, (2) a homogeneous sphere. Both are idealisations: in one case the density falls away so rapidly from the centre that at a short distance the mass may be supposed concentrated at the centre; in the other case the density scarcely increases at all towards the centre of the system. In case (1) the gravitational acceleration is of the Newtonian type, $F = A/r^2$, while in case (2) it is quasi-elastic, i.e. it varies as the distance, $F = Br$. In either case the integrals of the motion may be found by elementary methods, and the orbits are ellipses, with a common focus in case (1) and a common centre in case (2).

For definiteness, let us consider case (1), with a Newtonian force field and a Keplerian motion. For the quasi-elastic field an analogous discussion is possible. We know from the two-body problem that the orbit is entirely determined by five elements: the longitude Ω of the ascending node and the inclination i, which determine the position of the orbital plane; the distance ω of the pericentre from the node or the longitude $\varpi = \omega + \Omega$ of the pericentre, which determines the position of the orbit in that plane; and the major semiaxis a and eccentricity e, which give the size and shape of the orbit. The values of the elements are determined by the initial conditions, i.e. the coordinates and velocity components at an arbitrary initial instant. Denoting any one of the elements by E, we have $E = F(x_0, y_0, z_0, u_0, v_0, w_0)$, where the arguments of F are the initial values of the phase coordinates. But the choice of the initial instant is entirely arbitrary. Hence the above equation is equivalent to the condition that, in the motion in the phase path, the function $F(x, y, z, u, v, w) = E$ of the phase coordinates remains constant. That is, in a Keplerian motion the elements of the orbit are five independent one-valued integrals of the motion.

Let us now consider some examples of fields which are nearly but not exactly Newtonian.

(a) Let the field be still "spherical" but not Newtonian, as for example if a small spherically symmetrical distributed mass is added to the central condensation. From celestial mechanics it is known (cf., for example, [128]) that in this case there is a perturbing force R along the radius vector. The two elements Ω and i which determine the plane of the orbit remain unchanged, i.e. they remain one-valued integrals, but the elements a, e and ω are no longer integrals, and their perturbations are periodic. The perturbed motion is given by the differential equations

$$\left.\begin{aligned} \frac{da}{dt} &= \frac{2a^{3/2}\,e}{\sqrt{[GM(1-e^2)]}}\,R\sin v, \\ \frac{de}{dt} &= \frac{\sqrt{[a(1-e^2)]}}{\sqrt{(GM)}}\,R\sin v, \\ \frac{d\omega}{dt} &= -\frac{\sqrt{[a(1-e^2)]}}{e\sqrt{(GM)}}\,R\cos v, \end{aligned}\right\} \tag{5.39}$$

where G is the universal gravitational constant, M the mass concentrated at the centre, and v the true anomaly (the angular distance from the pericentre).

The period of revolution of the apse line is "almost always" (in the sense of the metric theory of manifolds) incommensurable with the period of the Keplerian motion, and so the perturbed orbit is not closed. It passes an infinity of times through every point within a certain plane annulus about the centre. Such a motion is called in celestial mechanics a *periplegmatic* motion.

Strictly speaking, such an orbit passes an infinity of times through an arbitrarily small neighbourhood of each point, but there is no physical difference between the two statements, since any "infinitesimal" body (a star) has a finite volume.

According to Jeans' theorem, the phase density remains constant along a phase path. In the case under consideration, the phase path fills a two-dimensional region.

Hence the phase density is constant throughout that region. From this it is clear that only four integrals of the motion are here sufficient to determine the phase path in six-dimensional phase space. The fifth integral of the motion still exists, of course, but it must be many-valued: for every integral represents an equation between the six phase coordinates, $I_k(x, y, z, u, v, w) = C_k$, and in this case the fifth integral must be many-valued in order to represent the fact that to each point x, y, z in the annular region there correspond at least two values of the velocities u, v, w.

Thus, for a spherically symmetrical gravitational field, there are only four one-valued integrals of the motion, and the phase density in general depends on them. They are the energy integral and the three angular momentum integrals, which we obtained in § 4.

(b) Next, let the field possess rotational symmetry. For this we assume that the distributed mass which in case (a) was added to the central concentrated mass has such symmetry. Then, if the unperturbed Keplerian orbit has an inclination $i \neq 0$, there will be, besides, the radial perturbing force R, other components S and W, one in the plane of the orbit perpendicular to the radius vector and the other normal to the plane of the unperturbed orbit. The perturbations of the elements are [128]

$$\left.\begin{aligned}
\frac{\mathrm{d}a}{\mathrm{d}t} &= \frac{2a^{3/2}}{\sqrt{[GM(1-e^2)]}}\,[eR\sin v + S(1 + e\cos v)],\\
\frac{\mathrm{d}e}{\mathrm{d}t} &= \frac{\sqrt{[a(1-e^2)]}}{\sqrt{(GM)}}\,[R\sin v + S(\cos v + \cos E)],\\
\frac{\mathrm{d}\varpi}{\mathrm{d}t} &= \frac{\sqrt{[a(1-e^2)]}}{e\sqrt{(GM)}}\left[-R\cos v + S\,\frac{2 + e\cos v}{1 + e\cos v}\,\sin v\right] + \frac{\mathrm{d}\Omega}{\mathrm{d}t}\sin^2\tfrac{1}{2}i,\\
\frac{\mathrm{d}i}{\mathrm{d}t} &= \frac{r}{\sqrt{[GMa(1-e^2)]}}\,W\cos(v + \varpi - \Omega),\\
\frac{\mathrm{d}\Omega}{\mathrm{d}t} &= \frac{r}{\sqrt{[GMa(1-e^2)]}\sin i}\,W\sin(v + \varpi - \Omega),
\end{aligned}\right\} \tag{5.40}$$

where in the second formula we have used the eccentric anomaly E, defined by $\cos E = (e + \cos v)/(1 + e\cos v)$.

It is certain that a stellar system having rotational symmetry will also have a considerable rotation. Hence, unlike what happens in the case of spherical symmetry, the star orbits will not be much different from circles (see Chapter X).

Let us consider, for definiteness, a system such as the Galaxy, i.e. one which has a rapid rotation (the circular velocity being much greater than the mean residual velocity) and is markedly flattened towards its equatorial plane.† In such a system not

† More precisely, we shall consider a plane sub-system.

only the eccentricities but also the inclinations of the orbits are small, and so, neglecting in (5.40) the small terms in e and i, we have

$$\left.\begin{aligned} \frac{\mathrm{d}a}{\mathrm{d}t} &= \frac{2a^{3/2}}{\sqrt{(GM)}} S, \\ \frac{\mathrm{d}e}{\mathrm{d}t} &= \frac{\sqrt{a}}{\sqrt{(GM)}} (R \sin v + 2S \cos v), \\ \frac{\mathrm{d}\varpi}{\mathrm{d}t} &= \frac{\sqrt{a}}{e\sqrt{(GM)}} (-R \cos v + 2S \sin v), \\ \frac{\mathrm{d}i}{\mathrm{d}t} &= \frac{r}{\sqrt{(GMa)}} W \cos(v + \varpi - \Omega), \\ \frac{\mathrm{d}\Omega}{\mathrm{d}t} &= \frac{r}{\sqrt{(GMa)} \sin i} W \sin(v + \varpi - \Omega). \end{aligned}\right\} \tag{5.41}$$

The right-nand sides of the third and fifth equations have small denominators and are therefore large. These equations describe the rapid rotation of the apse line and the line of nodes.

Thus, in consequence of the perturbations in the longitude ϖ of the pericentre and the longitude Ω of the ascending node, the orbit fills the whole volume obtained by rotating the plane annulus about an axis not normal to its plane. In phase space this volume is a three-dimensional sub-space. This means that to every point x, y, z in the volume there correspond several points u, v, w in velocity space, and each set of velocity coordinates corresponds to one branch (loop) of the orbit passing through the point x, y, z.

However, except in the central regions, where r is very small in comparison with a/e (the distance from the centre of the ellipse to the directrix), the rate of change $\mathrm{d}i/\mathrm{d}t$ of the inclination is of the same order as $\mathrm{d}\varpi/\mathrm{d}t$. As a result of the oscillations of the orbital plane, the above-mentioned volume is "spread out", and at each point x, y, z there appears, besides the multiplicity of values u, v, w, another independent multiplicity, since at each x, y, z, the values u, v, w are determined not only by the branch chosen but also by the inclination of the orbital plane. The phase path thus occupies a four-dimensional volume.

For a unique determination of the four-dimensional sub-space in six-dimensional space, it is necessary and sufficient to have two one-valued integrals of the form $I_k(x, y, z, u, v, w) = C_k$. Thus, for a stellar system with a rotationally symmetrical mass distribution, there are only two one-valued integrals of the motion on which the phase density depends. These integrals may be taken as the energy and the angular momentum about the axis of symmetry:

$$I_1 \equiv u^2 + v^2 + w^2 - 2U = C_1, \quad I_2 \equiv xv - yu = C_2, \tag{5.42}$$

or any two independent combinations of them. Then the phase density has the form

$$f = f(I_1, I_2). \tag{5.43}$$

(c) Let us finally consider the last of the cases studied in § 4, that of an arbitrary field of force. Here the perturbations of the orbital elements are given by the general eqns. (5.40). The perturbing forces R, S, W are arbitrary and, since none of them is distinguished in any way, we must assume that they are all of the same order of magnitude.

Not all five perturbations of the elements, however, but only four, lead to many-valuedness of the motion, since the perturbations of the semiaxis a and the eccentricity e depend on the same argument v and so are synchronous. The simultaneous perturbation of a and e causes a periodic deformation of the orbit which leads to one part of the many-valuedness; the change in ϖ rotates the orbit in the plane, and leads to a second part. The change in the inclination i of the orbit gives a third part, and finally the change in the longitude Ω of the node, that is, the rotation of the plane of the orbit at a fixed inclination, brings about a fourth part of the many-valuedness.

Thus the phase path in this case fills a five-dimensional volume, which is a subspace of six-dimensional phase space, and is uniquely defined by the value of one integral of the motion, namely the energy integral I_1:

$$I_1 \equiv u^2 + v^2 + w^2 - 2U. \tag{5.44}$$

Hence the phase density is a function of this integral only:

$$f = f(I_1). \tag{5.45}$$

Thus, for a gravitational field possessing no symmetry there is in general only a single one-valued integral of the motion on which the phase density depends.

Of course, the above arguments make no pretence of either mathematical rigour or physical generality. Their value is to have shown, for relatively special and particular examples, how the many-valued integrals of the motion naturally arise in determining the phase density.

According to Jeans' theorem the phase density is a function of the integrals of the motion. The preceding discussion shows that the reason why some of the integrals do not appear as arguments of the phase density is that they are many-valued. This discussion, though not rigorous, renders extremely plausible the hypothesis that the classical integrals, i.e. those which can be found by Jeans' theorem, are the only one-valued integrals.

The resulting situation is entirely analogous to that in statistical physics, where, to establish a statistical theory of equilibrium states of systems, it is necessary to introduce the *ergodic hypothesis*, according to which the phase density depends only on the energy integral I_1. For any ergodic system the energy integral is the only one-valued integral of the motion.†

It is easy to show that no system whose phase density depends only on the energy integral can have any internal motions, since the centroid velocity in it is zero at every point. For, in such a system, we have $f = f(u^2 + v^2 + w^2 - 2U)$, whence we see

† See, for example, J.E. Mayer and M.G. Mayer, *Statistical mechanics*, Wiley, New York, 1940.

that the phase density is an even function of each component of the velocity. If α is any of u, v, w, the corresponding centroid velocity component $\bar{\alpha}$ is

$$\bar{\alpha} = \int_{-\infty}^{\infty} \int_{-\infty}^{\infty} \int_{-\infty}^{\infty} f(u^2 + v^2 + w^2 - 2U)\,\alpha\, \mathrm{d}u\, \mathrm{d}v\, \mathrm{d}w.$$

The integration is taken over all velocity space. The existence of the integral is ensured by the sufficiently rapid approach to zero of the phase density. The integrand is an odd function, and therefore $\bar{\alpha} = 0$.

Hence it is clear that ordinary statistical physics is concerned only with systems having no internal motions (for example, a gas in a vessel at rest). This treatment is evidently quite inadequate for stellar dynamics, in which the most important systems are those having internal (i.e. macroscopic) motions.

By analogy with statistical physics we may apply the term *quasi-ergodic* to any stellar system for which the classical integrals, i.e. those obtained from Jeans' theorem, are the only one-valued integrals. In particular, in the important case of rotationally symmetrical systems (galaxies) the quasi-ergodic property signifies that the phase density depends on only two independent one-valued integrals of the motion: the energy integral and the angular momentum integral.

§ 6. Jeans' Problem. The Symmetry of the Distribution in Velocity Space

In § 5 we have seen that to each of the types of mass distribution considered there correspond certain properties of the phase density function. Since this function gives the distribution of points in six-dimensional phase space, we can also say that the distribution of mass in a stellar system determines the distribution of velocities (and of coordinates). The same property is expressed by the fundamental equation of stellar dynamics if it is regarded as an equation to determine the distribution function f for a given potential U. The problem of finding f from U will be referred to as *Jeans' problem*. It corresponds to the most natural and logical approach to the problem of stellar dynamics, since we know the specific factors which control the spatial distribution of stars and, in particular, tend to give the system various kinds of symmetry. They are the force of gravitation and such dynamical quantities as the total number (or mass) of stars, the rate of rotation (angular momentum), the dispersion of residual velocities ("temperature"), and others which we shall encounter in Chapter X. By the action of these dynamical parameters the system tends to acquire more or less rapidly the dynamic equilibrium figure if it does not already possess that figure.

Jeans' problem has no general solution, however. It involves an indeterminacy, to resolve which we shall use, in subsequent chapters, the *synthetic method*, which combines the statistical and hydrodynamical approaches. For certain particular cases, nevertheless, Jeans himself elegantly proved that certain properties of symmetry of the mass distribution in the system (i.e. of the potential) correspond to certain properties of symmetry in the stellar velocity distribution. The higher the symmetry of the mass distribution, the lower is that of the velocity distribution, and vice versa.

We shall discuss in order the three cases of mass distribution which were considered in § 4.

1. *Arbitrary Potential*

The phase density here depends only on a single one-valued integral

$$I_1 \equiv u^2 + v^2 + w^2 - 2U = C_1,$$

the energy integral. Hence

$$f = f(u^2 + v^2 + w^2 - 2U).$$

The components u, v, w appear symmetrically in the expression for f. Consequently the phase density, and the phase points, have in velocity space the highest possible symmetry, namely spherical symmetry.

2. *Rotationally Symmetrical Potential*

In this case the phase density depends on two independent one-valued integrals, those of energy and angular momentum:

$$I_1 \equiv u^2 + v^2 + w^2 - 2U = C_1, \quad I_2 \equiv xv - yu = C_2,$$

and the phase density is

$$f = f(I_1, I_2).$$

To ascertain the symmetry of the phase density, we use instead of the rectangular coordinates x, y, z the cylindrical coordinates ϱ, θ, z and the corresponding velocity components P, Θ, Z; see Fig. 26 (Chapter IV, § 9). Evidently $x = \varrho \cos\theta, y = \varrho \sin\theta$; $u = \mathrm{d}x/\mathrm{d}t = \mathrm{P}\cos\theta - \Theta\sin\theta$, $v = \mathrm{d}y/\mathrm{d}t = \mathrm{P}\sin\theta + \Theta\cos\theta$. Then the integrals I_1 and I_2 become†

$$\begin{aligned} I_1 &\equiv \mathrm{P}^2 + \Theta^2 + Z^2 - 2U(\varrho, z) = C_1, \\ I_2 &\equiv \varrho\Theta = C_2, \end{aligned} \tag{5.46}$$

and the phase density is correspondingly

$$f = f(\mathrm{P}^2 + \Theta^2 + Z^2 - 2U, \quad \varrho\Theta). \tag{5.47}$$

Hence we see that the function f is symmetrical only as regards two velocity components, P and Z. Accordingly, in velocity space the phase density, and the phase points, have rotational symmetry about the Θ-axis. In the present case the potential U depends on ϱ and z only. From (5.47) it follows that the phase density also involves only these two coordinates.

† It may be noticed that the integral I_2 in the form $I_2 = \varrho\Theta = C_2$ is easily obtained from the equations of motion in cylindrical coordinates (5.13), since $\partial U/\partial\theta = 0$.

3. *Spherically Symmetrical Potential*

In this case there are four independent one-valued integrals:

$$I_1 \equiv u^2 + v^2 + w^2 - 2U = C_1,$$

$$I_2 \equiv yw - zv = C_2,$$

$$I_3 \equiv zu - xw = C_3,$$

$$I_4 \equiv xv - yu = C_4,$$

and accordingly

$$f = f(I_1, I_2, I_3, I_4).$$

Since the function f is related to U by Poisson's equation, and the potential depends only on r, the phase density depends on the coordinates x, y, z only through $r = \sqrt{(x^2 + y^2 + z^2)}$. The energy integral involves the coordinates only through the potential $U = U(r)$, and to find the form of the function f we must take a combination of the integrals I_2, I_3, I_4 which depends only on r.

The quantities I_2, I_3, I_4 are the components of the axial vector of the total angular momentum $\mathbf{I} = \mathbf{r} \times \mathbf{V}$, where $\mathbf{r}$ is the radius vector and $\mathbf{V}$ the total velocity. The only combination, formed from the components of a vector, which is a one-valued invariant with respect to an arbitrary rotation of the coordinate axes is the square of its modulus. Thus the invariant required is $I^2 = I_2^2 + I_3^2 + I_4^2$. By Lagrange's theorem, $|\mathbf{a} \times \mathbf{b}|^2 = a^2 b^2 - (\mathbf{a} \cdot \mathbf{b})^2$. Hence

$$I^2 = r^2 V^2 - (\mathbf{r} \cdot \mathbf{V})^2. \tag{5.48}$$

The energy integral is

$$I_1 = V^2 - 2U(r). \tag{5.49}$$

In the present case, therefore, the phase density is a function of two one-valued integrals:

$$f = f(I_1, I^2). \tag{5.50}$$

To find the symmetry of the phase density we use spherical coordinates r, θ, φ (Fig. 27, § 2) and the corresponding velocity components R, Θ, Φ. Then $V^2 = R^2 + \Theta^2 + \Phi^2$, $\mathbf{r} \cdot \mathbf{V} = rR$, and (5.48), (5.49) give

$$I^2 = r^2(\Theta^2 + \Phi^2), \tag{5.51}$$

$$I_1 = R^2 + \Theta^2 + \Phi^2 - 2U(r). \tag{5.52}$$

The expression (5.50) for the phase density thus becomes

$$f = f\{R^2 + \Theta^2 + \Phi^2 - 2U(r),\ r^2(\Theta^2 + \Phi^2)\}. \tag{5.53}$$

If the square root of (5.51) is taken, a two-valued integral is obtained, and then the system is not quasi-ergodic.

Hence we see that the function f is symmetrical in the components Θ and Φ. For a spherically symmetrical potential, therefore, the phase density, and also the phase points, have in velocity space rotational symmetry about an axis parallel to the radius vector of the point considered.

From (5.53) it follows also that the coordinates appear in the phase density f only through the radius r.

§ 7. The Inverse Jeans' Problem

The fundamental equation of stellar dynamics (the Boltzmann equation) may be regarded not only as an equation to determine the phase density for a given potential (Jeans' problem) but also as one to determine the potential for a given phase density (the *inverse Jeans' problem*). This is a mathematical, not a physical, problem, and the extent of its correspondence to actual stellar systems must be examined in each particular case. As regards astronomy, therefore, the inverse problem is much less promising than the direct problem, but it has played a certain part in the historical development of stellar dynamics, and was the subject of some notable work by Eddington [25, 27], Oort [91] and others. This line of investigation culminated in the researches of Chandrasekhar, the results of which are given in his book *Principles of stellar dynamics* [17].

Between 1910 and 1930 the consideration of the inverse problem was a natural stage in the development of the dynamics of stellar systems, since the observational data available to stellar astronomy at that period related only to stars within a comparatively small neighbourhood of the Sun, extending to no more than a few hundred parsecs. The scanty information available concerning the local distribution of star density did not permit a reliable extrapolation to the rest of the Galaxy. This is clearly shown by the "Kapteyn universe" [47], which, despite its author's ingenuity and skill, affords a far from true representation of the structure of the Galaxy. It was therefore entirely reasonable and logical to attempt an indirect determination of the distribution of mass in the Galaxy, beginning from the known local distribution of stellar velocities in the neighbourhood of the Sun.

Another reason which led to the consideration of the inverse Jeans' problem lay in the difficulty of solving Jeans' problem itself that resulted from the uncertainty as regards the number of independent integrals of the motion I_k on which the phase density depends. In general, a system in a steady state must always have five independent integrals of the motion, which determine the phase path. However, general arguments concerning the symmetry of the system give at most four integrals, and in the case of rotational symmetry, which is of particular interest as regards applications to galaxies, only two integrals are obtained. It has been shown in § 5 that the remaining integrals should be many-valued (the quasi-ergodic hypothesis), and so they cannot be used to solve Jeans' problem.

In the inverse Jeans' problem the phase density, or at least the velocity function, is supposed known. Observations have shown that the distribution of residual velocities of stars in the neighbourhood of the Sun is given with sufficient accuracy by Schwarzschild's law $f = e^{-T}$, where T is a complete positive-definite quadratic form in the velocity components:

$$T = au^2 + bv^2 + cw^2 + 2fvw + 2gwu + 2huv + pu + qv + rw + s. \quad (5.54)$$

The coefficients $a, b, \ldots, s$ are functions of the coordinates, and also of the time if non-steady states are considered.

If we assume that the point in the Galaxy where the Sun (or the "observer") is located is "normal", i.e. is not a singular point, we can postulate that Schwarzschild's ellipsoidal law holds at every point in the Galaxy. There then arises the following mathematical problem, which we shall consider only for a system in a steady state.

A stellar system has rotational symmetry, and at every point in it the stellar velocities are distributed according to Schwarzschild's law. It is required to find an expression for the potential, or at least to ascertain, for example, its symmetry properties.

The Boltzmann eqn. (5.10) is, for a steady state,

$$D(f) \equiv u\frac{\partial f}{\partial x} + v\frac{\partial f}{\partial y} + w\frac{\partial f}{\partial z} + \frac{\partial U}{\partial x}\frac{\partial f}{\partial u} + \frac{\partial U}{\partial y}\frac{\partial f}{\partial v} + \frac{\partial U}{\partial z}\frac{\partial f}{\partial w} = 0. \quad (5.55)$$

First of all, we may notice that the properties of the potential which are derived from this equation do not depend on the particular manner in which f depends on the quadratic form T. For, if α denotes any of the six phase coordinates, we have $\partial f/\partial \alpha = (\mathrm{d}f/\mathrm{d}T)(\partial T/\partial \alpha)$ and, since (5.55) is homogeneous in the partial derivatives of f, $D(f) = (\mathrm{d}f/\mathrm{d}T)\, D(T)$. Hence eqn. (5.55) is exactly equivalent to the condition $D(T) = 0$, i.e. T is an integral of the motion.

This result is not at all unexpected, since Jeans' theorem shows that any function of an integral of the motion is also an integral of the motion, and so, instead of assuming Schwarzschild's law, we could equally well solve the inverse Jeans' problem for the "generalised" Schwarzschild's law

$$f = f(T), \quad (5.56)$$

which, using the terminology of Chapter II, we shall call the *ellipsoidal law*.

The usual procedure in solving the inverse Jeans' problem (see [41]) is to substitute the expression (5.54) for T in the fundamental eqn. (5.55). In the resulting cubic polynomial the coefficients of the various powers of u, v, w are equated to zero. Thus twenty partial differential equations to determine the coefficients in the quadratic form T are obtained.

This procedure will be discussed in greater detail in § 10. We may observe, however, that it is not necessary to write out these equations here (they are given, for example, in [17], p. 93), since the results of integrating them can be obtained more directly by the use of Jeans' theorem. If the phase density is given in the form (5.56) and (5.54), this is equivalent to assuming that the Boltzmann eqn. (5.55) has an integral of the motion which is quadratic in the velocity components. Since the stellar system is assumed to have rotational symmetry, this integral may conveniently be written in cylindrical coordinates:

$$\begin{aligned} T = a(\mathrm{P} - \mathrm{P}_0)^2 + b(\Theta - \Theta_0)^2 + c(Z - Z_0)^2 \\ + 2f(\Theta - \Theta_0)(Z - Z_0) + 2g(Z - Z_0)(\mathrm{P} - \mathrm{P}_0) \\ + 2h(\mathrm{P} - \mathrm{P}_0)(\Theta - \Theta_0) + \sigma, \end{aligned} \tag{5.57}$$

where $a, b, \ldots, h, \sigma, \mathrm{P}_0, \Theta_0, Z_0$ are functions of the coordinates; $\mathrm{P}_0, \Theta_0, Z_0$ are the centroid velocity components. According to Jeans' theorem, the phase density $f(T)$ must be a function of the integrals of the motion. In the case of rotational symmetry here considered, the equations of motion give two integrals:

$$\begin{aligned} I_1 &= \mathrm{P}^2 + \Theta^2 + Z^2 - 2U, \\ I_2 &= \varrho\Theta. \end{aligned} \tag{5.58}$$

Hence T must be equal to the most general expression, quadratic in the velocities, which can be formed from I_1 and I_2. This is

$$T = aI_1 + 2bI_2 + cI_2^2, \tag{5.59}$$

where a, b, c are constant coefficients. Substituting I_1 and I_2 from (5.58) and (5.59) and making the resulting quadratic form homogeneous, we have

$$T = h^2\mathrm{P}^2 + h^2\lambda^2(\Theta - \Theta_0)^2 + h^2Z^2 + \sigma, \tag{5.60}$$

where

$$h^2 = a = \text{constant}, \tag{5.61}$$

$$\lambda^2 = 1 + k_2\varrho^2, \tag{5.62}$$

$$\Theta_0 = -k_1\varrho/(1 + k_2\varrho^2), \tag{5.63}$$

$$\begin{aligned} k_1 &= b/a, \quad k_2 = c/a, \\ \sigma &= -2h^2[U + \tfrac{1}{2}\lambda^2\Theta_0^2]. \end{aligned} \tag{5.64}$$

Formulae (5.61)–(5.63) contain the fundamental results of the dynamical theory of rotationally symmetrical stellar systems in a steady state with an ellipsoidal velocity distribution. In what follows we shall call this theory, for brevity, the *ellipsoidal dynamics* of the Galaxy.†

A comparison of (5.60) and (5.57) shows that two components of the centroid velocity are zero: $\mathrm{P}_0 = Z_0 = 0$, and the circular velocity Θ_0 depends only on the coordinate ϱ. Thus follows the first important result: the only macroscopic motion which can occur in a stellar system in a steady state, with a rotationally symmetrical mass distribution and an ellipsoidal velocity distribution, is a barotropic rotation about the axis of symmetry.

It is easy to show that the coefficient k_2 cannot be negative. If it were negative, the circular velocity Θ_0 (5.63) would increase away from the axis of rotation and would become infinite at a finite distance, which is physically impossible.

† This term is proposed with due realisation of its inadequacy, but it is, we hope, sufficiently clear. Other terms appear either insufficiently accurate or too lengthy.

The sign of k_1, according to (5.63), determines the direction of the barotropic rotation. We shall suppose for definiteness that $k_1 > 0$. From (5.63) we have

$$\frac{d\Theta_0}{d\varrho} = -\frac{k_1(1 - k_2\varrho^2)}{(1 + k_2\varrho^2)^2}. \quad (5.65)$$

Hence the circular velocity at first increases away from the axis of rotation, reaching a maximum value when $\varrho = \sqrt{(1/k_2)}$, and then decreases to zero.

The position and shape of the velocity ellipsoid are evidently determined by the quadratic form T. It follows from (5.60) that the three axes of the velocity ellipsoid are respectively parallel to the cylindrical radius vector, the circular velocity, and the axis of symmetry. The value of λ in (5.60) gives the ratio of the semiaxes of the velocity ellipsoid:

$$\lambda = a/c, \quad (5.66)$$

where now a denotes the semiaxis in the P and Z directions and c that in the Θ direction; these should not be confused with the coefficients a and c in (5.59) and (5.61)–(5.64). Since

$$k_2 \geqslant 0, \quad (5.67)$$

it follows from (5.62) that

$$\lambda \geqslant 1. \quad (5.68)$$

From (5.60), (5.61), (5.68), (5.62) and (5.67) we derive a second important result: in a stellar system in a steady state, with rotational symmetry and an ellipsoidal velocity distribution, the velocity ellipsoid is an oblate spheroid; the semiaxes in the meridian plane are equal and constant, and the semiaxis perpendicular to that plane decreases away from the axis of symmetry. We see that the equatorial plane of the velocity ellipsoid, i.e. its plane of symmetry, coincides with the meridian plane through the axis of symmetry and the point considered, in complete agreement with the result previously derived from Jeans' theorem.

Let us now compare formulae (5.60)–(5.63) with the observational results. The manner of variation of the circular velocity of the centroid with distance appears to be in satisfactory agreement with observation for normal rotating galaxies. Figure 28 shows the results of measurements [70] of the circular velocity as a function of distance from the centre of the galaxy M33, corrected for the effect of foreshortening. The circles represent the results of the individual measurements, and the vertical lines represent the error in each determination. The continuous curve is the result of an empirical smoothing of the points; the broken curve is the theoretical result given by formula (5.63), in which the coefficients k_1 and k_2 are chosen by the method of least squares. We see that the agreement between theory and observation is satisfactory.

The measurement of the velocities of rotation of galactic nuclei and of elliptical galaxies has shown that they are in rigid rotation, i.e. their motions are not in accordance with (5.63). There is a discontinuity between the nuclei and the main peripheral parts of normal galaxies. This indicates a dynamical similarity between the elliptical galaxies and the nuclei of the normal galaxies. We shall return to this subject in more detail in Chapter X.

In order to apply formulae (5.60)–(5.64) to our Galaxy, we first express the coefficients k_1 and k_2 in terms of Oort's constants A and B. In the present notation, formulae (3.47) give

$$(\Theta_0/\varrho)_\odot = B - A, \quad (d\Theta_0/d\varrho)_\odot = B + A, \tag{5.69}$$

where the suffix $_\odot$ signifies that the quantities have their values for the distance of the Sun from the axis of the Galaxy ($\varrho_\odot = 7{\cdot}5$ kps).

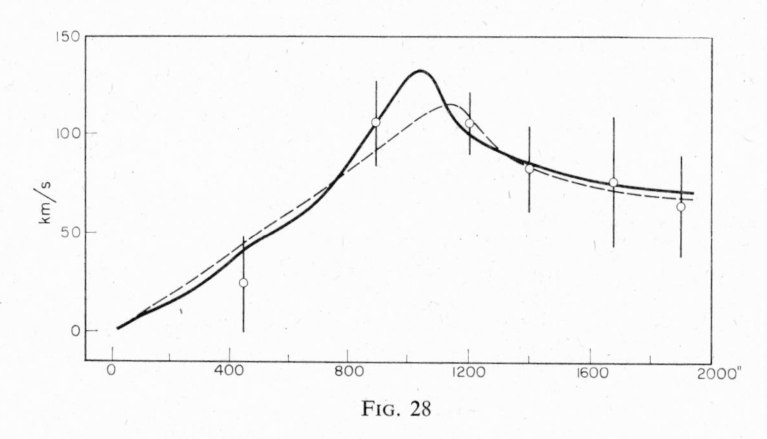

FIG. 28

From (5.69), (5.65) and (5.63) we obtain

$$\begin{aligned} B - A &= -k_1/(1 + k_2\varrho_\odot^2), \\ B + A &= -k_1/(1 - k_2\varrho_\odot^2)/(1 + k_2\varrho_\odot^2)^2. \end{aligned} \tag{5.70}$$

Elimination of k_1 gives

$$k_2 = -A/B\varrho_\odot^2, \tag{5.71}$$

and then

$$k_1 = -(B - A)^2/B. \tag{5.72}$$

Substitution of these values of k_1 and k_2 in (5.63) gives

$$\Theta_0 = [(B - A)^2 \varrho_\odot x/B] \div [1 - Ax^2/B],$$

where $x = \varrho/\varrho_\odot$ is the dimensionless radius. The numerical values $A = 20$ km/s per kps, $B = -13$ km/s per kps and

$$\varrho_\odot = 7{\cdot}5 \text{ kps} \tag{5.73}$$

give

$$\Theta_0 = 628x/(1 + 1{\cdot}54x^2). \tag{5.74}$$

This function has a maximum for $x = 0{\cdot}806$, corresponding to a distance $\varrho = 6$ kps from the galactic centre. The maximum velocity is 253 km/s. At the Sun's distance ($x = 1$) we have $\Theta_0 = 247$ km/s. Thus the Sun is in the region where the circular velocity of rotation of the Galaxy is decreasing, but not far from the maximum.

Figure 29 shows the function (5.74), and we see that the circular velocity of the centroid diminishes comparatively slowly. For $x = 4$ it is still 100 km/s, which is equal to the root-mean-square residual velocity of objects in the spherical sub-systems. This will have to be taken into account later.

The value of λ^2, which characterises the form of the velocity ellipsoid, is given, according to (5.62) and (5.71), by

$$\lambda^2 = 1 - (\varrho^2/\varrho_\odot^2)\, A/B. \tag{5.75}$$

Putting $\varrho = \varrho_\odot$, we obtain Lindblad's formula giving λ in terms of Oort's constants [66]:

$$\lambda^2 = 1 - A/B = (B - A)/B. \tag{5.76}$$

Since

$$B - A = \omega, \tag{5.77}$$

we can also write Lindblad's formula as

$$1/\lambda^2 = 1 + A/\omega. \tag{5.78}$$

This formula makes possible an indirect verification of the theory by comparison with observation. Taking the values given above for A and B, we find

$$1/\lambda = 0{\cdot}63. \tag{5.79}$$

According to Parenago ([107], pp. 130 and 147), the mean of a large number of determinations of the velocity ellipsoid gives for the ratio $1/\lambda$ the same value 0·63, in remarkable agreement.

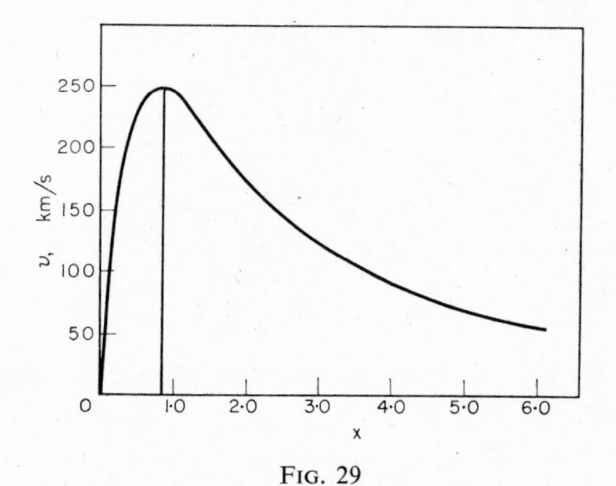

FIG. 29

However, formula (5.60) of the ellipsoidal dynamics of the Galaxy differs in two important respects from the observational results.

(1) As has been mentioned in Chapter II, the observations show that the ratios of the axes of the velocity ellipsoid are 8:5:4, whereas the theory requires an ellipsoid with two equal axes, the approximate ratios being 3:2:3.

(2) One axis of the velocity ellipsoid should, according to the theory, be directed precisely to the centre of the Galaxy,† but observation shows that the direction of the greatest axis deviates considerably from that of the galactic centre (the *deviation of the vertex*).

It may be mentioned in passing that formulae (5.61) and (5.62) for arbitrary ϱ are not at present capable of observational confirmation, since we can determine the velocity ellipsoid only for the neighbourhood of the Sun.

Let us now consider the relation between the potential U and the parameters appearing in the ellipsoidal distribution law (5.60). First of all, it should be noted that, so long as the actual form of the ellipsoidal law†† $f = f(T)$ remains indefinite, the only parameters of the distribution are h^2, λ^2 and Θ_0, which define the position and size of the velocity ellipsoid. The quantity σ is not yet a parameter of the distribution. For, as we saw in Chapter II, § 8, when the function $f(T)$ is given the position of the level surfaces $f(T) = C$ depends on σ; but if the actual form of $f(T)$ is not given, σ can be prescribed arbitrarily and then a function $f(T)$ can still be chosen such as to obtain any desired level surfaces.

We can therefore draw the following important conclusions from (5.61)–(5.64). Firstly, the parameters h^2, λ^2 and Θ_0 in the general ellipsoidal velocity distribution are independent of the potential U, i.e. of the distribution of mass in the system. This means that the general ellipsoidal velocity function obtained is determined not by the physical parameters of the problem but by the assumed mathematical form of the integral (5.57). Conversely, we see from (5.60) and (5.64) that the potential U is independent of the distribution parameters h^2, λ^2 and Θ_0.

Thus, for a general ellipsoidal distribution law of no specific form, the inverse Jeans' problem as formulated at the beginning of this section cannot be solved.

We have already mentioned that the actual form of the velocity function $f(T)$ remains indeterminate in the present theory, since the solution of the Boltzmann eqn. (5.55) does not depend on the actual form of $f(T)$, which must therefore be determined by other means. For this purpose it is most convenient to use some physical theory, or failing that, observational data.

It has been shown in Chapter II that the residual velocities of stars in the Galaxy (according to observational data for the neighbourhood of the Sun) are distributed approximately according to Schwarzschild's law:

$$f = e^{-T} = e^{-Q-\sigma}. \tag{5.80}$$

Let us consider the effect on the solution of the inverse Jeans' problem of specifying the actual form of the ellipsoidal law, for example in the form (5.80). If the function $f(T)$ is given, we may assume that the phase distribution is given, whereas previously only the velocity distribution could be regarded as given. The phase distribution determines the star density $\nu = \int f(T)\, d\Omega$, where $d\Omega$ is a volume element in velocity space. Hence, if the actual form of $f(T)$ is given, we can use formulae (5.60)–(5.63) to establish the actual relation between the parameters h^2, λ^2, Θ_0, σ of the phase distribu-

† As has been mentioned earlier, the Sun may be taken to lie in the galactic plane.

†† P. T. Reznikovskiĭ has given great assistance with regard to the problem of the part played by the actual form of the ellipsoidal law.

tion and the star density ν. Since σ is related to the potential U by (5.64), the way is thus open to a solution of the inverse Jeans' problem.

This method of solving the inverse Jeans' problem is not related to the Boltzmann equation, since in our statement of the problem any conclusions which might be drawn from that equation have already been derived previously.

Let us assume that Schwarzschild's law holds throughout the Galaxy. Since this law is a particular case of the general ellipsoidal law, the quadratic form in the exponent of (5.80) must be the quadratic form T given by formulae (5.60)–(5.64). In this case, therefore,

$$f = f_0 \exp[-h^2 \mathrm{P}^2 - h^2 \lambda^2 (\Theta - \Theta_0)^2 - h^2 Z^2], \tag{5.81}$$

where

$$f_0 = \exp(2h^2 U + h^2 \lambda^2 \Theta_0^2). \tag{5.82}$$

The quantity f_0 essentially determines the star density ν. For a comparison of eqn. (5.81) with (2.7) shows that

$$f_0 = h^3 \lambda \nu / \pi^{3/2}; \tag{5.83}$$

then from (5.82) and (5.83) we have the following relation between the star density ν and the potential U:

$$U = \frac{1}{2h^2} \log \frac{h^3 \lambda \nu}{\pi^{3/2}} - \tfrac{1}{2} \lambda^2 \Theta_0^2. \tag{5.84}$$

Formulae (5.84) and (5.61)–(5.63) essentially give the solution of the inverse Jeans' problem. For, if the phase density (the velocity distribution from Schwarzschild's law and the star density ν as a function of ϱ and z) is specified, these formulae determine the potential U at any point in the system. This solution is achieved only by assuming that the residual velocities are distributed in accordance with Schwarzschild's law.

In order to apply formula (5.84) it may usefully be transformed by introducing the derivatives of the star density ν with respect to ϱ and z. Differentiating (5.84) and using (5.62), (5.63), we have after some simple transformations

$$\frac{\partial (\log \nu)}{\partial \varrho} = -\frac{1}{\varrho}\left(1 - \frac{1}{\lambda^2}\right) + \frac{2h^2}{\varrho} \Theta_0^2 + 2h^2 \frac{\partial U}{\partial \varrho}, \tag{5.85}$$

$$\frac{\partial (\log \nu)}{\partial z} = 2h^2 \frac{\partial U}{\partial z}. \tag{5.86}$$

In terms of the velocity variances, given by

$$2h^2 = 1/\sigma_\varrho^2 = 1/\sigma_z^2, \quad 1/\lambda^2 = \sigma_\theta^2/\sigma_\varrho^2, \tag{5.87}$$

we have finally

$$\frac{\partial (\log \nu)}{\partial \varrho} = -\frac{\sigma_\varrho^2 - \sigma_\theta^2}{\varrho \sigma_\varrho^2} + \frac{\Theta_0^2}{\varrho \sigma_\varrho^2} + \frac{1}{\sigma_\varrho^2} \frac{\partial U}{\partial \varrho}, \tag{5.88}$$

$$\frac{\partial (\log \nu)}{\partial z} = \frac{1}{\sigma_\varrho^2} \frac{\partial U}{\partial z} = \frac{1}{\sigma_z^2} \frac{\partial U}{\partial z}. \tag{5.89}$$

We shall see later that these formulae are much used in applications of the ellipsoidal dynamics of the Galaxy.

In conclusion, it may be noted that the results here given of the ellipsoidal dynamics must be regarded as lacking adequate theoretical justification, since in a stellar dynamics which neglects stellar encounters the reason for the occurrence of Schwarzschild's law, and of any ellipsoidal law, is unexplained. In this theory Schwarzschild's law appears only as a fact derived from observations at a single point in the Galaxy, and the general ellipsoidal law is merely a postulate.

Furthermore, it is not difficult to see that from eqn. (5.84) the mass of the stellar system must necessarily be infinite, which is of course physically impossible. On substituting $\Theta_0 = k_1 \varrho/\lambda^2$, eqns. (5.82) and (5.83) give

$$\nu = (c/\lambda) \exp[(h^2 k_1^2 \varrho^2/\lambda^2) + 2h^2 U], \tag{5.90}$$

where c is a constant. When $\varrho \to \infty$ the exponent tends to $h^2 k_1^2/k_2$ = constant, since U tends to zero. For large ϱ, therefore, we have

$$\nu \approx c'/\lambda \approx c''/\varrho, \tag{5.91}$$

where c' and c'' are appropriate constants. When the density decreases so slowly with increasing distance, the integral

$$M = 2\pi \int_0^\pi \sin\theta \, d\theta \int_0^\infty \nu \varrho^2 \, d\varrho \int_{-\infty}^\infty dz, \tag{5.92}$$

which gives the total mass of the system, diverges.

§ 8. The Importance of Using One-valued Integrals of the Motion

In order to show the importance of the condition that the integrals of the motion which appear in the phase function should be one-valued, we may take as an example the mathematically very elegant theorem demonstrated by S. Chandrasekhar ([17], §§ 3.3, 3.5; see also [107]), who arrives at the conclusion that "for stellar systems in steady states and with differential motions the potential U must necessarily be characterized by helical symmetry".†

This theorem apparently contradicts the observational results, since no case of helical symmetry is observed among the galaxies.

We shall show that this contradiction exists only in so far as the proof of Chandrasekhar's theorem, which is a solution of the inverse Jeans' problem, overlooks the difference between one-valued and many-valued integrals.

In order to avoid the labour of integrating the twenty partial differential equations to which Chandrasekhar reduces the problem, we shall modify the mathematical formulation in a slight and unimportant way so as to deal with only seven differential equations. A consideration of these will suffice for a complete elucidation of the origin of the apparent disagreement mentioned above.

† Chandrasekhar's term "differential motions" has the same significance as our term *macroscopic* (or *internal*) *motions*.

Following G.G. Kuzmin [58], let us suppose for a moment that, instead of taking an integral of the motion Q as a quadratic form in the velocity components, we seek it as an infinite series of powers of these components:

$$Q = Q_0 + Q_1 + Q_2 + Q_3 + \dots, \tag{5.93}$$

where Q_k is a homogeneous form of degree k. The solution of the problem may be imagined as a sequence of successive approximations obtained by taking a gradually increasing number of terms in (5.93).

If we substitute such polynomials in the fundamental differential eqn. (5.55) and equate to zero the coefficients in the resulting polynomial, then, the more terms are taken in (5.93), the more conditions on the potential are obtained, since the result of substituting a polynomial of degree n in the equation is a polynomial of degree $n+1$. Since a homogeneous polynomial of degree n with three arguments has $\frac{1}{2}(n+1)(n+2)$ coefficients, we obtain each time $n+2$ excess conditions on the coefficients of the polynomial, which involve also the derivatives of the potential (i.e. the regular forces) which appear in the fundamental equation.

Thus we find that, the more general the form taken for the integral, the less general is the form obtained for the potential. Hence, if we wish to solve general problems of stellar dynamics, we should strive to begin from the simplest, not the most general, expressions for the integral Q, imposing as few as possible conditions on the potential, and therefore on the distribution of mass within the stellar system.

On these grounds the most satisfactory integral is the energy integral $I_1 = u^2 + v^2 + w^2 - 2U$, which, as we easily see by direct substitution, imposes no restriction on the potential. But this integral, as we have seen, has at the same time the decisive disadvantage that the phase density derived from it does not admit macroscopic (internal) motions. The next most simple form of the integral Q is the linear homogeneous expression

$$Q = au + bv + cw. \tag{5.94}$$

Substitution of (5.94) in the fundamental eqn. (5.55) gives

$$u\left(u\frac{\partial a}{\partial x} + v\frac{\partial b}{\partial x} + w\frac{\partial c}{\partial x}\right) + v\left(u\frac{\partial a}{\partial y} + v\frac{\partial b}{\partial y} + w\frac{\partial c}{\partial y}\right)$$
$$+ w\left(u\frac{\partial a}{\partial z} + v\frac{\partial b}{\partial z} + w\frac{\partial c}{\partial z}\right) + a\frac{\partial U}{\partial x} + b\frac{\partial U}{\partial y} + c\frac{\partial U}{\partial z} = 0,$$

which leads to the seven equations

$$\left.\begin{aligned}
&\frac{\partial a}{\partial x} = 0, \quad \frac{\partial b}{\partial y} = 0, \quad \frac{\partial c}{\partial z} = 0, && \text{(I)}\\
&\frac{\partial b}{\partial z} + \frac{\partial c}{\partial y} = 0, \quad \frac{\partial c}{\partial x} + \frac{\partial a}{\partial z} = 0, \quad \frac{\partial a}{\partial y} + \frac{\partial b}{\partial x} = 0, && \text{(II)}\\
&a\frac{\partial U}{\partial x} + b\frac{\partial U}{\partial y} + c\frac{\partial U}{\partial z} = 0. && \text{(III)}
\end{aligned}\right\} \tag{5.95}$$

Differentiating each of (II) with respect to each of x, y, z and using (I), we get

$$\partial^2 a/\partial y^2 = \partial^2 a/\partial z^2 = \partial^2 b/\partial x^2 = \partial^2 b/\partial z^2 = \partial^2 c/\partial x^2 = \partial^2 c/\partial y^2 = 0$$

and

$$\frac{\partial^2 b}{\partial x \partial z} + \frac{\partial^2 c}{\partial x \partial y} = 0, \quad \frac{\partial^2 c}{\partial x \partial y} + \frac{\partial^2 a}{\partial y \partial z} = 0, \quad \frac{\partial^2 a}{\partial y \partial z} + \frac{\partial^2 b}{\partial x \partial z} = 0,$$

and then by combining the last three equations we conclude that all the second-order partial derivatives of the coefficients a, b, c are zero, and therefore they are linear functions of the coordinates.

The eqns. (I) show that a, b, c do not depend on x, y, z respectively. Hence

$$\left.\begin{aligned} a &= c_1 y + c_2 z + a_0, \\ b &= a_1 z + a_2 x + b_0, \\ c &= b_1 x + b_2 y + c_0, \end{aligned}\right\} \tag{5.96}$$

where $a_0, a_1, a_2, \ldots, c_2$ are constants. Substituting in (II), we have

$$a_1 + b_2 = 0, \quad b_1 + c_2 = 0, \quad c_1 + a_2 = 0. \tag{5.97}$$

Replacing a_2, b_2, c_2 accordingly in (5.96) gives

$$\left.\begin{aligned} a &= c_1 y - b_1 z + a_0, \\ b &= a_1 z - c_1 x + b_0, \\ c &= b_1 x - a_1 y + c_0. \end{aligned}\right\} \tag{5.98}$$

Denoting by $\mathbf{A}$, $\mathbf{A}_0$, $\mathbf{A}_1$ and $\mathbf{r}$ the vectors

$$\mathbf{A}(a, b, c), \quad \mathbf{A}_0(a_0, b_0, c_0), \quad \mathbf{A}_1(a_1, b_1, c_1), \quad \mathbf{r}(x, y, z), \tag{5.99}$$

we can write eqns. (5.98) in the vector form

$$\mathbf{A} = \mathbf{r} \times \mathbf{A}_1 + \mathbf{A}_0. \tag{5.100}$$

Let $|\mathbf{A}_1| = k$. We rotate the coordinate axes so that the z-axis is along $\mathbf{A}_1$. In the new coordinates we have

$$\mathbf{A}_1(0, 0, k); \quad \mathbf{A}_0(\alpha, \beta, \gamma), \tag{5.101}$$

where α, β, γ are the new components of $\mathbf{A}_0$. Substituting these vectors in (5.100) and retaining for simplicity the same notation (a, b, c) and (x, y, z) for the components of $\mathbf{A}$ and $\mathbf{r}$, we obtain

$$a = ky + \alpha, \quad b = -kx + \beta, \quad c = \gamma. \tag{5.102}$$

According to (5.95) (III), U must be a function of the one-valued integrals of the equations $\mathrm{d}x/a = \mathrm{d}y/b = \mathrm{d}z/c$, or, in the above coordinate system,

$$\mathrm{d}x/(ky + \alpha) = \mathrm{d}y/(-kx + \beta) = \mathrm{d}z/\gamma. \tag{5.103}$$

Chandrasekhar transfers the origin to the point $(\beta/k, -\alpha/k, 0)$ and so obtains, instead of (5.102) and (5.103),

$$a = ky, \quad b = -kx, \quad c = \gamma; \tag{5.104}$$

$$\mathrm{d}x/ky = \mathrm{d}y/-kx = \mathrm{d}z/\gamma. \tag{5.105}$$

The first two members of (5.105) give $x\,\mathrm{d}x + y\,\mathrm{d}y = 0$, whence

$$\varrho^2 = x^2 + y^2 = C_1. \tag{5.106}$$

The first and third members of (5.105) give $\gamma\,\mathrm{d}x - ky\,\mathrm{d}z = 0$; substituting from (5.106) $y = \sqrt{(\varrho^2 - x^2)}$ and integrating, we have

$$z + (\gamma/k)\cos^{-1}[x/\sqrt{(x^2 + y^2)}] = C_2. \tag{5.107}$$

This may be written in cylindrical coordinates

$$z + \gamma\theta/k = C_2. \tag{5.108}$$

The potential U is a function of the integrals of (5.105). Thus (5.106), (5.108) give formally

$$U = U(\varrho, z + \gamma\theta/k). \tag{5.109}$$

This formula shows the helical symmetry of the potential U. For, by (5.109), U is unchanged when θ is increased by 2π (a complete turn round the z-axis) and z is decreased by $2\pi\gamma/k$. The quantity $2\pi\gamma/k$ is the pitch of the helix.

Chandrasekhar's theorem is thus formally proved; but the vital point is that the function $\theta = \cos^{-1}[x/\sqrt{(x^2 + y^2)}]$ is infinitely many-valued, and the helical symmetry is simply an expression of this many-valuedness. The same thing can be expressed by saying that this solution is not quasi-ergodic.

Stellar systems having helical symmetry do not in fact exist, since their masses would be infinite. Accordingly, Chandrasekhar was obliged to put $\gamma = 0$, eliminating the many-valuedness *a posteriori*. The quasi-ergodic condition (one-valuedness) which we have imposed on stellar systems automatically excludes such solutions *a priori*, and signifies that we regard as unreal all many-valued solutions of this kind. By abandoning the quasi-ergodic condition we obtain mathematically correct solutions, but these are physically unreal.

The same kind of inaccuracy is involved in replacing the integral (5.106), as an argument of the potential (5.109), by its square root ϱ. This would eventually appear when more particular solutions were considered.

Let us now apply the quasi-ergodic condition to the system and see what are the consequences as regards the proof of Chandrasekhar's theorem.

The integral (5.107) is one-valued only if $\gamma = 0$. Then $U = U(\varrho, z)$, and the symmetry is rotational. With $\gamma = 0$ we find from (5.104) and (5.94) $Q = k(yu - xv)$. This is the usual area integral (angular momentum integral). Thus, if we apply the natural condition that the phase density should be one-valued (quasi-ergodic system), we can deduce from the proof of Chandrasekhar's theorem only the following result: if a stellar system in a steady state with internal motions has a one-valued linear integral of the form (5.94), then the mass distribution in the system has rotational symmetry, and the integral (5.94) is the integral of angular momentum about the axis of symmetry.

§ 9. Quasi-integrals of the Motion. Oort's Quasi-integral

As we have seen, it is usually assumed that in the Galaxy there are no one-valued integrals of the motion except the energy integral and the area integral. The ellipsoidal dynamics of the Galaxy is generally based on these classical integrals. However, the resulting picture exhibits some important discrepancies with the observational results. Two of the most noticeable of these are, as already mentioned, the deviation of the vertex (i.e. the direction of the greatest axis of the velocity ellipsoid) from the direction of the centre of the Galaxy, and the inequality of the three axes of the observed ellipsoid.

For this reason, attempts have been made to find functions of the coordinates and velocities which should be *locally constant*, i.e. remain approximately constant under certain known conditions within a fairly large region of the Galaxy. Functions having this property will be called *quasi-integrals*, following Lindblad [68] and Kuzmin [58]. The problem of the quasi-integrals was first considered in a general form by Kuzmin [58].

Historically, the first quasi-integral was used by Oort [92] to study the distribution of star density at various distances from the galactic plane (i.e. along the z-axis). Oort's quasi-integral is obtained if we assume that, within some restricted region of the Galaxy, the gravitational potential $U(\varrho, z)$ can be resolved into two functions, one depending only on ϱ and the other only on z:†

$$U(\varrho, z) = U_1(\varrho) + U_2(z). \tag{5.110}$$

In this case the equations of motion for a free particle are

$$\left.\begin{aligned} \ddot{\varrho} - \varrho\dot{\theta}^2 &= U_1'(\varrho), \\ \varrho^2\dot{\theta} &= h = \text{constant}, \\ \ddot{z} &= U_2'(z), \end{aligned}\right\} \tag{5.111}$$

where the primes denote derivatives.

We see that the third equation now involves only the coordinate z. This means that the motion in the z-direction is independent of that in the coordinates ϱ and θ. The third equation in (5.111) is easily integrated and gives an independent integral of the motion in the z-direction:

$$I_3 = Z^2 - 2U_2(z) = 2H_1, \tag{5.112}$$

† This is true, in particular, if the Galaxy is a homogeneous spheroid. Then $U = A - B\varrho^2 - Dz^2$, where A, B and D are positive constants depending on the shape and size of the spheroid and on the density of the matter in it.

in addition to the two already known:

$$I_1 \equiv \mathrm{P}^2 + \Theta^2 + Z^2 - 2U = 2H, \quad I_2 \equiv \varrho\Theta = h.$$

The integral (5.112) will be called *Oort's integral.* The energy constant H_1 in the z-direction may be determined by substituting in (5.112) $z = 0$, $Z = Z_0$. This gives

$$Z^2 - 2[U_2(z) - U_2(0)] = Z_0^2. \tag{5.113}$$

Here Z_0 is clearly the "initial" velocity of the star, i.e. the velocity which it has whenever it crosses the galactic plane.

Hence we see that, if the distribution of the velocity component Z_0 at the galactic plane, where the Sun is located, is known, we can find the distribution function for the velocity Z at any distance from the galactic plane within the region where the fundamental assumption (5.110) concerning the potential holds good. For, let $f(Z_0^2)$ be the distribution function for the component Z_0. Then, from eqn. (5.113), the distribution function for the component Z is

$$F(Z^2) = f\{Z^2 - 2[U_2(z) - U_2(0)]\}. \tag{5.114}$$

In particular, if $f(Z_0^2)$ is a normal distribution, i.e. if the components Z_0 are distributed according to

$$f(Z_0^2) = [\nu_0/\sigma_z\sqrt{(2\pi)}]\exp(-Z_0^2/2\sigma_z^2) \tag{5.115}$$

(the mean value of Z_0 must evidently be zero), then, by (5.114), $F(Z^2)$ is also a normal distribution:

$$F(Z^2) = \nu_0\exp\{[U_2(z) - U_2(0)]/\sigma_z^2\}\cdot[1/\sigma_z\sqrt{(2\pi)}]\exp(-Z^2/2\sigma_z^2). \tag{5.116}$$

Hence the variance of the z-velocities σ_z^2 is a constant, independent of z.† The variances of the other two velocity components σ_ϱ^2 and σ_θ^2 are in general functions of ϱ, θ and z. This shows that the axes of the velocity ellipsoid must be unequal in the region where Oort's integral is valid.

From the velocity distribution (5.114) we find by integration the star density

$$\nu(z) = \int_{-\infty}^{\infty} f\{Z^2 - 2[U_2(z) - U_2(0)]\}\,\mathrm{d}Z. \tag{5.117}$$

The Z distribution function is $f\{Z^2 - 2[U_2(z) - U_2(0)]\}\,Z/Z_0$, and the times spent by each star in layers of thickness $\mathrm{d}z$ at z and at $z = 0$ are in the ratio Z_0/Z. Thus the star density $\nu(z)$ is given by (5.117).

If, in particular, $f(Z_0^2)$ is the normal distribution (5.115), integration of (5.116) gives

$$\nu(z) = \nu_0\exp\{[U_2(z) - U_2(0)]/\sigma_z^2\}.$$

If we take only points with a given ϱ, then $U_2(z) - U_2(0)$ in the exponent can be replaced by $U(\varrho, z) - U(\varrho, 0)$, giving

$$\nu(z) = \nu_0\exp\{[U(\varrho, z) - U(\varrho, 0)]/\sigma_z^2\}. \tag{5.118}$$

† In general, σ_z^2 is a function of z given by Jeans' second equation (see Chapter VIII, § 6):

$$\partial(\nu\sigma_z^2)/\partial z = \nu(z)\,\partial U/\partial z.$$

This relation was used in 1932 by Oort [92] to find the star density distribution in the z-direction.

Let us now consider what are the dimensions of the region in which Oort's quasi-integral is valid. Whatever the form of the gravitational potential in the Galaxy, it can be represented, in a sufficiently small region surrounding a non-singular point (ϱ_0, z_0), by a linear expression

$$U(\varrho, z) = U(\varrho_0, z_0) + (\partial U/\partial \varrho)_0 (\varrho - \varrho_0) + (\partial U/\partial z)_0 (z - z_0). \quad (5.119)$$

This expression has formally the additive property given by (5.110), and therefore the question naturally arises: is there any dynamical significance in a quasi-integral whose constancy is unrelated to the mass distribution in the system and follows simply from such purely mathematical properties of functions as continuity and differentiability? If a sufficiently small region is considered, then practically any function will be constant within that region. Such a function evidently need bear no physical relation to the system. Starting from eqn. (5.119), Kuzmin ([58], p. 350) concludes that Oort's quasi-integral is applicable only to stars belonging to the plane sub-systems, for which $|z| \leqslant 100$ ps.

The dispersion of the residual velocities of stars in the plane sub-systems is known to be very small in comparison with the circular velocity, and the circular velocity of the centroid is almost equal to that of a particle. The stars in the plane sub-systems all move in quasi-circular orbits, therefore, executing very small oscillations about the centroid, and each star remains, during its motion, within a region where the additive form (5.119) of the potential is valid. Thus the third integral (the quasi-integral) is applicable to the plane sub-systems.

It is not difficult to show, however, that the region where this integral is applicable is in fact much larger, and, in particular, extends to values of z far beyond the limits of the plane sub-systems. For the preceding discussion has been based only on a very general mathematical property of the potential (that it can be expanded in series), and so is valid for any form of the potential. We can make some statements *a priori* about properties of the potential which can be used to extend the range of validity of the quasi-integral derived above. If the Galaxy is of the shape of a normal spiral galaxy of type Sb or Sc, it must have a small dense nucleus and a main peripheral part considerably flattened at the poles. The Sun must be in the principal plane of symmetry, and in the peripheral part at a considerable distance from the centre.

In this case, as has been shown in Chapter III, § 7, the attraction of the Galaxy at any point C in the neighbourhood of the Sun can be approximately regarded as consisting of the attraction of two masses: (a) that of the nucleus, whose dimensions may be supposed small compared with the distance of the point C from the centre of the Galaxy, (b) that of a homogeneous flattened spheroid.†

Then the potential of the Galaxy can be written as a sum: $U = U_1 + U_2$; if the nucleus is assumed to be spherically symmetrical, then $U_1 = GM_1/r = GM_1/\sqrt{(\varrho^2 + z^2)}$, where M_1 is the mass of the nucleus, that is, the concentrated mass of the Galaxy.

† It may be recalled that the mass of the Galaxy was estimated by Oort using this approximate treatment (see Chapter III, § 7).

The potential of the homogeneous spheroid (the distributed mass of the Galaxy) is $U_2 = A - B\varrho^2 - Dz^2$, where A, B and D are positive constants. Thus

$$U = \frac{GM_1}{\sqrt{(\varrho^2 + z^2)}} + A - B\varrho^2 - Dz^2. \tag{5.120}$$

If z is small in comparison with ϱ, the first term can be expanded in powers of z/ϱ. For example, if

$$z < 0{\cdot}12\,\varrho, \tag{5.121}$$

we have, with a relative error of about 1%,

$$U = \left(\frac{GM_1}{\varrho} + A - B\varrho^2\right) - Cz^2. \tag{5.122}$$

Thus the potential of the Galaxy, on the above assumptions, has the additive Oort form (5.110).

The inequality (5.121) describes a volume in the Galaxy lying within a conical surface symmetrical about the galactic plane and the axis of rotation. The region in which Oort's quasi-integral is valid includes only that part of this volume which lies outside the central nucleus, and sufficiently far both from the nucleus and from the edge of the Galaxy.

If we take $\varrho_\odot = 8$ kps, then by (5.121) the upper limit of z in the neighbourhood of the Sun may be taken as about 1 kps. It should be noted that this limit gives only the distance, perpendicular to the galactic plane, to the corresponding point of the conical surface. In directions parallel to the galactic plane the distance is considerably greater.

Thus the range of validity of Oort's quasi-integral can be extended far into the region of intermediate and even spherical sub-systems.

The existence of a third independent integral makes possible, as mentioned above, an explanation of the inequality of the axes of the velocity ellipsoid. We see that by means of the third integral this anomaly in the general theory of the steady-state Galaxy can be explained for stars of the plane and intermediate sub-systems and even those of the spherical sub-systems. For the first two types the result is evident because they lie entirely within the limits of $|z|$ given above. As regards the spherical sub-systems, our data concerning the motions of stars with $|z| > 1000$ ps are as yet so meagre that practically all the information available about the velocity distribution in these sub-systems relates to stars with $|z| < 1000$ ps, and so they also are in practice within the range where Oort's integral is valid. It should be noted that, even when the time comes that the majority of the stellar motions used in determining the velocity ellipsoid belong to stars with $|z| > 1000$ ps, those stars lying at less than this distance from the galactic plane will still affect the result, and the axes of the ellipsoid will still be unequal. In order to resolve the problem it will be necessary to select those stars for which $|z| > 1000$ ps, a task which must be left to the fairly distant future.

§ 10. The Problem of the Third Integral for Systems with Rotational Symmetry

The discrepancies between the properties of the observed velocity ellipsoid and the results of the ellipsoidal theory of a steady-state Galaxy (the deviation of the vertex, the inequality of the axes, etc.) have been a continual stimulus to attempts to generalise the classical theory. Such generalisations are possible in three respects.

The first is the abandonment of the requirement of galactic stability. On this aspect there is an extensive literature initiated by Chandrasekhar [17]. We shall not, however, pause to consider these investigations in detail, not because they do not merit it but because present theories of systems not in a steady state necessarily involve the introduction of additional unknown functions of time, the determination of which will be possible only when we know at least the essential features of the physical process of the evolution of galaxies. For the time being, the theory must be confined to systems in a quasi-steady state, with a view to considering the problem of the possible time dependence of the parameters of the system.

The second approach may be described as dealing with the problem of local instability. As Chandrasekhar has shown, the solution of the problem of non-steady-state motion furnishes a simple explanation of both the inequality of the axes of the ellipsoid and the deviation of the vertex. However, these results, which are based on the ellipsoidal theory, necessarily assume that the above-mentioned anomalies of the velocity ellipsoid occur at every point in the Galaxy. In addition, they lead to a quite artificial dependence of the elements of the velocity ellipsoid on the coordinates. It is therefore of interest to consider the anomalies of the velocity ellipsoid as a *local* phenomenon, occurring only in the neighbourhood of the Sun. One possible treatment of this kind is given in Chapter VII, where it is shown that all these anomalies of the ellipsoid, as well as the K term in the radial velocities of some early-type stars, may be accounted for in a uniform manner as being the result of the dissipation of the local cluster into the surrounding star field.

There remains a third procedure, in which the Galaxy is, as before, treated as a system in a steady state, but in addition to the energy and angular momentum integrals I_1 and I_2, say, there exists a third integral which is to be determined. So long as the quasi-ergodic property remains a hypothesis, the search for a third integral is entirely legitimate. The literature here is again fairly extensive. Various authors have defined the third integral in different ways, using somewhat abstract mathematical methods, e.g. with the condition that the third integral should be quadratic in the velocity components, that the mass of the stellar system should be finite, and so on.

There are then two possible continuations. One is to determine, with the help of the three integrals, the gravitational potential of the stellar system, and hence the star density distribution. This can then be compared with the results of observation of both our own and other galaxies. The other approach is to regard the third integral only as a quasi-integral akin to the well-known Oort–Lindblad quasi-integral of the motion in the z-direction.

Space does not allow a fuller discussion of the interesting work of Chandrasekhar [17], Heckmann and Strassl [37], Fricke [34a] and Contopoulos [21a]. Here we shall consider only the quadratic integral explicitly introduced by Camm [16a], Fricke [34a] and in particular by G.G. Kuzmin [58, 61].†

It is to be noted, however, that all such investigations involve difficulties due to the fact that it is unknown whether stellar systems starting from reasonable initial states may after a physical process of evolution acquire a phase distribution which is a function of the third integral concerned.

The quasi-ergodic hypothesis denies this possibility. The resolution of the matter must await the development of a physical theory of the evolution of stellar systems.

To obtain a third integral, Kuzmin [58] considers a quadratic form in the velocities:

$$I_3 = a_{20}\mathrm{P}^2 + 2a_{11}\mathrm{P}Z + a_{02}Z^2 + a_{10}\mathrm{P} + a_{01}Z + a_{00}, \tag{5.123}$$

where a_{20}, a_{11}, a_{02} are functions of the coordinates ϱ and z; a_{10}, a_{01} linear functions of Θ with coefficients depending on ϱ and z; and a_{00} a similar quadratic function of Θ. Thus I_3 is a quadratic form in all three velocities P, Θ, Z, in which the component Θ does not appear explicitly. We shall eliminate Θ from the subsequent discussion by expressing it in terms of the area integral $I_2 = \varrho\Theta$.

In § 6 it has been shown that, for a system in a steady state with rotational symmetry, both the potential U and the phase density f depend on the coordinates ϱ and z only. Hence

$$\partial U/\partial\theta = 0, \quad \partial f/\partial\theta = 0, \quad \partial f/\partial t = 0, \tag{5.124}$$

and the Boltzmann eqn. (5.15) becomes

$$\mathrm{P}\frac{\partial f}{\partial\varrho} + Z\frac{\partial f}{\partial z} + \left(\frac{\partial U}{\partial\varrho} + \frac{\Theta^2}{\varrho}\right)\frac{\partial f}{\partial\mathrm{P}} - \frac{\mathrm{P}\Theta}{\varrho}\frac{\partial f}{\partial\Theta} + \frac{\partial U}{\partial z}\frac{\partial f}{\partial Z} = 0. \tag{5.125}$$

This equation can be transformed by expressing Θ in terms of the area integral $I_2 = \varrho\Theta$. Then

$$\mathrm{P}\frac{\partial f}{\partial\varrho} + Z\frac{\partial f}{\partial z} + \frac{\partial\Phi}{\partial\varrho}\frac{\partial f}{\partial\mathrm{P}} + \frac{\partial\Phi}{\partial z}\frac{\partial f}{\partial Z} = 0, \tag{5.126}$$

where we have put for brevity

$$\Phi = U - I_2^2/2\varrho^2. \tag{5.127}$$

Substituting I_3 from (5.123) on the left of (5.126), we have a cubic algebraic form in the velocities. If I_3 is an integral, this form must be identically zero. Equating to zero the coefficients of the various powers of the velocities, we therefore obtain ten linear partial differential equations for the coefficients a_{ij}, which, as usual in such

† In the Russian edition of this book (1958) this integral was called Kuzmin's integral. Since that time, Contopoulos [21a] in his paper of 1960 has pointed out that the same integral was earlier used by Chandrasekhar [16b, 17, 18], Camm [16a] and van Albada [2a]. The author is glad to correct this error.

cases, can be written in a compact form in four groups:

$$\left.\begin{aligned}
&\frac{\partial a_{20}}{\partial \varrho} = \frac{\partial a_{20}}{\partial z} + 2\frac{\partial a_{11}}{\partial \varrho} = 2\frac{\partial a_{11}}{\partial z} + \frac{\partial a_{02}}{\partial \varrho} = \frac{\partial a_{02}}{\partial z} = 0; && \text{(I)}\\
&\frac{\partial a_{00}}{\partial \varrho} + 2a_{20}\frac{\partial \Phi}{\partial \varrho} + 2a_{11}\frac{\partial \Phi}{\partial z} = 0, && \\
&\frac{\partial a_{00}}{\partial z} + 2a_{11}\frac{\partial \Phi}{\partial \varrho} + 2a_{02}\frac{\partial \Phi}{\partial z} = 0; && \text{(II)}\\
&\frac{\partial a_{10}}{\partial \varrho} = \frac{\partial a_{10}}{\partial z} + \frac{\partial a_{01}}{\partial \varrho} = \frac{\partial a_{01}}{\partial z} = 0; && \text{(III)}\\
&a_{10}\frac{\partial \Phi}{\partial \varrho} + a_{01}\frac{\partial \Phi}{\partial z} = 0. && \text{(IV)}
\end{aligned}\right\} \quad (5.128)$$

In eqn. (IV) we substitute linear expressions for a_{10} and a_{01} in terms of I_2, and also use the expression (5.127) for Φ in terms of I_2. Since the values of I_2 may be specified arbitrarily, and eqn. (IV) must be satisfied identically, we obtain

$$a_{10} = a_{01} = 0. \quad (5.129)$$

We now eliminate Φ and a_{00} from eqns. (II). In these equations the quantities a_{20}, a_{11} and a_{02} do not depend on I_2, and Φ is given in terms of I_2 by (5.127). The same arguments as led to (5.129) give the result

$$a_{00} = a_1 + a_2 I_2^2, \quad (5.130)$$

where a_1 and a_2 are functions of the coordinates ϱ and z only.

Substituting (5.130) in eqns. (II) and equating to zero the coefficients of I_2^2 and the terms not involving I_2, we have

$$\left.\begin{aligned}
\frac{\partial a_1}{\partial \varrho} + 2a_{20}\frac{\partial U}{\partial \varrho} + 2a_{11}\frac{\partial U}{\partial z} = 0,\\
\frac{\partial a_1}{\partial z} + 2a_{11}\frac{\partial U}{\partial \varrho} + 2a_{02}\frac{\partial U}{\partial z} = 0,
\end{aligned}\right\} \quad (5.131)$$

$$\left.\begin{aligned}
\frac{\partial a_2}{\partial \varrho} + 2\frac{a_{20}}{\varrho^3} = 0,\\
\frac{\partial a_2}{\partial z} + 2\frac{a_{11}}{\varrho^3} = 0.
\end{aligned}\right\} \quad (5.132)$$

Cross-differentiation of eqns. (5.132) and elimination of $\partial^2 a_2/\partial\varrho\,\partial z$ gives

$$\varrho\frac{\partial a_{20}}{\partial z} - \varrho\frac{\partial a_{11}}{\partial \varrho} + 3a_{11} = 0. \quad (5.133)$$

By differentiation of the eqns. (I) in (5.128) it is easy to see that a_{20} must be a quadratic function of z, a_{02} a quadratic function of ϱ, and a_{11} a bilinear function of ϱ and z. Then, using (5.133), we find

$$\left.\begin{aligned} a_{20} &= b_1 + b_2 z^2, \\ a_{02} &= c_1 + b_2 \varrho^2, \\ a_{11} &= -b_2 \varrho z, \end{aligned}\right\} \qquad (5.134)$$

where b_1, b_2, c_1 are constant coefficients.

If these expressions are substituted in (5.132), a determination of a_2 by integration gives $a_2 = (b_1 + b_2 z^2)/\varrho^2 + d_1$, where d_1 is another constant. Substitution of this value of a_2 in (5.130) gives, since $I_2 = \varrho\Theta$,

$$a_{00} = a_1 + d_1 I_2^2 + b_1 \Theta^2 + b_2 z^2 \Theta^2. \qquad (5.135)$$

When (5.134), (5.135), (5.129) are substituted as the coefficients in the original formula (5.123) for I_3, we have, omitting the constant term $d_1 I_2^2$ and dividing by b_2,

$$I_3 = (\varrho Z - z\mathrm{P})^2 + z^2\Theta^2 + (b_1/b_2)(\mathrm{P}^2 + Z^2) + (c_1/b_2) Z^2 + a_1/b_2. \qquad (5.136)$$

We eliminate $\mathrm{P}^2 + Z^2$ by means of the energy integral $I_1 = \mathrm{P}^2 + \Theta^2 + Z^2 - 2U$, and use the notation

$$\begin{aligned} z_0^2 &= (c_1 - b_1)/b_2, \\ (a_1 + 2b_1 U)/b_2 &= -2z_0^2 U^*; \end{aligned} \qquad (5.137)$$

omitting the constant term $b_1 I_1/b_2$, we have finally

$$I_3 = (\varrho Z - z\mathrm{P})^2 + z^2\Theta^2 + z_0^2(Z^2 - 2U^*). \qquad (5.138)$$

This is the third integral.

To determine the function U^*, from (5.131) with (5.134), (5.135) and (5.137) we have the equations

$$\begin{aligned} z_0^2 \frac{\partial U^*}{\partial \varrho} &= z^2 \frac{\partial U}{\partial \varrho} - \varrho z \frac{\partial U}{\partial z}, \\ z_0^2 \frac{\partial U^*}{\partial z} &= (\varrho^2 + z_0^2) \frac{\partial U}{\partial z} - \varrho z \frac{\partial U}{\partial \varrho}. \end{aligned} \qquad (5.139)$$

Hence we see, *inter alia*, that z_0 has the dimensions of length, and U^* those of potential.

From (5.139) we can derive a condition which must be satisfied by the potential in a stellar system in order that the third integral of the motion should exist. This condition is obtained by cross-differentiating these equations and eliminating U^*. Then

$$3\left(z\frac{\partial U}{\partial \varrho} - \varrho\frac{\partial U}{\partial z}\right) - (\varrho^2 + z_0^2 - z^2)\frac{\partial^2 U}{\partial \varrho\, \partial z} + \varrho z\left(\frac{\partial^2 U}{\partial \varrho^2} - \frac{\partial^2 U}{\partial z^2}\right) = 0. \qquad (5.140)$$

This is the required condition. It appears somewhat involved, and in particular it has no evident physical significance. It is in fact simply the condition for the existence of an integral of the motion which is a quadratic form in the velocities in a two-dimensional problem of dynamics.†

If the condition (5.140) is satisfied, the phase density is a function of three integrals: $f = f(I_1, I_2, I_3)$, where I_1 and I_2 are again the energy integral and the area integral: $I_1 = \mathrm{P}^2 + \Theta^2 + Z^2 - 2U$; $I_2 = \varrho\Theta$, and I_3 is the integral derived above. It is easy to show that in this case the velocity ellipsoid has unequal axes, and its greatest axis is not parallel to the galactic plane. These two properties of the velocity ellipsoid distinguish the case where there are three integrals from the classical case where there are only two. For, let the coordinate axes be rotated so that the integral I_3 is transformed into a sum of squares. Since the velocity Θ already appears in I_3 only as Θ^2, we need only rotate the axes in the PZ (meridian) plane. Denoting the velocity components along the new axes by v_1 and v_2, we have

$$I_3 = z_0^2(x_1 v_1^2 + x_2 v_2^2) + z^2\Theta^2 - 2z_0^2 U^*, \tag{5.141}$$

where x_1 and x_2 are the roots of the equation

$$z_0^2 x^2 - (\varrho^2 + z_0^2 + z^2)x + z^2 = 0, \tag{5.142}$$

and are always positive.

It is evident that this rotation of the axes leaves unchanged both the energy integral and the area integral. Thus the phase density is an even but unsymmetrical function of the two new velocity components. This means that the corresponding ellipsoid has in general three unequal axes, and no diametral plane coinciding with the galactic plane.

The angle α through which the coordinate axes are rotated in the above transformation is given by

$$\tan 2\alpha = 2\varrho z/(\varrho^2 + z_0^2 - z^2); \tag{5.143}$$

α is also the angle between one of the diametral planes of the ellipsoid and the galactic plane.

From (5.143) we see that $\alpha = 0$ if either $\varrho = 0$ (i.e. on the axis of rotation) or $z = 0$ (i.e. in the galactic plane). Unfortunately, we have at present no means of verifying whether the velocity ellipsoid is in fact inclined to the galactic plane, since the phenomenon becomes apparent only at considerable distances from that plane.

In general it should be noted that any attempt to obtain quadratic integrals of the motion for the solution of problems in the dynamics of stellar systems necessarily leads to very artificial conclusions. This is well seen in the work of Chandrasekhar [18], who solved a similar problem by assuming that the phase density is a "generalised" ellipsoidal function

$$f = f(T), \tag{5.144}$$

† See E. T. Whittaker, *A treatise on the analytical dynamics of particles and rigid bodies*, Cambridge University Press, 1937, p. 331.

where T is a complete quadratic function of the velocities, with coefficients depending on the coordinates:

$$
\begin{aligned}
T = a(\mathrm{P} - \mathrm{P}_0)^2 &+ b(\Theta - \Theta_0)^2 + c(Z - Z_0)^2 \\
&+ 2f(\Theta - \Theta_0)(Z - Z_0) + 2g(Z - Z_0)(\mathrm{P} - \mathrm{P}_0) \\
&+ 2h(\mathrm{P} - \mathrm{P}_0)(\Theta - \Theta_0) + \sigma; \qquad (5.145)
\end{aligned}
$$

$a, b, \ldots, h, \sigma, \mathrm{P}_0, \Theta_0, Z_0$ are to be found as functions of the coordinates.

Since, by Jeans' theorem, the phase density is an integral of the motion, the problem is evidently equivalent to finding the quadratic integral of the motion T.

Substituting (5.144), (5.145) in the Boltzmann eqn. (5.125), we obtain on the left-hand side a complete cubic polynomial in the velocity components P, Θ, Z, consisting of twenty terms. Since eqn. (5.125) must be satisfied identically, we obtain, by equating the various coefficients to zero, twenty first-order partial differential equations in the unknown coefficients in the quadratic form T.

We shall not give the calculations involved in the integration of these equations (see [17], pp. 93 ff.), but merely note that they naturally fall into four groups:

(I) ten equations for the coefficients,

(II) six equations for the centroid velocity components,

(III) one equation for the potential,

(IV) three "compatibility equations", which serve to determine the coefficient σ and from which we obtain two conditions of compatibility analogous to eqn. (IV) of (5.128).

A feature to which attention may be drawn is that the equations in groups (I) and (II) make it possible to complete the integration, although they do not involve the potential, i.e. the only function which represents the mass distribution and the force field. Thus the coefficients of the ellipsoid and the elements of the centroid motion are entirely determined (apart from some of the arbitrary constants of integration) by the formal condition that the required integral of the motion has a specific mathematical form.† Since there are no physical arguments to show that the integral of the motion must be quadratic, the result has no physical significance.

The same is true as regards the third integral. The coefficients a_{20}, a_{11}, a_{02} in the quadratic terms can be found by solving (5.128) (I), (II), and the coefficients a_{10}, a_{01} by solving (5.128) (III). These equations do not involve the potential U or the function Φ which replaces it, and so the principal coefficients in the integral I_3 are independent of the distribution of mass in the system.

Thus we can hardly extend the integral I_3 to the entire Galaxy, i.e. regard it as an exact integral. However, using the fact that (as Kuzmin has remarked) the expression (5.138) for I_3 involves the arbitrary constant z_0, we can choose the latter in such a way that I_3 remains approximately constant within a certain region of the Galaxy. In other words, I_3 may be used as a quasi-integral.

† The same conclusion was reached in § 7 by a different method of solution.

§ 11. The Potential Energy of Stellar Systems

In §§ 11 and 12 we shall consider some general topics in the statistical mechanics of stellar systems.

An important dynamical characteristic of a stellar system is the *potential energy*. The significance of this quantity may be elucidated as follows. Let us first consider a system of two stars, of masses m_1 and m_2, at a distance r_{12} apart. The force of mutual attraction between the stars is given by the vector $Gm_1m_2\boldsymbol{r}_{12}/r_{12}^3$, where $\boldsymbol{r}_{12}$ is the relative radius vector. When the stars undergo a relative vector displacement $\mathrm{d}\boldsymbol{r}_{12}$, the work done is

$$\mathrm{d}A_{12} = -Gm_1m_2\mathbf{r}_{12}\cdot\mathrm{d}\mathbf{r}_{12}/r_{12}^3 = -Gm_1m_2\,\mathrm{d}r_{12}/r_{12}^2,$$

since the scalar product $\boldsymbol{r}_{12}\cdot\mathrm{d}\boldsymbol{r}_{12} = +r_{12}\,\mathrm{d}r_{12}$. The minus sign signifies that, when r_{12} increases, the displacement is against the action of the force. The right-hand side is the differential $-Gm_1m_2\,\mathrm{d}r_{12}/r_{12}^2 = \mathrm{d}(Gm_1m_2/r_{12})$.

The function whose differential is equal to the infinitesimal work is called the *potential*, and the potential with reversed sign is called the *gravitational potential energy* or *Newtonian energy* of the system: $W = -Gm_1m_2/r_{12}$.

It is shown in textbooks on theoretical mechanics that the work done by gravitational forces in an arbitrary finite displacement between two arbitrary configurations p and q is independent of the paths of the particles m_1 and m_2, and is minus the increment of the potential energy: $A_{pq} = -(W_p - W_q)$. Evidently the potential energy is determined by these conditions only to within an arbitrary additive constant. This constant is usually chosen so that the potential energy satisfies the condition $\lim W = 0$ for $r_{12}\to\infty$. Then W is minus the "work of separation", i.e. the work which must be done against the force of mutual attraction to remove one of the stars to infinity from an initial distance r_{12}. The "work of separation" is equal and opposite to the work which must be done by the force of gravitation to bring the masses m_1 and m_2 from an initially infinite distance apart to the given distance r_{12}. In brief, the *Newtonian potential energy* is a quantity equal to the work done by the force of gravitation in bringing the system from a state of infinite separation to the configuration considered.

To determine the potential energy of a given system of particles m_i ($i = 1, 2, \ldots, N$) with mutual distances r_{ij} we proceed as follows. Let us imagine an initial state of infinite separation of every pair of particles. This can be reached by first making the particles coincide and then moving them radially away to infinity. We bring the particle m_1 from its initial position to that which it occupies in the given system. The work done by the gravitational forces is zero. Next we bring the particle m_2 from its original infinite distance to its final position. This displacement occurs in the gravitational field of m_1 alone, and so the corresponding work is $-Gm_1m_2/r_{12}$. When m_3 is brought from infinity to its final position in the system, the work done is $-Gm_1m_3/r_{13} - Gm_2m_3/r_{23}$. The work required to bring m_i into position is

$$-G\sum_{j=1}^{i-1} m_jm_i/r_{ij},$$

and so on.

Thus, when all the particles have reached their final positions, the potential energy of the system is found to be

$$W = -G \sum_{i=1}^{N} \sum_{j=1}^{i-1} \frac{m_i m_j}{r_{ij}} \tag{5.146}$$

or

$$W = -\frac{1}{2} G \sum_{i=1}^{N} \sum_{j=1}^{N} \frac{m_i m_j}{r_{ij}} \quad (i \neq j),$$

since each pair of particles appears twice in the latter sum.

The above argument shows that the process of building up the quantity W takes account only of the relative position of the stars in the system, and entirely neglects the energy of their motion.

In the successive movements of stars to their final positions the stars previously placed are supposed to be immovable in their positions, whereas in reality the forces of gravitation act simultaneously on all the stars and give them accelerations. But since the work done by gravitational forces depends only on the final positions of the particles, and not on the path taken, the potential energy W is the *energy of position.*

The case of stellar systems in a *steady state* is of particular importance. This means that the phase distribution of stars in the system does not vary with time. In such a system a star leaving any point is replaced by another. In other words, in a steady state the stars can be regarded as in fixed positions if we make no attempt to identify them, i.e. if we apply the principle of indistinguishability discussed in § 1.

Let us now establish the relation between the potential energy of the whole system and those of the individual stars. The potential energy of the ith star is evidently

$$W_i = -G m_i \sum_{j=1}^{N} \frac{m_j}{r_{ij}} \quad (i \neq j).$$

This formula and the second form of W given above show that the potential energy of the whole system is half the sum of the potential energies of the individual stars:

$$W = \tfrac{1}{2} \sum_{i=1}^{N} W_i. \tag{5.147}$$

This shows that the potential energy of a stellar system is *not additive*, i.e. the sum of the energies of the individual components (the stars) is not equal to the energy of the system.

The expression (5.146) for the potential energy may be simplified if we assume that all the stars have the same mass, i.e. $m_i = m$, $M = mN$, where N is the total number of stars in the system. Then

$$W = -Gm^2 \sum_{i>j} \frac{1}{r_{ij}} = -Gm^2 \frac{N(N-1)}{2\alpha},$$

where $1/\alpha = \overline{1/r_{ij}}$ is the mean reciprocal distance between pairs of stars. In the numerator of the last member we may omit unity as being small in comparison with N. Thus

$$W \approx -GM^2/2\alpha. \tag{5.148}$$

An important particular case is that of a spherical system. Here the level surfaces in the mass distribution (star density) are concentric spheres. From the theory of the Newtonian potential it is known that a homogeneous spherical shell exercises no attraction at an internal point. Using the fact that the work done by the force of gravitation is independent of the path of the particle, we can determine the potential energy of the system as follows. To bring the stars into an initial position at infinity we remove from the system successive spherical shells without changing their shape or concentricity. The stars are then returned, in order to calculate the work done by the force of gravitation, not separately but in shells, and each internal layer attracts those outside it as if its whole mass were concentrated at the centre of the system. The outer shells do not attract the mass within them. Hence we see that the work done by the forces of gravitation in bringing each shell from infinity to the finite radius corresponding to its position in the system is simply the sum of the amounts of work required to bring the individual stars into position.

Let us calculate the potential energy of a spherical system of radius a and mass M. We assume that a sphere of radius r and mass $M(r)$ has been formed by the successive application of spherical shells. Then the work done by gravitation in bringing an infinitesimal shell of mass $\mathrm{d}M(r)$ from infinity to radius r is $-[GM(r)/r]\,\mathrm{d}M(r)$. According to the above discussion the total energy of the system W is just the sum of the work done in the successive movement of the spherical shells. Hence the total potential energy is

$$W = -G\int_0^M \frac{M(r)}{r}\,\mathrm{d}M(r). \tag{5.149}$$

The mass of a shell is evidently $\mathrm{d}M(r) = 4\pi m\nu(r)\,r^2\,\mathrm{d}r$, where m is the mass of a star and $\nu(r)$ the star density. Hence (5.149) becomes

$$W = -4\pi Gm\int_0^a M(r)\,\nu(r)\,r\,\mathrm{d}r. \tag{5.150}$$

As an example, let us calculate the energy of a homogeneous sphere in which the star density is ν_0. Here $M(r) = \frac{4}{3}\pi m\nu_0 r^3$, $\nu(r) = \nu_0$, and formula (5.150) gives $W = -16\pi^2 Gm^2\nu_0^2 a^5/15$. This expression may be simplified by putting $M = \frac{4}{3}\pi m\nu_0 a^3$, which leads to the well-known formula

$$W = -\tfrac{3}{5}GM^2/a. \tag{5.151}$$

In the more general case of a polytropic sphere of index n (see Chapter IX, § 2) the corresponding formula is ([28], pp. 86–7)

$$W = -\frac{3}{5-n}\,GM^2/a.$$

Hence we see, *inter alia*, that $W > 0$ when $n > 5$. The sphere is therefore dynamically stable only if $n < 5$. When $n = 0$ we obtain (5.151).

§ 12. The Virial Theorem. Poincaré's Theorem on the Limiting Angular Velocity of Rotation

In this section we shall consider an important theorem of statistical mechanics, demonstrated in 1911 by Henri Poincaré [111] and known as the *virial theorem.* It is very general, and is frequently useful in obtaining approximate relations between the dynamical parameters of stellar systems.

Following Eddington [26], we take an arbitrary stellar system, and for complete generality we do not even assume that it is in a steady state. We shall, however, suppose that its size, shape and internal distribution of mass vary only slowly. The significance of this supposition will be seen later.

Let r_i denote the radius vector of the ith star relative to the centre of mass of the system. The quantity

$$J = \sum_{i=1}^{N} m_i r_i^2 = \sum_{i=1}^{N} m_i (x_i^2 + y_i^2 + z_i^2) \tag{5.152}$$

is the moment of inertia of the system about its centre of mass. We assume that

$$\mathrm{d}^2 J/\mathrm{d}t^2 = 0, \tag{5.153}$$

i.e. that J either remains constant, corresponding to a steady state, or varies at a constant rate, i.e.

$$\mathrm{d}J/\mathrm{d}t = \text{constant}. \tag{5.154}$$

In particular, this condition is approximately valid when J varies sufficiently slowly. In that case, expanding J in powers of t and keeping only the first two terms, we have

$$J \approx a + bt. \tag{5.155}$$

A system which satisfies the condition (5.155) will be called for brevity a *linearly non-steady system*. This term is due to P. P. Parenago [107]. In particular, any quasi-steady system may be regarded as linearly non-steady.

Differentiating (5.155) twice with respect to t, we have

$$\frac{1}{2}\frac{\mathrm{d}^2 J}{\mathrm{d}t^2} \equiv \sum_{i=1}^{N} m_i(\dot{x}_i^2 + \dot{y}_i^2 + \dot{z}_i^2) + \sum_{i=1}^{N} m_i\,(x_i\ddot{x}_i + y_i\ddot{y}_i + z_i\ddot{z}_i) = 0. \tag{5.156}$$

Here the dots above the letters denote time derivatives. The first term on the right is evidently twice the kinetic energy of the system:

$$\sum_{i=1}^{N} m_i v_i^2 = 2T, \tag{5.157}$$

where v_i is the velocity of the ith star. The second term is called the *virial*, whence the name of the theorem.

Let W be the potential energy of the system:

$$W = -G \sum_{i>j}^{N} \frac{m_i m_j}{r_{ij}}, \tag{5.158}$$

where $r_{ij} = \sqrt{[(x_i - x_j)^2 + (y_i - y_j)^2 + (z_i - z_j)^2]}$ is the distance between the ith and jth stars. Then the equations of motion of the ith star are

$$m_i \ddot{x}_i = -\partial W/\partial x_i, \quad m_i \ddot{y}_i = -\partial W/\partial y_i, \quad m_i \ddot{z}_i = -\partial W/\partial z_i, \tag{5.159}$$

and so the virial is

$$-\sum_{i=1}^{N} \left(x_i \frac{\partial W}{\partial x_i} + y_i \frac{\partial W}{\partial y_i} + z_i \frac{\partial W}{\partial z_i} \right).$$

Since W is a homogeneous function of degree -1 in the coordinates, Euler's theorem on homogeneous functions shows that

$$\sum_{i=1}^{N} \left(x_i \frac{\partial W}{\partial x_i} + y_i \frac{\partial W}{\partial y_i} + z_i \frac{\partial W}{\partial z_i} \right) = -W.$$

Thus in this case the virial is the potential energy W of the system, and eqns. (5.156) and (5.157) give

$$2T + W = 0. \tag{5.160}$$

This equation expresses the virial theorem: in any steady or linearly non-steady stellar system, the sum of the potential energy and twice the kinetic energy is zero.

Since, for a closed system, the law of conservation of energy gives

$$T + W = H = \text{constant}, \tag{5.161}$$

eqn. (5.160) can also be written

$$T = -H \tag{5.162}$$

or

$$W = 2H. \tag{5.163}$$

As an example, let us apply the virial theorem to a non-rotating stellar system. We have

$$2T = \sum_{i=1}^{N} m_i v_i^2 = M\hat{v}^2, \tag{5.164}$$

where M is the mass of the whole system and $\hat{v}$ the root-mean-square stellar velocity. Let the system be a homogeneous sphere of radius a. Then, as shown in § 11, the potential energy is

$$W = -\tfrac{3}{5} GM^2/a. \tag{5.165}$$

From (5.160), (5.164) and (5.165) we find

$$\hat{v}^2 = \tfrac{3}{5} GM/a, \tag{5.166}$$

which is a relation between the mass, radius and root-mean-square velocity of a homogeneous spherical stellar system. This relation shows that, during the evolution of a quasi-steady non-rotating system (not involving an appreciable decrease in mass), the mean square velocity is inversely proportional to the radius of the system. As an example, substituting the values $M = 1{\cdot}6 \times 10^5\, m_\odot$ and $a = 35$ ps for globular clusters, we find $\hat{v} = 4$ km/s.

For systems in rigid rotation with angular velocity ω we have

$$2T = M\hat{v}^2 + J\omega^2, \tag{5.167}$$

where M is the mass of the system, $\hat{v}$ the root-mean-square residual stellar velocity with respect to coordinates rotating with the system, and J the moment of inertia about the axis of rotation.

Substituting (5.167) in (5.160) and omitting the positive quantity $M\hat{v}^2$, we obtain *Poincaré's inequality* for discrete gravitating systems:

$$\omega^2 < -W/J. \tag{5.168}$$

This inequality is a necessary condition for the rigid rotation of the system to be a steady state (equilibrium). It may be made more particular by assuming that all the stars in the system are of equal mass. Then, substituting in (5.168) the expression (5.148) for the potential energy and using the fact that $J = Ms^2$, where s is the radius of gyration, we have

$$\omega^2 < GM/2\alpha s^2. \tag{5.169}$$

Here $1/\alpha$ is the mean reciprocal distance between pairs of stars.

The inequality (5.168) can also be applied to systems with a continuous distribution of mass, and for systems of a particular type its significance may be made more evident. For a homogeneous ellipsoid with semiaxes a, b, c we have

$$W = 8\pi^2 G\delta^2 a^2 b^2 c^2 I/15, \tag{5.170}$$

where δ is the density and I an elliptic integral:

$$I = \int_0^\infty \frac{dx}{\Delta}, \quad \text{where} \quad \Delta^2 = (a^2 + x)(b^2 + x)(c^2 + x).$$

Let us assume that the ellipsoid is rotating about its least axis. Taking the z-axis as the axis of rotation, we have for the moment of inertia

$$J = 4\pi\delta abc(a^2 + b^2)/15, \tag{5.171}$$

and the inequality (5.168) becomes

$$\omega^2 < 2\pi G\delta abcI/(a^2 + b^2). \tag{5.172}$$

For a spheroid ($a = b$) the result is

$$\omega^2 < 2\pi G\delta \frac{c}{\sqrt{(a^2 - c^2)}} \cos^{-1}\frac{c}{a}. \tag{5.173}$$

For a highly oblate spheroid ($c \ll a$), this is approximately

$$\omega^2 < 2\pi G\delta \frac{\pi}{2}\frac{c}{a}. \tag{5.174}$$

Finally, for a sphere ($a = b = c$)

$$\omega^2 < 2\pi G\delta, \tag{5.175}$$

which agrees exactly with Poincaré's inequality as usually formulated for liquids ([127], § 51).

In general, for equilibrium ellipsoids, one can write Poincaré's inequality (5.168) as

$$\omega^2 < 2\pi G\delta K, \tag{5.176}$$

where $K \leqslant 1$ depends only on the shape of the ellipsoid. This follows from the above formulae. The equality does not occur in practice, since it corresponds to a sphere, and then $\omega = 0$ for dynamical reasons.

As an example, let us consider the Galaxy, taking $c/a = 0{\cdot}1$, the star density in the neighbourhood of the Sun as $\nu = 3{\cdot}41 \times 10^{-57}$ per cm^3, and the mean mass of a star as $m = \frac{1}{2}m_\odot = 10^{33}$ g. Then the limiting angular velocity is $0''{\cdot}0044$ per y. This is less than the observed angular velocity of the rotation of the Galaxy in the Sun's neighbourhood, which is $\omega = 0''{\cdot}0068$ per y. For equilibrium it is necessary that the star density should be at least 2·4 times that observed. This result may be regarded as confirming the known fact that the Sun is outside the main dynamically stable part of the Galaxy.

It is important to note that ω^2 must be the smaller, the greater the omitted term $M\hat{v}^2$ in comparison with $J\omega^2$, i.e. the greater the energy of the irregular (residual) motions of the stars in comparison with that of their regular (rotational) motion.

In normal spirals, which include the Galaxy, the omitted term is small. Hence the angular velocity of such systems cannot be much different from Poincaré's limit, and Poincaré's inequality may be replaced by an approximate equality, by which the mean stellar velocity may be determined from the angular velocity or vice versa.

Poincaré's inequality is of importance in stellar dynamics. We have deliberately laid emphasis on the case of rotating homogeneous ellipsoids because, as will be seen in Chapter X, the great majority of galaxies are homogeneous ellipsoids to a first approximation. It will also be found that Poincaré's condition can be fulfilled in galaxies only by the formation of central nuclei and other structural features such as annular structure, i.e. by the attainment of a density distribution which is not homogeneous.

CHAPTER VI

REGULAR ORBITS OF STARS

§ 1. Circular and Almost Circular Orbits in a Stellar System with Rotational Symmetry and a Plane of Symmetry

IN THIS chapter we shall consider the motions of stars in the absence of irregular forces. Each star in a stellar system in a steady state describes, under the action of regular forces only, a regular orbit, which does not vary with time. Thus the dynamics of the system is determined by the distribution of regular orbits, and this is equivalent to specifying the coordinates and velocities of every star in the system (that is, the phase distribution) at some "initial" instant t_0.

However, a discussion of star orbits makes possible some general deductions concerning the nature of the motions in stellar systems. This chapter is therefore devoted to a discussion of the regular force field and the regular orbits of stars, especially since, as we have seen in Chapter V, the regular orbits still exist in systems where irregular forces act, and the stars merely change from one orbit to another under the action of these forces.

The nature of the regular orbits depends mainly on that of the gravitational field, which is determined by the size and shape of the system and by the distribution of mass within it. However, as we shall show, there exists even in the most general cases a class of orbits which are of particular importance in the dynamics of stellar systems. In systems having rotational symmetry, such as galaxies, this class includes circular orbits in the principal plane of symmetry, and these we shall now consider.

Our discussion of circular and almost circular orbits will be mainly based on that given by Lindblad [68]. Let us assume that the distribution of mass in a stellar system is rotationally symmetrical about some axis Oz, and symmetrical about some plane $z = 0$ perpendicular to that axis. Then the potential U is a function only of the cylindrical coordinates ϱ and z:

$$U = U(\varrho, z), \tag{6.1}$$

and

$$U(\varrho, z) = U(\varrho, -z). \tag{6.2}$$

Differentiation, using the fact that U is an even function of z, gives

$$(\partial U/\partial z)_{z=0} = 0, \quad (\partial^2 U/\partial \varrho\, \partial z)_{z=0} = 0. \tag{6.3}$$

The differential equations of motion in cylindrical coordinates are

$$\left.\begin{aligned}\frac{d^2\varrho}{dt^2} &= \varrho\left(\frac{d\theta}{dt}\right)^2 + \frac{\partial U}{\partial \varrho},\\ \frac{d}{dt}\left(\varrho^2\frac{d\theta}{dt}\right) &= \frac{\partial U}{\partial \theta},\\ \frac{d^2 z}{dt^2} &= \frac{\partial U}{\partial z}.\end{aligned}\right\} \tag{6.4}$$

From (6.1) we thus have the area integral

$$\varrho^2\, d\theta/dt = h. \tag{6.5}$$

We shall show that there are possible orbits lying entirely in the plane of symmetry. Let a star be, at some instant t_0, in the plane $z = 0$ and moving parallel to this plane, i.e. the initial position and velocity vectors be in the plane $z = 0$. The first eqn. (6.3) shows that the force acting on the star at time t_0 is also in that plane. Hence the acceleration of the star cannot take it out of the plane $z = 0$. Thus a star which is at any time moving in the plane $z = 0$ with a velocity lying in that plane will describe an orbit in the plane.†

In stellar systems such as galaxies, where the circular velocity of rotation Θ_0 is much greater than the root-mean-square residual velocity $\hat{v}$, so that $\gamma^{-1} = \hat{v}/\Theta_0 \ll 1$, the regular orbits of individual stars are not much different from circles (at least for stars of the plane sub-systems). Hence we should expect that circular and nearly circular orbits are possible in systems like the Galaxy. We shall show that plane circular orbits with $\varrho = \varrho_0 =$ constant, $z \equiv 0$ are possible. Since the condition $z \equiv 0$ may be supposed satisfied, from the above discussion, we substitute $\varrho = \varrho_0$ in the equations of motion. Denoting the angular velocity by ω_0, we obtain from (6.5) with $\varrho = \varrho_0$

$$\omega_0 = (d\theta/dt)_0 = h/\varrho_0^2 = \text{constant}. \tag{6.6}$$

Putting

$$\varrho_0 (d\theta/dt)_0 = h/\varrho_0 = \Theta_c, \tag{6.7}$$

we have from the first eqn. (6.4)

$$\frac{\Theta_c^2}{\varrho_0} + \left(\frac{\partial U}{\partial \varrho}\right)_0 = 0. \tag{6.8}$$

Hence Θ_c is just the circular velocity of a particle moving in a gravitational field given by the potential U. This shows that circular orbits are possible, and from (6.6) it follows that the circular motion must be uniform.

From (6.6), (6.7) and (6.8) we find two useful expressions for the area integral in circular motion:

$$h = \omega_0 \varrho_0^2, \tag{6.9}$$

$$h^2 = -\varrho_0^3 (\partial U/\partial \varrho)_0. \tag{6.10}$$

The suffix zero in (6.10) signifies that the derivative is taken for $z = 0$, $\varrho = \varrho_0$.

† An analytical proof is given by, for example, P.P. Parenago ([107], pp. 402–403).

Let us now find and investigate motions in almost circular orbits. If a star is moving in an orbit which, at some instant, differs only slightly from a circular orbit, its coordinates may be written

$$\left.\begin{aligned} \varrho &= \varrho_0 + \delta\varrho, \\ \theta = \theta_0 + \delta\theta &= \omega_0(t - t_0) + \delta\theta, \\ z &= \delta z, \end{aligned}\right\} \tag{6.11}$$

where ϱ_0 = constant is the radius of some circular orbit lying entirely in the plane $z = 0$, and θ_0 is the azimuthal angle corresponding to motion in this orbit.

The last eqn. (6.11) shows that the orbit deviates only slightly from the plane $z = 0$. For brevity we shall distinguish this orbit from the circular one by calling it a *varied* or *almost circular* orbit, and shall call $\delta\varrho$, $\delta\theta$, δz the *variations* of the corresponding coordinates. So far ϱ_0 has been arbitrary. We now define it so that the area integral h for the corresponding circular orbit is equal to its value for the varied orbit considered. This can always be done, since for given h the relation (6.10) can be used to find ϱ_0. The quantity $-(\partial U/\partial\varrho) = F(\varrho)$ is the magnitude of the regular gravitational force. Hence eqn. (6.10) becomes $h^2 = \varrho_0^3 F(\varrho_0)$. The right-hand side of this equation is an increasing function of ϱ_0 if F does not decrease more rapidly than $1/\varrho^3$. Such a rapid decrease is impossible for any reasonable distribution of mass within the system, since even if the mass is concentrated at the centre of the system F decreases only as $1/\varrho^2$. For any other mass distribution the force must decrease even less rapidly.†

Thus, for points within the system the quantity $-\varrho^3\,\partial U/\partial\varrho = \varrho^3 F(\varrho)$ is an increasing function of ϱ, and so eqn. (6.10) has only one solution. Substituting the resulting value of ϱ_0 in (6.6), we obtain the constant angular velocity ω_0 of circular motion, and thence the azimuthal angle $\theta_0 = \omega_0(t - t_0)$.

In terms of the area integral (6.5) we can write the differential equations of motion (6.4) as

$$\frac{d^2\varrho}{dt^2} = \frac{h^2}{\varrho^3} + \frac{\partial U}{\partial\varrho}, \quad \varrho^2\frac{d\theta}{dt} = h, \quad \frac{d^2 z}{dt^2} = \frac{\partial U}{\partial z}.$$

For the varied orbit, from the above condition, h and h^2 are found from (6.9) and (6.10). Hence the differential equations of the varied motion may be written

$$\left.\begin{aligned} \frac{d^2\varrho}{dt^2} &= -\frac{\varrho_0^3(\partial U/\partial\varrho)_0}{\varrho^3} + \frac{\partial U}{\partial\varrho}, \\ \frac{d\theta}{dt} &= \frac{\varrho_0^2\omega_0}{\varrho^2}, \\ \frac{d^2 z}{dt^2} &= \frac{\partial U}{\partial z}. \end{aligned}\right\} \tag{6.12}$$

† This conclusion holds for any point within the system. For external points the result may be different, as has been shown by Lindblad [68], but such problems will not be discussed here.

The potential $U = U(\varrho, z)$ can be expanded as a series of powers $\varrho - \varrho_0$ and z, keeping only terms of up to the second order of smallness. Using (6.3), we obtain

$$U \approx U_0 + (\partial U/\partial \varrho)_0 (\varrho - \varrho_0) + \tfrac{1}{2}(\partial^2 U/\partial z^2)_0 z^2 + \tfrac{1}{2}(\partial^2 U/\partial \varrho^2)_0 (\varrho - \varrho_0)^2.$$

Using this expansion, eqns. (6.12) can be written

$$\left.\begin{aligned} \frac{d^2\varrho}{dt^2} &= \left[1 - \left(\frac{\varrho}{\varrho_0}\right)^{-3}\right]\left(\frac{\partial U}{\partial \varrho}\right)_0 + \left(\frac{\partial^2 U}{\partial \varrho^2}\right)_0 (\varrho - \varrho_0), \\ \frac{d\theta}{dt} &= \omega_0 \left(\frac{\varrho}{\varrho_0}\right)^{-2}, \\ \frac{d^2 z}{dt^2} &= \left(\frac{\partial^2 U}{\partial z^2}\right)_0 z. \end{aligned}\right\} \tag{6.13}$$

Substituting in these equations the coordinates of a point (6.11) on the varied orbit and retaining only the first power of $\delta\varrho$ in expanding $(\varrho/\varrho_0)^{-3}$ and $(\varrho/\varrho_0)^{-2}$, we obtain the following equations for the variations of the coordinates:

$$\left.\begin{aligned} \frac{d^2\,\delta\varrho}{dt^2} &= \left(\frac{\partial^2 U}{\partial \varrho^2} + \frac{3}{\varrho}\frac{\partial U}{\partial \varrho}\right)_0 \delta\varrho, \\ \frac{d\,\delta\theta}{dt} &= -\,2\,\frac{\omega_0}{\varrho_0}\,\delta\varrho, \\ \frac{d^2\,\delta z}{dt^2} &= \left(\frac{\partial^2 U}{\partial z^2}\right)_0 \delta z. \end{aligned}\right\} \tag{6.14}$$

The first and third eqns. (6.14) are linear equations with constant coefficients. The nature of the almost circular varied motion depends on the signs of these coefficients. We shall show that both are negative.

Firstly,

$$\frac{\partial^2 U}{\partial \varrho^2} + \frac{3}{\varrho}\frac{\partial U}{\partial \varrho} = \frac{1}{\varrho^3}\frac{\partial}{\partial \varrho}\left(\varrho^3 \frac{\partial U}{\partial \varrho}\right); \tag{6.15}$$

we have seen that $-\varrho^3\,\partial U/\partial\varrho = \varrho^3 F(\varrho)$ is an increasing function of ϱ, at least for points within the system. Hence $\varrho^3\,\partial U/\partial\varrho$ is a decreasing function of ϱ, and (6.15) is negative. We can therefore put

$$\left(\frac{\partial^2 U}{\partial \varrho^2} + \frac{3}{\varrho}\frac{\partial U}{\partial \varrho}\right)_0 = -\,\varkappa_1^2, \tag{6.16}$$

where $\varkappa_1$ is a real positive quantity.

Secondly,

$$(\partial^2 U/\partial z^2)_0 < 0: \tag{6.17}$$

the z-component of the gravitational force for $z > 0$ is towards negative z, and for $z < 0$ it is in the opposite direction. Hence $\partial U/\partial z < 0$ for $z > 0$ and vice versa. Thus

$\partial U/\partial z$ is a decreasing function of z at $z = 0$, and its derivative with respect to z is negative there. We can therefore put

$$(\partial^2 U/\partial z^2)_0 = -\varkappa_2^2. \tag{6.18}$$

We substitute (6.16) and (6.18) in (6.14). Then the integration of the first and third eqns. (6.14) gives immediately

$$\begin{aligned} \delta\varrho &= a \sin \varkappa_1 (t - t_1), \\ \delta z &= b \sin \varkappa_2 (t - t_2), \end{aligned} \tag{6.19}$$

where a, b, t_1, t_2 are constants of integration. The quantities a and b are the amplitudes of the oscillations (about a circular orbit) executed by a particle moving in an almost circular orbit; t_1 and t_2 are such that the particle crosses the cylinder $\varrho = \varrho_0$ at $t = t_1$ and the plane $z = 0$ at $t = t_2$.

Substituting $\delta\varrho$ in the second eqn. (6.14) and integrating, we obtain

$$\delta\theta = 2(\omega_0/\varkappa_1 \varrho_0)\, a \cos \varkappa_1 (t - t_1), \tag{6.20}$$

where the further constant of integration has, for simplicity, been taken as zero.

Since the variations $\delta\varrho$, $\delta\theta$, δz of the coordinates are the deviations of the varied motion from the circular motion, it is necessary to choose a definite "comparison point" moving in a circular orbit. This choice does not affect $\delta\varrho$ and δz; for $\delta\theta$ it is most naturally made in such a way that the oscillations are symmetrical in the positive and negative directions. This corresponds to our choice of the new constant of integration as zero, since then $\overline{\delta\theta} = 0$.

In order to elucidate the nature of the motion in the varied orbit, we use a moving system of rectangular coordinates ξ, η, ζ, with the origin at the comparison point, the ζ-axis parallel to the axis of rotation, the ξ-axis along the radius of the circular orbit, and the η-axis tangential to this orbit in the positive direction of the circular motion.† The coordinates of a point on the varied orbit are thus

$$\left.\begin{aligned} \xi &= \delta\varrho = a \sin\varkappa_1 (t - t_1), \\ \eta &= \varrho_0\, \delta\theta = 2(\omega_0/\varkappa_1) a \cos \varkappa_1 (t - t_1), \\ \zeta &= \delta z = b \sin \varkappa_2 (t - t_2). \end{aligned}\right\} \tag{6.21}$$

Hence it follows that the particle describes the ellipse

$$\frac{\xi^2}{a^2} + \frac{\eta^2}{4\omega_0^2 a^2/\varkappa_1^2} = 1 \tag{6.22}$$

in the $\xi\eta$-plane about the comparison point. The period of this motion is $2\pi/\varkappa_1$. At the same time it executes a harmonic oscillation of period $2\pi/\varkappa_2$ parallel to the ζ-axis.

† Here it may not be superfluous to remind the reader that the rotation of the Galaxy is in the *negative* direction. This is because the positive direction of the z-axis is customarily taken to be to the northern pole of the Milky Way, and the rotation of the Galaxy as seen from that pole is clockwise. In stellar dynamics it would certainly be more logical to take the positive direction of the axis of the Galaxy towards the southern pole. This might, however, lead to misunderstanding by departing from he firmly established practice, and we shall therefore adhere to the conventional procedure.

As an example, let us consider the Galaxy in the neighbourhood of the Sun, and find the periods of the two motions. For convenience of calculations we may transform formula (6.16) by expressing $\varkappa_1$ in terms of Oort's constants A and B. To do so, we use the fact that $\partial U/\partial \varrho = -V_c^2/\varrho$, where V_c is the circular velocity at distance ϱ, and hence by differentiation

$$\frac{\partial^2 U}{\partial \varrho^2} = -2\frac{V_c}{\varrho}\frac{\partial V_c}{\partial \varrho} + \frac{V_c^2}{\varrho^2}$$

and

$$\frac{\partial^2 U}{\partial \varrho^2} + \frac{3}{\varrho}\frac{\partial U}{\partial \varrho} = -2\frac{V_c}{\varrho}\left(\frac{\partial V_c}{\partial \varrho} + \frac{V_c}{\varrho}\right). \tag{6.23}$$

From (3.47), (6.16) and (6.23) we find

$$\varkappa_1^2 = 4B(B - A), \tag{6.24}$$

or, using the numerical values (3.37) for A and B,

$$\varkappa_1 = 1{\cdot}34 \times 10^{-15}\ \mathrm{s}^{-1}, \tag{6.25}$$

so that the corresponding period is

$$2\pi/\varkappa_1 = 4{\cdot}67 \times 10^{15}\ \mathrm{s} = 1{\cdot}50 \times 10^8\ \mathrm{y}. \tag{6.26}$$

This is 80% of the period of rotation of the Galaxy, which is $1{\cdot}86 \times 10^8$ y (see (3.40)).

The quantity $\varkappa_2^2$ defined by (6.18) is called the *dynamical parameter* and denoted by C^2. Using the eqn. (5.90) derived in the ellipsoidal dynamics of the Galaxy, we have†

$$\varkappa_2^2 = -\sigma_z^2\frac{\partial^2 \log \nu}{\partial z^2} = C^2.$$

If the velocity variance σ_z^2 and the z-component of the star density variation $\partial^2 \log\nu/\partial z^2$ are known from observation, $\varkappa_2$ may be found.

The value of $\varkappa_2$ has been determined by Oort [92], P. P. Parenago [106, 108] and G. G. Kuzmin [60].†† The best value, according to Parenago's data, is

$$\varkappa_2 = 2{\cdot}37 \times 10^{-15}\ \mathrm{s}^{-1}, \tag{6.27}$$

and the corresponding period is

$$2\pi/\varkappa_2 = 2{\cdot}65 \times 10^{15}\ \mathrm{s} = 0{\cdot}84 \times 10^8\ \mathrm{y}. \tag{6.28}$$

This is little more than half the period (6.26) of the elliptical motion, and about 45% of the period of rotation of the Galaxy.

Thus motion in an almost circular orbit can be represented as a circular motion of period $2\pi/(A - B)$ combined with an epicyclic motion in a small ellipse with period

† The same formula was obtained by Oort by the use of eqn. (5.118), derived from the third quasi-integral.

†† Parenago [105] determined $\partial^2 \log\nu/\partial z^2$ for stars of various types (sub-systems), i.e. its dependence on σ_z^2.

$2\pi/\varkappa_1$ and a harmonic oscillation perpendicular to the galactic plane with period $2\pi/\varkappa_2$.

An important property of circular motion is its *dynamical stability:* any star whose orbit at some initial instant differs only slightly from a circle will continue for all time to have an orbit which differs only slightly from this circle. The stability of circular motion signifies that, if the variations of the coordinates are sufficiently small at the initial instant, they remain uniformly bounded for all time.

The study of the stability of a periodic motion usually involves the consideration of a varied (almost periodic) orbit. In the present case, however, the analysis given above cannot be directly applied as it stands. The reason is that, in examining the stability of a given circular motion, we must regard ϱ_0 as fixed and consider all possible almost circular orbits close to the given one. Previously our procedure was the opposite one: for each almost circular varied orbit we took a circular orbit close to it.

In discussing the stability of a given circular motion, we regard ϱ_0 in all the above formulae as fixed, and consider the varied motion (6.11). In the above analysis we considered only those varied orbits for which the area integral h has the same value as in the circular orbit. For such orbits the variations of the coordinates must, by (6.5) and (6.9), be such that

$$(\varrho_0 + \delta\varrho)^2 (\omega_0 + \delta\omega) = h, \tag{6.29}$$

where

$$\delta\omega = \mathrm{d}\,\delta\,\theta/\mathrm{d}t = \delta(\mathrm{d}\,\theta/\mathrm{d}t). \tag{6.30}$$

In discussions of stability it is usually assumed that the variations of the coordinates can take arbitrary values. The problem is then said to be one of *unconditional stability* (or simply one of *stability*). If the variations are related in any way, the problem is one of *conditional stability.*

Thus, in the terminology of Lyapunovs' theory of stability, we are considering the conditional stability of circular motion: the variations of coordinates and velocities are not arbitrary, but are related by (6.29).

Substituting the varied coordinates (6.11) in eqns. (6.12) with the condition (6.29), we obtain the differential eqns. (6.14) in the first approximation, when only the first powers of the variations are taken into account. It follows from the form of the solution (6.19), (6.20) that the circular motion $\varrho = \varrho_0$, $\omega = \omega_0$ is conditionally stable in the first approximation. For a final decision concerning the conditional stability, according to Lyapunov's theory, it is necessary to consider the second, and perhaps still higher, approximations.

§ 2. Determination of the Potential of the Galaxy

In § 1 we have considered the orbits of stars on the assumption that the stellar system is symmetrical about an axis (has rotational symmetry) and about a plane. Accordingly the potential U was a function of ϱ and z only, and was an even function of z. For a more detailed study of star orbits and a solution of other problems of galactic dynamics, P. P. Parenago derived an explicit analytical expression for the potential $U(\varrho, z)$.

Parenago's work [104, 106] was based on the theoretical expression (5.63) giving the circular velocity of the centroid as a function of the distance from the axis of rotation in a rotating Galaxy in a steady state:

$$V_c = k_1\varrho/(1 + k_2\varrho^2). \tag{6.31}$$

Parenago regarded this expression as being fairly well confirmed by observations of the long-period Cepheids (a plane sub-system) in the Galaxy and by spectroscopic measurements of the nearest other galaxies, M 31 in Andromeda and M 33 in Triangulum.

The analysis assumes that the circular velocity of the centroid in plane sub-systems is almost identical with that of a particle, given by

$$V_c^2 = -\varrho\, \mathrm{d}U/\mathrm{d}\varrho, \tag{6.32}$$

where $U(\varrho)$ is the potential in the galactic plane. From (6.31) and (6.32) we have

$$\mathrm{d}U/\mathrm{d}\varrho = -k_1^2\varrho/(1 + k_2\varrho^2)^2,$$

and hence by integration

$$U = \frac{k_1^2}{2k_2}\frac{1}{1 + k_2\varrho^2} + C. \tag{6.33}$$

The constant of integration C may be approximately determined from the condition that the potential outside the Galaxy must be $U = GM/\varrho$. Let ϱ_1 be the equatorial radius of the Galaxy. Since the potential (like any primitive of a bounded function) must be continuous at $\varrho = \varrho_1$, we have the condition

$$\frac{k_1^2}{2k_2}\frac{1}{1 + k_2\varrho_1^2} + C = \frac{GM}{\varrho_1}.$$

Substituting the values $k_1 = 72$ km/s per kps $= 2{\cdot}34 \times 10^{-15}$ s^{-1}, $k_2 = 0{\cdot}0237$ kps^{-2} $= 2{\cdot}49 \times 10^{-45}$ cm^{-2}, $\varrho_1 = 13$ kps $= 4{\cdot}0 \times 10^{22}$ cm, $M = 1{\cdot}3 \times 10^{11}\, m_\odot = 2{\cdot}6 \times 10^{44}$ g, we find

$$C = 0{\cdot}69 \times 10^{-14}\ \mathrm{cm}^2/\mathrm{s}^2. \tag{6.34}$$

The first term in the potential (6.33) is, for $\varrho = \varrho_0 = 7{\cdot}2$ kps,

$$U_0 = 4{\cdot}93 \times 10^{-14}\ \mathrm{cm}^2/\mathrm{s}^2.$$

Thus C is comparatively small at all points in the Galaxy, but its value is somewhat indefinite on account of the indefiniteness of the equatorial radius ϱ_1 and mass M of the Galaxy. Parenago points out that the mass must in reality be less than the value used above. This means that C must in reality be even less than (6.34). In a first approximation, therefore, C may be neglected, and

$$U = U_c/(1 + k_2\varrho^2), \tag{6.35}$$

where

$$U_c = k_1^2/2k_2 = 11{\cdot}0 \times 10^{-14}\ \mathrm{cm}^2/\mathrm{s}^2. \tag{6.36}$$

For the potential outside the galactic plane Parenago assumed the expression

$$U(\varrho, z) = U_c \varphi(z)/(1 + k_2 \varrho^2), \tag{6.37}$$

where the factor $\varphi(z)$ is a positive function which decreases monotonically in both directions from $z = 0$ and satisfies the conditions $\varphi(0) = 1$, $\varphi(\pm\infty) = 0$, $\varphi'(0) = \varphi'(\pm\infty) = 0$; $\varphi''(0) < 0$.

To find the actual form of $\varphi(z)$ we can use eqn. (5.89) of the ellipsoidal dynamics of the Galaxy:

$$\frac{\partial(\log \nu)}{\partial z} = \frac{1}{\sigma_z^2} \frac{\partial U}{\partial z}, \tag{6.38}$$

where σ_z^2 is the constant variance of the z-component of velocity and $\nu = \nu(\varrho, z)$ is the star density. Integrating (6.38), we have†

$$U(\varrho, z) - U(\varrho, 0) = \sigma_z^2 \log[\nu(\varrho, z)/\nu(\varrho, 0)], \tag{6.39}$$

where ϱ is regarded as a constant. The right-hand side of (6.39) can be obtained directly from observations for various sub-systems. In particular, values have been calculated for several sub-systems of variable stars (long-period Cepheids, short-period Cepheids, and Mira-type variables), for the neighbourhood of the Sun, by B.V. Kukarkin [52]. The quantity $U(\varrho, 0)$ for this neighbourhood, i.e. for $\varrho = \varrho_0$, is just the quantity

$$U_0 = U_c/(1 + k_2 \varrho_0^2) = 4{\cdot}93 \times 10^{-14}\ \text{cm}^2/\text{s}^2 \tag{6.40}$$

mentioned previously. Thus (6.39) for $\varrho = \varrho_0$ gives the ratio $U(\varrho_0, z)/U(\varrho_0, 0) = \varphi(z)$. As a check, a similar calculation may be made for $\varrho = 0$, i.e. for the axis of the Galaxy.

In this way Parenago ([106], p. 249) concluded that very similar values of $\varphi(z)$ are obtained in the two cases, and that they can be satisfactorily approximated for $0 < z < 2$ to 3 kps by the expression

$$\varphi(z) = \exp(-\lambda z^2), \tag{6.41}$$

where

$$\lambda = 0{\cdot}056\ \text{kps}^{-2} = 5{\cdot}9 \times 10^{-45}\ \text{cm}^{-2}. \tag{6.42}$$

From (6.37) and (6.41) we have finally for the potential of the Galaxy

$$U(\varrho, z) = [U_c/(1 + k_2 \varrho^2)] \exp(-\lambda z^2). \tag{6.43}$$

The constants U_c and λ in this formula have the values (6.36) and (6.42);

$$k_2 = 0{\cdot}0237\ \text{kps}^{-2} = 2{\cdot}49 \times 10^{-45}\ \text{cm}^{-2}. \tag{6.44}$$

§ 3. Some Properties of Regular Galactic Orbits

In § 1 we have considered the motion of stars in almost circular orbits in systems such as the Galaxy, assuming that the stars can depart slightly from the galactic plane $z = 0$. Now, using Parenago's expression for the potential, we may discuss another

† P. P. Parenago [106], in deriving (6.39), used Jeans' hydrodynamic equation (see Chapter VIII, §6) $\partial(\nu\sigma_z^2)/\partial z = \nu\, \partial U/\partial z$, putting $\sigma_z^2 =$ constant. In our exposition, however, it is better to use eqn. (5.89).

class of motions, namely those in orbits lying entirely in the galactic plane, but otherwise arbitrary. We shall follow the method of Parenago ([106], § 7).

Let ϱ, θ be the galactic polar coordinates of a star, and P, Θ the corresponding velocity components. Then the differential equations of the two-dimensional motion are

$$\dot{\mathrm{P}} = \frac{\Theta^2}{\varrho} + \frac{\mathrm{d}U}{\mathrm{d}\varrho}, \quad \varrho\Theta = h, \tag{6.45}$$

where the potential U may be regarded as depending only on ϱ.

Eliminating Θ from these equations, multiplying by $2\,\mathrm{P}\mathrm{d}t = 2\,\mathrm{d}\varrho$ and integrating, we have the energy integral

$$\mathrm{P}^2 = -\frac{h^2}{\varrho^2} + 2(U + H), \tag{6.46}$$

which is easily brought to the usual form if h is eliminated by means of the area integral. Then (6.46) becomes

$$V^2 = 2(U + H). \tag{6.47}$$

Evidently H is the energy constant. For stars whose velocity does not exceed the velocity of escape, the total energy must be negative ($H < 0$).

Equation (6.46) may be used for a qualitative investigation of the orbits. Let us determine the points corresponding to extremum values of ϱ. Putting $\mathrm{P} = 0$, we have from (6.46)

$$h^2 = 2(U + H)\varrho^2. \tag{6.48}$$

The potential U in the galactic plane is given by (6.35). For simplicity we shall replace k_2 in the denominator of (6.35) by k. Then

$$U = U_c/(1 + k\varrho^2). \tag{6.49}$$

In terms of a new variable

$$\xi = k\varrho^2, \tag{6.50}$$

the potential is simply

$$U = U_c/(1 + \xi). \tag{6.51}$$

Substituting (6.51) in (6.48), we obtain a quadratic equation for the extrema of ξ, and therefore of ϱ:

$$2H\xi^2 + [2(U_c + H) - kh^2]\xi - kh^2 = 0. \tag{6.52}$$

The discriminant of this equation is

$$\begin{aligned} D &= (U_c + H - \tfrac{1}{2}kh^2)^2 + 2kh^2 H \\ &= (U_c - \tfrac{1}{2}kh^2)^2 + H^2 + 2H(U_c + \tfrac{1}{2}kh^2). \end{aligned} \tag{6.53}$$

Hence $D > 0$ if $H > H_1$, where

$$H_1 = -[\sqrt{U_c} - \sqrt{(\tfrac{1}{2}kh^2)}]^2. \tag{6.54}$$

When $H > H_1$ the roots of eqn. (6.52) are real. Denoting them by ξ_1 and ξ_2, we have by Vieta's theorem

$$\begin{aligned} \xi_1 + \xi_2 &= -[2(U_c + H) - kh^2]/2H, \\ \xi_1 \xi_2 &= -kh^2/2H. \end{aligned} \tag{6.55}$$

We shall show that

$$2(U_c + H) - kh^2 > 0. \tag{6.56}$$

The area integral (6.45) gives, with (6.50),

$$kh^2 = \xi \Theta^2, \tag{6.57}$$

and, since $2H$ on the left of (6.56) can be written in terms of $\mathrm{P}^2 + \Theta^2 - 2U$ by means of the energy integral, (6.51) and (6.57) give

$$2(U_c + H) - kh^2 = \frac{2\xi U_c}{1+\xi} + \mathrm{P}^2 - (\xi - 1)\Theta^2 = \frac{2\xi U_c}{1+\xi} + \xi \mathrm{P}^2 - (\xi - 1)V^2.$$

Replacing V in the last term by the parabolic velocity V_p and using the equation $V_p^2 = 2U = 2U_c/(1+\xi)$, we obtain the quantity $\xi \mathrm{P}^2 + V_p^2$, which is evidently positive. This proves (6.56).

From (6.54), (6.55) and (6.56) it follows that when $H < 0$ eqn. (6.52) has two real positive roots ξ_1 and ξ_2. For definiteness we assume that $\xi_1 < \xi_2$. When $H = 0$ there is only one root, $\xi_1 = kh^2/(2U_c - kh^2)$, the other becoming infinite. When $H > 0$ one root (ξ_1) is positive and the other (ξ_2) is negative. The first case ($H_1 < H < 0$) is called *quasi-elliptic*, the second ($H = 0$) *quasi-parabolic*, and the third ($H > 0$) *quasi-hyperbolic*.

In the quasi-elliptic case, a star with a negative energy of not too large magnitude ($H_1 < H < 0$) moves in an orbit which lies entirely between two concentric circles of radii $\varrho_1 = \sqrt{(\xi_1/k)}$ and $\varrho_2 = \sqrt{(\xi_2/k)}$. In the quasi-parabolic and quasi-hyperbolic cases, the whole orbit lies outside the circle of radius $\varrho_1 = \sqrt{(\xi_1/k)}$; $\varrho_2 = \infty$.

It should be mentioned, however, that these deductions remain valid only if ξ does not exceed unity, i.e. for points nearer to the centre than the Sun is. For greater ξ, the value of the constant term C in the expression (6.33) for the potential becomes important: at the Sun's distance it is already about 14% of the potential.

The inaccuracy of the expression used for the potential is seen from the fact *(inter alia)* that, between the two negative roots of the equation $D = 0$:

$$H_1 = -[\sqrt{U_c} - \sqrt{(\tfrac{1}{2}kh^2)}]^2 \quad \text{and} \quad H_2 = -[\sqrt{U_c} + \sqrt{(\tfrac{1}{2}kh^2)}]^2,$$

the two roots ξ_1 and ξ_2 of eqn. (6.52) become imaginary, and this clearly has no physical meaning. Even an attempt to retain the additive constant in the potential would not escape error, however, since the direct joining, at the outer boundary of the Galaxy, of Parenago's potential to that of a point mass is extremely arbitrary and really demands a knowledge of the distribution of mass in the Galaxy. Unfortunately, also, the error is the greater, the closer to the edge of the Galaxy is the region considered since there the individual orbits of stars are of greatest importance (see the discussion of the corona in the Introduction).

Let us now consider the integration of the differential equations of a galactic orbit. Substituting the potential U (6.49) in the energy integral (6.46), and putting $d\varrho/dt$ for P, we have

$$\left(\frac{d\varrho}{dt}\right)^2 = -\frac{h^2}{\varrho^2} + \frac{2U_c}{1+k\varrho^2} + 2H.$$

Integration gives

$$t + C = \int \frac{\sqrt{(1+k\varrho^2)}\varrho\, d\varrho}{\sqrt{[2Hk\varrho^4 + (2U_c + 2H - kh^2)\varrho^2 - h^2]}},$$

or, with $\xi = k\varrho^2$,

$$t + C = \frac{1}{2\sqrt{k}} \int \frac{\sqrt{(1+\xi)}\, d\xi}{\sqrt{[2H\xi^2 + (2U_c + 2H - kh^2)\xi - kh^2]}}.$$

Since the denominator contains just the polynomial whose roots are ξ_1 and ξ_2, we can rewrite this as

$$t + C = \frac{1}{2\sqrt{(-2kH)}} \int \frac{\sqrt{(1+\xi)}\, d\xi}{\sqrt{[(\xi_2 - \xi)(\xi - \xi_1)]}}. \tag{6.58}$$

The arbitrary constant C can be eliminated by measuring time from the instant when the star passes through the apocentre (the apogalacticon). At this point $\xi = \xi_2$, and so

$$t = \frac{1}{2\sqrt{(-2kH)}} \int_{\xi}^{\xi_2} \frac{\sqrt{(1+\xi)}\, d\xi}{\sqrt{[(\xi_2 - \xi)(\xi - \xi_1)]}}. \tag{6.59}$$

By the substitution

$$\xi = x^2 - 1 \tag{6.60}$$

the right-hand side of (6.59) is reduced to an elliptic integral of the second kind $E(\varphi, \theta)$, and we have

$$t = \frac{\sqrt{(1+\xi_2)}}{\sqrt{(-2kH)}} E(\varphi, \theta), \tag{6.61}$$

where

$$\varphi = \sin^{-1} \sqrt{[(\xi - \xi_1)/(\xi_2 - \xi_1)]} \tag{6.62}$$

and

$$\sin\theta = \sqrt{[(\xi_2 - \xi_1)/(1 + \xi_2)]}.$$

The time between successive passages through the pericentre (perigalacticon) may be called the *anomalistic* period. It is clearly equal to twice the time from ξ_1 to ξ_2. Denoting it by P_a, we have from (6.62) and (6.61)

$$P_a = 2\frac{\sqrt{(1+\xi_2)}}{\sqrt{(-2kH)}} E(\tfrac{1}{2}\pi, \theta). \tag{6.63}$$

For a circular orbit, $\xi = \xi_1 = \xi_2$; $\theta = 0$; $E(\frac{1}{2}\pi, 0) = \frac{1}{2}\pi$. Since $V_c^2 = -\varrho\, \partial U/\partial\varrho$, we obtain from (6.47), (6.49) and (6.50)

$$-2H = 2U - V_c^2 = \frac{2U_c}{1+\xi} - \frac{2\xi U_c}{(1+\xi)^2} = 2U_c/(1+\xi)^2.$$

For an almost circular orbit, therefore, we have approximately

$$P_a = \frac{\pi}{\sqrt{(2kU_c)}} (1 + \xi)^{3/2}. \tag{6.64}$$

The period of the absolute orbital motion is

$$P = 2\pi/\omega_c = 2\pi\varrho_0/V_c = 2\pi(1 + \xi)\sqrt{(2kU_c)}. \tag{6.65}$$

Formulae (6.64) and (6.65) give

$$P_a/P = \tfrac{1}{2}\sqrt{(1 + \xi)}. \tag{6.66}$$

Thus the anomalistic period of revolution is not in general equal to the absolute period, and consequently there is a rotation of the apse line.

The condition $P_a = P$ holds if $\xi = 3$, or $\varrho = 10{\cdot}2$ kps. The two periods are equal at this distance only. For $\xi < 3$, $P_a < P$ and therefore the line of nodes rotates against the direction of motion of the star; for $\xi > 3$ it rotates in the same direction. At the Sun's distance $P_a = 150$ My, and $P = 190$ My.

Parenago has given a table and a graph for the Sun's neighbourhood, whereby ξ_1, ξ_2 and P_a (or, what is the same thing, ϱ_1, ϱ_2 and P_a) may easily be found from P and Θ, or conversely.

Subsequently R.M.Dzigvashvili [23] constructed similar tables and graphs for $\varrho = 2, 3, 4, 5, 6, 9, 10, 11$ and 12 kps, and from them he drew some conclusions of cosmogonical interest. It was found that stars of the plane and intermediate sub-systems, in their motion in the Galaxy, remain always at approximately the same distance from the centre, so that their orbits are in general not much different from circles. The stars of the spherical sub-systems, however, move at very variable distances from the centre of the Galaxy. Almost all of them, whatever their position in the Galaxy, pass through perigalacticon very close to the centre (not more than 1 kps). This result confirms the view that the stars of the plane and intermediate sub-systems were formed in or near those regions of the Galaxy where they now are, namely in a flat layer adjoining the galactic plane.

The place of origin of the stars of the spherical sub-systems remains undecided. They may have been formed in the nucleus of the Galaxy, or at various points in the spherical volume which they now occupy. Although Dzigvashvili used Parenago's calculations out to $\varrho = 12$ kps, his qualitative conclusions are probably not invalidated by the inexactitude of the potential assumed.

§ 4. The Distribution of Mass in the Galaxy and Some Related Problems

The problem of the distribution of mass in our own and other galaxies can undoubtedly be solved only by means of a physical theory of the structure and motion of stellar systems. In Chapter X we shall give the results of at least a first approximation to such a theory.

Until this physical theory of galactic structure exists, the only possible way of proceeding is to attempt to determine by semi-empirical trial the distribution of mass in galaxies, using the data on galactic structure and motions obtained from direct observation and elementary theory, and thereby to lay the experimental foundations for the development of the physical theory. Such studies will, of course, be mainly carried out with respect to our own Galaxy, for which the observational data are most extensive. Data concerning other galaxies (the nearest ones) will serve for purposes of comparison.

The pioneer in this subject was Oort, who attacked it soon after developing his theory of galactic rotation. In the first paper [92] Oort represented the Galaxy by a homogeneous spheroid with a central nucleus, and thus found the mass of the Galaxy. The essential results of this investigation have been given in Chapter III, § 7. Later the work was continued by Oort and van Woerkom [93], Chandrasekhar [18], Takase [129] and others. The subject has been especially developed in recent years by Soviet astronomers, including Parenago [103, 106], Kuzmin [56], Kukarkin [52], Safronov [112], Idlis [40] and others.

In each of these papers some more or less arbitrary additional hypotheses are of necessity introduced, and so they may be generally described as using the *method of models.* In this method a more or less natural scheme of the structure of the Galaxy, with parameters determined from observation, is used. The greater the agreement between the resulting general picture and the results of observations of the mass distribution and law of rotation of the Galaxy, the more satisfactory is the scheme thought to be.

Such investigations are carried out in two general ways. One way, to which belongs the work of Chandrasekhar [18], Safronov [112], Kuzmin [56] and others, continues to develop and improve Oort's original scheme. These authors begin from the representation of the Galaxy as some combination of interpenetrating homogeneous ellipsoids. The other way, which has been taken chiefly by Parenago, is to find the dynamical characteristics of the Galaxy from the analytical expression (6.43) for the potential.

Let us first consider the work of Parenago [103, 106]. Knowing the potential U, we can find the value of the density $\delta(\varrho, z)$ by means of Poisson's equation in cylindrical coordinates:

$$-4\pi G\delta = \frac{\partial^2 U}{\partial \varrho^2} + \frac{1}{\varrho}\frac{\partial U}{\partial \varrho} + \frac{\partial^2 U}{\partial z^2}. \tag{6.67}$$

Substitution of the potential (6.43) gives

$$\delta(\varrho, z) = \frac{U_c}{2\pi G(1 + k_2\varrho^2)}\left[\frac{2k_2(1 - k_2\varrho^2)}{(1 + k_2\varrho^2)^2} - (2\lambda^2 z^2 - \lambda)\right] \exp(-\lambda z^2). \tag{6.68}$$

In particular, the density in the neighbourhood of the Sun is

$$\delta_0 = \delta(\varrho_0, 0) = 5{\cdot}88 \times 10^{-24}\ \mathrm{g/cm^3}. \tag{6.69}$$

Subtracting from δ_0 the observed value of the star density near the Sun, which is $2{\cdot}8 \times 10^{-24}$ g/cm³, we find the density of diffuse gas and dust, namely $3{\cdot}1 \times 10^{-24}$ g/cm³.

Putting $\varrho = z = 0$ in (6.68), we find for the density at the centre of the Galaxy

$$\delta_c = U_c(2k_2 + \lambda)/2\pi G = 32{\cdot}3 \times 10^{-24} \text{ g/cm}^3 = 5{\cdot}5\, \delta_0.$$

Parenago [106] and Kuzmin [56, 57, 60] have proposed another way of determining the dynamical density of the Galaxy in the neighbourhood of the Sun. Using formula (6.32), which relates the potential U to the circular velocity, and formulae (3.46) for Oort's constants, we find

$$\left(\frac{\partial^2 U}{\partial \varrho^2} + \frac{1}{\varrho}\frac{\partial U}{\partial \varrho}\right)_0 = 2(A^2 - B^2),$$

where the suffix zero, as usual, denotes the values of quantities in the neighbourhood of the Sun.

Substituting this in (6.67) and using the dynamical parameter $C^2 = -(\partial^2 U/\partial z^2)_0$, we obtain

$$4\pi G\delta_0 = C^2 - 2(A^2 - B^2).$$

The values given by eqn. (3.36) may be used for Oort's constants; the determination of C^2 as in § 1 gives (according to Parenago) $C^2 = 5330 \text{ km}^2/\text{s}^2$ per $\text{kps}^2 = 5{\cdot}60 \times 10^{-30} \text{ s}^{-2}$. Thus the dynamical density of the Galaxy in the neighbourhood of the Sun is $\delta_0 = 6{\cdot}23 \times 10^{-24} \text{ g/cm}^3$, which is close to the value (6.69).

By means of the expression (6.35) for the potential we can calculate a number of quantities which are of interest in galactic dynamics. These include the velocity of escape V_p and the circular velocity V_c in the galactic plane for various values of ϱ:

$$\begin{aligned} V_p &= \sqrt{(2U)} = \sqrt{(2U_c)}/\sqrt{(1 + k_2\varrho^2)}, \\ V_c &= \sqrt{(-\varrho\,\partial U/\partial \varrho)} = \sqrt{(2k_2 U_c)}\varrho/(1 + k_2\varrho^2). \end{aligned} \qquad (6.70)$$

Table VI gives the results of calculating various dynamical quantities for the Galaxy. In this table ϱ is the distance from the centre of the Galaxy in kiloparsecs, U the potential (6.35) in the galactic plane, A and B Oort's constants of galactic rotation, ω the angular velocity of rotation of the Galaxy at various distances from the centre, $P = 2\pi/\omega$ the period of rotation, V_c and V_p the circular velocity of the centroid and the velocity of escape, calculated from formulae (6.70), P_z the period of oscillation in the z-direction for a star moving in an almost circular orbit, δ the density of matter in the Galaxy, in grams per cubic centimetre, and δ/δ_0 the density relative to its value near the Sun.

The logarithmic density gradient $\partial(\log_{10}\delta)/\partial\varrho$ given in the last column is one of the most important characteristics whereby a given group of stars may be assigned to one of the three main components of the stellar population of the Galaxy. In the neighbourhood of the Sun the logarithmic density gradient, usually denoted by m, has an average value of $-0{\cdot}11$ for plane, $-0{\cdot}20$ for intermediate and $-0{\cdot}26$ for spherical sub-systems. Hence it follows that the star density in plane sub-systems decreases much more slowly than that in spherical sub-systems in a direction parallel to the galactic plane. In other words, the spherical sub-systems are much more concentrated towards the centre of the Galaxy than are the plane sub-systems.

It must be borne in mind that all the quantities in Table VI, except the potential U and the logarithmic gradient m, pertain to the plane sub-systems, since they have been calculated on the assumption that the circular velocity of the centroid is not greatly different from that of a particle. The diffuse matter also forms a plane sub-system in the Galaxy, and so the logarithmic gradient of the density δ of stellar matter is found to be the same as that of the star density ν for plane sub-systems.

We may now give a brief account of the results of obtaining the mass distribution by representing the Galaxy as a superposition of several homogeneous spheroids. This approach to the problem will be referred to as *Oort's method.* In describing this method we shall follow the work of Safronov [112], one of the most recent and most general treatments.

The Galaxy is regarded as an assembly of homogeneous spheroids with a common centre and a common axis of symmetry. Let us consider an arbitrary point in the common equatorial plane; in what follows this point will be identified with the position of the Sun. Let F_i be the gravitational acceleration at the point due to the ith spheroid. If the point is outside the spheroid we write F_i', and if it is inside F_i''. The spheroids are said to be *internal* and *external* respectively (relative to the point considered, of course). If the point is taken to be at the position of the Sun in the Galaxy, the mass of the internal spheroids is due mainly to the stars of the spherical sub-systems, on account of their marked concentration towards the centre of the Galaxy. The external spheroids must contain predominantly the stars of the plane sub-systems and the diffuse matter, and must be more flattened than the internal spheroids.

Thus we have the equation

$$\sum_i F_i' + \sum_i F_i'' = V^2/\varrho, \tag{6.71}$$

where V is the circular velocity of a particle at distance ϱ, which can be replaced by the corresponding centroid velocity for plane sub-systems.

Differentiating eqn. (6.71) with respect to ϱ and using Oort's constant A instead of $\mathrm{d}V/\mathrm{d}\varrho$ by means of the first formula (3.46), we obtain a second equation:

$$\sum_i \frac{\mathrm{d}F_i'}{\mathrm{d}\varrho} + \sum_i \frac{\mathrm{d}F_i''}{\mathrm{d}\varrho} = \frac{V}{\varrho}\left(4A + \frac{V}{\varrho}\right). \tag{6.72}$$

The quantity V/ϱ can also be expressed in terms of Oort's constants by using the first formula (3.47), but this is better not done because the constant B is not accurately known.

It is known from potential theory that the force of attraction of a homogeneous spheroid at an external point is given by

$$F_i' = 2\pi G \delta_i a_i^2 c_i \varrho \int_{\lambda_i}^{\infty} \frac{\mathrm{d}s}{(a_i^2 + s)^2 \sqrt{(c_i^2 + s)}}, \tag{6.73}$$

where λ_i is the positive root of the equation

$$\frac{\varrho^2}{a_i^2 + \lambda} + \frac{z^2}{c_i^2 + \lambda} = 1.$$

TABLE VI. THEORETICAL VALUES OF DYNAMICAL QUANTITIES IN THE GALAXY (AFTER PARENAGO)

ϱ (kps)	U (10^{14} cm^2/s^2)	A (km/s per kps)	B (km/s per kps)	ω (10^{-15}s^{-1})	P (10^6 y)	V_c (km/s)	V_p (km/s)	P_z (10^6 y)	δ (10^{-24} g/cm^3)	δ/δ_0	$(\partial \log_{10} \delta)/\partial \varrho$ (kps^{-1})
0	11·3	0·0	−80·7	2·62	76	0	475	55	32·3	5·5	0·000
1	11·0	2·2	−76·2	2·54	78	78	469	55	30·1	5·1	−0·061
2	10·1	7·4	−64·9	2·34	85	145	450	58	24·7	4·2	−0·107
3	9·0	13·2	−50·9	2·08	96	192	424	61	18·7	3·2	−0·132
4	7·7	17·4	−37·8	1·79	111	221	393	66	13·8	2·3	−0·136
5	6·6	19·6	−27·3	1·52	131	235	362	72	10·2	1·7	−0·126
6	5·6	20·2	−19·4	1·29	154	238	333	78	7·7	1·3	−0·111
7	4·7	19·6	−13·9	1·09	183	234	306	85	6·1	1·0	−0·095
7·2	4·53	19·4	−13·0	1·05	190	233	301	86	5·88	1·00	−0·092
8	4·0	18·4	−10·0	0·92	216	227	282	92	5·0	0·8	−0·082
9	3·4	16·9	−7·3	0·78	254	218	260	100	4·2	0·7	−0·072
10	2·9	15·4	−5·4	0·68	295	208	241	108	3·6	0·6	−0·065

From formula (6.73) F_i'' is obtained by putting $\lambda_i = 0$. The derivative $\mathrm{d}F_i'/\mathrm{d}\varrho$ is easily calculated from these formulae.† Since F_i'' involves ϱ only as a factor, $\mathrm{d}F_i''/\mathrm{d}\varrho = F_i''/\varrho$.

The mass of the ith spheroid $M_i = 4\pi\delta_i a_i^2 c_i/3$, and we can therefore write (6.73) more briefly as

$$F_i' = M_i' P_i', \tag{6.74}$$

where P_i' denotes the product of the remaining factors. If M' is the total mass of the internal spheroids, and b_1 the mean value of the coefficients P_i':

$$b_1 = \sum_i M_i' P_i'/M', \tag{6.75}$$

then

$$\sum_i F_i' = b_1 M'. \tag{6.76}$$

Similarly, for the attraction of an external spheroid,

$$F_i'' = \delta_i P_i'', \tag{6.77}$$

where P_i'' is the product of the other factors. Let δ_0 be the sum of the densities of the external spheroids. Since, in the neighbourhood of the Sun, the mass of stars and diffuse matter belongs mainly to the plane sub-systems, δ_0 must be almost equal to the galactic density near the Sun, which is well known from observation; its value has been given above in (6.69). Using (6.77), we have

$$\sum_i F_i'' = b_2 \delta_0, \tag{6.78}$$

where b_2 is the mean value of P_i'':

$$b_2 = \sum_i \delta_i P_i''/\delta_0. \tag{6.79}$$

Substituting (6.76) and (6.78) in (6.71) and using the same arguments for eqn. (6.72), we obtain two equations:

$$\begin{aligned} b_1 M' + b_2 \delta_0 &= V^2/\varrho, \\ b_3 M' + b_2 \delta_0 &= \frac{V^2}{\varrho}\left(4\frac{\varrho}{V}A + 1\right), \end{aligned} \tag{6.80}$$

where b_3 is given by the formula

$$\sum_i \varrho \frac{\mathrm{d}F_i'}{\mathrm{d}\varrho} = b_3 M'$$

analogous to (6.76).

It is easy to see that, if

$$b_2' = \sum_i M_i'' P_i''/M'',$$

† To save space we omit the formula for $\mathrm{d}F_i'/\mathrm{d}\varrho$.

where M'' is the total mass of the external spheroids, we can replace (6.80) by the similar equations

$$b_1 M' + b_2' M'' = V^2/\varrho,$$
$$b_3 M' + b_2' M'' = \frac{V^2}{\varrho}\left(4\frac{\varrho}{V}A + 1\right). \tag{6.81}$$

Here M' is practically the mass of the spherical sub-systems and M'' that of the plane and intermediate sub-systems.

To find b_1 and b_3, Safronov assumes the following distribution of the density ν'' of the spherical sub-systems, from the results of Parenago's work:

ϱ (kps)	0	2	4	6	8
ν''/ν_8''	105	72	23	4·5	1

Here the unit of density is taken as ν_8'', the value of ν'' for $\varrho = 8$ kps. By taking five internal spheroids with major semiaxes $a' = 2, 3, 4, 6, 7{\cdot}2$ kps and axis ratios $a'/c' = 2, 3, 5, 7, 10$, Safronov finds b_1 and b_3.

To find b_2, it is assumed that for the plane sub-systems there is only one "mean" spheroid, with axis ratio $a''/c'' = 18$ and 14 from Oort's results in two different investigations (1932 and 1941).

The values obtained for the masses lie within the limits

$$M' = 0{\cdot}32 \text{ to } 0{\cdot}75, \quad M'' = 0{\cdot}27 \text{ to } 0{\cdot}71, \tag{6.82}$$

and the total mass of the Galaxy is

$$M = 1{\cdot}1. \tag{6.83}$$

All these are expressed in units of 10^{11} solar masses. The value of the total mass is satisfactory, but M' and M'' are not sufficiently reliable to give any indication of the relative masses of spherical and plane sub-systems in the Galaxy.

§ 5. Some Properties of Star Orbits in Spherical Systems

In this section we shall consider star orbits in a system for which the mass distribution is spherically symmetrical; such are the globular clusters, and the elliptical galaxies of type E 0.

The form of the individual orbits depends, of course, on the mass distribution as a function of r, but some general conclusions about the nature of the orbits can be drawn merely from the spherical symmetry of the system.

If a star is within the system at a distance r from the centre, it is acted on by a force $GM(r)/r^2$ towards the centre, where $M(r)$ is the mass within a sphere of radius r. The gravitational potential is thus

$$U(r) = \int_r^\infty \frac{GM(r)}{r^2}\,\mathrm{d}r. \tag{6.84}$$

We naturally use spherical coordinates r, θ, φ and velocities R, Θ, Φ (see Fig. 27, Chapter V, § 2), with the origin at the centre of the system. The potential U depends on r only, and the differential equations of motion (5.19) become

$$\left.\begin{aligned} \dot{R} &= \frac{\Theta^2 + \Phi^2}{r} + \frac{\mathrm{d}U}{\mathrm{d}r}, \\ \dot{\Theta} &= -\frac{R\Theta}{r} + \frac{\Phi^2}{r}\cot\theta, \\ \dot{\Phi} &= -\frac{R\Phi}{r} - \frac{\Theta\Phi}{r}\cot\theta. \end{aligned}\right\} \tag{6.85}$$

Multiplying the second equation by Θ and the third by Φ and adding, replacing R by $\dot{r}$ and integrating, we obtain the area integral (angular momentum integral) about the centre:

$$r^2(\Theta^2 + \Phi^2) = l^2. \tag{6.86}$$

Putting

$$T^2 = \Theta^2 + \Phi^2, \tag{6.87}$$

where T is the transverse velocity, in the first eqn. (6.85) and (6.86), we have

$$\dot{R} = \frac{T^2}{r} + \frac{\mathrm{d}U}{\mathrm{d}r}, \tag{6.88}$$

$$r^2 T^2 = l^2. \tag{6.89}$$

Multiplying the differential eqn. (6.88) by $R = \dot{r}$ and using (6.89), we have the energy integral:

$$R^2 + T^2 - 2U = 2H. \tag{6.90}$$

The two integrals (6.89) and (6.90) determine, as we shall see, the general properties of the star orbits.

In Chapter V, § 4, we saw that, for a spherical system, there exists the angular momentum vector integral $\mathbf{r} \times \mathbf{V} = \mathbf{l}$, whose components along the coordinate axes are I_2, I_3, I_4 given by formulae (5.34). Hence it follows that each star orbit is a plane curve, and the plane of the orbit is perpendicular to the vector $\mathbf{l} = \mathbf{r}_0 \times \mathbf{V}_0$, where $\mathbf{r}_0$ and $\mathbf{V}_0$ are the initial radius vector and velocity. The orientation of the orbit is unimportant here, since the system is spherically symmetrical.

We shall now show that circular orbits are possible. For a circular orbit of radius r_c we have $r \equiv r_c$, $R = \dot{r} = 0$. Then (6.88) gives

$$T_c^2/r_c = -(\mathrm{d}U/\mathrm{d}r)_{r=r_c}, \tag{6.91}$$

where T_c is the circular velocity. Substitution of the potential U (6.84) gives

$$T_c^2 = GM(r_c)/r_c. \tag{6.92}$$

Thus the circular velocity is a uniquely defined function of the coordinates, and if it is given for all r the law of mass distribution in the system is known. From (6.92), moreover, the circular motion is uniform.

It is easy to see that straight lines through the centre of the system are also possible orbits. The orbit is a straight line if the velocity at some instant $t = t_0$ is along the radius vector, i.e. $T_0 = 0$. Then, according to the area integral (6.89), $l = 0$, and therefore $r^2 T^2 = 0$ at every instant. Hence $T \equiv 0$, and the path is therefore a straight line.

It may be shown that, in any case except the two limiting ones of circular and rectilinear orbits, the orbit lies entirely between two concentric spheres of radii equal to the apocentric and pericentric distances.†

This property can be derived from the integrals of the motion (6.89) and (6.90). Eliminating the transverse velocity T from these integrals, we have

$$R^2 = -f(r), \tag{6.93}$$

where

$$f(r) \equiv \frac{l^2}{r^2} - 2U(\ ) - 2H.$$

If the motion can exist, the condition $-R^2 = f(r) \leqslant 0$ must hold. The greatest and least roots of the equation

$$f(r) \equiv \frac{l^2}{r^2} - 2U(r) - 2H = 0 \tag{6.94}$$

therefore give the apocentric and pericentric distances.

Let us first suppose that the constant energy H is negative, i.e. the velocity is less than the velocity of escape. There are three possible cases.

(1) $R^2 \equiv 0$ at every instant. Then $(\mathrm{d}r/\mathrm{d}t)^2 \equiv 0$, $r = r_c =$ constant, and the orbit is circular. Circular orbits are possible at all distances from the centre, as has already been shown.

(2) $R^2 \not\equiv 0$, $l = 0$. The orbit is a straight line, and eqn. (6.94) becomes

$$U(r) = -H. \tag{6.95}$$

Equation (6.84) shows that $U(r)$ is a monotonically decreasing function of r. For $H < 0$, therefore, eqn. (6.95) has a single positive root, which gives the apocentric distance r_a. Since R is not zero for any value of r except r_a, the star oscillates along a diameter, moving symmetrically about the centre of the system.

(3) $R^2 \not\equiv 0$, $l \neq 0$. The path clearly includes points for which $f(r) = -R^2 < 0$. From (6.94) and (6.84) we find

$$f'(r) = -\frac{l^2}{r^3} + 2\frac{GM(r)}{r^2}.$$

Hence the equation $f'(r) = 0$ can have not more than one root. From (6.94) with $H < 0$ we obtain $f(0) = +\infty, f(\infty) = -2H > 0$. It follows that the function $f(r)$

† The derivation of the properties of circular orbits given here and on the next page is by P. T. Reznikovskii, to whom the author's thanks are due for permission to use this material before publication.

has the form shown in Fig. 30, and in the present case the equation $f(r) = 0$ has two real positive roots r_1 and r_2.

Thus the orbits of almost all stars whose velocities do not exceed the velocity of escape ($H < 0$) are more or less elongated ovals, and the apse line may precess.

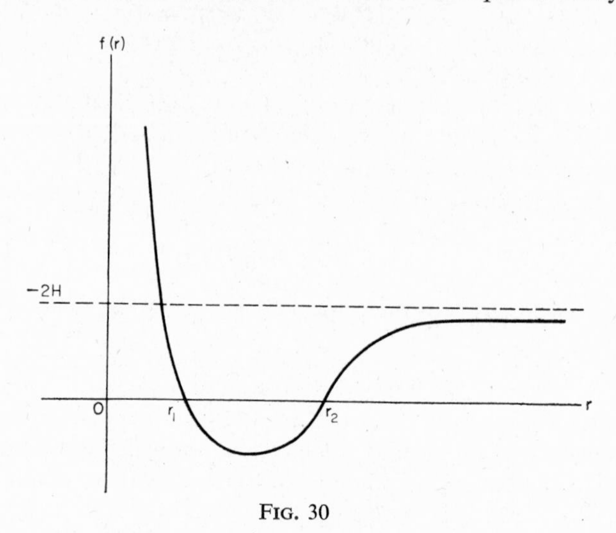

FIG. 30

A similar discussion for $H \geqslant 0$ shows that the whole orbit lies outside the circle of radius r_1, and r_1 (the pericentric distance) is the only positive root of the equation $f(r) = 0$.

For a more detailed analysis of the properties of the orbits, it is necessary to know the actual form of the potential U, i.e. the distribution of mass in the system. In particular, if the potential is known, the problem reduces to quadratures. For, eliminating the transverse velocity from the integrals (6.89) and (6.90), we have

$$R^2 = (\mathrm{d}r/\mathrm{d}t)^2 = 2H - l^2/r^2 + 2U, \tag{6.96}$$

whence t is obtained as a function of r by integration. The azimuthal angle φ can then be found from the area integral (6.89) by using the relation $T = r\,\mathrm{d}\varphi/\mathrm{d}t$.

Some approximate results of a more specific nature, however, may be obtained even if U is unknown. The orbit and the motion in it are determined at each point by the local properties of the force field, since in any regular force field the motion is uniquely determined by the initial coordinates and velocities.

In particular, in the outermost parts of a system, where the mass of the stars lying still further from the centre may be neglected, the motion and the nature of the orbit are quasi-Keplerian under the Newtonian potential

$$U = GM/r. \tag{6.97}$$

Substituting this potential in (6.94), we obtain a quadratic equation:

$$l^2u^2 - 2GMu - 2H = 0, \tag{6.98}$$

where

$$u = 1/r. \tag{6.99}$$

If the whole orbit lies in the region of the quasi-Newtonian potential, it is approximately elliptical and eqn. (6.98) determines the position of the apocentre and the pericentre. If only the apocentre lies in this region, the apocentric distance corresponds to the smaller root of eqn. (6.98).

It is easy to see that the field in the central parts of the system is quasi-elastic. The star density $\nu(r)$, by continuity, must satisfy the condition $d\nu/dr = 0$ for $r = 0$. Hence it must be almost constant for small r: $\nu(r) \approx \nu_0$. Then $M \approx \frac{4}{3}\pi r^3 m\nu_0$, where m is the mass of one star, and from (6.84) we obtain the Hookean potential

$$U = -\alpha r^2, \tag{6.100}$$

in which α is given in terms of the central density ν_0 by the relation

$$\alpha = 4\pi G m\nu_0/3. \tag{6.101}$$

In a quasi-elastic force field, the orbits are ellipses with centre at the origin, and the motion takes place with an angular velocity $\omega = \sqrt{\alpha}$ which is constant and the same for every orbit. Substituting the potential (6.100) in (6.94), we obtain the biquadratic equation

$$2\alpha r^4 - 2Hr^2 + l^2 = 0.$$

If the whole orbit lies in the central region, this equation gives the semiaxes of the quasi-Hookean ellipse.

It is evident that, by considering only the individual orbits of stars, we can draw no conclusions concerning the structure of the stellar system as a whole (shape, mass distribution, etc.). To do so, it is necessary to examine the distribution of the orbits, or rather of the orbital elements. This will be done for spherical systems in Chapter IX.

CHAPTER VII

THE PROBLEM OF LOCAL DYNAMICS

§ 1. Statement of the Problem

IN PREVIOUS chapters we have discussed the general methods used in the dynamics of stellar systems in steady states. The problem was to investigate the motions of such a stellar system as a whole. Assumptions were made concerning the structure of stellar systems, smoothing out all structural non-uniformities. For example, the Galaxy was regarded as a system having rotational symmetry. Nevertheless, stellar systems of even the most regular forms have a complex structure: for instance, the Galaxy includes spiral arms, star clouds lying more or less along these arms, and star clusters. According to modern ideas, the star clouds are the centres of the continuing formation of young stars. These are mainly O and B stars, which (according to V.A. Ambartsumyan) are formed in the O associations, and subgiants and dwarfs, which are formed in the T associations and subsequently dispersed into the surrounding space. In this way the patchy distribution of stars corresponding to the clouds of the Milky Way is set up. The mean age of the stars in the associations is only 10^7 to 10^8 y. We have seen in Chapter IV that the stellar condensations such as star clouds (and also condensations of diffuse matter) are of importance in the dynamics of the Galaxy.

In the chapters on the general dynamics of stellar systems it has been assumed that the systems are in a quasi-steady state, i.e. a state of dynamic equilibrium which changes only very slowly under the action of the irregular forces. This viewpoint again corresponds to a smoothed structure of stellar systems, devoid of any marked fluctuations in star density.

If, however, the star distribution includes appreciable condensations, the dynamical picture may be considerably changed. The condensations must in general be unstable and be dispersed fairly rapidly among the surrounding stars or *field stars*. Besides the natural loss of stars by dynamical escape at velocities exceeding the velocity of escape, every condensation tends to be disrupted by the tidal action of the regular force field of the whole system. Such forces will be particularly effective in rapidly rotating systems such as the Galaxy.

It is known that, in the neighbourhood of the Sun, there is a large cluster of stars of class O and the early B sub-classes, mainly from B 0 to B 5 inclusive. This cluster is probably an offshoot from the spiral structure of the Galaxy in the region of the Sun, and is usually called the *local cluster*. The discovery of the associations makes it natural to suppose that the local cluster results from the accumulation of early-type stars, formed in O (and possibly T) associations, which have not yet mingled with the general stellar population.

It is important to note that the motions of the stars near the Sun exhibit certain features which do not accord with the general theory of a quasi-steady Galaxy, and for the early-type very bright O and B stars these features are particularly marked.

In stellar dynamics there are two essentially different methods of investigating these peculiarities of stellar motions, and a choice between them can be made only on the basis of extensive data regarding stellar motions, not only for the immediate neighbourhood of the Sun but also for a region comprising an appreciable fraction of the entire Galaxy.

The first method starts from the assumption that the observed anomalies in stellar motions are general, i.e. that they occur throughout the Galaxy, varying continuously from point to point. With this approach, we must evidently attempt to modify the dynamical theory as a whole. Following such a method, we can seek to go from a theory of systems in a steady state to one for non-steady systems, to add quasi-integrals to the exact integrals of the motion, and so on.

The second method regards the observed deviations from the general theory as being due to properties of the structure of the stellar system in the neighbourhood of the point concerned; in particular, the properties of the system in the neighbourhood of the Sun are to be interpreted as being due to the presence of an extensive condensation, the local cluster, the "framework" of which is formed by the *complex* of early-type (O and B) stars—a term due to A. F. Torondzhadze.

The first method remains within the scope of the general problem of stellar dynamics, which is to examine the structure and motions of the stellar system as a whole. The second method, however, involves the study of only a comparatively small region within the system, for example that occupied by a single star cloud. We shall call this *local dynamics*.

In a certain sense the problem of local dynamics may be considered independently of the general problem. The local condensation considered may be regarded as being in a given field of external force due to the system of field stars distributed in a given way. The local condensation may be dynamically stable or unstable, independently of the stability of the whole system. We can therefore consider local departures from a steady state within a system which, as a whole, is in a steady state, and this will be our statement of the problem.

The choice between the two methods must at present remain open, but we may justify a discussion of local dynamics, on two grounds.

(1) The existence of the local cluster must have some effect on the observed properties of stellar motions. In other words, the problem of local dynamics is meaningful provided that the local cluster actually axists.

(2) The observed peculiarities of stellar motions (the K effect and the deviation of the vertex) pertain as a rule to only the brightest, and therefore to the nearest, stars. For faint stars at distances exceeding 500 ps, these properties do not occur, and the velocity distribution approaches the normal distribution obtained from the theory of systems in a steady state.

In this chapter we shall consider the problem of local dynamics for the neighbourhood of the Sun. We regard the local cluster as a dynamically unstable system within a stellar system in a steady state and a constant field of force, into which it is gradually dispersing. We shall show that all the main observed features and anomalies of the motions of the local cluster can then be interpreted.

§ 2. The Local Cluster and its Motion in the Galaxy

Let us first see what is known of the detailed structure of the Galaxy in the neighbourhood of the Sun.

The Sun lies almost exactly in the galactic plane, parallel to which is the spiral structure of the Galaxy. Studies of our own and other galaxies show that the spiral arms are not at all homogeneous, but consist of more or less continuous chains of clouds and agglomerations of stars with many side-branches and even separate pieces. The most detailed and reliable information on the distribution of stars in the spirals of the Galaxy in the neighbourhood of the Sun is due to B. A. Vorontsov-Vel'yaminov [135, 136]. The work of Vorontsov-Vel'yaminov and others has shown that the clouds in galactic spirals consist mainly of early-type giants.

In recent years the study of the spiral structure of the Galaxy has been considerably advanced by the use of radio astronomy. The results have shown that the Sun is either inside or at the edge of one of the clouds in a spiral arm. This cloud is elongated in the direction of the arm, which passes near the Sun in the direction $l = 242°$ (R. V. Kunitskiĭ [54]). The presence of the cloud was discovered in 1879 by Gould, and is most evident in the distribution of the O and B stars, and to some extent in the dwarf M stars. Gould later discovered that the bright B stars are concentrated not along the galactic equator but along a great circle called *Gould's belt*, at an angle of 17° to the galactic equator. This indicated the existence of a star cloud whose plane of maximum concentration is at an angle of 17° to the galactic plane.

The cloud was named by Shapley the *local cluster*. Probably stars of other types (dwarfs and, especially, giants) also occur in the local cluster, although this would be more difficult to establish, because almost all the white giants are in star clouds, but only a restricted number of dwarfs belong to these clouds.

It follows from the above that the local cluster belongs entirely to the plane component of the stellar population of the Galaxy.

It may also be noted that the existence of the local cluster as an independent dynamical entity was established from radial velocities by Mineur [72, 73] in 1929 and confirmed in 1950 by R. B. Shatsova [117, 118] from a detailed study of proper motions given in Boss's *General Catalogue* and of radial velocities.

Mineur [72, 73] noted that the galactic rotation term, involving the longitude, in the radial velocity $v_r = Ar \sin 2(l - l_0)$, where l_0 is the longitude of the centre and A is Oort's constant ($= -\frac{1}{2}R\, \mathrm{d}\omega/\mathrm{d}R$, where R is the distance of the observer from the centre and ω the angular velocity of rotation), admits four solutions: the principal one corresponds to the centre of the Galaxy $l_0 = 325°$, and another to the anticentre $l_1 = 325° - 180° = 145°$, and there are also $l_3 = l_0 + 90° = 55°$ and $l_4 = l_0 - 90° = 235°$, for which the sign of the constant A must be reversed. It is evident that the angular velocity cannot increase in absolute value away from the centre. Hence a change in the sign of A must mean a change in the sign of ω, i.e. a reversal of the direction of rotation. Mineur pointed out that the direction l_4 is very close to that of the centre of the local cluster ($l = 242°$). By dividing the stars with known radial velocities into six groups according to distance, Mineur found that the stars nearest the Sun tend to revolve about the centre of the local cluster, but the more distant stars tend to revolve about the centre of the Galaxy. The transition from one tendency to the other takes place continuously as the distance increases.

In 1930 Bok [15] noted that the second Oort's constant B must always have the opposite sign to that of A, and so Mineur's conclusions can be valid only if the constant B for the nearest stars also has the sign opposite to that which it has for more distant stars. Since $A > 0$ and $B < 0$ for the distant stars, $A < 0$ and $B > 0$ for the near stars. The truth of the last inequality cannot yet be tested with any reliability, since the value of B for the near stars is too uncertain.

In 1950 R.B.Shatsova [117] carried out an extensive investigation embodying a different approach to the problem. She made use of the fact that the bright O and B stars which form the "framework" of the local cluster are seen as Gould's belt, which is inclined to the plane of the Milky Way. If the local cluster really exists, then, no matter what its motion, unless it is purely translational, it must involve rotation, since the laws of kinematics show that any motion of a rigid body can be represented as a combination of translational and rotational motions. The non-rigidity of the local cluster leads to a third component of the motion, namely a deformation.

Shatsova's investigation was based on the proper motions of 33,342 stars in Boss's *General Catalogue*. She assumed that each star in the neighbourhood of the Sun participates in two rotations, that of the Galaxy and that of the local cluster. The direction of the axis of rotation of the local cluster was taken to be perpendicular to the plane of Gould's belt, and the most probable value of the angle between the axes of the local cluster and of the Galaxy was taken as 10°. This angle is sufficient for a kinematic separation of two rotations about non-parallel axes. Among Shatsova's results are the following.

(1) The direction of the centre of the local cluster is $l = 253°$.

(2) The distance of the Sun from the axis of rotation of the local cluster is $r = 150$ ps.

(3) The angular velocity of rotation of the local cluster at the Sun's distance is $\omega_0 = -1''\cdot63$ per century, which is 2·4 times the rate of rotation of the Galaxy at this point and is in the same direction. Thus Shatsova's results contradict those of Mineur.

(4) Oort's constant $A = -\frac{1}{2}R\,\mathrm{d}\omega_0/\mathrm{d}R$ for the rotation of the local cluster is found to be, in Oort's units, $A = +0\cdot289$ km/s per ps, which is 15 times its value for the galactic rotation. The high value of A is evidently due to a high gradient of the angular velocity.

Using the rotation of the local cluster, Shatsova was able to explain for the first time such phenomena as the north–south galactic asymmetry of proper motions: there are systematic differences in the mean proper motions of stars in the northern and southern galactic hemispheres. She also accounted for the presence of third-order longitude harmonics (called the *Dyson terms*) in the mean proper motions.

Subsequently [118] these results were confirmed by her analysis of 3049 radial velocities. The present author [81] has also obtained similar conclusions from entirely different arguments.

Finally, Torondzhadze has pointed out in a recent paper [131] that the numerical results of Shatsova's work can equally well be interpreted as showing an expansion of the local cluster parallel to the galactic plane. In the general case they could be regarded as arising from a simultaneous expansion and rotation. In subsequent sections of this chapter we shall show that a number of peculiarities in the motions of stars near the Sun may be explained in terms of an expansion of the local cluster parallel to the galactic plane. Thus all these kinematic properties are explained in terms of local dynamics.

Unfortunately, a more detailed account of this work would take us too far afield, since it would necessitate the quoting of lengthy formulae and a discussion of algebraic details and initial data concerning stellar motions. This would hardly be within the scope of this book, whose main purpose is to discuss fundamentals. The reader is therefore referred to the original works of the authors named above. Here we shall merely note that a first attempt to take account of the rotational motion of the local cluster has already led to an explanation of two peculiar features of stellar motions, which cannot be explained by the general theory of galactic rotation.

§ 3. Observed Anomalies of Stellar Motions in the Neighbourhood of the Sun

Let us now consider the observed anomalies in the motions of stars, and attempt to interpret them on the hypothesis of the dissipation of the local cluster. It is evident that this dissipation can (and in general must) co-exist with rotation of the local cluster if the cluster is dynamically unstable.

The most important anomalies with respect to the predictions of the steady-state theory are (1) the inequality of the axes of the velocity ellipsoid, (2) the deviation of the vertex from the direction of the centre of the Galaxy, (3) the positive K effect for the O and B stars. We shall consider these in turn.

The problem of the unequal axes of the velocity ellipsoid has already been discussed and solved in Chapter V. In that chapter it was shown, on the basis of the work of Oort, Lindblad, Kuzmin and others, that the phenomenon can be explained by the existence of a third integral (quasi-integral) of the motion, without going outside the steady-state theory. We saw that Oort's integral I_3 is valid only in a region defined by the condition that ϱ should be large compared with z, and not throughout the Galaxy. This condition defines a conical region adjoining the galactic plane. In the vicinity of the Sun the half-thickness of this region may be estimated from the condition that $|z| < 0{\cdot}12\,\varrho$. The recent value $\varrho = 7200$ ps gives the upper limit of $|z|$ as about 800 ps.

We shall not analyse in detail the observational results concerning the deviation of the vertex, but simply give the general results; for further information the reader may consult, for example, P. P. Parenago's book ([107], pp. 128–131).

(1) Except for B stars and M giants, the velocity ellipsoid depends only slightly on the spectral type and apparent distribution of the stars.

(2) The position of the vertex depends on the absolute magnitude of the stars considered: for giants the mean longitude of the vertex is $l = 344°$, and for dwarfs it is 332°, the deviation of the vertex being $\Delta l = 19°$ and 7° respectively.

(3) For B stars and M dwarfs the longitude of the vertex is not accurately known, because the axes of the ellipsoid which lie in the galactic plane are almost equal, and so the ellipsoid degenerates to a spheroid with axis of rotation perpendicular to the galactic plane.

In 1939 Blaauw [11] determined the velocity ellipsoid on the basis of a very large number (18,000) of the most accurately known proper motions. By means of an analysis of dispersions, a statistical method superior to those generally used, Blaauw concluded that the deviation of the vertex is observed, as a rule, for the brightest stars, i.e. on average for the nearest stars. For faint stars the deviation of the vertex is very slight: Blaauw's value for the longitude of the vertex is $l = 322°$.

Leaving aside for the moment the B and dwarf M stars, we can summarise these results thus: the deviation of the vertex is characteristic mainly of the giants and of the nearest stars.

Attempts at a theoretical interpretation of the deviation of the vertex have been made by the use of both general and local galactic dynamics. In 1935 Shiveshwarkar [119] considered the problem on the assumption that, at each point in the Galaxy, there is besides the circular velocity Θ_0 a radial velocity $P_0 > 0$ of the centroid, corresponding to an expansion of the Galaxy parallel to the galactic plane. He obtained for the deviation angle Δl the formula $\tan 2\Delta l = m/(k^2 - h^2)$, where h^2, k^2 and m are the coefficients in the quadratic form giving Schwarzschild's law in two dimensions:

$$f = f_0 \exp[-h^2(P - P_0)^2 - m(P - P_0)(\Theta - \Theta_0) - k^2(\Theta - \Theta_0)^2].$$

The theoretical value of Δl is of the order of 11°.

It is difficult, however, to assent to the fundamental assumptions underlying this work: despite the expansion, the Galaxy is assumed to be in a steady state, which is an obvious contradiction. Moreover, doubts may arise concerning the modification of the general theory of the Galaxy in order to explain one isolated phenomenon, however important in itself, while neglecting the effect of this modification on the whole complex of phenomena observed in the part of the Galaxy accessible to observation.

Thus the attempts made to explain the deviation of the vertex in terms of the general theory of the Galaxy have not proved successful. It is certain that any attempt to relate the deviation of the vertex to an expansion of the whole Galaxy will require a solution of the dynamical problem for a system in a non-steady state. Such a theory cannot be based merely on isolated facts, without taking account of the physical process which brings about the non-steady motion of the system.

In 1934 Heckmann and Strassl [37] carried out an investigation of fundamental importance, considering the problem of local dynamics in the following form. In a Galaxy which is rotationally symmetrical and in a steady state, large numbers of fluctuations of star density are continually formed and dispersed, each of which occupies only a very small part of the Galaxy and exists for a time very short in comparison with the fundamental unit of time, the period of rotation of the Galaxy. Since the fluctuations are small and random, the potential of the regular forces, which is the integral of the ratio of density to distance, must be independent of time as a result of the averaging which occurs in integration. This corresponds to the hypothesis that the Galaxy as a whole is in a steady state. But an observer attached to some fluctuation, who observes the motions of stars in a small region and for a time of order not exceeding the lifetime of a fluctuation, will find a local phase distribution which is not steady. Thus the situation considered by Heckmann and Strassl may be described as "locally non-steady".

To solve this problem, they assumed that the local cluster is such a fluctuation of star density. They supposed that the velocity distribution in the local cluster is Schwarzschildian, with coefficients depending on coordinates and time. Expanding the potential in powers of the small quantities $\xi = x - x_0$, $\eta = y - y_0$, $\zeta = z - z_0$ (where x_0, y_0, z_0 are the coordinates of the observer), they found theoretical expressions for twenty parameters which define the local fluctuation in terms of twenty-one quantities (sixteen coefficients in the expansion of the centroid velocity in powers of ξ, η, ζ up to the second order, and five independent elements of the velocity ellipsoid). The properties of the theoretical velocity distribution in the local cluster include a deviation of the vertex. Unfortunately, the large number of unknowns which have to be found by comparing theory and observation, and the insufficient accuracy of the spatial velocities of stars, made it impossible to derive any numerical results.

The *K* effect is that the observed mean radial velocity of bright O and B stars is positive, and equals 4·3 km/s. This may be illustrated by the following data, taken from Parenago's book ([107], p. 106).

Spectral type	Mean apparent magnitude $\overline{m}$	*K* effect (km/s)
O to B2	4·4	+6·2
O to B2	6·9	+0·9
B3 to B5	4·8	+4·6
B3 to B5	6·8	+0·0

Thus the *K* effect for bright O and B stars decreases with increasing distance and for $\overline{m} = 6$, i.e. at distances of 400 to 500 ps, becomes almost zero. For stars of other spectral classes, the *K* effect is absent, as we see from the following table, based on the 1936 data of Smart and Green [123]:

Spectrum	B	A	F	G	K	M
K (km/s)	+4·9	+1·7	+0·3	−0·2	+0·3	+0·7

The hypotheses which have been advanced at various times to explain the *K* effect cannot be regarded as satisfactory. They include the following.

(1) The *relativistic K effect* in the atmospheres of massive stars. This is given by $K = GM/Rc$, where M is the mass of the star, R its radius, and c the velocity of light. According to the calculations of Plaskett and Pearce [109], the relativistic *K* effect varies from +2·62 km/s for O stars to +1·0 km/s for B3 to B5 stars. Thus it cannot entirely explain the observed *K* effect, and also should be independent of distance.

The relativistic K effect may, however, be of some importance, and should not be ignored. The fact that the observed K effect becomes imperceptible at large distances may indicate that it does not depend on any one cause, but results from the interaction of several causes.

(2) The *kinematic K effect*, discussed in Chapter III (formula (3.27)), corresponds to one form of hydrodynamical deformation of a volume element in the system in accordance with Helmholtz' theorem. Such an effect should, however, increase in magnitude in proportion to r, i.e. should behave in the opposite manner to that observed.

(3) The existence of a stream of bright B stars in the southern hemisphere (in the constellations of Scorpius and Centaurus) does not differ essentially from the hypothesis of a local cluster, since it likewise envisages the existence of a large accumulation of bright B stars in the neighbourhood of the Sun. The assumption of such a stream involves an indeterminacy, since its existence or otherwise depends largely on the assumed elements of the solar motion.

The only difference from the hypothesis of a local cluster is that the Sun is assumed to lie outside, rather than inside, the agglomeration of B stars.

The Scorpius–Centaurus stream hypothesis was put forward in the 1930's by Plaskett and Pearce [109, 110]. Its shortcomings were made evident in 1936 by the detailed work of Smart [121] and especially by the extensive investigation of Nordström [76]. The same conclusion was reached more recently by A. F. Torondzhadze [130]. These authors showed that the stream motion is simulated by the incorrect value of the Sun's velocity used by Plaskett and Pearce. In addition, Nordström showed that, if the stars whose membership of the stream is dubious are rejected (as is usually done in such cases), a repetition of the calculation gives the same value of $+4{\cdot}2$ km/s. This shows directly that the Scorpius–Centaurus stream theory proves nothing.†

The arguments in favour of the existence of the local cluster have been given at the beginning of this chapter. Here we may add the fact that, according to recent discoveries by Soviet and other astronomers, the O and B stars, and also the supergiants of various classes, form isolated clouds along the spiral arms of the Galaxy. The concepts put forward by V. A. Ambartsumyan of the formation of O and B stars in associations, which are subsequently dispersed and perhaps form the star clouds observed in the arms of the Galaxy, are similar in nature. It may be supposed that the local cluster is one such cloud. We do not know to what extent this cloud contains an agglomeration of late-type stars, but it is possible that the star clouds are concentrations mainly of O and B stars rather than of the other classes. These stars therefore serve to mark the clouds, and are easily detected on account of their rarity in the Galaxy as a whole and their high luminosity.

† In [107] P. P. Parenago retained Plaskett and Pearce's explanation of the K effect, but he gave no new data to justify this.

§ 4. The Linearised Equations of Motion in Lindblad's System of Rotating Coordinates

The above discussion leads to the dynamical problem of investigating the effect of the local cluster on the kinematics of stellar motions as regards an observer within the cluster.

A typical procedure in problems of this type is to consider a small region within the Galaxy, comprising some neighbourhood of the observer's centroid. The observer's centroid participates in the galactic rotation, and it is therefore convenient, in investigating the local dynamics, to use a rotating coordinate system with its origin at the observer's centroid. The coordinates of points within the region considered may be regarded as small compared with the distance of the observer from the centre of the Galaxy. In this way, as we shall see, the problem may be linearised and its solution thereby considerably simplified.

In the present section we shall derive Lindblad's equations for the relative motion of stars near an arbitrary centroid. We shall assume that the centroid C_0 moves in a circular orbit in the galactic plane; let ϱ_0 be the radius of this orbit and ω_0 the constant angular velocity. The corresponding quantities for a general centroid will be denoted by ϱ and ω.

We use a fixed system of galactocentric rectangular coordinates, with the Z-axis along the axis of the Galaxy and the X-axis passing through the position of the centroid C_0 at time $t = 0$ (Fig. 31, where the Z-axis is out of the paper). Then $\theta = \omega_0 t$. In addition, we define a moving system of rectangular coordinates x, y, z, with its origin at the centroid C_0, the z-axis parallel to the Z-axis, and the x-axis along OC_0.

Let P be an arbitrary point in the neighbourhood of C_0. Its coordinates and velocity components in the two systems are related by

$$\left.\begin{aligned} X &= (\varrho_0 + x)\cos\theta - y\sin\theta, \\ Y &= (\varrho_0 + x)\sin\theta + y\cos\theta, \\ Z &= z; \end{aligned}\right\} \tag{7.1}$$

$$\left.\begin{aligned} \dot{X} &= \dot{x}\cos\theta - \dot{y}\sin\theta - \omega_0 Y, \\ \dot{Y} &= \dot{x}\sin\theta + \dot{y}\cos\theta + \omega_0 X, \\ \dot{Z} &= \dot{z}. \end{aligned}\right\} \tag{7.2}$$

The kinetic energy of a particle of unit mass expressed in terms of the moving coordinates is therefore

$$\begin{aligned} T = \tfrac{1}{2}(\dot{x}^2 + \dot{y}^2 + \dot{z}^2) - \omega_0 \dot{x} y + \omega_0(\varrho_0 + x)\dot{y} \\ + \tfrac{1}{2}\omega_0^2[(\varrho_0 + x)^2 + y^2]. \end{aligned} \tag{7.3}$$

Substituting in Lagrange's equations of motion, we obtain

$$\left.\begin{aligned} \ddot{x} - 2\omega_0\dot{y} - \omega_0^2(\varrho_0 + x) &= \partial U/\partial x + \partial U_1/\partial x, \\ \ddot{y} + 2\omega_0\dot{x} - \omega_0^2 y &= \partial U/\partial y + \partial U_1/\partial y, \\ \ddot{z} &= \partial U/\partial z + \partial U_1/\partial z, \end{aligned}\right\} \tag{7.4}$$

where U is the potential of the regular forces in the Galaxy, and U_1 the potential of the local cluster. These equations are exact, and represent the motions of stars in the neighbourhood of the observer's centroid.

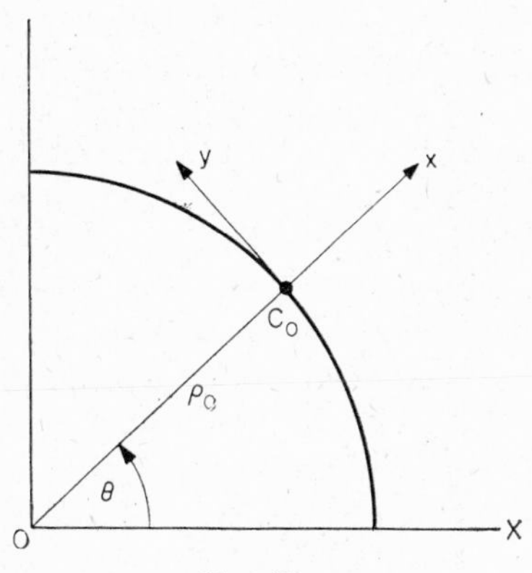

FIG. 31

Since the dimensions of the local cluster are small compared with those of the Galaxy, the coordinates x, y, z may be taken as small compared with ϱ_0 for points whose distance from C_0 does not exceed the dimensions of the local cluster ($\sim$ 500 ps). In this way eqns. (7.4) may be linearised.

In order to expand the right-hand sides in powers of x, y, z as far as the first-order terms, it is sufficient to represent the potential U of the Galaxy as a series of powers of x, y, z as far as the second order:

$$\begin{aligned} U = U_0 &+ (\partial U/\partial x)_0 x + (\partial U/\partial y)_0 y + \tfrac{1}{2}(\partial^2 U/\partial x^2)_0 x^2 \\ &+ \tfrac{1}{2}(\partial^2 U/\partial y^2)_0 y^2 + \tfrac{1}{2}(\partial^2 U/\partial z^2)_0 z^2 + (\partial^2 U/\partial x\,\partial y)_0 xy \\ &+ (\partial^2 U/\partial y\,\partial z)_0 yz + (\partial^2 U/\partial z\,\partial x)_0 zx. \end{aligned} \tag{7.5}$$

The suffix zero denotes the values of the derivatives at the point $x = y = z = 0$. To simplify this expression we use the fact that the potential U has rotational symmetry about the axis of the Galaxy and symmetry about the galactic plane. Hence

$$\begin{aligned} U &= U(\varrho, z) = U(\sqrt{\{[\varrho_0 + x]^2 + y^2\}}, z), \\ U(\varrho, -z) &= U(\varrho, z), \end{aligned} \tag{7.6}$$

and therefore

$$\left.\begin{aligned}
\frac{\partial U}{\partial x} &= \frac{\varrho_0 + x}{\varrho}\frac{\partial U}{\partial \varrho}, \quad \frac{\partial U}{\partial y} = \frac{y}{\varrho}\frac{\partial U}{\partial \varrho},\\
\frac{\partial U(\varrho, z)}{\partial z} &= -\frac{\partial U(\varrho, -z)}{\partial z},\\
\frac{\partial^2 U}{\partial x^2} &= \frac{(\varrho_0 + x)^2}{\varrho^2}\frac{\partial^2 U}{\partial \varrho^2} + \frac{y^2}{\varrho^3}\frac{\partial U}{\partial \varrho},\\
\frac{\partial^2 U}{\partial y^2} &= \frac{y^2}{\varrho^2}\frac{\partial^2 U}{\partial \varrho^2} + \frac{(\varrho_0 + x)^2}{\varrho^3}\frac{\partial U}{\partial \varrho},\\
\frac{\partial^2 U}{\partial x \partial y} &= \frac{(\varrho_0 + x)y}{\varrho^2}\frac{\partial^2 U}{\partial \varrho^2} - \frac{(\varrho_0 + x)y}{\varrho^3}\frac{\partial U}{\partial \varrho},\\
\frac{\partial^2 U(\varrho, z)}{\partial \varrho \partial z} &= -\frac{\partial^2 U(\varrho, -z)}{\partial \varrho \partial z}.
\end{aligned}\right\} \quad (7.7)$$

Putting in eqns. (7.7) $x = y = z = 0$, $\varrho = \varrho_0$, and substituting the results in (7.5), we find

$$U = U_0 + (\partial U/\partial \varrho)_0 x + \tfrac{1}{2}(\partial^2 U/\partial \varrho^2)_0 x^2 + \tfrac{1}{2}(1/\varrho_0)(\partial U/\partial \varrho)_0 y^2 + \tfrac{1}{2}(\partial^2 U/\partial z^2)_0 z^2. \quad (7.8)$$

Since the local cluster is predominantly seen in the distribution of giant stars, i.e. is part of the plane component of the galactic population, the velocity of the centroid is equal to the circular velocity of a particle to within a fraction of a kilometre per second. Thus we can put

$$\omega_0^2 = -(1/\varrho_0)(\partial U/\partial \varrho)_0. \quad (7.9)$$

Substituting (7.8) and (7.9) in eqns. (7.4), we easily obtain

$$\left.\begin{aligned}
\ddot{x} - 2\omega_0 \dot{y} + \alpha_1 x &= \partial U_1/\partial x,\\
\ddot{y} + 2\omega_0 \dot{x} &= \partial U_1/\partial y,\\
\ddot{z} + \alpha_3 z &= \partial U_1/\partial z,
\end{aligned}\right\} \quad (7.10)$$

where the constant coefficients α_1 and α_3 are given by

$$\alpha_1 = \left(\frac{1}{\varrho}\frac{\partial U}{\partial \varrho} - \frac{\partial^2 U}{\partial \varrho^2}\right)_0, \quad \alpha_3 = -\left(\frac{\partial^2 U}{\partial z^2}\right)_0. \quad (7.11)$$

Let us now examine the effect of the attraction of the local cluster, assuming for simplicity that the latter is a homogeneous ellipsoid with its principal axes parallel to the moving coordinate axes. Then the theory of attractions shows that the force at an internal point is given by

$$\partial U_1/\partial x = -\beta_1 x, \quad \partial U_1/\partial y = -\beta_2 y, \quad \partial U_1/\partial z = -\beta_3 z, \quad (7.12)$$

where $\beta_1, \beta_2, \beta_3$ are positive constants depending on the axes of the ellipsoid and proportional to the star density within the local cluster. Equations (7.10) become

$$\left.\begin{aligned} \ddot{x} - 2\omega_0 \dot{y} + (\alpha_1 + \beta_1)x = 0, \\ \ddot{y} + 2\omega_0 \dot{x} + \beta_2 y = 0, \\ \ddot{z} + (\alpha_3 + \beta_3)z = 0. \end{aligned}\right\} \tag{7.13}$$

These are the linearised equations of motion for stars in the neighbourhood of the observer's centroid.

The following remarks should be made concerning the use of eqns. (7.13). They are valid only in a small neighbourhood of the point C_0, which moves in the galactic plane on a circle of radius $\varrho_0 = \sqrt{(x_0^2 + y_0^2)}$ with angular velocity ω_0 about the centre of the Galaxy. The limits of this neighbourhood are given by the condition that the apparent magnitude of a B0 star is $6^m{\cdot}0$. This gives the distance $r_0 = 400$ to 500 ps $= 0{\cdot}05$ to $0{\cdot}06$ times ϱ_0. A star moving with velocity V relative to C_0 traverses a distance equal to r_0 in a time $T_0 = r_0/V$. In what follows we shall have occasion to use eqns. (7.13) to study the motions of stars belonging to the plane component of the galactic population. For these stars the value of V may be taken as the root-mean-square residual velocity: $V = 20$ km/s, and then $T_0 = 2{\cdot}5 \times 10^7$ y. For O and B stars V has only half this value, and so T_0 is correspondingly greater. However, as we shall see later, the low value of V is itself a consequence of the dissipation of the local cluster, since outside the local cluster the O and B stars apparently have residual velocities equal to those of the other stars in the plane component of the Galaxy.

There is a further point which requires discussion. If we take the point C_0 as centroid, then eqns. (7.13) show that we consider the velocities of all stars as residual velocities relative to the same centroid C_0. We are therefore, strictly speaking, justified in applying these equations only within a single macroscopic volume element about the point C_0, i.e. only at distances less than 100 ps. This would mean that we could include only stars with $m < 3^m$ (assuming that this means $M < 3^m$), of which there are only a few dozen in the whole sky. On going beyond the macroscopic volume element we should have to take into account the change in the centroid velocity, i.e. make a correction for the differential galactic rotation. This correction is of the order of Ar_0, where $A = 0{\cdot}019$ km/s per ps, and for $r_0 = 500$ ps this is 10 km/s, which is comparable with the root-mean-square residual velocities. We shall not do so, however, for the following reasons.

Firstly, by taking into account the differential galactic rotation, we do not avoid the difficulty, since the change in the parameters of the velocity ellipsoid is not allowed for. By forming the second-order moments of the residual velocities we tacitly assume that the velocity ellipsoid is the same at all points in the region considered; but both the general theory and specific calculations made by the author in 1934 [78] show that there is a systematic change in the elements of the ellipsoid, depending on the size of the region within which the stellar velocities are considered.

Secondly, the elimination of the differential galactic rotation terms would involve corrections to each individual velocity, and therefore a knowledge of the distance of each star, which is at present affected by a large error. An attempt to make the necessary corrections after calculating the second-order moments of the residual velocities is, of course, possible in principle, but is not of practical use, because the relative error in the galactic rotation term Ar_0 is $(\overline{r^2} - \bar{r}^2)/\overline{r^2}$ and is difficult to calculate because of the difficulty of finding $\overline{r^2}$ (see, for example, Nordström's investigation of radial velocities [76]).

Lastly, no correction to allow for the distances of the individual stars is made in practice in calculating the elements of the velocity ellipsoid.

Our results, in which all stellar velocities are referred to the same centroid, therefore differ both in principle and in practice from those obtained by referring the velocities to the corresponding topocentric centroids; see, for example, Lindblad's derivation of the ellipsoidal velocity distribution [66].

§ 5. Conditions for the Dynamical Stability of Stellar Condensations

The equations of relative motion in the form (7.13) were first used by B. J. Bok in 1934 [14] to examine the conditions for the stability of moving clusters of stars in the Galaxy. We may mention his results in passing, although we are actually interested in the case of instability, i.e. the opposite case to that considered by Bok.

The signs of the coefficients $\alpha_1, \alpha_3, \beta_1, \beta_2, \beta_3$ in eqns. (7.13) are of importance as regards stability. As we have already stated, the coefficients $\beta_1, \beta_2, \beta_3$ are always positive. The coefficient α_1 defined by the first eqn. (7.11) is always negative. This may be seen as follows. Firstly, $\partial U/\partial \varrho$ is always negative, since $\partial U/\partial \varrho = -F$, where F is the absolute magnitude of the gravitational force. Secondly, the magnitude of F decreases with increasing ϱ, and therefore $\partial U/\partial \varrho$ increases with ϱ, so that its derivative $\partial^2 U/\partial \varrho^2$ must be positive. Thus both terms in the expression for α_1 are negative. More clearly still, perhaps, α_1 may be written

$$\alpha_1 = -\left[\varrho \frac{\partial}{\partial \varrho}\left(\frac{1}{\varrho}\frac{\partial U}{\partial \varrho}\right)\right]_0 = \left[\varrho \frac{\partial}{\partial \varrho}\left(\frac{F}{\varrho}\right)\right]_0,$$

which shows that α_1 is negative unless F/ϱ increases with ϱ, which is impossible in reality, since it would mean that F increases faster than the first power of ϱ.

The coefficient α_3 is positive, since $\partial^2 U/\partial z^2 < 0$. This follows because near the galactic plane the distribution of mass in the Galaxy is not much different from plane-parallel layers whose density increases towards the galactic plane. Hence $\partial U/\partial z < 0$ for $z > 0$ and $\partial U/\partial z > 0$ for $z < 0$. Thus we can put

$$\alpha_3 + \beta_3 = \varkappa_3^2. \tag{7.14}$$

Then the third eqn. (7.13) can be integrated immediately to give the result

$$z = C \cos \varkappa_3 (t - t_0), \tag{7.15}$$

which we already know: apart from the notation, it is the same as the second formula (6.19). The numerical values of $\varkappa_3$ here and $\varkappa_2$ there differ because the former includes the attraction of the local cluster as well as the action of the regular force of the Galaxy.

To find x and y we substitute in the first two eqns. (7.13)

$$x = A\,e^{ikt}, \quad y = B\,e^{ikt}. \tag{7.16}$$

Then we obtain, as usual in such cases, the homogeneous algebraic equations

$$(\alpha_1 + \beta_1 - k^2)\,A - 2i\omega_0 k B = 0,$$

$$2i\omega_0 k A + (\beta_2 - k^2)\,B = 0.$$

These have non-zero solutions for A and B only if k satisfies the *characteristic equation*

$$k^4 - (\alpha_1 + \beta_1 + \beta_2 + 4\omega_0^2)\,k^2 + (\alpha_1 + \beta_1)\,\beta_2 = 0. \tag{7.17}$$

In order that the motions of the stars in the local cluster should be finite, k must be real (when the solutions for x and y will contain only trigonometrical terms).

Equation (7.17) may be written

$$k^4 - pk^2 + q = 0,$$

where

$$p = \alpha_1 + \beta_1 + \beta_2 + 4\omega_0^2, \qquad q = (\alpha_1 + \beta_1)\,\beta_2. \tag{7.18}$$

Hence we see that the conditions $p > 0$, $q > 0$, $\frac{1}{4}p^2 \geqslant q$ must hold if the solution is bounded. The second condition, from (7.18), is

$$\alpha_1 + \beta_1 > 0, \tag{7.19}$$

since β_2 is always positive. If this equation holds, the first condition is automatically satisfied, since β_2 and $4\omega_0^2$ in (7.18) are positive. From (7.18) we easily see also that $p^2 > 4q$ if (7.19) holds.

Thus the condition (7.19) is sufficient to ensure that the motions of all stars in the local cluster are finite, i.e. that the local cluster is stable. Physically, the condition (7.19) evidently signifies that the attraction of the local cluster exceeds the disruptive tidal force due to the attraction of the whole Galaxy.

If $\alpha_1 + \beta_1 < 0$, the product of the roots of the characteristic eqn. (7.17) is negative. Thus this equation has two conjugate imaginary roots and two real roots of opposite signs. Then at least one particular solution for x and y (7.16) will be an indefinitely increasing function of time. The general solution will therefore also increase indefinitely, and the cluster will be unstable.

The following point should be mentioned. These conclusions regarding the stability have been based on the linearised eqns. (7.13), i.e. the equations of the first approximation. A. M. Lyapunov [64] has shown that stability in the first approximation does not in general imply stable solutions of the full equations; the nature of the latter solutions requires a consideration of the second-order, or even higher, terms. If, however, the solutions of the first-approximation equations are unstable, then those of the full equations are necessarily unstable also.

With regard to the present discussion this means that, when $\alpha_1 + \beta_1 < 0$, the solution of the linearised eqns. (7.13) and that of the exact eqns. (7.4) are both unstable; if $\alpha_1 + \beta_1 > 0$, the solution of the linearised equations is stable, but this does not imply the stability of the solutions of eqns. (7.4).

The first-approximation stability condition (7.19) may be put in a form whose significance is more evident. We begin by expressing α_1 in terms of Oort's constants A and B. The first eqn. (7.11) can clearly be written

$$\alpha_1 = -\left[\varrho \frac{\partial}{\partial \varrho}\left(\frac{1}{\varrho}\frac{\partial U}{\partial \varrho}\right)\right]_0. \tag{7.20}$$

Omitting the suffix zero in (7.9), we have $(1/\varrho)\,\partial U/\partial \varrho = -\omega^2$, and therefore

$$\alpha_1 = 2\omega_0\omega_0'\varrho_0, \tag{7.21}$$

where $\omega_0' = (\mathrm{d}\omega/\mathrm{d}\varrho)_0$. Formulae (3.11) and (7.21) give

$$\alpha_1 = 4A(B - A). \tag{7.22}$$

The constant β_1 can be written as

$$\beta_1 = G\delta\beta_0, \tag{7.23}$$

where δ is the density of matter and β_0 a constant whose value has been calculated in Chapter III (called α' in (3.55)). Its approximate value is

$$\beta_0 = \pi^2 c/a, \tag{7.24}$$

where a and c are the semiaxes of the meridian section of the flattened spheroid. Substituting these expressions for α_1 and β_1 in (7.19), we obtain

$$\delta \geqslant 4A(A - B)/G\beta_0. \tag{7.25}$$

This is the explicit form of the condition for stability of stellar condensations. The quantity

$$\delta_0 = 4A(A - B)/G\beta_0 \tag{7.26}$$

may be called the *critical density*, since, if the density $\delta > \delta_0$, the condensation is stable (at least in the first approximation), and if $\delta < \delta_0$ it is unstable.

For a given point in the Galaxy the critical density depends, according to (7.24) and (7.26), only on the ratio c/a of the semiaxes of the cluster. The values of these quantities for the neighbourhood of the Sun are given in Table VII, taken from P. P. Parenago's book [107].

TABLE VII

c/a	0·1	0·2	0·3	0·4	0·5	0·6	0·8	1·0
β_0	0·278	0·499	0·677	0·824	0·946	1·048	1·211	1·333
δ_0	0·71	0·40	0·29	0·24	0·21	0·19	0·16	0·15

Here δ_0 is expressed in solar masses per cubic parsec. For comparison, we may mention that the actual value of the total star density in the neighbourhood of the Sun is 0·1 star per ps^3 = 0·04 solar mass per ps^3, taking the mean stellar mass as about 0·4 solar mass (V. S. Safronov, 1951 [113]).

Table VII shows that the critical density increases rapidly with decreasing c/a, reaching a value almost 18 times the galactic star density for the most flattened clusters ($c/a = 0{\cdot}1$). Since the local cluster belongs to the plane sub-system and its thickness is less than its dimensions in the galactic plane by at least a factor of 5 to 10, it must be extremely unstable and rapidly dissipated into the surrounding space. According to the previous discussion, this conclusion is valid whatever the form of the higher-order terms in the series expansions of the right-hand sides of eqns. (7.4).

§ 6. The Quasi-precession of the Velocity Ellipsoid

Let us now return to the first two eqns. (7.13). As has just been shown, $\alpha_1 + \beta_1$ must be negative in the local system, and so we can put

$$\alpha_1 + \beta_1 = -\varkappa_1^2. \tag{7.27}$$

Since $\beta_2 > 0$, we can put

$$\beta_2 = \varkappa_2^2. \tag{7.28}$$

Then the first two eqns. (7.13) become

$$\begin{aligned} \ddot{x} - 2\omega_0 \dot{y} - \varkappa_1^2 x &= 0, \\ \ddot{y} + 2\omega_0 \dot{x} + \varkappa_2^2 y &= 0. \end{aligned} \tag{7.29}$$

Before doing any actual calculations, let us make a qualitative analysis of eqns. (7.29), and examine the associated features of the dynamical problem.

Up to now we have emphasised the validity of the principle of indistinguishability of stars in the dynamics of stellar systems. It is easy to see that this principle does *not* hold in local dynamics. If we select some macroscopic volume element in an arbitrary system in a steady state with no density fluctuations, then the number of stars in this volume element and their velocity distribution will be invariable, since the stars which leave it are continually replaced by an equal number with the same velocities, which exactly make up the loss. Under these conditions the stars are indistinguishable. If the local cluster is not in a steady state, however, it is being dissipated into the surrounding space, where the star density is less, and the stars entering a macroscopic volume element will not make up for those leaving it. A particularly important point is that we shall observe outside the initial volume mainly the stars which have left the cluster, because there are more of them than of the field stars. This property will be the more marked, the greater the excess density in the cluster. The stars observed will be of two kinds, the "ordinary" field stars and the "marked" stars which have come from the cluster. It is not necessary that all stars should be considered. The same result holds for any particular species of star (for example, those of a particular spectral class).

This point must be taken into account in discussing the problem of local dynamics. We shall shortly see that the Coriolis acceleration, which occurs in any motion in rotating coordinates, has an effect on the motion of the "marked" stars. In systems without density fluctuations, the Coriolis acceleration is unimportant. It still affects the movement of individual stars, of course, but each star leaving a macroscopic volume element enters another one and is thus "lost", in the sense that it becomes indistinguishable from the other stars with the same velocity which were earlier or later in the same volume.†

Let us now examine eqns. (7.29). These are the equations of harmonic oscillations in two dimensions, where the "elastic force" has components $\varkappa_1^2 x$ and $-\varkappa_2^2 y$ and the "resistance of the medium" (Coriolis force) has components $2\omega_0 \dot{y}$ and $-2\omega_0 \dot{x}$. To determine the action of these forces we shall omit each in turn and examine the effect of the other alone.

Omitting the "elastic force", we have

$$\begin{aligned} \ddot{x} - 2\omega_0 \dot{y} &= 0, \\ \ddot{y} + 2\omega_0 \dot{x} &= 0. \end{aligned} \tag{7.30}$$

Since we are here interested in the change of the velocities, not that of the coordinates, we obtain by integration

$$\dot{x} = V \sin 2\omega_0 (t - t_0), \quad \dot{y} = V \cos 2\omega_0 (t - t_0), \tag{7.31}$$

where V and t_0 are constants. The constant V is given by the condition $V^2 = \dot{x}_0^2 + \dot{y}_0^2$, i.e. it is the initial residual velocity. The constant t_0 is evidently the instant when the vector $\mathbf{V}$ is along the y-axis. It is easy to see that, when the argument $\tau = 2\omega_0(t - t_0)$ increases from 0 to $\frac{1}{2}\pi$, $\dot{x}$ changes from 0 to V and $\dot{y}$ from V to 0, and so on. This corresponds to a rotation of the vector $\mathbf{V}$ in the direction opposite to that of the rotation ω_0.

We see that the action of the Coriolis forces does not change the magnitude of the velocity, but causes the vector $\mathbf{V}$ to rotate about the origin with an angular velocity $2\omega_0$, which is independent of the velocity of the star, and in the direction opposite to that of the galactic rotation.

Let us suppose that at some instant the velocity ellipsoid has its vertex towards the centre of the Galaxy ($\Delta l = 0$), in accordance with the steady-state theory. Then, in the course of time, the velocity vectors of all stars will remain unchanged in magnitude but will rotate through equal angles in the direction opposite to that of galactic rotation. Since the latter is clockwise as seen from the northern pole of the Galaxy, the vertex deviates in the direction of increasing galactic longitude, i.e. in the observed direction. This rotational motion may be called the *quasi-precession* of the velocity ellipsoid. Thus an observer within a star cluster will always observe mainly the same stars, and so will detect a quasi-precession of the ellipsoid. If a system such as the

† These problems are closely related to that of the intermingling of statistical systems in phase space, which was investigated by a notable young physicist, the late N.S. Krylov [51]. However, a discussion of them is beyond the scope of this book.

local cluster is of finite dimensions, the natural dilution effect (i.e. the decrease in density owing to the radial dissipation of stars) will cause the proportions of "marked" stars to decrease away from the boundaries of the cluster. The quasi-precession also must therefore be a clearly local effect. If, in determining the velocity ellipsoid, we select stars at a considerable distance from the cluster, no quasi-precession will be found. This result is confirmed by observation, since a determination by Blaauw of the ellipsoid for remote stars has given, as mentioned above, no deviation of the vertex.

§ 7. The Quasi-tidal Effect on the Velocity Ellipsoid

We now omit the Coriolis force in eqns. (7.29), obtaining

$$\begin{aligned} \ddot{x} - \varkappa_1^2 x &= 0, \\ \ddot{y} + \varkappa_2^2 y &= 0. \end{aligned} \tag{7.32}$$

These equations can be integrated independently, and give

$$\begin{aligned} x &= x_0 \cosh \varkappa_1 t + (\dot{x}_0/\varkappa_1) \sinh \varkappa_1 t, \\ y &= y_0 \cos \varkappa_2 t + (\dot{y}_0/\varkappa_2) \sin \varkappa_2 t; \end{aligned} \tag{7.33}$$

hence the velocity components are

$$\begin{aligned} \dot{x} &= x_0 \varkappa_1 \sinh \varkappa_1 t + \dot{x}_0 \cosh \varkappa_1 t, \\ \dot{y} &= -y_0 \varkappa_2 \sin \varkappa_2 y + \dot{y}_0 \cos \varkappa_2 t. \end{aligned} \tag{7.34}$$

These formulae show that, as t increases, $\dot{y}$ remains bounded, but $\dot{x}$ increases without limit. The expressions (7.34) are too complex to give a clear idea of the phenomenon, and we shall simplify them by using the fact that the star density in the local cluster is small. The effect of its attraction on the motions of stars therefore cannot be great compared with that of the gravitational forces of the whole Galaxy. Hence we put $\beta_1 = \beta_2 = 0$. Then (7.27), (7.22) and (7.28) give

$$\varkappa_1^2 = -\alpha_1 = 4A(A - B), \qquad \varkappa_2 = 0. \tag{7.35}$$

Equations (7.34), with the conditions (7.35), become

$$\begin{aligned} \dot{x} &= x_0 \varkappa_1 \sinh \varkappa_1 t + \dot{x}_0 \cosh \varkappa_1 t, \\ \dot{y} &= \dot{y}_0. \end{aligned} \tag{7.36}$$

The second-order moments of the velocities of stars in a small circle of radius r_0 are

$$\left.\begin{aligned} \overline{\dot{x}^2} &= \overline{\dot{x}_0^2} \cosh^2 \varkappa_1 t + \overline{x_0^2} \varkappa_1^2 \sinh^2 \varkappa_1 t \\ &= \overline{\dot{x}_0^2} + (\overline{\dot{x}_0^2} + \overline{x_0^2} \varkappa_1^2) \sinh^2 \varkappa_1 t, \\ \overline{\dot{x}\dot{y}} &= \overline{\dot{x}_0 \dot{y}_0} \cosh \varkappa_1 t, \\ \overline{\dot{y}^2} &= \overline{\dot{y}_0^2}, \end{aligned}\right\} \tag{7.37}$$

since evidently $\overline{x_0 \dot{x}_0} = \overline{x_0 \dot{y}_0} = 0$; the bar denotes averaging.

Let us suppose that, in accordance with the steady-state theory of the Galaxy, the principal axes of the velocity ellipsoid are at some instant along the x, y, z axes, i.e. $\overline{\dot{x}_0 \dot{y}_0} = 0$, and that two principal axes are equal: $\overline{\dot{x}_0^2} = \overline{\dot{z}_0^2}$. Then, by (7.37), we have for any subsequent instant

$$\left.\begin{aligned} \overline{\dot{x}^2} &= \overline{\dot{x}_0^2} \left[1 + \left(1 + \frac{\varkappa_1^2 \overline{x_0^2}}{\overline{\dot{x}_0^2}}\right) \sinh^2 \varkappa_1 t\right], \\ \overline{\dot{y}^2} &= \overline{\dot{y}_0^2}, \\ \overline{\dot{x}\dot{y}} &= 0. \end{aligned}\right\} \qquad (7.38)$$

Also $\overline{\dot{y}\dot{z}} = \overline{\dot{z}\dot{x}} = 0$, since the motion in the z-direction is independent of that in the other two directions.

Equation (7.15) may be written

$$z = z_0 \cos \varkappa_3 t + (\dot{z}_0/\varkappa_3) \sin \varkappa_3 t,$$

whence

$$\dot{z} = -\varkappa_3 z_0 \sin \varkappa_3 t + \dot{z}_0 \cos \varkappa_3 t.$$

Thus

$$\overline{\dot{z}^2} = \overline{z_0^2} \varkappa_3^2 \sin^2 \varkappa_3 t + \overline{\dot{z}_0^2} \cos^2 \varkappa_3 t$$

or

$$\overline{\dot{z}^2} = \overline{\dot{z}_0^2} \left[1 - \left(1 - \frac{\varkappa_3^2 \overline{z_0^2}}{\overline{\dot{z}_0^2}}\right) \sin^2 \varkappa_3 t\right].$$

The coefficient of $\sin^2 \varkappa_3 t$ is very small in the neighbourhood of the Sun. Neglecting the attraction of the local cluster, we have by (7.14) and (7.11) $\varkappa_3^2 = \alpha_3 = -(\partial^2 U/\partial z^2)_0$. This quantity has been calculated in Chapter VI, § 1, where it was called $\varkappa_2^2$: its value was found to be $(2{\cdot}37)^2 \times 10^{-30}\ \mathrm{s}^{-2} = 5{\cdot}62 \times 10^{-30}\ \mathrm{s}^{-2}$.

For the star density distribution in the z-direction we use the "barometric formula" $\nu = \nu_0 \exp(-z/\beta)$, where ν_0 is the density in the galactic plane. Then the mean square is

$$\overline{z_0^2} = \int_0^\infty z^2 \exp(-z/\beta)\, \mathrm{d}z/\beta = 2\beta^2.$$

For the value of β for O and B stars we take that given by Parenago ([107], p. 291, Table 55): 58 and 46 ps, or on average $\beta = 50$ ps $= 1{\cdot}54 \times 10^{15}$ km. Thus $\varkappa_3^2 \overline{z_0^2} = 26{\cdot}7\ \mathrm{km^2/s^2}$. For the velocity variance in the z-direction $\overline{\dot{z}_0^2} = \sigma_z^2$ we take Nordström's value for O5–B5 stars brighter than the sixth apparent magnitude ([76], p. 162, Table 37): $\overline{\dot{z}_0^2} = (5{\cdot}4)^2 = 29{\cdot}2\ \mathrm{km^2/s^2}$. Thus the factor in question is $1 - \varkappa_3^2 \overline{z_0^2}/\overline{\dot{z}_0^2} = 1 - 26{\cdot}7/29{\cdot}2 = 0{\cdot}103$. This may be taken as zero within the limits of the natural indeterminacy of the initial data, since it would be sufficient to reduce σ_z by only 0·2 km/s to make the factor zero, whereas the mean error in σ_z is 0·8 km/s. Thus we have $\overline{\dot{z}^2} = \overline{\dot{z}_0^2} =$ constant. This result might have been foreseen, since in a flat rotating Galaxy the dispersion in the z-direction must be small.

Returning now to eqns. (7.38), we deduce from them the following conclusions: (1) the directions of all three axes of the velocity ellipsoid remain unchanged, (2) $b^2 = \overline{\dot{y}^2}$ and $c^2 = \overline{\dot{z}^2}$ remain unchanged, (3) $a^2 = \overline{\dot{x}^2}$ is extended in the manner given by

$$\frac{\overline{\dot{x}^2} - \overline{\dot{x}_0^2}}{\overline{\dot{x}_0^2}} = \left(1 + \frac{\varkappa_1^2 \overline{x_0^2}}{\overline{\dot{x}_0^2}}\right) \sinh^2 \varkappa_1 t. \tag{7.39}$$

If the attraction of the local cluster is neglected, formula (7.37) shows that

$$\begin{aligned}\varkappa_1 &= \sqrt{|\alpha_1|} = 2\sqrt{[A(A - B)]} \\ &= 1{\cdot}67 \times 10^{-15}\ \mathrm{s}^{-1} \\ &= 5{\cdot}26 \times 10^{-8}\ \mathrm{y}^{-1}.\end{aligned} \tag{7.40}$$

The expression (7.39) may be put in a somewhat more convenient form if we assume, for simplicity, that the star density is constant within the local cluster and that the volume occupied by this cluster at the initial instant has the form of a circular cylinder of radius r_0 and small height, not exceeding the present thickness of the plane component of the Galaxy. Then it is easily shown that

$$\overline{x_0^2} = \tfrac{1}{4} r_0^2, \tag{7.41}$$

and the relative increase in the velocity variance in the x-direction (the square of the greatest semiaxis of the velocity ellipsoid) is given by

$$\frac{\overline{\dot{x}^2} - \overline{\dot{x}_0^2}}{\overline{\dot{x}_0^2}} = \left(1 + \frac{\varkappa_1^2 r_0^2}{4\overline{\dot{x}_0^2}}\right) \sinh^2 \varkappa_1 t. \tag{7.42}$$

This deformation of the velocity ellipsoid, consisting of an extension along the line to the galactic centre, is called the *quasi-tidal effect*.

Thus an observer situated within a star cloud such as the local cluster will observe two properties of the velocity ellipsoid: (1) the quasi-tidal effect, whereby the ellipsoid is elongated in the direction of the galactic centre, (2) the quasi-precession, whereby the greatest axis of the ellipsoid rotates in the direction opposite to the rotation of the Galaxy.

The combined effect depends on the relative magnitude of these two characteristics. If the quasi-precession is slow, the ellipsoid is simply elongated along the vertex line. If it is fast, the ellipsoid turns through 180° before any given group of young stars can escape from the local cluster. Then the quasi-tidal effect will tend to equalise the ellipsoid axes in the galactic plane, i.e. to reduce the eccentricity of this cross-section of the ellipsoid, bringing it nearer to a circular form.

§ 8. The Application of the Theory to the Observational Results

Let us see what is obtained when the formulae derived above are applied to observational results, considering first the quasi-precession. We shall suppose that, at the initial instant $t = t_0$, the greatest axis of the velocity ellipsoid passes exactly through the centre of the Galaxy. The action of the Coriolis forces will cause the ellipsoid to rotate with angular velocity $2\omega_0$. If the total angle turned through is Δl, the "age" of the velocity ellipsoid is

$$t - t_0 = \Delta l / 2\omega_0. \tag{7.43}$$

To determine the deviation Δl of the vertex, we use the results of Nordström's determination of the velocity ellipsoid from radial velocities of stars brighter than the sixth apparent magnitude [76]. The majority of these stars are giants, and their apparent brightness indicates that most of them are within the local cluster.

Nordström's value for the longitude of the vertex is $l = 340°$; taking for the centre of the Galaxy $l_0 = 325°$, we find the deviation of the vertex $\Delta l = 15°$.

The angular velocity of the rotation of the Galaxy is, according to P.P. Parenago ([107], p. 158), $\omega_0 = 0''{\cdot}0068$ per y. Then (7.43) gives $t - t_0 = 4{\cdot}0 \times 10^6$ y. Thus the age of the velocity ellipsoid is unexpectedly small. Of course, this value is not very accurate, and probably gives only the order of magnitude. But it certainly shows that the bright stars forming the local cluster are relatively young.†

The age would be larger if the deviation of the vertex were greater than the value used above. The difficulty as regards our theory is that the observed deviation is too small. The age is somewhat increased if the calculations are based on the velocity ellipsoid for the B stars which form the "framework" of the local cluster. According to Nordström's data, the longitude of the vertex for B stars brighter than the sixth magnitude is $l = 236°$. This is very different from the mean longitude of the vertex used above ($l = 340°$). If we suppose that the value $l = 236°$ is correct, we should add to it 180°, since the Coriolis force must rotate the vertex in the direction of increasing longitudes. Then $l = 416°$, and the deviation of the vertex is $\Delta l = 91°$. From (7.43) we then obtain $t - t_0 = 2{\cdot}4 \times 10^7$ y. However, if such a large deviation of the vertex is assumed correct, the possibility of even greater deviations must be taken into account.

The time after which the deviation $\Delta l = 180°$ is

$$P = 4{\cdot}8 \times 10^7 \text{ y}. \tag{7.44}$$

We may take this as giving, in order of magnitude, the age of the stars in the local cluster. As a check, we may compute the mean distance traversed by stars in the local cluster in time P. Putting $V = \sigma = 10$ km/s, we find $s = VP = 490$ ps. This is approximately equal to the radius of the local cluster. Hence the value of P given by (7.44) is an upper limit to the age of the stars in the local cluster, since after this time the stars must have left the cluster and mingled with the general galactic population.

Let us now consider the quasi-tidal effect. Nordström's data [76] give for the greatest semiaxis $\sigma_1 = a$ and the semiaxis $\sigma_3 = c$ perpendicular to the galactic plane

$$a = 12{\cdot}4 \text{ km/s}, \quad c = 5{\cdot}1 \text{ km/s}, \tag{7.45}$$

whence

$$\overline{\dot{x}^2} = 154 \text{ km}^2/\text{s}^2, \quad \overline{\dot{x}_0^2} = \overline{\dot{z}^2} = 26 \text{ km}^2/\text{s}^2.$$

† The small value of Δl is doubtless explained, at least in part, by the mixing of stars from the local cluster with other stars.

Although the existence of a quasi-integral in the z-direction, discused in Chapter V, § 10, renders the semiaxes c and a independent, we may still use in a first approximation the conclusions of the steady-state theory of the Galaxy. Since the semiaxis c must remain constant, and at the initial instant must, according to the steady-state theory, equal a, we can put $\overline{\dot{z}^2} = \overline{\dot{x}_0^2}$. Then the observed relative change in dispersion in the x-direction is

$$(\overline{\dot{x}^2} - \overline{\dot{z}^2})/\overline{\dot{z}^2} = 4{\cdot}90. \tag{7.46}$$

Putting in (7.42) $\overline{\dot{x}_0^2} = \overline{\dot{z}^2}$, we obtain the relative change in velocity dispersion in the x-direction as

$$(\overline{\dot{x}^2} - \overline{\dot{z}^2})/\overline{\dot{z}^2} = (1 + \varkappa_1^2 r_0^2/4\overline{\dot{z}^2}) \sinh^2 \varkappa_1 t. \tag{7.47}$$

In order to use this formula it is necessary to know the "initial" radius r_0 of the local cluster. Unfortunately, this quantity cannot be found directly from observation and is largely arbitrary, since the local cluster is ill-defined and its boundaries are not accurately known. The value of r_0 is rather that of some "effective" radius, i.e. the radius of an ideal local cluster which has a uniform structure and a star density equal to its mean value for the actual local cluster, and occupies a circular cylinder, and gives the observed manner of change of the velocity ellipsoid.

We make the assumption, for want of better, that the local system has not appreciably changed in size. Then r_0 may be taken as the present value of the radius, given by the condition that the boundary of the local system corresponds to $m = 6^m$ for O and B stars:

$$r_0 = 500 \text{ ps}. \tag{7.48}$$

The coefficient of r_0^2 in (7.47) is small: $\varkappa_1^2/4\overline{\dot{z}^2} \approx 2{\cdot}17 \times 10^{-6}$. When r_0 has the value (7.48), the term $\varkappa_1^2 r_0^2/4\overline{\dot{z}^2}$ does not exceed 0·3. In a first approximation, we therefore obtain from (7.47) the age of the velocity ellipsoid as

$$t = \sqrt{\frac{\overline{\dot{x}^2} - \overline{\dot{z}^2}}{\overline{\dot{z}^2}}} \cdot \frac{1}{\varkappa_1}. \tag{7.49}$$

Hence, using the numerical values (7.46) and (7.40), we find the age of the velocity ellipsoid from the quasi-tidal effect:

$$t = 9{\cdot}7 \times 10^7 \text{ y}. \tag{7.50}$$

This value is of the same order of magnitude as the age (7.44) determined above from the quasi-precession.

Thus we can suppose that the age of the velocity ellipsoid, as determined from the bright giants which constitute the local cluster, is of the order of 10^7 y. This is in good agreement with the conclusion of V. A. Ambartsumyan, B. A. Vorontsov-Vel'yaminov and others that these are young stars. When t is less than this value the linearised equations of motion (7.29) may be used.

§ 9. The Combined Quasi-precession and Quasi-tidal Effects

In the general case, the linearised equations of motion in the galactic plane are

$$\ddot{x} - 2\omega_0 \dot{y} + (\alpha_1 + \beta_1) x = 0, \qquad \ddot{y} + 2\omega_0 x + \beta_2 y = 0. \tag{7.51}$$

To simplify the problem, we shall assume that the distribution of mass in the local cluster has rotational symmetry about an axis parallel to the z-axis. Then

$$\beta_1 = \beta_2 = \beta. \tag{7.52}$$

Let us also assume that the excess star density due to the local cluster (that is, the star density in the local cluster minus the galactic field star density) is

$$\delta = \lambda \delta_0, \tag{7.53}$$

where δ_0 is the critical density given by (7.26) and λ the *instability coefficient*, which satisfies the condition

$$0 \leqslant \lambda \leqslant 1. \tag{7.54}$$

When $\lambda = 0$ we have complete instability and density $\delta = 0$; when $\lambda = 1$ we have the limit of stability and $\delta = \delta_0$.

Let us now express the dynamical parameters of the problem in terms of Oort's constants. From (7.23) and (7.26), $\beta_1 = 4A(A - B)\lambda = \varkappa_1^2 \lambda$. Hence (7.22), (7.52) and (3.45) give

$$\alpha_1 = -\varkappa_1^2, \quad \beta = \varkappa_1^2 \lambda, \quad \omega_0^2 = (A - B)^2. \tag{7.55}$$

Substituting in eqn. (7:17), we obtain the solution

$$k^2 = \tfrac{1}{2}\varkappa_1^2 \{2\lambda - g \pm \sqrt{[(2\lambda - g)^2 + 4\lambda(1 - \lambda)]}\}.$$

It is useful to write this in the dimensionless form

$$s^2 = \tfrac{1}{2}(2\lambda - g) \pm \tfrac{1}{2}\Delta, \tag{7.56}$$

where

$$s = k/\varkappa_1, \quad g = B/A, \quad \Delta = \sqrt{[(2\lambda - g)^2 + 4\lambda(1 - \lambda)]}. \tag{7.57}$$

It is evident that, for $\lambda < 1, \Delta > 2\lambda - g$, and therefore the four roots of the equation include two real and two imaginary. In terms of the two real positive constants s_1^2 and s_2^2 given by

$$s_1^2 = \tfrac{1}{2}[\Delta + (2\lambda - g)] \quad \text{and} \quad s_2^2 = \tfrac{1}{2}[\Delta - (2\lambda - g)], \tag{7.58}$$

we have the general solution of (7.51):

$$x = A_1 \cos s_1\tau + A_2 \sin s_1\tau + B_1 \cosh s_2\tau + B_2 \sinh s_2\tau,$$

$$y = l_1 A_2 \cos s_1\tau - l_1 A_1 \sin s_1\tau - l_2 B_2 \cosh s_2\tau - l_2 B_1 \sinh s_2\tau,$$

where $\tau = \varkappa_1 t$ is the dimensionless time and

$$l_1 = \frac{s_2^2 + \lambda}{\lambda\sqrt{(1-g)}}\, s_1, \quad l_2 = \frac{s_1^2 - \lambda}{\lambda\sqrt{(1-g)}}\, s_2. \tag{7.59}$$

The values of s_1 and s_2 for various values of λ are given in Table VIII. We take $A = 0{\cdot}020$ and $B = -0{\cdot}013$ km/s per ps, whence $g = -13/20 = -0{\cdot}65$.

TABLE VIII

λ	0·0	0·1	0·2	0·3	0·4	0·5	0·6	0·7	0·8	0·9	1·0
s_1	0·8062	0·9722	1·0886	1·1832	1·2649	1·3378	1·4042	1·4655	1·5228	1·5768	1·6279
s_2	0·0000	0·3085	0·3674	0·3873	0·3873	0·3738	0·3490	0·3127	0·2627	0·1903	0·0000

Again denoting by $x_0, y_0, \dot{x}_0, \dot{y}_0$ the coordinates and velocities at $t = 0$, we obtain the following equations to determine the coefficients A_1, A_2, B_1, B_2:

$$\left.\begin{aligned} x_0 &= A_1 + B_1, \\ y_0 &= l_1 A_2 - l_2 B_2, \\ \dot{y}_0 &= -\varkappa_1 l_1 s_1 A_1 - \varkappa_1 l_2 s_2 B_1, \\ \dot{x}_0 &= \varkappa_1 s_1 A_2 + \varkappa_1 s_2 B_2, \end{aligned}\right\} \tag{7.60}$$

whence

$$\left.\begin{aligned} A_1 &= -\frac{(\Delta - g)(\Delta + g - 2\lambda)\, x_0 + 4\sqrt{(1-g)}\,\lambda \dot{y}_0/\varkappa_1}{4\lambda\Delta}, \\ A_2 &= \frac{2\lambda\sqrt{(1-g)}\, y_0 + (\Delta - g)\, \dot{x}_0/\varkappa_1}{2\Delta\sqrt{[\lambda(1-\lambda)]}}\, s_2, \\ B_1 &= \frac{(\Delta + g)(\Delta - g + 2\lambda)\, x_0 + 4\sqrt{(1-g)}\,\lambda\, \dot{y}_0/\varkappa_1}{4\lambda\Delta}, \\ B_2 &= -\frac{2\lambda\sqrt{(1-g)}\, y_0 + (\Delta + g)\, \dot{x}_0/\varkappa_1}{2\Delta\sqrt{[\lambda(1-\lambda)]}}\, s_1. \end{aligned}\right\} \tag{7.61}$$

A detailed examination of the solutions for various values of λ is not necessary here, since the comparison of theory and observation is made difficult not only by the fact that we do not know the correct value of λ but also because we ignore a number of factors which may considerably affect the observed velocity distribution of O and B stars. These factors will be discussed later.

As an example, we may take the single value $\lambda = 0{\cdot}5$, since the results vary only slightly with λ except when it is close to 0 or 1. We shall show that, in accordance with the cases of quasi-precession and quasi-tidal effect already considered, the changes in the shape of the velocity ellipsoid and in the direction of the vertex are of the order of magnitude which would be expected from the limiting cases described previously.

The velocity components are

$$\begin{aligned}\dot{x} &= -\varkappa_1 s_1 A_1 \sin s_1\tau + \varkappa_1 s_1 A_2 \cos s_1\tau \\ &\quad + \varkappa_1 s_2 B_1 \sinh s_2\tau + \varkappa_1 s_2 B_2 \cosh s_2\tau, \\ \dot{y} &= \varkappa_1 l_1 s_1 A_1 \cos s_1\tau - \varkappa_1 l_1 s_1 A_2 \sin s_1\tau \\ &\quad - \varkappa_1 l_2 s_2 B_1 \cosh s_2\tau + \varkappa_1 l_2 s_2 B_2 \sinh s_2\tau.\end{aligned} \tag{7.62}$$

To simplify the problem further we assume that the initial coordinates of all stars are the same: $x_0 = y_0 = 0$. Then (7.61) become

$$\left.\begin{aligned} -\varkappa_1 s_1 A_1 &= \sqrt{(1-g)}\, s_1 \dot{y}_0/\Delta, \\ \varkappa_1 s_1 A_2 &= (\Delta - g)\, \dot{x}_0/2\Delta, \\ \varkappa_1 s_2 B_1 &= \sqrt{(1-g)}\, s_2 \dot{y}_0/\Delta, \\ -\varkappa_1 s_2 B_2 &= (\Delta + g)\, \dot{x}_0/2\Delta. \end{aligned}\right\} \tag{7.63}$$

Substituting (7.63) in (7.62) and averaging, we obtain for the second-order velocity moments $\overline{\dot{x}^2}$, $\overline{\dot{x}\dot{y}}$, $\overline{\dot{y}^2}$ linear functions of their initial values $\overline{\dot{x}_0^2}$, $\overline{\dot{x}_0\dot{y}_0}$, $\overline{\dot{y}_0^2}$, with coefficients easily expressed in terms of those in (7.62). We shall show only that the precession and tidal effects on the velocity ellipsoid are perceptible after a time of the order of 5×10^7 y. We may therefore assume that at the initial instant $t = 0$ the principal axes of the velocity ellipsoid are in the directions of the coordinate axes in agreement with the steady-state theory of the Galaxy, i.e. we put $\overline{\dot{x}_0\dot{y}_0} = 0$. Then the second-order moments are

$$\left.\begin{aligned} \overline{\dot{x}^2} &= (\alpha_{11} + \alpha_{12}\cos 2s_1\tau + \alpha_{13}\cos s_1\tau \cosh s_2\tau \\ &\quad + \alpha_{14}\cosh 2s_2\tau)\,\overline{\dot{x}_0^2} + (\beta_{11} + \beta_{12}\cos 2s_1\tau \\ &\quad + \beta_{13}\sin s_1\tau \sinh s_2\tau + \beta_{14}\cosh 2s_2\tau)\,\overline{\dot{y}_0^2}, \\ \overline{\dot{x}\dot{y}} &= (\alpha_{21}\sin 2s_1\tau + \alpha_{22}\cos s_1\tau \sinh s_2\tau \\ &\quad + \alpha_{23}\sin s_1\tau \cosh s_2\tau + \alpha_{24}\sinh 2s_2\tau)\,\overline{\dot{x}_0^2} \\ &\quad + (\beta_{21}\sin 2s_1\tau + \beta_{22}\cos s_1\tau \sinh s_2\tau \\ &\quad + \beta_{23}\sin s_1\tau \cosh s_2\tau + \beta_{24}\sinh 2s_2\tau)\,\overline{\dot{y}_0^2}, \\ \overline{\dot{y}^2} &= (\alpha_{31} + \alpha_{32}\cos 2s_1\tau + \alpha_{33}\sin s_1\tau \sinh s_2\tau \\ &\quad + \alpha_{34}\cosh 2s_2\tau)\,\overline{\dot{x}_0^2} + (\beta_{31} + \beta_{32}\cos 2s_1\tau \\ &\quad + \beta_{33}\cos s_1\tau \cosh s_2\tau + \beta_{34}\cosh 2s_2\tau)\,\overline{\dot{y}_0^2}. \end{aligned}\right\} \tag{7.64}$$

To find the coefficients α_{ij} and β_{ij} we need only substitute (7.59) and (7.63) in (7.62), multiply and average (the last process is denoted by the bars). The functions of double arguments are used where possible to facilitate numerical calculations. The expressions for α_{ij} and β_{ij} are cumbersome, and on that account will not be given here.

The results of calculations of the second-order moments are shown in Table IX. We have taken $\overline{\dot{x}_0^2} = 1$ and, by Lindblad's formula, $\overline{\dot{y}_0^2}/\overline{\dot{x}_0^2} = B/(B - A) = 13/33$. The result, rounded, is $\overline{\dot{y}_0^2} = 0{\cdot}4$. For $\lambda = 0{\cdot}5$, according to Table VIII, $s_1 = 1{\cdot}3378$, $s_2 = 0{\cdot}3738$, whence $\varkappa_1 s_1 = 4{\cdot}47 \times 10^{-7}\,\mathrm{s}^{-1}$, $\varkappa_1 s_2 = 1{\cdot}25 \times 10^{-7}\,\mathrm{s}^{-1}$.

TABLE IX

τ_1'	τ_2'	$\overline{\dot{x}^2}$	$\overline{\dot{x}\,\dot{y}}$	$\overline{\dot{y}^2}$	Δl (deg)	a	c	t (10^7 y)
0·1	0·0279	0·998	0·057	0·405	2·7	0·999	0·635	0·24
0·2	0·0559	0·989	0·114	0·418	5·6	0·997	0·642	0·47
0·3	0·0838	0·975	0·168	0·440	8·7	0·994	0·653	0·71
0·4	0·1118	0·957	0·220	0·469	12·1	0·990	0·667	0·94
0·5	0·1397	0·936	0·268	0·506	16·0	0·986	0·682	1·18
0·6	0·1676	0·908	0·312	0·547	20·4	0·981	0·701	1·42
0·7	0·1956	0·876	0·350	0·594	25·6	0·980	0·714	1·65
0·8	0·2235	0·843	0·384	0·646	31·4	0·979	0·726	1·89
0·9	0·2515	0·806	0·411	0·699	37·6	0·982	0·734	2·12
1·0	0·2794	0·767	0·432	0·753	44·0	0·988	0·737	2·36

The first two columns in Table IX give the time arguments $\tau_1' = \varkappa_1 s_1 t$ and $\tau_2' = \varkappa_1 s_2 t$. The corresponding times in units of 10^7 y are shown in the last column. Columns 3 to 5 give the velocity moments, column 6 the angle of rotation of the vertex line in degrees (deviation of the vertex), and columns 7 and 8 the semiaxes of the velocity ellipsoid.

We see that the deviation of the vertex is 25° after only $1{\cdot}65 \times 10^7$ y, i.e. the quasi-precession has a very marked effect. The quasi-tidal effect elongates the velocity ellipsoid, and the value of c^2/a^2 changes in this time from 0·4 to $(0{\cdot}714/0{\cdot}980)^2 = 0{\cdot}531$, i.e. by 33%.

The angle Δl continues to increase with time, and after a further $2{\cdot}60 \times 10^7$ y it has increased to more than 70°. The flattening of the ellipsoid reaches a maximum value for $t = 2{\cdot}12 \times 10^7$ y and then decreases again, returning to the initial value 0·400 at $t = 3{\cdot}75 \times 10^7$ y.

The further changes in the velocity ellipsoid are more difficult to judge, because as t increases we approach the limit of applicability of the linear approximation. Another difficulty arises because the local cluster is embedded in the general galactic star population, and so there are some stars within the cluster which do not belong to it. These stars will evidently have a different velocity ellipsoid, which is almost the same as that given by the steady-state theory of the Galaxy. Thus the observed velocity ellipsoid is the result of the superposition of two ellipsoids, one belonging to the local cluster and the other to the general population. The appropriate formulae are easily derived, but their application would require an estimate of the relative numbers of the two kinds of star in the neighbourhood of the Sun. This is a difficult problem.

Finally, a third complicating factor is that the formation of new stars in the associations within the local cluster appears to be still in progress, and therefore the velocity ellipsoid of the O and B stars is a combination of a continuous sequence of ellipsoids corresponding to successively formed groups of O and B stars. Let some group of these stars be formed at time t, and let us assume that, at this instant, the ellipsoid of the group is that given by the steady-state theory of the Galaxy. After a time dt the ellipsoid of this group has undergone a change by the quasi-precession and quasi-tidal processes mentioned above. At the same time a new group of O and B stars is formed, whose ellipsoid will be that which belonged to the first group at time t. Thus the observed ellipsoid is a superposition of these two. In the course of time a continuous sequence of ellipsoids will be superposed. The velocity ellipsoid actually observed in the neighbourhood of the Sun therefore contains the entire history of the local cluster from the time when it came into existence.

All these problems require separate investigation.

§ 10. The *K* Effect as a Result of the Dissipation of the Local Cluster

Let us now regard the K effect as the result of the dissipation into the surrounding space of the local cluster, a star cloud whose mass is so small that its attraction may be neglected. For simplicity we shall again consider the two-dimensional problem, since the only stars for which the K effect is observed, namely the O and B stars, are typical representatives of the plane sub-systems.

We assume that, at the initial instant $t = 0$, the cloud occupies a circle of radius r_0 and is of unit density everywhere (Fig. 32). Let ϱ, φ be polar coordinates in the galactic plane, with origin at the centre of the local cluster and initial line in a fixed direction; let ψ be the angle between a star's velocity $\mathbf{v}$ and the initial line.

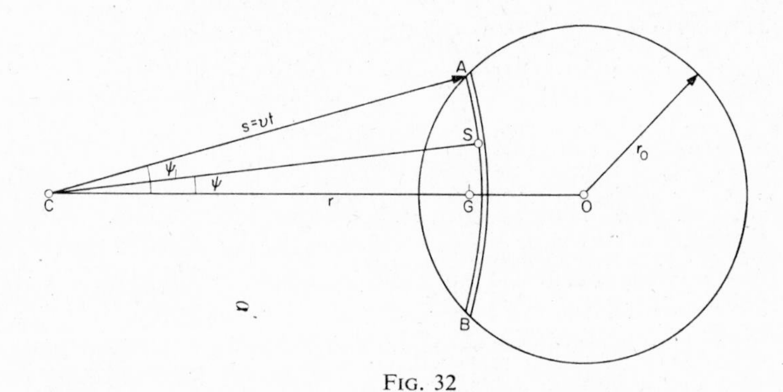

FIG. 32

Since the local cluster is small compared with the Galaxy, we neglect the gravitational acceleration of both the Galaxy and the local cluster itself within the cluster, and assume that all stars move in straight lines at constant velocities.

Let us consider at time $t > 0$ a small region $d\tau$ centred at a point C whose distance from the origin $r > r_0$, i.e. which lies outside the local cluster. The stars with velocities between v and $v + dv$ which at time t are in the area $d\tau$ were at $t = 0$ within an annulus of width $dr = t\,dv$ bounded by an arc ASB with centre C and radius $s = vt$, the distance traversed by stars of velocity v in time t. Figure 32 shows the initial position S of one such star.

We assume as a first approximation that the stellar velocity distribution in the cloud is Maxwellian. Then

$$dn = C \exp(-v^2/2\sigma^2)\, v\, dv\, d\psi, \tag{7.65}$$

where $v\,dv\,d\psi$ is an area element in velocity space, and C a constant factor, which is here unimportant. Thus, for given t, the mean radial velocity $\bar{v}_r = \overline{v \cos\psi}$ in the region $d\tau$ is

$$\bar{v}_r = \frac{\int_{v_1}^{v_2} \exp(-v^2/2\sigma^2)\, v^2\, dv \int_{-\psi_1}^{\psi_1} \cos\psi\, d\psi}{\int_{v_1}^{v_2} \exp(-v^2/2\sigma^2)\, v\, dv \int_{-\psi_1}^{\psi_1} d\psi}$$

$$= \frac{\int_{v_1}^{v_2} \exp(-v^2/2\sigma^2)\, v^2 \sin\psi_1\, dv}{\int_{v_1}^{v_2} \exp(-v^2/2\sigma^2)\, v\psi_1\, dv}, \tag{7.66}$$

where $v_1 = (r - r_0)/t$, $v_2 = (r + r_0)/t$, and ψ_1 is the maximum value of ψ for given values of v and t (Fig. 32). Formula (7.66) may also be written

$$t\bar{v}_r = \frac{\int_{r-r_0}^{r+r_0} \exp(-s^2/2\sigma^2 t^2)\, s^2 \sin\psi_1\, ds}{\int_{r-r_0}^{r+r_0} \exp(-s^2/2\sigma^2 t^2)\, s\psi_1\, ds}. \tag{7.67}$$

We know from elementary statics that $s_G = (s/\psi_1)\sin\psi_1$ is the distance CG, where G is the centre of mass of the arc ASB, and $2\psi_1\, s\, ds$ is the area of the infinitesimal annulus bounded by ASB. Thus we have a geometrical interpretation of the quantity $\bar{v}_r$: the product $\bar{v}_r t = s_0$, or $\bar{v}_r = s_0/t$, where s_0 is the distance from C to the centre of mass of a circle of centre O and radius r_0, whose density along any arc ASB is constant and proportional to $\exp(-s^2/2\sigma^2 t^2)$.

Since these calculations are only approximate, it is hardly worthwhile making an exact calculation of $\bar{v}_r$ from formula (7.67). We shall take only the case where $v = s/t$ is small in comparison with σ. Then $\exp(-s^2/2\sigma^2 t^2) = 1$. The mass distribution within the circle representing the local cluster is therefore uniform; hence $s_0 = r$, and

$$\bar{v}_r = r/t. \tag{7.68}$$

This result was first derived by Milne [71] to interpret the red shift in the spectra of the galaxies as a result of dissipation of a gravitationally unstable cluster of galaxies. For brevity we shall call the process corresponding to eqn. (7.68) a *Milnean expansion.*

The velocity is seen to be proportional to the distance and inversely proportional to the time.

It remains to show that, in the conditions prevailing in the Galaxy, a Milnean expansion leads to a K effect which decreases with increasing distance. The radial velocity given by formula (7.68) will be observed only if the star cloud is dissipated into a vacuum. Otherwise, the mean radial velocity (K effect) observed at a distance r is a weighted mean of radial velocities of field stars and of cluster stars undergoing dissipation. We shall suppose that the cloud is at rest relative to the field stars, and so the velocities of the "partial" centroids of the field stars will be zero.

Since, according to (7.68), the mean radial velocity of the stars in the cloud is r/t, we obtain for the K effect, i.e. the radial velocity of the centroid (here a resultant value, because of the mingling of the cloud stars and field stars),

$$K(r) = \{\nu(r)/[\nu(r) + \nu_1]\}\, r/t, \tag{7.69}$$

where ν_1 is the density of field stars and $\nu(r)$ the star density in the cloud. Within and near the cloud, i.e. for $r \approx r_0$, the density $\nu(r)$ exceeds ν_1, and so formula (7.69) gives approximately an ordinary Milnean expansion: $K(r) \approx r/t$. Outside the cloud the star density decreases rapidly as r increases (the dilution effect).

At a distance r, the star density $\nu(r)$ is

$$\nu(r) = (2\nu_0/\sigma^2) \int\limits_{(r-r_0)/t}^{(r+r_0)/t} \exp\,(-v^2/2\sigma^2)\, v\psi_1\, \mathrm{d}v, \tag{7.70}$$

where ν_0 is the initial density within the cloud. This expression is easily obtained from the denominator of (7.66) if we substitute expressions for v_1 and v_2 and for the constant factor $C = \nu_0/\sigma^2$ in the Maxwellian distribution function.

Let us consider the case $r \gg r_0$. Then formula (7.70) is much simplified. Since $\sin \psi_1 < r_0/r$, we can put $\sin \psi_1 \approx \psi_1 < r_0/r$ for $r \gg r_0$. Hence

$$\begin{aligned} \nu(r) &< 2\nu_0 \frac{r_0}{r} \int\limits_{(r-r_0)/t}^{(r+r_0)/t} \exp\,(-v^2/2\sigma^2)\, v\, \mathrm{d}v \\ &< 2\nu_0 (r_0/r) \exp\,[-(r - r_0)^2/2\sigma^2 t^2] \int\limits_{(r-r_0)/t}^{(r+r_0)/t} v\, \mathrm{d}v \\ &= 4\nu_0 (r_0^2/t^2) \exp[-(r - r_0)^2/2\sigma^2 t^2]. \end{aligned} \tag{7.71}$$

Since σt is the mean distance traversed by cloud stars in time t, the region where $r > \sigma t$ may be taken as "outside" the cluster. Formula (7.71) shows that in this region the star density tends rapidly to zero with increasing r.

It follows that at large distances $\nu(r)$ becomes much less than ν_1. Then formula (7.69) gives $K(r) = [\nu(r)/\nu_1]\, r/t$, or, from (7.71),

$$K(r) < 4 \frac{\nu_0}{\nu_1} \frac{r_0^2}{t^2} \exp\,[-(r - r_0)^2/2\sigma^2 t^2]\, r/t.$$

Hence we see that, for $r > \sigma t$, the K effect falls rapidly to zero with increasing r, as is in fact observed.

§ 11. The Dynamics of the Local Cluster

It is a commonly accepted view that the velocity distribution of B stars presents several unexplained anomalies.

(1) The residual velocities are extremely small, the root-mean-square value (Nordström) being $\sigma = 9{\cdot}9$ km/s, whereas the average value for all stars is 17·8 km/s, which is almost twice as great.

(2) The longitude of the vertex is $l = 236°$, whereas its value for all stars brighter than the sixth magnitude is $l = 340°$.

(3) The projection of the velocity ellipsoid on the galactic plane has an axis ratio $\sigma_2/\sigma_1 = 0{\cdot}86$, which is very large compared with the value for all other types of star: the mean value for all stars is $\sigma_2/\sigma_1 = 0{\cdot}73$, which corresponds to a less nearly circular ellipse.

(4) The ratio of the axis perpendicular to the galactic plane is $\sigma_3/\sigma_1 = 0{\cdot}41$, which is very small, the value for all stars being $\sigma_3/\sigma_1 = 0{\cdot}65$.

All these anomalies can be naturally explained if we assume that the Sun is located in one of the galactic star clouds in the local cluster. This cluster undergoes dissipation into the surrounding stellar population, and stars which leave it become indistinguishable from the rest.

To end this chapter we may deduce the following conclusions.

(1) Although the stars which are involved in the dissipation are young, their average age being of the order of 10^7 y, the stars which have high velocities are younger still, since they rapidly leave the cloud and are lost among the surrounding stars. The slowly moving stars remain in the cloud for a comparatively long time, some of them permanently as a result of being captured by the force field of the cloud. By "captured" we mean that, however weak the gravitational field of the local cluster itself, there exists a velocity of escape, and any star whose velocity is less than this remains within the local cluster until, as a result of further dissipation, its velocity becomes greater than the velocity of escape. This process may take a very long time if the rate of star formation in the local cluster is sufficiently high. The low-velocity stars are therefore older, on the average, than the high-velocity stars. This easily explains the fact that the deviation of the vertex is greater for the low-velocity stars, as noted by Blaauw [12]. The deviation of the vertex is accounted for as a result of the action of the Coriolis force, and gives the mean age of the O and B stars within 300 ps of the Sun as of the order of 5×10^7 y. During this time the vertex can turn through 180°.

(2) The small dispersion of the residual velocities of O and B stars ($\sigma = 9{\cdot}9$ km/s, which is little more than half the mean value for all other stars in the plane sub-system) is explained by the preferential dissipation of the more rapidly moving stars into the surrounding star field. There is thus a deficiency of high-velocity stars, and the observed dispersion of residual velocities is considerably less than the normal dispersion for stars in the plane sub-system. It may be supposed that the true velocity dispersion for O and B stars is the same as the normal dispersion for the stars (mostly giants) in the plane sub-system.

(3) The escape of high-velocity stars just described leads to a velocity ellipsoid less elongated in the direction of the vertex. On account of the rapid precessional rotation of the velocity ellipsoid (with twice the angular velocity of the galactic rotation), the tidal action of the galactic field, which tends to elongate the ellipsoid in the direction from the observer to the galactic centre, leads to a uniform expansion of the ellipsoid in all directions parallel to the galactic plane. This may account for the extremely low dispersion of the velocities of B stars in the z-direction as compared with the other two directions.

Calculation shows that the value $\sigma_3/\sigma_1 = 0{\cdot}41$ at present observed would be reached after $4{\cdot}2 \times 10^7$ y, a value which is in agreement with the mean age of B stars as estimated from the deviation of the vertex.

In an interesting and detailed paper A.F. Torondzhadze [131] considers the same problem by another method. He examines the individual relative orbits of O and B stars in the neighbourhood of the observer's centroid, and relates the K effect for these stars to other anomalies in their kinematics. In particular, he notes a lack of agreement between the solar velocity and the residual velocity dispersion, and assesses this discrepancy by means of Strömberg's formula for the asymmetry of stellar velocities, $v_y = -p\sigma^2 + \beta$. It is found that the velocity dispersion and the solar velocity observed for the O and B star complex do not satisfy this relation. By means of Strömberg's formula, Torondzhadze finds corrections to the elements of the solar motion, taking $\sigma = 17$ km/s in accordance with the results of Mineur [74]. The age of the O and B star complex forming the local cluster is found to be 22×10^7 y, which does not differ too greatly from the value obtained above.

It should be mentioned, however, that the procedure used by Torondzhadze – the validity of which he himself has questioned because of the results of Parenago ([107], p. 155) – is in a sense opposite to ours. Torondzhadze takes the velocity dispersion of the O and B stars to be normal and corrects the solar motion accordingly. Our assumption is that the dissipation of the O and B complex must be accompanied by a selective loss of high-velocity stars, and so the observed velocity dispersion should be reduced, as in fact it is. The O and B star dispersion $\bar{\sigma}^2 = \frac{1}{3}(\sigma_1^2 + \sigma_2^2 + \sigma_3^2)$ is taken from Nordström's results ($\sigma = 9{\cdot}9$ km/s). If Torondzhadze's collation of data is used, the values of $\bar{\sigma}$ are as shown in Table X.

TABLE X. VELOCITY DISPERSION IN km/s FOR BRIGHT O AND B STARS

Author	σ_1	σ_2	σ_3	$\bar{\sigma}$	Reference
Strömberg	7·8	10·8	9·5	9·4	(a)
Lindblad	10·9	9·8	4·5	8·8	(b)
Mineur	14·2	12·0	4·4	11·0	(c)
Mineur	17·4	12·6	7·2	13·1	(c)

(a) *Astrophysical Journal* **61,** 363, 1925.

(b) *Monthly Notices of the Royal Astronomical Society* **90,** 503, 1930.

(c) *Bulletin Astronomique, Mémoires et Variétés,* [2] **7,** 321, 1933.

It is difficult to agree with the value taken, $\bar{\sigma} = 17$ km/s, which is close to only the value $\sigma_1 = 17{\cdot}4$ km/s for the direction of the vertex given in Mineur's second treatment.

If we use the above-mentioned hypothesis that the O and B star complex is in reality a kinematically normal group of bright stars of the plane sub-system, with values of all the kinematic parameters differing only slightly from the mean values for all stars, we can suppose that the value of $\bar{\sigma}$ taken by Torondzhadze is simply the normal dispersion, which according to Nordström, for example, is 17·8 km/s. In this case the corrections found for the solar velocity components may be regarded as valid and equal to the velocity component of the whole O and B star complex in a direction which the author takes to be that of the spiral arm of the Galaxy passing near the Sun and forming the geometrical locus of the centres of the associations which continually produce O and B stars. This velocity is 4·3 km/s.

What is the nature of the local cluster? The future progress of stellar astronomy will no doubt modify our conception of the local agglomeration called by this name. We do not propose to relate the entire history of the subject, but it may be mentioned that the local cluster was originally regarded as a chance aggregation of stars, resembling the star clouds of the Milky Way. It was thought to be most evident in the distribution of O and B, and probably dwarf M, stars. A similar distribution of other types of star was supposed to be present, leading to the idea of a large mass of the local cluster. The existence of star clouds and the local cluster was related to no particular structural features of the Galaxy such as the spiral structure, because at that time (1920–1930) the details of this structure were entirely unknown.

The investigations of recent years have shown conclusively that the local cluster appears only as a partial increase in density, of the type represented by the associations. Such formations may be called *star complexes*, a term due to Torondzhadze. These complexes are closely related to the associations: they are a direct result of the dissipation of the latter. Being dynamically unstable, they are in turn continually dissipated among the surrounding stars, and at the same time they are continually renewed by the young stars leaving the associations.

The extent to which the complexes are related to the spiral arms is a matter for future investigation, since it is at present not at all clear whether the spiral arms of the Galaxy extend as far as the neighbourhood of the Sun. The hypothesis may be advanced, however, that the star complexes in the neighbourhood of the Sun are perhaps the remains of spiral arms disrupted by the differential rotation of the Galaxy—an idea due to Vorontsov-Vel'yaminov.

The existence of the local cluster is certain, because otherwise there would be no explanation of the phenomenon of Gould's belt. We now see, moreover, that the presence of the local cluster has a marked effect on the kinematics of large groups of stars, whose kinematic properties must be studied in order to eliminate the effect of local factors on the observed properties of stellar motions.

CHAPTER VIII

DYNAMICS OF CENTROIDS

§ 1. Macroscopic Motions in Stellar Systems

IT HAS been pointed out in the preceding chapters that stellar systems, despite the extremely low star density in them, possess the property of continuity which we usually associate with the concept of a material medium occupying the whole of some volume of space, such as a gas or a liquid. The existence of the centroid illustrates the continuity of motion in stellar systems. The centroid velocity $\mathbf{V}_0$, with components u_0, v_0, w_0, is, as we have seen in Chapter II, the mean stellar velocity in a given macroscopic volume element:

$$\left.\begin{aligned} u_0 &= \bar{u} = \int uf\,\mathrm{d}\Omega/\nu, \\ v_0 &= \bar{v} = \int vf\,\mathrm{d}\Omega/\nu, \\ w_0 &= \bar{w} = \int wf\,\mathrm{d}\Omega/\nu, \end{aligned}\right\} \tag{8.1}$$

where f is the phase density, ν the star density and $\mathrm{d}\Omega$ a volume element in velocity space.

To find the internal (macroscopic) motions, an exact knowledge of the phase density is necessary. However, certain general conclusions can be drawn on the assumption that the system possesses some symmetry. We shall assume that the system is in a steady state, and consider the three cases discussed in Chapter V, § 5.

(1) The stellar system has no symmetry. It has been shown in Chapter V, § 5, that in this case all three components of the centroid velocity are zero, and therefore there are no macroscopic motions in the system.

(2) The system has rotational symmetry. In this case, according to Chapter V, § 6, the phase density is a function of the energy integral and the area integral: $f = f(\mathrm{P}^2 + \Theta^2 + Z^2 - 2U, \varrho\Theta)$, and is an even function of P and Z. Hence

$$\left.\begin{aligned} \mathrm{P}_0 &= \int f\mathrm{P}\,\mathrm{d}\Omega/\nu = 0, \\ Z_0 &= \int fZ\,\mathrm{d}\Omega/\nu = 0, \\ \Theta_0 &= \int f\Theta\,\mathrm{d}\Omega/\nu = \bar{\Theta}(\varrho,z). \end{aligned}\right\} \tag{8.2}$$

Thus we see that the centroid velocity at any point in the system is perpendicular to the meridian (ϱz) plane. The line of motion of the centroid may be called a *streamline* by analogy with hydrodynamics. The streamlines are in general determined by the differential equations $\mathrm{d}\varrho/\mathrm{P}_0 = \varrho\,\mathrm{d}\,\theta/\Theta_0 = \mathrm{d}z/Z_0$. In the present case we see that the streamline of every centroid is a circle. For, since $\mathrm{P}_0 = Z_0 = 0$, it follows that $\mathrm{d}\varrho = \mathrm{d}z = 0$, or ϱ = constant, z = constant. In the coordinates ϱ, θ, z these equations represent circles, which in this case are coordinate lines.

At each point on the circle the direction of the tangent is the same as that of the centroid velocity. This velocity, of magnitude Θ_0, will be called *the circular velocity of the centroid*; it is in general different from the circular velocity Θ_c of a particle. Unlike the centroids, the particles can execute a circular motion only in the equatorial plane (plane of symmetry) of the system, and moreover the value of Θ_c is always given by the relation $\Theta_c^2 = -\varrho\,\partial U/\partial\varrho$.

Thus the only macroscopic motion possible in a rotationally symmetrical stellar system in a steady state is rotation about the axis of symmetry. In particular, if the velocity distribution in the system is ellipsoidal, then

$$f = f(\mathrm{P}^2 + \Theta^2 + Z^2 - 2U + k_2\varrho^2\Theta^2 + k_1\varrho\Theta).$$

It is easily seen from (8.2) that in this case

$$\mathrm{P}_0 = 0, \quad Z_0 = 0, \quad \Theta_0 = \Theta_0(\varrho) = -k_1\varrho/(1 + k_2\varrho^2). \tag{8.3}$$

Thus we return to the conclusion derived in another manner in Chapter V, § 7: for a stellar system in a steady state with rotational symmetry and an ellipsoidal velocity distribution, the centroid velocity field is one of barotropic rotation.

(3) The system has spherical symmetry. In this case the phase density f is a function of the energy integral and the square of the angular momentum about the centre of symmetry (Chapter V, § 6):

$$f = f[R^2 + \Theta^2 + \Phi^2 - 2U, \quad r^2(\Theta^2 + \Phi^2)].$$

Thus the phase density is an even function of R, Θ and Φ. Hence

$$\left.\begin{aligned} R_0 &= \int f R\,\mathrm{d}\Omega/\nu = 0, \\ \Theta_0 &= \int f\Theta\,\mathrm{d}\Omega/\nu = 0, \\ \Phi_0 &= \int f\Phi\,\mathrm{d}\Omega/\nu = 0. \end{aligned}\right\} \tag{8.4}$$

In a spherically symmetrical stellar system in a steady state, therefore, no macroscopic motions are possible.

The above discussion shows that the steady-state condition imposes very considerable restrictions on the internal motions in the stellar system.

§ 2. Some Properties of the Kinematics of Centroids

The hydrodynamic aspect of the motions in a stellar system is somewhat different from the corresponding situation in a fluid. The difference is due mainly to the fact that stellar systems are extremely rarefied media, i.e. the mean free path of the particles is very long.

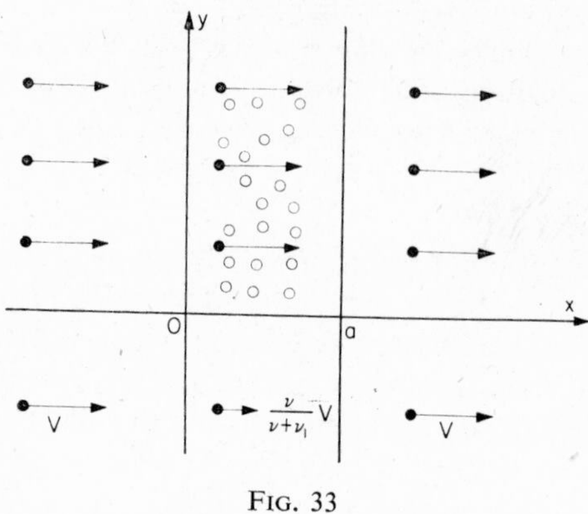

FIG. 33

In order to investigate these properties of motion in rarefied media, let us consider two simple examples.

EXAMPLE I: Linear motion through a plane layer at rest. Suppose that a current of identical particles is moving parallel to the x-axis, with spatial density ν, and that across its path lies an infinite homogeneous plane-parallel layer of similar particles at rest, with spatial density ν_1 (Fig. 33). Let the surfaces of the layer be $x = 0$, $x = a$. For simplicity we shall suppose that the particles do not interact and, in particular, do not undergo collisions.

This situation corresponds approximately to the motion of an open cluster such as the Pleiades or the Hyades in the Galaxy, or the collision of masses of gas whose relative velocity is large compared with the thermal velocities of the molecules.

The velocity of the medium where $x < 0$ is determined by the movement of the particles. The macroscopic motion of these particles is therefore characterised by the centroid velocity V. The medium in the layer $0 < x < a$ consists of two kinds of particles, those moving with mean velocity V and those at rest. Thus at each point in this layer the centroid velocity is equal to the weighted mean of these velocities, $(\nu V + \nu_1 0)/(\nu + \nu_1) = \nu V/(\nu + \nu_1)$. This is evidently less than V, and there is a retardation of the motion, which in the absence of external forces is a purely kinematic phenomenon; we shall call it a *kinematic retardation.*

At first sight this result might seem paradoxical, since the kinetic energy of the motion is reduced and yet no work is done. For, if we ignore the particles at rest, the kinetic energy per unit volume is $\frac{1}{2}\nu V^2$. When both kinds of particle are taken into account, however, it is $\frac{1}{2}(\nu + \nu_1)\,[\nu V/(\nu + \nu_1)]^2 = \frac{1}{2}\nu^2 V^2/(\nu + \nu_1) < \frac{1}{2}\nu V^2$.

In reality there is no paradox, because the missing energy is converted into the energy of irregular thermal motion. Relative to the centroid, each particle in the current has velocity $V - \nu V/(\nu + \nu_1) = \nu_1 V/(\nu + \nu_1)$, and each particle in the fixed layer has velocity $0 - \nu V/(\nu + \nu_1) = -\nu V/(\nu + \nu_1)$. The "lost" energy of the irregular motion is therefore

$$\frac{1}{2}\nu\left(\frac{\nu_1 V}{\nu + \nu_1}\right)^2 + \frac{1}{2}\nu_1\left(\frac{\nu V}{\nu + \nu_1}\right)^2 = \frac{1}{2}\frac{\nu\nu_1}{\nu + \nu_1}V^2.$$

It is easily seen that, by adding this expression for the energy of the irregular motion to that for the energy of the regular motion, we obtain the original energy:

$$\frac{1}{2}\frac{\nu\nu_1}{\nu + \nu_1}V^2 + \frac{1}{2}\frac{\nu^2}{\nu + \nu_1}V^2 = \frac{1}{2}\nu V^2.$$

For $x > a$ we have again a parallel stream of particles moving with velocity V, and so we observe a kinematic acceleration of the motion of the centroid.

Thus the simple example discussed above shows that, in the motion of centroids, positive and negative kinematic accelerations are observed, accompanied by the change of part of the kinetic energy of regular motion into that of irregular motion, and vice versa.

Example II: Let us now consider another example, also very simple, which shows that the motion of centroids is not subject to the laws of motion of particles.

Let all the particles move in the xy-plane, and let the axis Oy be the source of a stream of identical particles moving with velocity V_1 parallel to the axis Ox (Fig. 34);

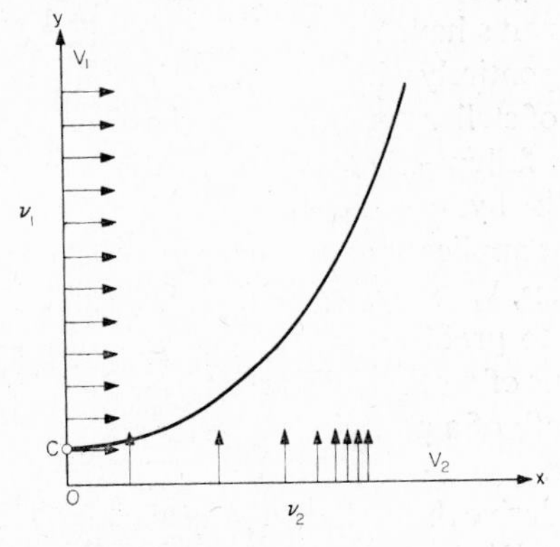

Fig. 34

let the spatial density of these particles be ν_1 = constant along the axis Oy. Likewise, let the axis Ox be the source of similar particles moving parallel to the axis Oy with velocity V_2, but let the spatial density ν_2 of these particles be a function of x, e.g. $\nu_2 = \lambda x$.

The components u and v of the centroid velocity along the coordinate axes are equal to the weighted means of the velocity components of the two streams:

$u = \nu_1 V_1/(\nu_1 + \lambda x), v = \lambda x V_2/(\nu_1 + \lambda x)$. For $x \to 0$ we have $u = V_1, v = 0$; for $x \to \infty$, $u = 0, v = V_2$, in accordance with the relative proportions of particles from the first and second streams for various values of x.

The equations of motion of the centroid are evidently $dx/dt = \nu_1 V_1/(\nu_1 + \lambda x)$, $dy/dt = \lambda x V_2/(\nu_1 + \lambda x)$, whence $dy/dx = kx$, with $k = \lambda V_2/\nu_1 V_1$. Integration gives the equation of a family of parallel parabolas: $y = \frac{1}{2}kx^2 + C$, each of which is the path of a centroid given by the initial condition $y = C$ for $x = 0$.

It is easy to see that, in the general case $\nu_1 = f_1(x, y), \nu_2 = f_2(x, y)$, we obtain the differential equation of the path of the centroid as

$$\frac{dy}{dx} = \frac{V_2}{V_1}\frac{f_2(x, y)}{f_1(x, y)}.$$

Hence it is clear that, by appropriate choice of $f_1(x, y)$ and $f_2(x, y)$, we can cause the centroid to move along any predetermined curve, simply on account of the relative strength of the two streams.

§ 3. Equations of Motion of Centroids in Rectangular Coordinates

It is possible to derive equations of motion for centroids, similar in all respects to the equations of motion of a compressible viscous fluid. These equations were first obtained by Jeans, and are usually called the *hydrodynamical equations* of stellar dynamics; they have, however, not been much used as yet. The reason for this neglect is certainly, at least in part, the one-sided conventional approach to the dynamics of stellar systems. These systems have usually been regarded only as discrete assemblies of individual stars moving entirely without interaction, and the attempt has been made to develop the dynamics of stellar systems on the basis of a study of the paths of these individual stars [65]. The failure of these attempts is in itself a reason to seek other approaches. The use of the hydrodynamical equations is undoubtedly attractive, because it makes possible the application to stellar dynamics of the extensively developed methods of fluid mechanics. We may expect that, in turn, the application of the equations of fluid mechanics to problems of stellar dynamics will make possible a consideration of new problems of motion in rarefied gases, since stellar systems are media in which the mean free path of a particle is large in comparison with any conceivable characteristic dimensions.

In this section we shall derive the hydrodynamical equations of motion of a centroid in rectangular coordinates. We have seen that, in the fundamental eqn. (5.3) of stellar dynamics, the right-hand side is zero if either the action of the irregular forces is negligibly small or the phase density is invariant with respect to these forces. In either case eqn. (5.3) becomes the Boltzmann eqn. (5.8). The latter may be written

$$\frac{\partial f}{\partial t} + u\frac{\partial f}{\partial x} + v\frac{\partial f}{\partial y} + w\frac{\partial f}{\partial z} + X\frac{\partial f}{\partial u} + Y\frac{\partial f}{\partial v} + Z\frac{\partial f}{\partial w} = 0, \quad (8.5)$$

where X, Y, Z are the components of the gravitational force.

The components u_0, v_0, w_0 of the centroid velocity are the mean velocities of the stars in the macroscopic volume element considered. To obtain the equations of motion of the centroid, therefore, we must "average" (integrate) eqn. (8.5) over all velocity space. Here it must be borne in mind that, in any actual stellar system, the stellar velocities are limited by the velocity of escape. As $V \to \infty$, therefore, the phase density must tend rapidly to zero. Let us assume that

$$\lim_{V\to\infty} f = 0, \quad \lim_{V\to\infty} Vf = 0. \tag{8.6}$$

We multiply eqn. (8.5) by the volume element $d\Omega = du\,dv\,dw$, and integrate over all velocity space. Since

$$\int f\,d\Omega = \nu,$$
$$\int uf\,d\Omega = \nu u_0, \quad \int vf\,d\Omega = \nu v_0, \quad \int wf\,d\Omega = \nu w_0, \tag{8.7}$$

where u_0, v_0, w_0 are the components of the centroid velocity, and, by (8.6),

$$\int \frac{\partial f}{\partial u}\,d\Omega = \int [f]_{-\infty}^{+\infty}\,dv\,dw = 0, \quad \int \frac{\partial f}{\partial v}\,d\Omega = 0, \quad \int \frac{\partial f}{\partial w}\,d\Omega = 0,$$

we have the first hydrodynamical equation:

$$\frac{\partial \nu}{\partial t} + \frac{\partial(\nu u_0)}{\partial x} + \frac{\partial(\nu v_0)}{\partial y} + \frac{\partial(\nu w_0)}{\partial z} = 0. \tag{8.8}$$

This is the same as the hydrodynamical *equation of continuity* if the centroid velocity is equated to the fluid velocity.

Next, we multiply eqn. (8.5) successively by $u\,d\Omega$, $v\,d\Omega$, $w\,d\Omega$ and again integrate over all velocity space. Then, using (8.6), (8.7) and the relations $\int u^2 f\,d\Omega = \nu\overline{u^2}$, $\int uvf\,d\Omega = \nu\overline{uv}$, $\int uwf\,d\Omega = \nu\overline{uw}$, etc., where the bar denotes the result of averaging, and also

$$\int u\frac{\partial f}{\partial u}\,d\Omega = \int [uf]_{-\infty}^{+\infty}\,dv\,dw - \int f\,d\Omega = -\nu,$$

$$\int u\frac{\partial f}{\partial v}\,d\Omega = 0, \quad \int u\frac{\partial f}{\partial w}\,d\Omega = 0,$$

etc., we obtain the three equations

$$\left.\begin{aligned}
\frac{\partial(\nu u_0)}{\partial t} + \frac{\partial(\nu\overline{u^2})}{\partial x} + \frac{\partial(\nu\overline{uv})}{\partial y} + \frac{\partial(\nu\overline{uw})}{\partial z} &= \nu X,\\
\frac{\partial(\nu v_0)}{\partial t} + \frac{\partial(\nu\overline{vu})}{\partial x} + \frac{\partial(\nu\overline{v^2})}{\partial y} + \frac{\partial(\nu\overline{vw})}{\partial z} &= \nu Y,\\
\frac{\partial(\nu w_0)}{\partial t} + \frac{\partial(\nu\overline{wu})}{\partial x} + \frac{\partial(\nu\overline{wv})}{\partial y} + \frac{\partial(\nu\overline{w^2})}{\partial z} &= \nu Z.
\end{aligned}\right\} \tag{8.9}$$

Equations (8.8) and (8.9) are the *equations of transfer;* eqns. (8.9) represent the transfer of momentum, and eqn. (8.8) the transfer of particle (star) density.

In this form the hydrodynamical equations, though concise, cannot be used for a comparison with the ordinary hydrodynamical equations of a compressible viscous fluid. Such a comparison can be made if the quantities $\overline{u^2}$, $\overline{uv}$, ... in eqn. (8.9) are written explicitly in terms of the centroid velocity components.

We write the stellar velocity components as

$$u = u_0 + u', \quad v = v_0 + v', \quad w = w_0 + w', \tag{8.10}$$

where u_0, v_0, w_0 are again the centroid velocity components and u', v', w' the residual velocities, which are such that $\overline{u'} = 0$, $\overline{v'} = 0$, $\overline{w'} = 0$. Then

$$\left.\begin{aligned} &\overline{u^2} = u_0^2 + \overline{u'^2}, \quad \overline{v^2} = v_0^2 + \overline{v'^2}, \quad \overline{w^2} = w_0^2 + \overline{w'^2}, \\ &\overline{vw} = v_0 w_0 + \overline{v'w'}, \quad \overline{wu} = w_0 u_0 + \overline{w'u'}, \\ &\overline{uv} = u_0 v_0 + \overline{u'v'}. \end{aligned}\right\} \tag{8.11}$$

In substituting these quantities in the first eqn. (8.9), we use the identities

$$\left.\begin{aligned} \frac{\partial(\nu u_0)}{\partial t} &= u_0 \frac{\partial \nu}{\partial t} + \nu \frac{\partial u_0}{\partial t}, \\ \frac{\partial(\nu u_0^2)}{\partial x} &= u_0 \frac{\partial(\nu u_0)}{\partial x} + \nu u_0 \frac{\partial u_0}{\partial x}, \\ \frac{\partial(\nu u_0 v_0)}{\partial y} &= u_0 \frac{\partial(\nu v_0)}{\partial y} + \nu v_0 \frac{\partial u_0}{\partial y}, \\ \frac{\partial(\nu u_0 w_0)}{\partial z} &= u_0 \frac{\partial(\nu w_0)}{\partial z} + \nu w_0 \frac{\partial u_0}{\partial z}. \end{aligned}\right\} \tag{8.12}$$

In the resulting equation the sum of the terms containing the factor u_0 is zero, by the equation of continuity (8.8). Thus we have finally

$$\begin{aligned} &\nu \left(\frac{\partial u_0}{\partial t} + u_0 \frac{\partial u_0}{\partial x} + v_0 \frac{\partial u_0}{\partial y} + w_0 \frac{\partial u_0}{\partial z} \right) \\ &\quad + \frac{\partial(\nu \overline{u'^2})}{\partial x} + \frac{\partial(\nu \overline{u'v'})}{\partial y} + \frac{\partial(\nu \overline{u'w'})}{\partial z} = \nu X. \end{aligned}$$

The same procedure is applied to $\nu \overline{v_0}$, $\nu \overline{v_0 u_0}$, $\nu \overline{v_0^2}$ and $\nu v_0 w_0$ and the terms proportional to v_0 are separated; the result is substituted in the second eqn. (8.9). The third equation is similarly treated, and the sum of the w_0 terms is likewise zero by the equation of continuity.

Finally, dividing the three equations so obtained by ν and putting for brevity

$$\begin{aligned} &p_{xx} = \nu \overline{u'^2}, \qquad p_{yy} = \nu \overline{v'^2}, \qquad p_{zz} = \nu \overline{w'^2}, \\ &p_{yz} = p_{zy} = \nu \overline{v'w'}, \quad p_{zx} = p_{xz} = \nu \overline{w'u'}, \quad p_{xy} = p_{yx} = \nu \overline{u'v'}, \end{aligned} \tag{8.13}$$

we obtain the hydrodynamical equations

$$\left.\begin{aligned}
&\frac{\partial u_0}{\partial t} + u_0 \frac{\partial u_0}{\partial x} + v_0 \frac{\partial u_0}{\partial y} + w_0 \frac{\partial u_0}{\partial z} \\
&\qquad + \frac{1}{\nu}\left(\frac{\partial p_{xx}}{\partial x} + \frac{\partial p_{xy}}{\partial y} + \frac{\partial p_{xz}}{\partial z}\right) = X, \\
&\frac{\partial v_0}{\partial t} + u_0 \frac{\partial v_0}{\partial x} + v_0 \frac{\partial v_0}{\partial y} + w_0 \frac{\partial v_0}{\partial z} \\
&\qquad + \frac{1}{\nu}\left(\frac{\partial p_{yx}}{\partial x} + \frac{\partial p_{yy}}{\partial y} + \frac{\partial p_{yz}}{\partial z}\right) = Y, \\
&\frac{\partial w_0}{\partial t} + u_0 \frac{\partial w_0}{\partial x} + v_0 \frac{\partial w_0}{\partial y} + w_0 \frac{\partial w_0}{\partial z} \\
&\qquad + \frac{1}{\nu}\left(\frac{\partial p_{zx}}{\partial x} + \frac{\partial p_{zy}}{\partial y} + \frac{\partial p_{zz}}{\partial z}\right) = Z.
\end{aligned}\right\} \tag{8.14}$$

To these must be added the equation of continuity (8.8):

$$\frac{\partial \nu}{\partial t} + \frac{\partial(\nu u_0)}{\partial x} + \frac{\partial(\nu v_0)}{\partial y} + \frac{\partial(\nu w_0)}{\partial z} = 0.$$

In this form the hydrodynamical equations of stellar dynamics are identical with the familiar Navier–Stokes equations of motion of a compressible viscous fluid.†

In deriving these equations we have used the fact that only regular forces act in the system. It is useful to note that eqns. (8.14) can also be obtained from much more general assumptions. Let us assume that both regular and irregular forces act, the latter being due to the interaction of stars in encounters, and impose two conditions on the irregular forces.

(1) Encounters between stars do not change the number of stars in any given macroscopic volume element. Then $\int (\partial f/\partial t)_i \, d\Omega = 0$. This condition always holds if the encounters are of the nature of collisions, i.e. if they take place at distances much less than the size of the macroscopic volume element and if the result of the encounter is to change the velocities but not the coordinates of the stars, as is true if the encounter radius is less than the radius of the macroscopic volume element.

(2) Encounters between stars do not change the total momentum vector. Then $\int u(\partial f/\partial t)_i \, d\Omega = 0$, $\int v(\partial f/\partial t)_i \, d\Omega = 0$, $\int w(\partial f/\partial t)_i \, d\Omega = 0$. This condition holds if the encounter radius is small compared with the mean distance between the stars. In that case the stars involved in the encounter may be regarded as forming an isolated system, for which the law of conservation of momentum holds.

It is easy to see that, when these two conditions are satisfied, the derivation of

† See, for example, H. Lamb, *Hydrodynamics*, 6th ed., p. 576, Cambridge University Press, 1932.

eqns. (8.14) remains valid even in the presence of irregular forces.† The concept of the centroid and the peculiarities of its dynamics apply to arbitrary gases and liquids also. Hence eqns. (8.14) may be called the equations of motion of an arbitrary continuous medium.

§ 4. The Dynamical Significance of the Velocity Moment Tensor

Let us now compare the motion of the centroids with the regular streaming motion of stars. For comparison with eqns. (8.14) we consider a star stream in which each star moves in a definite orbit in such a way that its velocity at any point is a definite function of the coordinates, and in general of the time: $u = f_1(x, y, z, t)$, $v = f_2(x, y, z, t)$, $w = f_3(x, y, z, t)$. In this case eqns. (8.14) are replaced by the ordinary equations of motion of a free particle: $du/dt = X$, $dv/dt = Y$, $dw/dt = Z$, or

$$\left.\begin{aligned} \frac{\partial u}{\partial t} + u\frac{\partial u}{\partial x} + v\frac{\partial u}{\partial y} + w\frac{\partial u}{\partial z} &= X, \\ \frac{\partial v}{\partial t} + u\frac{\partial v}{\partial x} + v\frac{\partial v}{\partial y} + w\frac{\partial v}{\partial z} &= Y, \\ \frac{\partial w}{\partial t} + u\frac{\partial w}{\partial x} + v\frac{\partial w}{\partial y} + w\frac{\partial w}{\partial z} &= Z. \end{aligned}\right\} \tag{8.15}$$

A comparison of eqns. (8.14) and (8.15) shows that the terms in brackets on the left of (8.14) correspond to the kinematic acceleration of the centroids. This acceleration is due to the diffusion of stars across the boundaries of the macroscopic volume element.

On comparing the expressions (8.13) for $p_{xx}, p_{xy}, \ldots, p_{zz}$ with those given in Chapter I for the velocity moments, we see that

$$\begin{aligned} p_{xx} &= \mu_{200}, \quad p_{yy} = \mu_{020}, \quad p_{zz} = \mu_{002}, \\ p_{yz} &= \mu_{011}, \quad p_{zx} = \mu_{101}, \quad p_{xy} = \mu_{110}. \end{aligned} \tag{8.16}$$

The moments μ_{ijk} form the *moment tensor*

$$\boldsymbol{M} = \begin{bmatrix} \mu_{200} & \mu_{110} & \mu_{101} \\ \mu_{110} & \mu_{020} & \mu_{011} \\ \mu_{101} & \mu_{011} & \mu_{002} \end{bmatrix}, \tag{8.17}$$

and so we can say that the derivatives of the coefficients of the Schwarzschild velocity ellipsoid with respect to the coordinates give the kinematic acceleration (positive or negative) of the centroids in their motion in the stellar system.

† The two conditions given are equivalent to the mathematical condition that $(\partial f/\partial t)_i$ is orthogonal to 1, u, v, w; see D. Hilbert, *Outline of a general theory of linear integral equations* (*Grundzüge einer allgemeinen Theorie der linearen Integralgleichungen*), ch. XXII, Teubner, Berlin, 1924.

To obtain a more direct idea of the physical significance of the moment tensor, let us consider an infinitesimal rectangular parallelepiped† in the system, with its sides parallel to the coordinate axes. We shall use a slightly different notation, denoting the coordinates by x_1, x_2, x_3 and the corresponding velocities by u_1, u_2, u_3. Then the volume of the parallelepiped is $\mathrm{d}x = \mathrm{d}x_1\,\mathrm{d}x_2\,\mathrm{d}x_3$, x denoting the radius vector of the point (x_1, x_2, x_3). Since the star density is small, the action of the irregular forces may be neglected over distances of the order of the dimensions of $\mathrm{d}x$.

Let us determine the momentum acquired by the volume $\mathrm{d}x$ in unit time from stars entering and leaving it. The number of stars entering $\mathrm{d}x$ through the face which is perpendicular to the x_1-axis and passes through the corner (x_1, x_2, x_3), with velocities lying in the rectangular parallelepiped $\mathrm{d}u = \mathrm{d}u_1\,\mathrm{d}u_2\,\mathrm{d}u_3$ in velocity space, is $\mathrm{d}n_1 = u_1 f(u, x)\,\mathrm{d}u\,\mathrm{d}x_2\,\mathrm{d}x_3$. The corresponding momentum components are $u_1^2 f(u, x)\,\mathrm{d}u\,\mathrm{d}x_2\,\mathrm{d}x_3$, $u_1 u_2 f(u, x)\,\mathrm{d}u\,\mathrm{d}x_2\,\mathrm{d}x_3$, $u_1 u_3 f(u, x)\,\mathrm{d}u\,\mathrm{d}x_2\,\mathrm{d}x_3$. The total momentum components are

$$\left.\begin{aligned} &\mathrm{d}x_2\,\mathrm{d}x_3 \int u_1^2 f(u, x)\,\mathrm{d}u, \\ &\mathrm{d}x_2\,\mathrm{d}x_3 \int u_1 u_2 f(u, x)\,\mathrm{d}u, \\ &\mathrm{d}x_2\,\mathrm{d}x_3 \int u_1 u_3 f(u, x)\,\mathrm{d}u, \end{aligned}\right\} \tag{8.18}$$

where the integration is taken over the half-space $u_1 > 0$. The stars which leave $\mathrm{d}x$ through the same face take with them a negative contribution, which is equivalent to a positive momentum in the opposite direction. Thus the total momentum acquired by $\mathrm{d}x$ in unit time from stars passing through it is equal to (8.18) with the integration taken over all velocity space.

Putting for brevity

$$p_{ik} = \int u_i u_k f(u, x)\,\mathrm{d}u \quad (i, k = 1, 2, 3), \tag{8.19}$$

we obtain the rate-of-momentum-change tensor $\boldsymbol{P}$ per unit volume: p_{ik} gives the rate of change of the momentum along the x_k-axis due to stars passing through unit area perpendicular to the x_i-axis. Evidently $p_{ik} = p_{ki}$, i.e. the tensor $\boldsymbol{P}$ is symmetrical. It is a function of the coordinates x_i. In our notation this may be written $p_{ik} = p_{ik}(x)$. The quantities (8.18) are

$$p_{11}(x)\,\mathrm{d}x_2\,\mathrm{d}x_3, \quad p_{12}(x)\,\mathrm{d}x_2\,\mathrm{d}x_3, \quad p_{13}(x)\,\mathrm{d}x_2\,\mathrm{d}x_3. \tag{8.20}$$

Let us now consider stars which cross the opposite face of the volume $\mathrm{d}x$. This face is parallel and equal to the one already mentioned, and is at a distance $\mathrm{d}x_1$ from it in the direction of x_1 increasing. The corresponding increments of the momentum components are

$$\begin{gathered} -p_{11}(x + \mathrm{d}x_1)\,\mathrm{d}x_2\,\mathrm{d}x_3, \quad -p_{12}(x + \mathrm{d}x_1)\,\mathrm{d}x_2\,\mathrm{d}x_3, \\ -p_{13}(x + \mathrm{d}x_1)\,\mathrm{d}x_2\,\mathrm{d}x_3. \end{gathered} \tag{8.21}$$

The minus sign signifies a loss of momentum.

† Here, and always in such cases, the "infinitesimal" volume is to be taken as of the order of the macroscopic volume element defined in Chapter II.

The net loss of momentum from stars crossing both the faces perpendicular to the x_1-axis is

$$-[p_{11}(x + \mathrm{d}x_1) - p_{11}(x)]\,\mathrm{d}x_2\,\mathrm{d}x_3$$
$$= -(\partial p_{11}/\partial x_1)\,\mathrm{d}x_1\,\mathrm{d}x_2\,\mathrm{d}x_3 = -(\partial p_{11}/\partial x_1)\,\mathrm{d}x.$$

The corresponding expressions for the increments of the momentum components along the other two axes are easily seen to be $-(\partial p_{12}/\partial x_1)\,\mathrm{d}x$ and $-(\partial p_{13}/\partial x_1)\,\mathrm{d}x$. On referring these to unit volume and making similar calculations for the faces perpendicular to the x_2 and x_3 axes, we find the components of the total rate of change of momentum to be

$$\left.\begin{aligned} -\left(\frac{\partial p_{11}}{\partial x_1} + \frac{\partial p_{21}}{\partial x_2} + \frac{\partial p_{31}}{\partial x_3}\right) &= -\frac{\partial p_{i1}}{\partial x_i}, \\ -\left(\frac{\partial p_{12}}{\partial x_1} + \frac{\partial p_{22}}{\partial x_2} + \frac{\partial p_{32}}{\partial x_3}\right) &= -\frac{\partial p_{i2}}{\partial x_i}, \\ -\left(\frac{\partial p_{13}}{\partial x_1} + \frac{\partial p_{23}}{\partial x_2} + \frac{\partial p_{33}}{\partial x_3}\right) &= -\frac{\partial p_{i3}}{\partial x_i}, \end{aligned}\right\} \tag{8.22}$$

where for brevity the rule of summation over a repeated suffix is used on the right-hand sides.

The expressions (8.22) are minus the components of the divergence of the tensor $\boldsymbol{P}$, which, apart from the notation, are the same as the terms in eqns. (8.14) giving the kinematic acceleration in stellar systems.

§ 5. Equations of Motion of Centroids in Cylindrical and Spherical Coordinates

Various coordinate systems are used in studying the motions occurring in various types of stellar system; the most usual are rectangular, cylindrical and spherical coordinates.

In this section we shall derive the hydrodynamical equations of stellar dynamics in cylindrical and spherical coordinates, assuming as before that the phase density tends rapidly to zero as $V \to \infty$. Let

$$\lim_{V\to\infty} f = 0, \quad \lim_{V\to\infty} Vf = 0, \quad \lim_{V\to\infty} V^2 f = 0. \tag{8.23}$$

The Boltzmann equation in cylindrical coordinates is eqn. (5.15):

$$\frac{\partial f}{\partial t} + \mathrm{P}\frac{\partial f}{\partial \varrho} + \frac{\Theta}{\varrho}\frac{\partial f}{\partial \theta} + Z\frac{\partial f}{\partial z} + \left(\frac{\partial U}{\partial \varrho} + \frac{\Theta^2}{\varrho}\right)\frac{\partial f}{\partial \mathrm{P}}$$
$$+ \left(\frac{1}{\varrho}\frac{\partial U}{\partial \theta} - \frac{\mathrm{P}\Theta}{\varrho}\right)\frac{\partial f}{\partial \Theta} + \frac{\partial U}{\partial z}\frac{\partial f}{\partial Z} = 0. \tag{8.24}$$

To simplify the integration in obtaining the hydrodynamical equations from (8.24) we write Λ for any one of the three velocity components in curvilinear coordinates and $E(\mathrm{M}, \mathrm{N})$ for an arbitrary integrable function of the other two components M, N. Then, using the conditions (8.23) and the definitions of star density and centroid velocity, we have the following formulae which will be of use:

$$\left.\begin{aligned}
\int f\,\mathrm{d}\Omega &= \nu,\\
\int \Lambda f\,\mathrm{d}\Omega &= \Lambda_0,\\
\int (\partial f/\partial \Lambda)\,\mathrm{d}\Omega &= \int [f]_{-\infty}^{\infty}\,\mathrm{dM}\,\mathrm{dN} = 0,\\
\int \Lambda(\partial f/\partial \Lambda)\,\mathrm{d}\Omega &= \int [f\Lambda]_{-\infty}^{\infty}\,\mathrm{dM}\,\mathrm{dN} - \int f\,\mathrm{d}\Lambda\,\mathrm{dM}\,\mathrm{dN} = -\nu,
\end{aligned}\right\}\quad (8.25)$$

$$\left.\begin{aligned}
\int E(\partial f/\partial \Lambda)\,\mathrm{d}\Omega &= \int E[f]_{-\infty}^{\infty}\,\mathrm{dM}\,\mathrm{dN} = 0,\\
\int Ef\,\mathrm{d}\Omega &= \nu\overline{E},\\
\int E\Lambda(\partial f/\partial \Lambda)\,\mathrm{d}\Omega &= \int E[f\Lambda]_{-\infty}^{\infty}\mathrm{dM}\,\mathrm{dN} - \int Ef\,\mathrm{d}\Lambda\,\mathrm{dM}\,\mathrm{dN}\\
&= -\nu\overline{E},\\
\int E\Lambda^2(\partial f/\partial \Lambda)\,\mathrm{d}\Omega &= \int E[f\Lambda^2]_{-\infty}^{\infty}\mathrm{dM}\,\mathrm{dN} - 2\int E\Lambda f\,\mathrm{d}\Lambda\,\mathrm{dM}\,\mathrm{dN}\\
&= -2\nu\overline{E\Lambda}.
\end{aligned}\right\}\quad (8.26)$$

We multiply eqn. (8.24) by the volume element in velocity space—in cylindrical coordinates $\mathrm{d}\Omega = \mathrm{dP}\,\mathrm{d}\Theta\,\mathrm{d}Z$—and integrate over all P, Θ and Z. Then, using (8.25) and (8.26), we obtain the equation of continuity in cylindrical coordinates:

$$\frac{\partial \nu}{\partial t} + \frac{\partial(\nu \mathrm{P}_0)}{\partial \varrho} + \frac{1}{\varrho}\frac{\partial(\nu\Theta_0)}{\partial \theta} + \frac{\partial(\nu Z_0)}{\partial z} + \frac{1}{\varrho}\nu \mathrm{P}_0 = 0. \quad (8.27)$$

Next, we multiply eqn. (8.24) successively by $\mathrm{P}\,\mathrm{d}\Omega$, $\Theta\,\mathrm{d}\Omega$ and $Z\,\mathrm{d}\Omega$, and again integrate, using (8.25) and (8.26). This gives the three equations of transfer in cylindrical coordinates:

$$\left.\begin{aligned}
&\frac{\partial(\nu \mathrm{P}_0)}{\partial t} + \frac{\partial(\nu\overline{\mathrm{P}^2})}{\partial \varrho} + \frac{1}{\varrho}\frac{\partial(\nu\overline{\mathrm{P}\Theta})}{\partial \theta} + \frac{\partial(\nu\overline{\mathrm{P}Z})}{\partial z} - \frac{\nu\overline{\Theta^2}}{\varrho} + \frac{\nu\overline{\mathrm{P}^2}}{\varrho} = \nu\frac{\partial U}{\partial \varrho},\\
&\frac{\partial(\nu\Theta_0)}{\partial t} + \frac{\partial(\nu\overline{\mathrm{P}\Theta})}{\partial \varrho} + \frac{1}{\varrho}\frac{\partial(\nu\overline{\Theta^2})}{\partial \theta} + \frac{\partial(\nu\overline{\Theta Z})}{\partial z} + \frac{2}{\varrho}\nu\overline{\mathrm{P}\Theta} = \frac{\nu}{\varrho}\frac{\partial U}{\partial \theta},\\
&\frac{\partial(\nu Z_0)}{\partial t} + \frac{\partial(\nu\overline{\mathrm{P}Z})}{\partial \varrho} + \frac{1}{\varrho}\frac{\partial(\nu\overline{\Theta Z})}{\partial \theta} + \frac{\partial(\nu\overline{Z^2})}{\partial z} + \frac{1}{\varrho}\nu\overline{\mathrm{P}Z} = \nu\frac{\partial U}{\partial z}.
\end{aligned}\right\}\quad (8.28)$$

To obtain the hydrodynamical equations we must proceed as in § 3 by expressing the total velocity moments $\overline{\mathrm{P}^2}$, $\overline{\mathrm{P}\Theta}$, ... in terms of the centroid velocities $\mathrm{P}_0, \Theta_0, Z_0$ and the residual velocities P', Θ', Z'. Since

$$\mathrm{P} = \mathrm{P}_0 + \mathrm{P}', \quad \Theta = \Theta_0 + \Theta', \quad Z = Z_0 + Z', \quad (8.29)$$

we have

$$\overline{P^2} = P_0^2 + \overline{P'^2}, \quad \overline{P\Theta} = P_0\Theta_0 + \overline{P'\Theta'}, \quad \text{etc.,} \tag{8.30}$$

for, by the definition of residual velocities, $\overline{P'} = \overline{\Theta'} = \ldots = 0$. Substitution of (8.30) in (8.28), using the relations

$$\left.\begin{aligned}
\frac{\partial(\nu P_0)}{\partial t} &= P_0 \frac{\partial \nu}{\partial t} + \nu \frac{\partial P_0}{\partial t},\\
\frac{\partial(\nu P_0^2)}{\partial \varrho} &= P_0 \frac{\partial(\nu P_0)}{\partial \varrho} + \nu P_0 \frac{\partial P_0}{\partial \varrho},\\
\frac{\partial(\nu P_0 \Theta_0)}{\partial \theta} &= P_0 \frac{\partial(\nu \Theta_0)}{\partial \theta} + \nu \Theta_0 \frac{\partial P_0}{\partial \theta}, \quad \ldots,
\end{aligned}\right\} \tag{8.31}$$

and the equation of continuity (8.27), shows that terms containing the factors P_0, Θ_0 and Z_0 cancel, leaving

$$\left.\begin{aligned}
&\frac{\partial P_0}{\partial t} + P_0 \frac{\partial P_0}{\partial \varrho} + \frac{\Theta_0}{\varrho} \frac{\partial P_0}{\partial \theta} + Z_0 \frac{\partial P_0}{\partial z} - \frac{\Theta_0^2}{\varrho}\\
&\qquad + \frac{1}{\nu}\left[\frac{\partial p_{\varrho\varrho}}{\partial \varrho} + \frac{1}{\varrho}\frac{\partial p_{\varrho\theta}}{\partial \theta} + \frac{\partial p_{\varrho z}}{\partial z} - \frac{p_{\theta\theta}}{\varrho} + \frac{p_{\varrho\varrho}}{\varrho}\right] = \frac{\partial U}{\partial \varrho},\\
&\frac{\partial \Theta_0}{\partial t} + P_0 \frac{\partial \Theta_0}{\partial \varrho} + \frac{\Theta_0}{\varrho} \frac{\partial \Theta_0}{\partial \theta} + Z_0 \frac{\partial \Theta_0}{\partial z} + \frac{1}{\varrho} P_0 \Theta_0\\
&\qquad + \frac{1}{\nu}\left[\frac{\partial p_{\varrho\theta}}{\partial \varrho} + \frac{1}{\varrho}\frac{\partial p_{\theta\theta}}{\partial \theta} + \frac{\partial p_{\theta z}}{\partial z} + \frac{2}{\varrho} p_{\varrho\theta}\right] = \frac{1}{\varrho}\frac{\partial U}{\partial \theta},\\
&\frac{\partial Z_0}{\partial t} + P_0 \frac{\partial Z_0}{\partial \varrho} + \frac{\Theta_0}{\varrho} \frac{\partial Z_0}{\partial \theta} + Z_0 \frac{\partial Z_0}{\partial z}\\
&\qquad + \frac{1}{\nu}\left[\frac{\partial p_{\varrho z}}{\partial \varrho} + \frac{1}{\varrho}\frac{\partial p_{\theta z}}{\partial \theta} + \frac{\partial p_{zz}}{\partial z} + \frac{1}{\varrho} p_{\varrho z}\right] = \frac{\partial U}{\partial z},
\end{aligned}\right\} \tag{8.32}$$

with the notation

$$\begin{aligned}
p_{\varrho\varrho} &= \nu \overline{P'^2}, \quad p_{\varrho\theta} = p_{\theta\varrho} = \nu \overline{P'\Theta'}, \quad p_{\varrho z} = p_{z\varrho} = \nu \overline{P'Z'},\\
p_{\theta\theta} &= \nu \overline{\Theta'^2}, \quad p_{\theta z} = p_{z\theta} = \nu \overline{\Theta' Z'}, \quad p_{zz} = \nu \overline{Z'^2}.
\end{aligned} \tag{8.33}$$

Equations (8.32) with the equation of continuity (8.27) form the hydrodynamical equations of motion of centroids in cylindrical coordinates.

It is easily shown, in the same way as in § 4, that the terms in brackets in eqns. (8.32) correspond to the kinematic acceleration of the centroids.

Let us now derive the hydrodynamical equations in spherical coordinates. The Boltzmann equation is in these coordinates (5.20)

$$\frac{\partial f}{\partial r} + R\frac{\partial f}{\partial r} + \frac{\Theta}{r}\frac{\partial f}{\partial \theta} + \frac{\Phi}{r\sin\theta}\frac{\partial f}{\partial \varphi} + \left(\frac{\Theta^2+\Phi^2}{r} + \frac{\partial U}{\partial r}\right)\frac{\partial f}{\partial R}$$
$$+ \left(-\frac{R\Theta}{r} + \frac{\Phi^2}{r}\cot\theta + \frac{1}{r}\frac{\partial U}{\partial \theta}\right)\frac{\partial f}{\partial \Theta}$$
$$+ \left(-\frac{R\Phi}{r} - \frac{\Theta\Phi}{r}\cot\theta + \frac{1}{r\sin\theta}\frac{\partial U}{\partial \varphi}\right)\frac{\partial f}{\partial \Phi} = 0. \tag{8.34}$$

We multiply this equation by the volume element in velocity space—in spherical coordinates $\mathrm{d}\Omega = \mathrm{d}R\,\mathrm{d}\Theta\,\mathrm{d}\Phi$—and integrate. Then, using integrals similar to (8.25) and (8.26), we obtain the equation of continuity in spherical coordinates:

$$\frac{\partial \nu}{\partial t} + \frac{\partial(\nu R_0)}{\partial r} + \frac{1}{r}\frac{\partial(\nu\Theta_0)}{\partial \theta} + \frac{1}{r\sin\theta}\frac{\partial(\nu\Phi_0)}{\partial \varphi} + \frac{2\nu R_0}{r} + \frac{\nu\Theta_0}{r}\cot\theta = 0. \tag{8.35}$$

Next, multiplying eqn. (8.34) successively by $R\,\mathrm{d}\Omega$, $\Theta\,\mathrm{d}\Omega$ and $\Phi\,\mathrm{d}\Omega$, and integrating with the use of (8.25) and (8.26), we have the following equations of transfer in spherical coordinates:

$$\left.\begin{aligned}
&\frac{\partial(\nu R_0)}{\partial t} + \frac{\partial(\nu\overline{R^2})}{\partial r} + \frac{1}{r}\frac{\partial(\nu\overline{\Theta R})}{\partial \theta} + \frac{1}{r\sin\theta}\frac{\partial(\nu\overline{\Phi R})}{\partial \varphi}\\
&\qquad - \frac{\nu}{r}(\overline{\Theta^2} + \overline{\Phi^2}) = \nu\frac{\partial U}{\partial r},\\
&\frac{\partial(\nu\Theta_0)}{\partial t} + \frac{\partial(\nu\overline{R\Theta})}{\partial r} + \frac{1}{r}\frac{\partial(\nu\overline{\Theta^2})}{\partial \theta} + \frac{1}{r\sin\theta}\frac{\partial(\nu\overline{\Phi\Theta})}{\partial \varphi}\\
&\qquad - \frac{\nu\overline{\Phi^2}}{r}\cot\theta + \frac{3}{r}\nu\overline{R\Theta} + \frac{\nu\overline{\Theta^2}}{r}\cot\theta = \frac{\nu}{r}\frac{\partial U}{\partial \theta},\\
&\frac{\partial(\nu\Phi_0)}{\partial t} + \frac{\partial(\nu\overline{R\Phi})}{\partial r} + \frac{1}{r}\frac{\partial(\nu\overline{\Theta\Phi})}{\partial \theta} + \frac{1}{r\sin\theta}\frac{\partial(\nu\overline{\Phi^2})}{\partial \varphi}\\
&\qquad + \frac{3}{r}\nu\overline{R\Phi} + \frac{2}{r}\nu\overline{\Theta\Phi}\cot\theta = \frac{\nu}{r\sin\theta}\frac{\partial U}{\partial \varphi}.
\end{aligned}\right\} \tag{8.36}$$

To obtain the hydrodynamic equations we express the total velocities as $R = R_0 + R'$, $\Theta = \Theta_0 + \Theta'$, $\Phi = \Phi_0 + \Phi'$, where R', Θ', Φ', are the residual-velocity components.

Then calculations analogous to those given above lead to the three equations

$$\left.\begin{aligned}
&\frac{\partial R_0}{\partial t} + R_0 \frac{\partial R_0}{\partial r} + \frac{1}{r}\Theta_0 \frac{\partial R_0}{\partial \theta} + \frac{1}{r \sin\theta}\Phi_0 \frac{\partial R_0}{\partial \varphi} - \frac{1}{r}(\Theta_0^2 + \Phi_0^2) \\
&\quad + \frac{2}{r} R_0^2 + \frac{R_0\Theta_0}{r}\cot\theta + \frac{1}{\nu}\left[\frac{\partial p_{rr}}{\partial r} + \frac{1}{r}\frac{\partial p_{r\theta}}{\partial \theta} + \frac{1}{r\sin\theta}\frac{\partial p_{r\varphi}}{\partial \varphi}\right. \\
&\quad \left. - \frac{1}{r}(p_{\theta\theta} + p_{\varphi\varphi}) + \frac{2p_{rr}}{r} + \frac{p_{r\theta}}{r}\cot\theta\right] = \frac{\partial U}{\partial r}, \\
&\frac{\partial \Theta_0}{\partial t} + R_0 \frac{\partial \Theta_0}{\partial r} + \frac{1}{r}\Theta_0 \frac{\partial \Theta_0}{\partial \theta} + \frac{1}{r\sin\theta}\Phi_0 \frac{\partial \Theta_0}{\partial \varphi} \\
&\quad - \frac{\Phi_0^2}{r}\cot\theta + \frac{1}{r}R_0\Theta_0 + \frac{1}{\nu}\left[\frac{\partial p_{r\theta}}{\partial r} + \frac{1}{r}\frac{\partial p_{\theta\theta}}{\partial \theta}\right. \\
&\quad \left. + \frac{1}{r\sin\theta}\frac{\partial p_{\varphi\theta}}{\partial \varphi} - \frac{p_{\varphi\varphi}}{r}\cot\theta + \frac{3}{2}p_{r\theta} + \frac{p_{\theta\theta}}{r}\cot\theta\right] = \frac{1}{r}\frac{\partial U}{\partial \theta}, \\
&\frac{\partial \Phi_0}{\partial t} + R_0 \frac{\partial \Phi_0}{\partial r} + \frac{1}{r}\Theta_0 \frac{\partial \Phi_0}{\partial \theta} + \frac{1}{r\sin\theta}\Phi_0 \frac{\partial \Phi_0}{\partial \varphi} + \frac{1}{r}R_0\Phi_0 \\
&\quad + \frac{1}{r}\Theta_0\Phi_0\cot\theta + \frac{1}{\nu}\left[\frac{\partial p_{\varphi r}}{\partial r} + \frac{1}{r}\frac{\partial p_{\varphi\theta}}{\partial \theta} + \frac{1}{r\sin\theta}\frac{\partial p_{\varphi\varphi}}{\partial \varphi}\right. \\
&\quad \left. + \frac{3}{r}p_{r\varphi} + \frac{2}{r}p_{\theta\varphi}\cot\theta\right] = \frac{1}{r\sin\theta}\frac{\partial U}{\partial \varphi},
\end{aligned}\right\} \quad (8.37)$$

where

$$\begin{aligned}
&p_{rr} = \nu\overline{R'^2}, \quad p_{r\theta} = p_{\theta r} = \nu\overline{R'\Theta'}, \quad p_{r\varphi} = p_{\varphi r} = \nu\overline{R'\Phi'}, \\
&p_{\theta\theta} = \nu\overline{\Theta'^2}, \quad p_{\theta\varphi} = p_{\varphi\theta} = \nu\overline{\Theta'\Phi'}, \quad p_{\varphi\varphi} = \nu\overline{\Phi'^2}.
\end{aligned} \quad (8.38)$$

These, together with (8.35), give the hydrodynamical equations of motion of centroids in spherical coordinates. As before, the terms in brackets in eqns. (8.37) represent the kinematic acceleration of the centroids.

§ 6. The Hydrodynamical Equations for a Rotationally Symmetrical System in a Steady State

The most interesting systems from the astronomical point of view are systems like the galaxies, in a steady state and possessing rotational symmetry. For such systems the Boltzmann equation has the form (5.125)

$$\mathrm{P}\frac{\partial f}{\partial \varrho} + Z\frac{\partial f}{\partial z} + \left(\frac{\partial U}{\partial \varrho} + \frac{\Theta^2}{\varrho}\right)\frac{\partial f}{\partial \mathrm{P}} - \frac{\mathrm{P}\Theta}{\varrho}\frac{\partial f}{\partial \Theta} + \frac{\partial U}{\partial z}\frac{\partial f}{\partial Z} = 0. \quad (8.39)$$

The phase density is a function of the energy integral and the area integral:

$$f = f[\mathrm{P}^2 + \Theta^2 + Z^2 - 2U(\varrho, z), \varrho\Theta]; \tag{8.40}$$

as in § 3, it is assumed to satisfy the conditions

$$\lim_{V\to\infty} f = 0, \qquad \lim_{V\to\infty} f V = 0. \tag{8.41}$$

To derive the hydrodynamical equations we must multiply eqn. (8.39) successively by 1, P, Θ and Z and integrate over all velocity space.† Then from (8.40) and (8.41) it easily follows that the first and third of these integrations lead to identities.

According to (8.40), f is an even function of P and Z. Hence $\int \mathrm{P}Zf\,\mathrm{d}\Omega = 0$, $\int \mathrm{P}\Theta Z(\partial f/\partial\Theta)\,\mathrm{d}\Omega = -\int \mathrm{P}Zf\,\mathrm{d}\Omega = 0$. Using these relations and formulae (8.25), (8.26), we obtain from the second and fourth integrations the equations

$$\begin{aligned} \frac{\partial(\nu\overline{\mathrm{P}^2})}{\partial\varrho} - \left(\nu\frac{\partial U}{\partial\varrho} + \frac{1}{\varrho}\nu\overline{\Theta^2}\right) + \frac{1}{\varrho}\nu\overline{\mathrm{P}^2} &= 0, \\ \frac{\partial(\nu\overline{Z^2})}{\partial z} &= \nu\frac{\partial U}{\partial z}. \end{aligned} \tag{8.42}$$

These equations could, as already stated, be obtained from the more general eqns. (8.28), since in the present case $\mathrm{P}_0 = Z_0 = \overline{\mathrm{P}\Theta} = \overline{\mathrm{P}Z} = \overline{\Theta Z} = 0$. Such a derivation differs formally from the one given in the text, in that the condition $\lim_{V\to\infty} f V^2 = 0$ is imposed as well as (8.41). From the symmetry of the function f with respect to P and Z (cf. (8.40)) it follows that $\overline{\mathrm{P}^2} = \overline{Z^2}$. Putting

$$\begin{aligned} p &= \int \mathrm{P}^2 f\,\mathrm{d}\Omega = \nu\overline{\mathrm{P}^2} = \nu\overline{Z^2}, \\ q &= \int \Theta^2 f\,\mathrm{d}\Omega = \nu\overline{\Theta^2}, \end{aligned} \tag{8.43}$$

we obtain from (8.42) the equations of transfer for a rotationally symmetrical stellar system in a steady state:

$$\begin{aligned} \frac{\partial p}{\partial\varrho} + \frac{p-q}{\varrho} &= \nu\frac{\partial U}{\partial\varrho}, \\ \frac{\partial p}{\partial z} &= \nu\frac{\partial U}{\partial z}. \end{aligned} \tag{8.44}$$

These extremely simple equations were first derived by Jeans [44] in 1922. They are often called "hydrodynamical equations", but are actually equations of transfer.

To obtain the hydrodynamical equations, the velocity of the centroid must be introduced in (8.44). It has been shown in § 1 that, for a steady-state system with rotational symmetry, the centroids move in circles, i.e. $\mathrm{P}_0 = Z_0 = 0$, $\Theta_0 = \Theta_0(\varrho, z)$. In the present case, therefore,

† Equations (8.27), (8.28) could be used, but we prefer to give a direct proof.

$$\mathrm{P} = \mathrm{P}', \quad Z = Z', \quad \Theta = \Theta_0 + \Theta', \tag{8.45}$$

and

$$\overline{\mathrm{P}^2} = \overline{\mathrm{P}'^2} = \sigma_\varrho^2, \quad \overline{Z^2} = \overline{Z'^2} = \sigma_z^2, \quad \overline{\Theta^2} = \overline{\Theta_0^2} + \overline{\Theta'^2} = \Theta_0^2 + \sigma_\theta^2, \tag{8.46}$$

where σ_ϱ^2, σ_θ^2, σ_z^2 are the velocity variances in the corresponding directions. From (8.45) and (8.46)

$$p = \nu\sigma_\varrho^2 = \nu\sigma_z^2, \quad q = \nu\Theta_0^2 + \nu\sigma_\theta^2. \tag{8.47}$$

The equal quantities σ_ϱ^2 and σ_z^2 characterise the distribution of residual velocities in the meridian plane, and σ_θ^2 that in the direction perpendicular to this plane. We shall therefore call σ_ϱ^2 the *radial variance* and σ_θ^2 the *transverse variance*. Substituting (8.47) in Jeans' eqns. (8.44), we have the hydrodynamical equations for a rotationally symmetrical stellar system in a steady state:

$$\begin{aligned} \frac{\partial(\nu\sigma_\varrho^2)}{\partial\varrho} + \frac{\nu(\sigma_\varrho^2 - \sigma_\theta^2)}{\varrho} &= \nu\frac{\Theta_0^2}{\varrho} + \nu\frac{\partial U}{\partial\varrho}, \\ \frac{\partial(\nu\sigma_\varrho^2)}{\partial z} &= \nu\frac{\partial U}{\partial z}. \end{aligned} \tag{8.48}$$

The quantities U and ν are related by Poisson's equation, which in cylindrical coordinates, with U independent of θ, is

$$\frac{1}{\varrho}\frac{\partial}{\partial\varrho}\left(\varrho\frac{\partial U}{\partial\varrho}\right) + \frac{\partial^2 U}{\partial z^2} = -4\pi G m\nu, \tag{8.49}$$

where G is the gravitational constant and m the average mass of a star.

Thus we have only three eqns. (8.48), (8.49) to determine the five unknowns ν, U, Θ_0, σ_ϱ^2, σ_θ^2, and the problem of solving the hydrodynamical equations is indeterminate. This indeterminacy is due to the important fact that, besides the hydrodynamical quantities ν, U and Θ_0, the equations involve also the essentially statistical quantities σ_ϱ^2 and σ_θ^2.

With a purely hydrodynamical approach, the quantities on the left-hand sides of eqns. (8.48) are unknowns. They express the kinematic acceleration of the centroids, and so the difficulty arises, in terms of mechanics, becausethe hydrodynamical problem involves an essentially statistical factor, the kinematic acceleration.

Thus the hydrodynamical approach to solving the problems of stellar dynamics is inadequate, and statistical methods also must be used. This we shall do in Chapters IX and X, in which the problem of the dynamics of stellar systems will be solved by combining the hydrodynamical and the statistical approach.

§ 7. The Explanation of the Asymmetry of Stellar Motions in the Galaxy

It has just been shown that a purely hydrodynamical approach does not suffice to solve completely the problem of the dynamics of stellar systems. Some general conclusions may, however, be obtained in this way.

Let us ascertain, for example, the effect of the kinematic acceleration on the motion of the centroids in a system with rotational symmetry. The circular velocity Θ_0 of the centroid appears only in the first of the hydrodynamical eqns. (8.48). In this equation we substitute for the derivative of the potential its expression in terms of the circular velocity Θ_c of a particle at distance ϱ:

$$\partial U/\partial \varrho = -\Theta_c^2/\varrho. \tag{8.50}$$

Then

$$\Theta_0^2 - \Theta_c^2 = \left[\frac{1}{\nu}\frac{\partial(\nu\sigma_\varrho^2)}{\partial \varrho} + \frac{\sigma_\varrho^2 - \sigma_\theta^2}{\varrho}\right]\varrho. \tag{8.51}$$

From this formula we see that the centroid velocity Θ_0 is in general not equal to the circular velocity Θ_c of a particle at the same distance ϱ, i.e. a centroid undergoes a kinematic acceleration (positive or negative) relative to a particle moving in a circular orbit.

This derivation is valid if the stellar population in the system considered is uniform, since then there will be in each macroscopic volume element a single centroid common to all stars, with velocity Θ_0. If, however, the system comprises several sub-systems with different residual-velocity dispersions, then the kinematic acceleration given by the right-hand side of formula (8.51) will be different for each sub-system. (The kinematic acceleration is, of course, $\Theta_0 - \Theta_c$, not $\Theta_0^2 - \Theta_c^2$, but the latter difference can equally well be used to characterise the acceleration.) Hence every sub-system will have its own centroid velocity. Thus, in a rotationally symmetrical steady-state system whose star population is kinematically inhomogeneous, the centroid velocity for each sub-system is determined by the stellar velocity dispersion in that sub-system.

As an example of the application of these results to specific problems of stellar dynamics, let us consider the theoretical interpretation of the asymmetry of stellar motions in the Galaxy. As has been stated in Chapter III, the asymmetry of stellar velocities is the phenomenon that the centroids of sub-systems with different residual-velocity dispersions are displaced relative to one another along the *axis of asymmetry*, which is a straight line perpendicular to the direction of the galactic centre. The value of Δv, the difference in the solar velocity for different sub-systems, is given by

$$\Delta v = a\sigma^2 + b, \tag{8.52}$$

where σ^2 is the residual-velocity variance for a given sub-system, and a and b are constants, first obtained observationally by Strömberg [125] in 1922.

Observation shows that the residual velocities of stars in the Galaxy are distributed approximately in accordance with an ellipsoidal law (more precisely, with Schwarzschild's law). In this case, as we have seen, the radial variance σ_ϱ^2 is constant, and the transverse variance σ_θ^2 is

$$\sigma_\theta^2 = \sigma_\varrho^2/(1 + k_2\varrho^2). \tag{8.53}$$

Hence the relation (8.51) for the Galaxy can be written

$$\Theta_0^2 - \Theta_c^2 = \varrho\left(\frac{\partial \log \nu}{\partial \varrho} + \frac{k_2\varrho}{1 + k_2\varrho^2}\right)\sigma_\varrho^2. \tag{8.54}$$

Denoting the ratio of the variances by

$$\lambda^2 = \sigma_\varrho^2/\sigma_\theta^2 = 1 + k_2\varrho^2, \tag{8.55}$$

we can write more concisely

$$\Theta_0^2 - \Theta_c^2 = \varrho\sigma_\varrho^2\,\partial \log(\lambda\nu)/\partial\varrho. \tag{8.56}$$

The right-hand side of this equation represents the kinematic acceleration of the centroids. The sign of the acceleration evidently depends on the rate at which the star density ν decreases with increasing distance ϱ. If $\lambda\nu$ is a decreasing function of ϱ, then by (8.56) the kinematic acceleration is negative. Thus we have the following result: if the star density in the Galaxy, or in any sub-system, decreases in such a way that the product $\lambda\nu$ is a decreasing function of ϱ, then the centroids will undergo a kinematic retardation.

Observation shows that this condition seems to hold for the whole Galaxy and also for the individual sub-systems. For any actual sub-system, therefore, $\Theta_0 < \Theta_c$. If we imagine a sequence of sub-systems with decreasing residual velocities, then the circular velocity of the centroid is, by (8.56), equal to Θ_c when the residual velocities are zero. Hence Θ_c is the limiting circular velocity for plane sub-systems, which have only small residual stellar velocities.

In stellar astronomy, the kinematic retardation is customarily measured by means of Strömberg's asymmetry of stellar motions $S = \Theta_c - \Theta_0$. This is easily expressed in terms of the right-hand side of (8.56). We shall suppose that S is small in comparison with the circular velocity Θ_c of a particle. Then we have, as far as the first-order terms in S/Θ_c,

$$\frac{\Theta_c^2 - \Theta_0^2}{\Theta_c^2} = \frac{2\Theta_0}{\Theta_c}\cdot\frac{S}{\Theta_c} + \left(\frac{S}{\Theta_c}\right)^2 \approx \frac{2S\Theta_0}{\Theta_c^2}, \tag{8.57}$$

i.e.

$$\Theta_c^2 - \Theta_0^2 \approx 2S\Theta_0. \tag{8.58}$$

From (8.56) and (8.58) we have an approximate formula for the kinematic retardation:

$$S = -\frac{\varrho}{2\Theta_0}\frac{\partial \log(\lambda\nu)}{\partial\varrho}\sigma_\varrho^2. \tag{8.59}$$

This formula gives a theoretical interpretation of the asymmetry of stellar motions.

At each point in the system ϱ, Θ_0, ν and λ are constants. Hence (8.59) becomes

$$S = a\sigma_\varrho^2. \tag{8.60}$$

This differs from the usual Strömberg asymmetry formula

$$S = a\sigma^2 + b \tag{8.61}$$

in that the term b is absent. This is easily explained by the fact that in the empirical formula (8.61) the deviation S is defined, not with respect to the circular velocity Θ_c of a particle as in the derivation of (8.56) and (8.60), but with respect to the velocity of the centroid in the neighbourhood of the Sun, which we denote by Θ_S.

If the deviation of the centroid velocity in formula (8.61) is denoted by S' to distinguish it from (8.60), then

$$S' = \Theta_0 - \Theta_S = (\Theta_0 - \Theta_c) + (\Theta_c - \Theta_S). \tag{8.62}$$

The expression in the first parenthesis is just S (8.60), and that in the second parenthesis is a constant at the point considered (the neighbourhood of the Sun). Putting therefore $\Theta_c - \Theta_S = b$, we obtain (8.61).

Thus the observed Strömberg asymmetry formula is entirely explained. It should be remembered, however, that our arguments are valid only for values of S which are small compared with Θ_c.

We may also derive two minor consequences of the fundamental formula (8.56). For plane sub-systems, $\Theta_c^2 - \Theta_0^2$ is small, and we can put approximately $\partial(\lambda\nu)/\partial\varrho = 0$, whence $\lambda\nu = c(z)$, or

$$\nu = c(z)/\sqrt{(1 + k_2\varrho^2)}. \tag{8.63}$$

Thus we conclude, in agreement with observation, that the star density for plane sub-systems decreases slowly away from the centre of the Galaxy.

Secondly, replacing Θ_c^2 in (8.56) by $-\varrho\,\partial U/\partial\varrho$, we can write this equation as

$$\sigma_\varrho^2 \frac{\partial \log(\lambda\nu)}{\partial\varrho} - \frac{\partial U}{\partial\varrho} = \frac{\Theta_0^2}{\varrho} > 0,$$

and we can put approximately

$$\sigma_\varrho^2 \frac{\partial \log(\lambda\nu)}{\partial\varrho} = \frac{\partial U}{\partial\varrho} \tag{8.64}$$

for the spherical sub-systems, where Θ_0 is small.

CHAPTER IX

DYNAMICS OF SPHERICAL STELLAR SYSTEMS

§ 1. The Basic Equations for Spherical Systems

THE simplest and also the most important type of stellar system is that in which the distribution of mass is spherically symmetrical. Such systems include the globular clusters and also some elliptical nebulae whose equatorial compression is zero (type E0). Evidently these systems can have no appreciable rotation.

To investigate such systems it is natural to use spherical coordinates r, θ, φ, with the origin at the centre of the system, and the corresponding linear velocities R, Θ, Φ as shown in Fig. 27 (Chapter V, § 2).†

In these coordinates the differential equation for the phase density (the Boltzmann equation) is given by formula (5.20). Assuming that the system is in a steady state ($\partial f/\partial t = 0$), we can write this equation as

$$R\frac{\partial f}{\partial r} + \frac{\Theta}{r}\frac{\partial f}{\partial \theta} + \frac{\Phi}{r\sin\theta}\frac{\partial f}{\partial \varphi} + \left(\frac{\Theta^2+\Phi^2}{r} + \frac{\partial U}{\partial r}\right)\frac{\partial f}{\partial R}$$
$$+ \left(\frac{\Phi^2}{r}\cot\theta - \frac{R\Theta}{r} + \frac{1}{r}\frac{\partial U}{\partial \theta}\right)\frac{\partial f}{\partial \Theta}$$
$$+ \left(-\frac{R\Phi}{r} - \frac{\Theta\Phi}{r}\cot\theta + \frac{1}{r\sin\theta}\frac{\partial U}{\partial \varphi}\right)\frac{\partial f}{\partial \Phi} = 0. \qquad (9.1)$$

In a spherically symmetrical system the phase density f and the potential U must be independent of θ and φ, i.e. $\partial f/\partial\theta = \partial f/\partial\varphi = 0$, $\partial U/\partial\theta = \partial U/\partial\varphi = 0$. Hence (9.1) becomes

$$R\frac{\partial f}{\partial r} + \left\{\frac{\mathrm{d}U}{\mathrm{d}r} + \frac{\Theta^2+\Phi^2}{r}\right\}\frac{\partial f}{\partial R}$$
$$+ \left\{\frac{\Phi^2}{r}\cot\theta - \frac{R\Theta}{r}\right\}\frac{\partial f}{\partial \Theta} + \left\{-\frac{\Theta\Phi}{r}\cot\theta - \frac{R\Phi}{r}\right\}\frac{\partial f}{\partial \Phi} = 0. \qquad (9.2)$$

We shall call this equation the fundamental equation of stellar dynamics for spherical systems.

† The term *radial velocity* which will be used to denote R should not be confused with the term used in observational astronomy to denote velocity in the line of sight.

According to the above assumption, the phase density must be independent of the coordinate θ, and, by equating the coefficient of $\cot\theta$ to zero, we obtain the two equations

$$\Phi \frac{\partial f}{\partial \Theta} - \Theta \frac{\partial f}{\partial \Phi} = 0,$$
$$R \frac{\partial f}{\partial R} + \left\{ \frac{dU}{dr} + \frac{\Theta^2 + \Phi^2}{r} \right\} \frac{\partial f}{\partial R} - \frac{R}{r} \left\{ \Theta \frac{\partial f}{\partial \Theta} + \Phi \frac{\partial f}{\partial \Phi} \right\} = 0. \tag{9.3}$$

It is easy to see that the first of these equations implies that f depends on the square of the transverse velocity

$$T^2 = \Theta^2 + \Phi^2; \tag{9.4}$$

for the general theory of such equations shows that f must be an arbitrary function of the integral of the ordinary equation $d\Theta/\Phi = d\Phi/(-\Theta)$ or $\Theta\, d\Theta + \Phi\, d\Phi = 0$, which gives on integration

$$\Theta^2 + \Phi^2 = \text{constant}. \tag{9.5}$$

Using Jeans' theorem, we conclude that, in spherical systems, the phase density is a function of one space coordinate (the radius r) and two velocities, the radial velocity R and the squared transverse velocity $T^2 : f = f(r, R, T^2)$.

Eliminating the variables Θ and Φ from the second eqn. (9.3) by means of the obvious relation

$$\Theta \frac{\partial f}{\partial \Theta} + \Phi \frac{\partial f}{\partial \Phi} = T \frac{\partial f}{\partial T}, \tag{9.6}$$

we obtain the fundamental equation in the concise form

$$R \frac{\partial f}{\partial r} + \left\{ \frac{dU}{dr} + \frac{T^2}{r} \right\} \frac{\partial f}{\partial R} - \frac{RT}{r} \frac{\partial f}{\partial T} = 0. \tag{9.7}$$

The corresponding ordinary differential equations are

$$\frac{dr}{R} = \frac{dR}{dU/dr + T^2/r} = \frac{dT}{-RT/r}. \tag{9.8}$$

Two integrable combinations are easily found. The first and third members of (9.8) give

$$\frac{dr}{r} + \frac{dT}{T} = 0. \tag{9.9}$$

Hence we have the angular momentum integral (area integral) about the centre of the system:

$$rT = h. \tag{9.10}$$

A composition of the second and third members gives

$$\frac{dr}{R} = \frac{R\,dR + T\,dT}{R\,dU/dr}, \tag{9.11}$$

or

$$R\,dR + T\,dT - (dU/dr)\,dr = 0. \tag{9.12}$$

The energy integral is thus obtained:

$$\tfrac{1}{2}(R^2 + T^2) - U = H. \tag{9.13}$$

Since, as shown above, f must depend on T^2, we have the result that in spherical systems the phase density is a function of only two integrals of the motion, the squared angular momentum and the energy:

$$f = f(R^2 + T^2 - 2U, r^2 T^2). \tag{9.14}$$

It is evident that the replacement of T^2 by T would involve taking the square root of the angular-momentum integral, and would therefore lead to many-valuedness.

The above proof is due to Shiveshwarkar [120]. The same result was derived directly from Jeans' theorem in Chapter V, § 6: formula (5.53) becomes (9.14) after the substitution $\Theta^2 + \Phi^2 = T^2$.

Since the phase density (9.14) is an even function of R and T, the mean values of R and T are zero. Thus we again conclude (cf. Chapter VIII, § 1) that, in a spherical stellar system in a steady state, no macroscopic motions are possible.

Hence, in particular, it follows that in this case the equations of transfer and the hydrodynamical equations are identical, and the equation of continuity is identically satisfied.

The hydrodynamical equation (equation of transfer) for spherical systems can be derived as a very simple particular case of the hydrodynamical equations in spherical coordinates (8.37), but it is also easily obtained by direct averaging of the fundamental eqn. (9.7). We shall suppose that the phase density satisfies the conditions

$$\lim_{V\to\infty} f = 0, \quad \lim_{V\to\infty} Vf = 0, \tag{9.15}$$

where V is the speed. To obtain the hydrodynamical equation we multiply eqn. (9.7) by R and integrate over all velocity space, using the properties of the phase density expressed by eqn. (9.14).

Denoting a volume element in velocity space by $\mathrm{d}\Omega$, we have

$$\int R^2 \frac{\partial f}{\partial r}\,\mathrm{d}\Omega = \frac{\partial}{\partial r}\int R^2 f\,\mathrm{d}\Omega = \frac{\mathrm{d}(\nu \overline{R^2})}{\mathrm{d}r}. \tag{9.16}$$

The integrals

$$\int R(\partial f/\partial R)\,\mathrm{d}\Omega \quad \text{and} \quad \int T^2 R(\partial f/\partial R)\,\mathrm{d}\Omega \tag{9.17}$$

also appear. Since by (9.15) $\lim Rf = 0$ for $R \to \pm\infty$ and by the definition of the star density $\int f\,\mathrm{d}\Omega = \nu$, we obtain, on integration by parts,

$$\left.\begin{aligned}\int R\frac{\partial f}{\partial R}\,\mathrm{d}\Omega &= \int R\frac{\partial f}{\partial R}\,\mathrm{d}R\,\mathrm{d}\Theta\,\mathrm{d}\Phi\\ &= \int [Rf]_{-\infty}^{\infty}\,\mathrm{d}\Theta\,\mathrm{d}\Phi - \int f\,\mathrm{d}\Omega = -\nu,\\ \int T^2 R\frac{\partial f}{\partial R}\,\mathrm{d}\Omega &= \int T^2\,[Rf]_{-\infty}^{\infty}\,\mathrm{d}\Theta\,\mathrm{d}\Phi - \int T^2 f\,\mathrm{d}\Omega = -\nu\overline{T^2}.\end{aligned}\right\} \tag{9.18}$$

Finally, the integral $\int R^2 T(\partial f/\partial T)\,\mathrm{d}\Omega$ is evaluated as follows. From (9.15), $\lim \Theta f = \lim \Phi f = 0$ for $\Theta, \Phi \to \pm\infty$ respectively; using (9.6), we therefore find

$$\int R^2 T \frac{\partial f}{\partial T}\,\mathrm{d}\Omega = \int R^2 \Theta \frac{\partial f}{\partial \Theta}\,\mathrm{d}\Omega + \int R^2 \Phi \frac{\partial f}{\partial \Phi}\,\mathrm{d}\Omega$$

$$= \int R^2\,[\Theta f]_{-\infty}^{\infty}\,\mathrm{d}R\,\mathrm{d}\Phi - \int R^2 f\,\mathrm{d}\Omega + \int R^2 [\Phi f]_{-\infty}^{\infty}\,\mathrm{d}R\,\mathrm{d}\Theta$$

$$- \int R^2 f\,\mathrm{d}\Omega = -2\nu\overline{R^2}. \tag{9.19}$$

Thus the hydrodynamical equation for spherical systems is

$$\frac{\mathrm{d}(\nu\overline{R^2})}{\mathrm{d}r} - \nu\frac{\mathrm{d}U}{\mathrm{d}r} - \frac{\nu\overline{T^2}}{r} + \frac{2\nu\overline{R^2}}{r} = 0. \tag{9.20}$$

Equation (9.20) can be obtained from the first eqn. (8.36) by using the following results valid in the present case: $\overline{R} = \overline{\Theta} = \overline{\Phi} = \overline{R\Theta} = \overline{R\Phi} = \overline{\Theta\Phi} = 0$, $\partial U/\partial\theta = \partial U/\partial\varphi = \partial f/\partial\theta = \partial f/\partial\varphi = 0$ and $\overline{\Theta^2} + \overline{\Phi^2} = \overline{T^2}$. It is easily seen that the remaining eqns. (8.36) and the equation of continuity (8.35) become identities. The direct derivation of eqn. (9.20) given in the text differs formally from that using (8.36) in that we do not here impose the condition $\lim V^2 f = 0$ as $V \to \infty$.

With Jeans' notation

$$p = \nu\overline{R^2}, \quad q = \nu\overline{T^2}, \tag{9.21}$$

eqn. (9.20) becomes finally [83]

$$\frac{\mathrm{d}p}{\mathrm{d}r} + \frac{2p - q}{r} = \nu\frac{\mathrm{d}U}{\mathrm{d}r}. \tag{9.22}$$

Since there are no internal motions in a stellar system under the conditions assumed, the quantities p and q given by (9.21) are proportional to the squared semiaxes of the velocity ellipsoid. The quantities $\overline{R^2}$ and $\overline{T^2}$ are just the variances of the radial and transverse residual velocities.

The hydrodynamical eqn. (9.22) involves four unknown functions: the density ν, the potential U, and Jeans' functions p and q. The density and the potential are related by Poisson's equation, which in our notation is

$$\frac{1}{r^2}\frac{\mathrm{d}}{\mathrm{d}r}\left(r^2\frac{\mathrm{d}U}{\mathrm{d}r}\right) = -4\pi G m\nu, \tag{9.23}$$

m being the mean mass of a star. Hence we see that, without additional information, the problem is indeterminate.

One possible method of solution is to use supplementary hypotheses concerning the nature of the unknown functions, and this has in fact been the procedure used up to the present time. We shall now consider some of the most general and most important results.

§ 2. The Use of the Theory of Polytropic Spheres

Let us assume that the velocity distribution is spherically symmetrical at every point in a stellar system. Then the dispersions of the velocities R, Θ and Φ will be the same. Since $p = \nu\overline{R^2}$ and $q = \nu\overline{T^2} = \nu(\overline{\Theta^2} + \overline{\Phi^2})$, it follows that $q = 2p$. Hence eqn. (9.22) becomes the usual equation of hydrostatic equilibrium:

$$\frac{1}{\nu}\frac{\mathrm{d}p}{\mathrm{d}r} = \frac{\mathrm{d}U}{\mathrm{d}r}. \tag{9.24}$$

The application of this equation to gas spheres was first investigated by Emden [31].

In spherical symmetry the gravitational force is

$$\mathrm{d}U/\mathrm{d}r = -GM(r)/r^2, \tag{9.25}$$

where $M(r)$ is the mass within a sphere of radius r, i.e.

$$M(r) = 4\pi m \int_0^r \nu(r)\, r^2\, \mathrm{d}r. \tag{9.26}$$

Substituting (9.25) and (9.26) in (9.24), and differentiating with respect to r in order to remove the integral, we obtain

$$\frac{\mathrm{d}}{\mathrm{d}r}\left(\frac{r^2}{\nu}\frac{\mathrm{d}p}{\mathrm{d}r}\right) = -4\pi G m \nu(r)\, r^2. \tag{9.27}$$

Thus Poisson's eqn. (9.23) is a simple consequence of eqn. (9.25).

In a spherical stellar system all physical parameters are one-valued functions of the radius. Hence any function can be expressed in terms of any other. In particular, with no loss of generality, the quantity p may be assumed to be a function of ν. Emden's method makes the particular assumption that

$$p = K\nu^{\gamma}, \tag{9.28}$$

where K and γ are constants. A medium in which the condition (9.28) holds is said to be *polytropic*.

Putting

$$\gamma = 1 + 1/n, \tag{9.29}$$

we introduce another parameter n, called the *index of the polytrope*. For convenience we use also a new unknown function θ defined by

$$\nu = \lambda\,\theta^n \quad (\lambda = \text{constant}). \tag{9.30}$$

Then

$$p = K\lambda^{(n+1)/n}\,\theta^{n+1}. \tag{9.31}$$

In terms of the dimensionless radius

$$\xi = r/\alpha, \tag{9.32}$$

where α is a constant having the dimensions of length whose value is to be chosen suitably, eqn. (9.27) becomes

$$\frac{1}{\xi^2}\frac{\mathrm{d}}{\mathrm{d}\xi}\left(\xi^2\frac{\mathrm{d}\theta}{\mathrm{d}\xi}\right)+\frac{4\pi G M\alpha^2}{\lambda^{(1-n)/n}K(n+1)}\theta^n=0.$$

If we take

$$\alpha^2 = K(n+1)\,\lambda^{(1-n)/n}/4\pi G m, \tag{9.33}$$

the equation takes the simple form

$$\frac{1}{\xi^2}\frac{\mathrm{d}}{\mathrm{d}\xi}\left(\xi^2\frac{\mathrm{d}\theta}{\mathrm{d}\xi}\right)+\theta^n=0. \tag{9.34}$$

This is called *Emden's equation.*

Let λ be taken as the density at the centre of the system; then, from (9.30), we have the boundary condition

$$\theta = 1 \quad \text{for} \quad \xi = 0. \tag{9.35}$$

For continuity we must have

$$\mathrm{d}\theta/\mathrm{d}\xi = 0 \quad \text{for} \quad \xi = 0. \tag{9.36}$$

Solutions of Emden's equation which satisfy the boundary conditions (9.35) and (9.36) are called *Emden functions.*

We shall not give a detailed discussion of the Emden functions; the theory of these functions is a mathematical problem having an extensive literature, to which the interested reader is referred. Here we shall simply note the most important results.

Evidently the only Emden functions having any physical significance are those without singularities. The following important theorem holds: every solution of Emden's equation finite at the origin has a derivative which is zero there. In other words, the second boundary condition (9.36) is a consequence of the condition of finiteness.

To prove this theorem, we use the function $\tau = \theta\xi$. If θ is finite for $\xi = 0$, then $\tau = 0$ there. We have

$$(\mathrm{d}\theta/\mathrm{d}\xi)_{\xi=0} = \lim_{\xi\to 0}\frac{\xi\,\mathrm{d}\tau/\mathrm{d}\xi - \tau}{\xi^2}.$$

But

$$\tau = (\mathrm{d}\tau/\mathrm{d}\xi)_{\xi=0}\,\xi + \tfrac{1}{2}(\mathrm{d}^2\tau/\mathrm{d}\xi^2)_{\xi=0}\,\xi^2 + \dots$$

and

$$\mathrm{d}\tau/\mathrm{d}\xi = (\mathrm{d}\tau/\mathrm{d}\xi)_{\xi=0} + (\mathrm{d}^2\tau/\mathrm{d}\xi^2)_{\xi=0}\,\xi + \dots.$$

Hence

$$(\mathrm{d}\theta/\mathrm{d}\xi)_{\xi=0} = \tfrac{1}{2}(\mathrm{d}^2\tau/\mathrm{d}\xi^2)_{\xi=0}.$$

Since

$$\mathrm{d}^2\tau/\mathrm{d}\xi^2 = \xi\,\mathrm{d}^2\theta/\mathrm{d}\xi^2 + 2\,\mathrm{d}\theta/\mathrm{d}\xi$$

$$= \frac{1}{\xi}\frac{\mathrm{d}}{\mathrm{d}\xi}\left(\xi^2\frac{\mathrm{d}\theta}{\mathrm{d}\xi}\right),$$

Emden's equation shows that $\mathrm{d}^2\tau/\mathrm{d}\xi^2 = -\xi\theta^n$. Since θ is assumed finite when $\xi = 0$, $\mathrm{d}^2\tau/\mathrm{d}\xi^2 = 0$ there. Hence $\mathrm{d}\theta/\mathrm{d}\xi = 0$ there. This completes the proof.

The theory of polytropic spheres shows that exact solutions of Emden's equation can be obtained only for $n = 0$, 1 and 5.

When $n = 0$, eqn. (9.34) becomes

$$\frac{1}{\xi^2} \frac{d}{d\xi} \left(\xi^2 \frac{d\theta_0}{d\xi} \right) = -1. \tag{9.37}$$

A repeated integration gives

$$\theta_0 = C_2 - \frac{C_1}{\xi} - \frac{1}{6} \xi^2, \tag{9.38}$$

where C_1 and C_2 are constants of integration. The condition that $\xi = 0$ is not a singularity gives $C_1 = 0$, and the boundary condition $\theta = 1$ for $\xi = 0$ gives $C_2 = 1$. The Emden function for $n = 0$ is therefore

$$\theta_0 = 1 - \frac{1}{6} \xi^2. \tag{9.39}$$

It is easily seen that the second boundary condition, $d\theta/d\xi = 0$ for $\xi = 0$, is automatically satisfied. When

$$\xi = \Xi_0 = \sqrt{6}, \tag{9.40}$$

the function θ_0 is zero, and for greater ξ it is negative. We may suppose that this point marks the outer boundary of the system, and Ξ_0 is the radius of the system.

To obtain a solution for $n = 1$, we change the unknown function by putting

$$\theta = \tau/\xi. \tag{9.41}$$

Emden's eqn. (9.34) then becomes

$$\frac{d^2\tau}{d\xi^2} + \frac{\tau^n}{\xi^{n-1}} = 0. \tag{9.42}$$

For $n = 1$ we have a simple linear equation,

$$\frac{d^2\tau}{d\xi^2} + \tau = 0. \tag{9.43}$$

The general integral of this is $\tau = C_1 \sin\xi + C_2 \cos\xi$. Hence $\theta = (C_1/\xi) \sin\xi + (C_2/\xi) \cos\xi$. For finiteness at $\xi = 0$ we must put $C_2 = 0$, and thus obtain *Ritter's solution:*

$$\theta_1 = (C_1/\xi) \sin\xi, \tag{9.44}$$

which is finite at $\xi = 0$. The first boundary condition, $\theta_1 = 1$ for $\xi = 0$, gives $C_1 = 1$. The radius Ξ of the system may be taken as the smallest positive root of the equation $\theta_1 = 0$. Evidently

$$\Xi = \pi. \tag{9.45}$$

To find a solution when $n = 5$, we use new variables defined by

$$\xi = e^{-s}, \quad \theta_5 = \frac{1}{\sqrt{2}} e^{\frac{1}{2}s} z. \tag{9.46}$$

Substitution in eqn. (9.34) with $n = 5$ gives

$$\frac{d^2 z}{ds^2} = \tfrac{1}{4} z\,(1 - z^4). \tag{9.47}$$

Multiplication by $2\,dz/ds$ and integration gives

$$(dz/ds)^2 = \tfrac{1}{4} z^2 (1 - \tfrac{1}{3} z^4) + C.$$

If the second boundary condition (9.36) is imposed, $C = 0$. In the resulting equation $2\,dz/z = \pm\sqrt{(1 - \tfrac{1}{3} z^4)}\,ds$, the plus sign must be taken, since z must increase with s. We now put

$$\tfrac{1}{3} z^4 = \sin^2 \zeta. \tag{9.48}$$

Then logarithmic differentiation gives $2\,dz/z = \cot\zeta\,d\zeta$, and we have the very simple equation $d\zeta/\sin\zeta = ds$, whence by integration

$$\tan\tfrac{1}{2}\zeta = C'\,e^s, \tag{9.49}$$

with C' a new constant of integration. To find θ_5 as a function of ξ we express $\sin\zeta$ in terms of $\cot\tfrac{1}{2}\zeta$ in (9.48). Then

$$\frac{z^2}{\sqrt{3}} = \frac{2\cot\tfrac{1}{2}\zeta}{1 + \cot^2\tfrac{1}{2}\zeta} = \frac{2C\,e^{-s}}{1 + C^2\,e^{-2s}} = \frac{2C\xi}{1 + C^2\xi^2},$$

where $C = 1/C'$.

Now substituting this expression in (9.46), we have *Schuster's solution:*

$$\theta_5 = (3C^2)^{\frac{1}{4}}/(1 + C^2\xi^2)^{\frac{1}{2}}. \tag{9.50}$$

The boundary condition $\theta_5 = 1$ for $\xi = 0$ gives $3C^2 = 1$, and the Emden function is

$$\theta_5 = 1/(1 + \tfrac{1}{3}\xi^2)^{\frac{1}{2}}. \tag{9.51}$$

This function decreases monotonically with increasing ξ and satisfies the boundary condition on the derivative. The radius of the system is infinite.

The corresponding star density

$$\nu_5 = \theta_5^5 = 1/(1 + \tfrac{1}{3}\xi^2)^{5/2} \tag{9.52}$$

is said to obey *Schuster's law*. In the first quarter of this century a considerable literature was written about Schuster's law and its application to globular clusters, in connection with the development at that time by Emden and others of the theory of equilibrium states of gas spheres. This approach to stellar dynamics was not very fruitful because, firstly, the application of the equation of hydrostatic equilibrium (9.22) and Emden's eqn. (9.34) to stellar systems was justified only by the purely outward analogy between star clusters and gas spheres and, secondly, Schuster's law is distinguished from an infinity of other solutions of Emden's equation only by the fact that for $n = 5$ the solution can be expressed in terms of elementary functions.

I. N. Minin [75] has recently shown that Schuster's law must hold in a spherical system which is in a stable state, with the rate of dissipation of stars at each point proportional to the star density. He has also derived a number of interesting conclusions concerning the evolution of spherical systems.

We now see that the hydrostatic equation can be rigorously applied in all cases, provided only that the dispersion of stellar velocities at each point in the system is the same in all directions (this condition is a little more general than the requirement of spherical symmetry of velocities). It is evident that Emden's theory can be approximately applied when the velocity distribution does not deviate too greatly from a spherical distribution.

We have seen that, when $n < 5$, configurations of finite radius are obtained. When $n = 5$ (Schuster's case) the radius becomes infinite, since θ_5 does not become zero for any finite value of ξ. The general theory shows that the radius is infinite for all $n > 5$ also.

When $n < 5$ the mass of the system can easily be calculated as a function of the parameters K, λ and n. The mass within a sphere of radius $\alpha\xi$ is, by (9.26), (9.30) and (9.32),

$$M(\alpha\xi) = 4\pi m \alpha^3 \lambda \int_0^\xi \xi^2 \theta^n \, d\xi,$$

or, from Emden's equation,

$$M(\alpha\xi) = -4\pi m \alpha^3 \lambda \int_0^\xi \frac{d}{d\xi}\left(\xi^2 \frac{d\theta}{d\xi}\right) d\xi = -4\pi m \alpha^3 \lambda \xi^2 \, d\theta/d\xi. \tag{9.53}$$

The total mass is, substituting the expression (9.33) for α,

$$M_n = -4\pi \left[\frac{K(n+1)}{4\pi G}\right]^{3/2} m^{-1/2} \lambda^{(3-n)/2n} [\xi^2 \, d\theta/d\xi]_{\xi=\Xi_n}, \tag{9.54}$$

where Ξ_n denotes the smallest root of the equation

$$\theta_n = 0. \tag{9.55}$$

When $n = 5$, Ξ_n becomes infinite, but it is easily shown that

$$\lim_{\xi\to\infty} \xi^2 \frac{d\theta_5}{d\xi} = -\sqrt{3}. \tag{9.56}$$

In this case, therefore, the mass is finite:

$$M_5 = 4\pi (3K/2\pi G)^{3/2} \sqrt{3} m^{-1/2} \lambda^{-1/5}. \tag{9.57}$$

When $n > 5$ the mass is found to be infinite. A detailed analysis would be too lengthy for inclusion here, and the reader must therefore be referred elsewhere.†

A comparison of the polytropic laws with data on the distribution of stars in globular clusters leads to uncertain results. For example, the data of von Zeipel in 1913 [139] give Table XI.

† See, for example, S. Chandrasekhar, *An introduction to the study of stellar structure*, University of Chicago, 1939.

TABLE XI

Cluster	n	Probable error	n_1
M 2	5·01	$\pm 0{\cdot}014$	5·15
M 3	6·40	$\pm 0{\cdot}011$	5·05
M13	5·48	$\pm 0{\cdot}010$	4·93
M15	5·10	$\pm 0{\cdot}022$	5·08
Mean	5·62		5·03

If we exclude from consideration the data concerning the central parts of the clusters, the somewhat smaller values given in the last column of the table are obtained.

From these data we might conclude that the globular clusters correspond to a polytropic law with $n = 5$. This conclusion is not reliable, however, because some clusters give values of $n > 5$. These values cannot be regarded as real, because they give an infinite mass. Furthermore, the hypothesis of spherical symmetry in the stellar velocity distribution, on which the above analysis is based, also needs confirmation. This will be discussed in later sections of the present chapter.

§ 3. The Most Probable Phase Distribution

So far, in discussing spherical stellar systems, we have regarded them as continuous media, a point of view which may reasonably be called "hydrodynamical". We have seen that this treatment is inadequate for the development of a complete dynamical theory. The difficulty arises on account of the discrete structure of stellar systems, because the fundamental problem of determining the quantity $2p - q$ is related to the nature of the stellar velocity distribution. We shall therefore now supplement the hydrodynamic analysis of the preceding sections by considering spherical stellar systems from the viewpoint of statistical physics, taking into account the fact that they consist of large numbers ($N \gg 1$) of uniform particles, the stars.

Let us examine the following general statistical problem. Suppose that an assembly of N stars of equal mass m is known to form a spherical system in a steady state. What is the most probable phase distribution?

To solve this problem we must determine the number of independent parameters which have to be specified to give a unique solution. A "parameter" is any physical quantity characteristic of the system. In the following discussion we shall use what are called *additive parameters*, i.e. those which are the sums of various quantities for each star separately.†

As the first additive parameter we naturally take the total mass of the system, equal to the sum of the masses of the stars:

$$M = \sum m_i = mN, \tag{9.58}$$

† These definitions are, of course, not at all precise. A more detailed account of the concept of additive parameters is given in Chapter X.

where m is the mean mass of one star and N the number of stars. If all the stars have the same mass and the total number of stars is known, then the mass of the system is known.

We shall suppose that the system is in a steady state, conservative, and dynamically isolated from other systems, and we shall show that in these conditions the sum of the total energies of the individual stars,

$$\sum H_i = \sum T_i + \sum W_i, \tag{9.59}$$

where H_i, T_i and W_i are respectively the total, kinetic and potential energies of the ith star, may be taken as an additive parameter. Evidently $\sum T_i = T$, where T is the total energy of the system. The difficulty is that the sum of the energies of the individual stars is not equal to the energy of the whole system.

It has been shown in Chapter V, § 11, that the sum of the potential energies of the individual stars is twice the potential energy W of the system: $\sum W_i = 2W$. Hence

$$\sum H_i = T + 2W. \tag{9.60}$$

Thus the total energy $H = T + W$ of the system is not an additive parameter, although it is an integral of the motion under the conditions assumed. It is easy to see that the virial theorem (Chapter V, § 12) can be used to prove the existence of two (or even an infinity of) energy integrals. From this theorem it follows that not only the total energy of the system but its kinetic and potential energy separately are conserved. Thus any combination of these three quantities is a possible parameter. According to eqn. (9.60) the second additive parameter may be taken for convenience as

$$E = 2(T + 2W). \tag{9.61}$$

This is twice the sum of the energies of the individual stars. We shall call E the *energy parameter* of the system.

The mass and energy parameters alone do not uniquely define a spherical stellar system: Jeans' theorem shows that an energy integral exists in any system, even if it is without symmetry. In order to define the system we must therefore use the integrals of the motion which exist only in spherical systems, namely the area (or angular momentum) integrals. It has been shown in § 1 that, when the force field is spherically symmetrical, there exists for every particle the angular momentum integral $r^2 T^2 = l^2$, where T is the transverse velocity. Hence the required third additive parameter for spherical systems must be of the form

$$Q = \sum r_i^2 T_i^2 = \sum l_i^2. \tag{9.62}$$

It is easy to discover the physical significance of the parameter Q, since the quantity

$$Q/N = \overline{r^2 T^2} = \overline{l^2} \tag{9.63}$$

is the variance of the angular momenta of the stars. Since the number N of stars is given by the mass of the system, it is evident that, if Q is given, the variance of the angular momenta is known.

The necessity of specifying the value of Q is seen as follows. From the argument given in Chapter VI, § 5, it follows that when $\overline{l^2} = 0$ almost all† stars must move exactly radially, oscillating along the radius vector. Such a system is dynamically possible for any law of mass distribution, although it could scarcely occur in reality. The greater $\overline{l^2}$, the more stars have oval orbits differing to some extent from circles. Some stars may have circular orbits, as we have seen in Chapter VI, § 5.

Another limiting case, which is also dynamically possible, is a system in which every star describes a circular orbit. All actual spherical systems must lie between these two extremes. It is therefore necessary to specify Q in order to define a spherical system. This also follows from the need to specify the division of stars into subsystems (Chapter X).

Thus we have the following statistical problem: what is the most probable law of phase distribution for an isolated spherical stellar system in a steady state for which the three additive parameters

$$\left.\begin{aligned} &(1)\ M = mN = \textstyle\sum m_i, \text{ the mass,} \\ &(2)\ E = 2(T + 2W) = 2\textstyle\sum H_i, \text{ twice the sum of the energies, and} \\ &(3)\ Q = \textstyle\sum l_i^2 = \sum r_i^2 T_i^2, \text{ the sum of the squared angular momenta,} \end{aligned}\right\} \quad (9.64)$$

are given?

To solve this problem we use the method of phase cells, which is widely employed in statistical physics.†† We suppose that the radius of the system is finite††† and that the stellar velocities do not exceed some finite limit. These two conditions can be combined in the requirement that the phase volume Γ corresponding to the system should be finite.

The limiting velocity may, for example, be taken as the maximum velocity of escape in the system, since, if the velocity of any star exceeds this limit, we have no reason to regard it as a member of the system. Such a high-velocity star remains only temporarily within the space occupied by the system, and rapidly passes beyond the system. The few high-velocity stars whose velocities exceed the velocity of escape at a given point but not its maximum value in the system cannot be of importance.

We divide the volume Γ into a very large number $S(1 \ll S \ll N)$ of phase cells $\gamma_s(s = 1, 2, \ldots, S)$, in such a way that the dimensions of each cell are very small $(\gamma_s \ll \Gamma)$, and yet the number n_s of phase points in each cell is large, so that the law of large numbers can be applied to every cell (or at least to the great majority of the cells).

In consequence of the smallness of the phase cells, the values of the coordinates and velocities of the stars will differ very little within a cell, and no great error is committed if the coordinates, velocities and all continuous functions of these are taken to have

† Here the words "almost all" are used as in set theory and signify "all except for a set of measure zero".

†† See, for example, J. E. Mayer and M. G. Mayer, *Statistical mechanics*, Wiley, New York, 1940.

††† By the radius of the system we mean the radius of the main part of it, where the star density is so large that a phase density may be defined. The remainder of the system constitutes the corona. The concept of the radius will be discussed more closely in Chapter X.

the same value for every star in a cell. Then the additive parameters (9.64) can be taken in the form

$$(1)\ N = \sum_s n_s, \quad (2)\ E = 2\sum_s n_s H_s, \quad (3)\ Q = \sum_s n_s l_s^2, \tag{9.65}$$

where n_s is the population of the sth cell.

Finally, let us assume that the stellar population of the system is statistically homogeneous, and therefore each star has the same *a priori* probability of being in any part of the phase volume. More precisely, we assume that the *a priori* probability p_s that a randomly chosen star will be in the cell γ_s is proportional to the volume of γ_s and is independent of its position in Γ.

The probability that the cell γ_1 contains n_1 stars, the cell γ_2 contains n_2 stars, and so on, is by the binomial law

$$P = \frac{N!}{n_1!n_2!\dots n_s!} p_1^{n_1} p_2^{n_2} \dots p_s^{n_s}, \tag{9.66}$$

if

$$(1)\ \sum_s n_s = N, \quad (2)\ 2\sum_s n_s H_s = E, \quad (3)\ \sum_s n_s l_s^2 = Q. \tag{9.67}$$

The problem of finding the most probable distribution is equivalent to finding the maximum of Lagrange's function

$$L = \log P + k_1 \sum_s n_s + 2k_2 \sum_s n_s H_s + k_3 \sum_s n_s l_s^2, \tag{9.68}$$

where k_1, k_2, k_3 are Lagrangian multipliers.

To find the maximum, we regard the quantities n_s as taking a continuous sequence of values. This treatment is necessary because we seek the phase function as a continuous function of its arguments. Then, to find n_s, we have from (9.68) and (9.66) the condition

$$\frac{\partial L}{\partial n_s} = -\frac{\mathrm{d} \log n_s!}{\mathrm{d} n_s} + \log p_s + k_1 + 2k_2 H_s + k_3 l_s^2 = 0. \tag{9.69}$$

Since the population of each cell is considerably greater than unity, the derivative of $n_s!$ can be replaced by putting

$$\frac{\mathrm{d} \log n_s!}{\mathrm{d} n_s} \approx \frac{\log (n_s + 1)! - \log n_s!}{(n_s + 1) - n_s} = \log n_s. \tag{9.70}$$

Substituting this value in eqn. (9.69) and omitting the suffix s, which is no longer needed, we have

$$n = p \exp (k_1 + 2k_2 H + k_3 l^2). \tag{9.71}$$

This formula gives the required most probable distribution, but it may usefully be transformed.

Replacing n in (9.71) by $\mathrm{d}n$, and p by the geometrical probability

$$p = \gamma/\Gamma = \mathrm{d}\omega\, \mathrm{d}\Omega/\Gamma, \tag{9.72}$$

where $d\omega$ and $d\Omega$ are volume elements in coordinate space and velocity space, we obtain

$$dn = (1/\Gamma)\, d\omega\, d\Omega \exp(k_1 + 2k_2 H + k_3 l^2). \tag{9.73}$$

Thus the distribution function is

$$f = (1/\Gamma) \exp(k_1 + 2k_2 H + k_3 l^2). \tag{9.74}$$

Substituting H and l^2 from the energy and area integrals

$$H = \tfrac{1}{2}(R^2 + T^2) + W, \quad l^2 = r^2 T^2, \tag{9.75}$$

we have the distribution law as a function of the three phase coordinates R, T, r:

$$f = f(R, T, r) = (1/\Gamma) \exp[k_1 + k_2(R^2 + T^2 + 2W) + k_3 r^2 T^2]. \tag{9.76}$$

Hence we see that k_2 must be negative, since otherwise the number of stars would increase with their velocity, and that k_3 must also be negative. In (9.76) we replace the Lagrangian multipliers k_1, k_2, k_3 by the more convenient quantities

$$K = (1/\Gamma) \exp k_1, \tag{9.77}$$

$$h^2 = -k_2, \tag{9.78}$$

$$\varkappa^2 = k_3/k_2. \tag{9.79}$$

Then the distribution function takes the final form

$$f(R, T, r) = K \exp(-2h^2 W) \exp\{-h^2[R^2 + (1 + \varkappa^2 r^2) T^2]\}. \tag{9.80}$$

Putting $T^2 = \Theta^2 + \Phi^2$, we can also write

$$f(R, \Theta, \Phi, r) = K \exp(-2h^2 W) \exp[-h^2 R^2 - h^2(1 + \varkappa^2 r^2)\Theta^2 - h^2(1 + \varkappa^2 r^2)\Phi^2]. \tag{9.81}$$

This formula shows that the distribution is a Schwarzschildian function of the velocities. Formula (9.81) gives the distribution with respect to the coordinates as well as the velocities. The distribution found above may therefore be called a *Schwarzschild–Boltzmann distribution.*

Thus the use of the method of phase cells gives the following conclusion: for a statistically homogeneous spherical stellar system in a steady state and bounded in phase space, the most probable distribution is an ellipsoidal Schwarzschild–Boltzmann distribution. In particular, when $\varkappa = 0$, (9.81) gives a spherical Maxwell–Boltzmann distribution.

The problem of how the most probable distributions tend to be established in stellar systems will be discussed in Chapter X, which concerns the evolution and lifetimes of various types of stellar system. Here we shall assume the result that in stellar systems the distribution is in fact the most probable one (or does not differ greatly therefrom). In the next three sections we shall seek to find from this the star density and the values of the parameters appearing in the distribution (9.80).

§ 4. The Truncation of the Phase Distribution. Determination of the Star Density

In finding the most probable distribution by the method of phase cells we have imposed a natural and very important restriction, namely that the phase volume is bounded. The phase distribution (9.80) obtained is consequently truncated in both coordinate space and velocity space.

Let us first consider the truncation of the distribution in velocity space. The values of the velocity components R and T must be limited by the condition that the total velocity of each star does not exceed the velocity of escape. The latter is given by the condition that a star goes to infinity. Taking the energy integral $R^2 + T^2 + 2W = 2H$ for a given star and putting $r = \infty$, we obtain (since $\lim W = 0$)

$$\lim_{r\to\infty} (R^2 + T^2) = 2H. \tag{9.82}$$

Hence the minimum velocity with which a star can escape is given by the condition $H = 0$, and the velocity of escape corresponds to zero total energy. The condition for a star to belong to the system is therefore

$$R^2 + T^2 + 2W < 0. \tag{9.83}$$

Following Ambartsumyan [5], we can make an approximate estimate of the relative number of stars escaping, i.e. an estimate of the truncation of the distribution in velocity space, as follows.

Since our purpose is merely to estimate the order of magnitude of the ratio of the number of stars escaping to the total number of stars, we assume for simplicity that the velocity distribution is Maxwellian. We can write the number of stars with velocities between v and $v + dv$ as $dn = n_0 \exp(-v^2/2\sigma^2)\, v^2\, dv$, where n_0 is a constant factor whose value is here unimportant and σ^2 the variance of the velocities in any direction. The required fraction of stars escaping is then

$$b = \int_{v_p}^{\infty} \exp(-v^2/2\sigma^2)\, v^2\, dv \Big/ \int_0^{\infty} \exp(-v^2/2\sigma^2)\, v^2\, dv,$$

where v_p denotes the velocity of escape. A calculation of the integrals gives

$$b = \frac{2}{\sqrt{\pi}} \frac{v_p}{\sqrt{2}.\sigma} \exp(-v_p^2/2\sigma^2) + 1 - \operatorname{erf}(v_p/\sigma),$$

where

$$\operatorname{erf} t = \frac{2}{\sqrt{\pi}} \int_0^t \exp(-t^2)\, dt.$$

Denoting by $\overline{v_p^2}$ and $\overline{\sigma^2}$ the mean values of the squared velocity of escape and the velocity variance for the whole system, we can write the mean value of b in the approximate form

$$\bar{b} = \frac{2}{\sqrt{\pi}} \sqrt{\frac{\overline{v_p^2}}{2\overline{\sigma^2}}} \exp(-\overline{v_p^2}/2\overline{\sigma^2}) + 1 - \operatorname{erf}[\sqrt{(\overline{v_p^2}/\overline{\sigma^2})}]. \tag{9.84}$$

The value of $\overline{v_p^2/2\sigma^2}$ can be found from the virial theorem (Chapter V, § 12). Since σ^2 gives the stellar velocity variance in any direction, $\overline{\sigma^2} = \frac{1}{3}\overline{v^2}$. Let T be the kinetic energy of the system, M its mass, and N the total number of stars. Then, assuming that every star has the same mass, we find that $\overline{v^2} = \sum v_i^2/N = 2\sum \frac{1}{2} m_i v_i^2/mN = 2T/M$. Hence

$$\overline{\sigma^2} = 2T/3M. \tag{9.85}$$

By the definition of the velocity of escape, $v_p^2 = -2W$, where W is the potential energy per unit mass at the point considered. Thus $\overline{v_p^2} = -2\overline{W} = -2\sum W_i m_i/M$. It has been shown in Chapter V, § 11, that the sum of the potential energies of the individual stars is twice the potential energy of the system. Hence

$$\overline{v_p^2} = -4W/M. \tag{9.86}$$

From (9.85) and (9.86) we find $\overline{v_p^2}/2\overline{\sigma^2} = -3W/T$, or, using the fundamental equation of the virial theorem $W + 2T = 0$ (5.160), $\overline{v_p^2}/2\overline{\sigma^2} = 6$. Substituting this in (9.84), we have the mean fraction of stars escaping:

$$\bar{b} = (2/\sqrt{\pi})\, e^{-6}\sqrt{6} + 1 - \operatorname{erf}\sqrt{12} = 0{\cdot}0073 < 1\%. \tag{9.87}$$

The above argument must be taken as showing that the Maxwellian distribution, and also the Schwarzschildian and similar distributions, which lead to a positive (not zero) probability of arbitrarily large stellar velocities, cannot occur in actual stellar systems. They must be regarded as truncated at a velocity equal to the velocity of escape, and the value found for $\bar{b}$ is a measure of the average truncation of the Maxwellian distribution for the whole system.

It is seen from the above estimate that the number of stars escaping is small compared with the total number of stars in the system. In a first approximation, therefore, the star density and the parameters of the distribution can be determined by neglecting the escaping stars, i.e. by taking the phase distribution not to be truncated in velocity space. This makes possible a considerable simplification of the subsequent calculations.

To find the star density, the phase density (9.81) should be integrated over the part of velocity space which corresponds to the phase volume occupied by the system, i.e. over all velocities which satisfy the condition (9.83). The expression thus obtained, however, is so involved that there is no point in quoting it.

To determine the star density in a first approximation, we multiply the phase density (9.81) by the volume element $d\Omega = dR\, d\Theta\, d\Phi$ and integrate over *all* velocity space. Then the expression for the star density is

$$\begin{aligned} \nu(r) &= \frac{\pi^{3/2} K}{h^3(1 + \varkappa^2 r^2)} \exp(-2h^2 W) \quad \text{for} \quad r \leqslant a, \\ &= 0 \text{ for } r > a. \end{aligned} \tag{9.88}$$

Hence it follows, *inter alia*, that the ratio $k_3/k_2 = \varkappa^2$ must be positive (i.e. $k_3 < 0$). Otherwise, the star density given by formula (9.88) would become infinite at a finite distance $r = \sqrt{(-1/\varkappa^2)}$, which is physically impossible.

The truncation of the velocity distribution reduces the mass of the system. In the central part of the system this decrease does not exceed a small fraction of one per cent of the density given by formula (9.88). If the phase distribution is known, the exact values of the truncation coefficient at various distances from the centre can easily be calculated, and so a correction can be applied to formula (9.88). We shall not do this here, since it has no practical importance.

The expression (9.88) for the star density makes it possible to estimate the truncation of the phase distribution in coordinate space, i.e. to estimate the radius of the system. Since $W \to 0$ when $r \to \infty$, (9.88) shows that the star density must decrease with increasing distance r from the centre when the latter is sufficiently large. Hence the radius a of the system may reasonably be taken as the value of r beyond which the density given by (9.88) is so small that a macroscopic volume element can no longer be defined.†

From the above discussion it follows that a stellar system has no sharp boundary, but this does not mean that it must extend to infinity. The star density distribution, which statistically is just the distribution law for the spatial coordinates of the stars, must be truncated, and the point of truncation is at what we may reasonably call the radius of the system, i.e. the radius of its main central part.

In this respect there is a close analogy between stellar systems and gas spheres such as the Sun and stars. One speaks without hesitation of the radius of a star, though knowing quite well that a star can have no sharp boundary. In particular, at the outer boundary of a star, neither the density nor the pressure can be zero; both must be positive, and their values must be determined from the conditions of hydrostatic equilibrium, taking into account radiation pressure and the atmospheric pressure.

§ 5. Discussion of the Expression Obtained for the Phase Density

It has already been pointed out that under the above conditions the phase distribution is an ellipsoidal Schwarzschild–Boltzmann distribution. From (9.81) we see that the axes of the velocity ellipsoid are in the same directions as those of the spherical coordinate system in velocity space.

The solution which we have obtained leads to a number of general conclusions concerning the properties of spherical stellar systems. As a first approximation, we shall neglect the truncation of the phase density in velocity space.

Multiplying the distribution function (9.81) successively by R^2, Θ^2 and Φ^2, integrating over all velocity space and dividing the results by the star density (9.88), we find the velocity variances:

$$\overline{R^2} = 1/2h^2, \tag{9.89}$$

$$\overline{\Theta^2} = \overline{\Phi^2} = 1/2h^2(1 + \varkappa^2 r^2). \tag{9.90}$$

† It should be mentioned that this definition of the radius is unsuitable in so far as, to find a, we must know the law of variation of the potential W with distance, which requires a complete solution of the problem.

Since $T^2 = \Theta^2 + \Phi^2$, the variance of the transverse velocity is

$$\overline{T^2} = 1/h^2(1 + \varkappa^2 r^2). \tag{9.91}$$

By means of formulae (9.89) and (9.91) we can find the ratio of the quantities p and q defined in (9.21):

$$p/q = \overline{R^2}/\overline{T^2} = \tfrac{1}{2}(1 + \varkappa^2 r^2). \tag{9.92}$$

Hence it follows that, when $\varkappa = 0$, $q = 2p$, which, as we have seen, corresponds to a spherical distribution.

From the above formulae we can draw the following conclusions.

(1) The parameter h determines the radial-velocity dispersion at every point in a spherical system. Thus the radial-velocity dispersion is the same throughout the system. The "equation of state" for spherical systems is therefore

$$p = \nu \overline{R^2} \tag{9.93}$$

and, since $\overline{R^2}$ is constant, this equation shows that the pressure is proportional to the density. Thus the equation of a spherical stellar system is the same as for a perfect gas in an isothermal state (Boyle's law). The quantity $p(r)$ for stellar systems corresponds to the pressure in ordinary hydrostatics, and the star density $\nu(r)$ corresponds to the mass density.

(2) The ratio of the radial-velocity variance to the velocity variance in either direction perpendicular to it is, by (9.89) and (9.90),

$$\overline{R^2}/\overline{\Phi^2} = \overline{R^2}/\overline{\Theta^2} = 1 + \varkappa^2 r^2. \tag{9.94}$$

Since the ratio of the squared semiaxes of the velocity ellipsoid is equal to the ratio of the corresponding variances, the velocity ellipsoid is a spheroid elongated in the radial direction. The radial semiaxis does not vary with distance from the centre of the system, and the ellipsoid becomes more elongated as the transverse variance decreases.

It is useful to note that the two limiting cases $\overline{R^2} \neq 0, \overline{T^2} = 0$ and $\overline{R^2} = 0, \overline{T^2} \neq 0$, mentioned in § 3 as being dynamically possible, do not occur for any real and finite values of $\varkappa$, by (9.92).

The former case (purely radial motion of almost all stars) occurs in the limit as $\varkappa \to \infty$; the second case (purely circular motion) is not even a limiting case. Thus not every case which is dynamically possible, i.e. does not violate the laws of theoretical mechanics, can be the statistically most probable one.

(3) The value of the parameter K determines the mass of the system. This follows from (9.88), which, on multiplication by $4\pi r^2$ and integration over r from 0 to a, gives

$$M = \frac{4\pi^{5/2}}{h^3} K \int_0^a \frac{r^2}{1 + \varkappa^2 r^2} \exp(-2h^2 W)\, dr. \tag{9.95}$$

The fact that the potential W itself depends on the mass distribution in the system does not essentially affect the situation.

(4) Since, by (9.21), (9.89) and (9.91), $p = \nu(r)/2h^2$, $q = \nu(r)/h^2(1 + \varkappa^2 r^2)$ and $U(r) = -W(r)$, we see that the star density given by (9.88) is formally an exact solution of the hydrodynamical eqn. (9.22).

Thus the two entirely different methods, hydrodynamical and statistical, have led to exactly the same result for the mass distribution in spherical systems. This must be regarded as confirming the general correctness of our procedure. Here it is important to note that the hydrodynamical equation has been derived with hardly any specialising assumptions.

(5) From the above formulae it is easy to derive an integral equation for the star density $\nu(r)$. By logarithmic differentiation of (9.88), we obtain

$$\frac{\mathrm{d}\log\nu(r)}{\mathrm{d}r} = -\frac{2\varkappa^2 r}{1 + \varkappa^2 r^2} - 2h^2\frac{\mathrm{d}W}{\mathrm{d}r}. \tag{9.96}$$

The quantity $\mathrm{d}W/\mathrm{d}r$ is (apart from sign) the force acting on a star of unit mass. Hence

$$\mathrm{d}W/\mathrm{d}r = GM(r)/r^2, \tag{9.97}$$

where $M(r)$ is the mass within a sphere of radius r. Evidently

$$M(r) = 4\pi m\int_0^r \nu(r)\, r^2\,\mathrm{d}r. \tag{9.98}$$

From (9.96), (9.97) and (9.98) we have the following integro–differential equation, in which the only unknown is the star density:

$$\frac{\mathrm{d}\log\nu(r)}{\mathrm{d}r} = -\frac{2\varkappa^2 r}{1 + \varkappa^2 r^2} - \frac{8\pi G h^2 m}{r^2}\int_0^r \nu(r)\, r^2\,\mathrm{d}r. \tag{9.99}$$

Of course, integral and differential equations may be obtained from (9.99) by integration and differentiation, but they are less convenient than (9.99).

The solution of eqn. (9.99) (for example, by successive approximations) gives $\nu(r)$. As an initial approximation we may take Schuster's law, or even $\nu(r) = \bar{\nu}$, a constant. We shall not discuss here the complete solution of eqn. (9.99), but only the first approximation.

Let us put $\nu(r) = \bar{\nu} =$ constant for $r \leqslant a$, $\nu(r) = 0$ for $r > a$. Then (9.99) gives

$$\nu(r) = \frac{\nu(0)}{1 + \varkappa^2 r^2}\exp\left(-\tfrac{4}{3}Gh^2\pi m\bar{\nu}r^2\right), \tag{9.100}$$

where $\nu(0)$ is the central density. Hence, putting $r = a$, we obtain a relation between the densities at the boundary and at the centre:

$$\nu(a) = \frac{\nu(0)}{1 + \varkappa^2 a^2}\exp\left(-h^2 GM/a\right), \tag{9.101}$$

where

$$M = M(a) = \tfrac{4}{3}\pi a^3 m\bar{\nu}. \tag{9.102}$$

The central density $\nu(0)$ may be eliminated from (9.100). Putting in (9.88) $r = a$ and comparing with (9.101), we have

$$\nu(0) = \frac{K\pi^{3/2}}{h^3} \exp\left\{h^2\left[\frac{GM}{a} - 2W(a)\right]\right\}.$$

Evidently $W(a) = -GM/a$. Hence

$$\nu(0) = \frac{K\pi^{3/2}}{h^3} \exp(3h^2 GM/a).$$

Substituting this quantity in (9.100), we have finally

$$\nu(r) = \frac{K\pi^{3/2} \exp(3h^2 GM/a)}{h^3(1 + \varkappa^2 r^2)} \exp\left(-\tfrac{4}{3} Gh^2 \pi m \bar{\nu} r^2\right). \tag{9.103}$$

Although this expression is only approximate, it is entirely reasonable. Further approximations are much more complex, and the calculations involved may not be worth while, because the whole theory is itself approximate.

§ 6. Determination of the Parameter h^2 of the Phase Distribution

In this section we shall examine the problem of determining the parameter h^2 of the phase distribution. This parameter is a constant which determines the dispersion of the radial components of stellar velocities at every point in a stellar system. However, there is a region in any spherical system where the velocity distribution is always spherical, namely the neighbourhood of the centre. To determine h^2 in this region it is sufficient to find the velocity variance σ_R^2 in any direction; then $h^2 = 1/2\sigma_R^2$.

It has been shown in Chapter VI, § 5, that, in the neighbourhood of the centre of a spherical system the force field is quasi-Hookean, and for small r it is very weak. We can also suppose that the number of stars which remain permanently in the immediate neighbourhood of the centre is very small. This property of orbit distribution will be demonstrated in § 7. It always holds if the star density at the centre of the system is bounded. Hence the velocity dispersion in the centre of the system will be determined by the stars which pass through the centre and oscillate along a diameter of the system.

Let us consider one such star. Its motion is described by the differential equation

$$\mathrm{d}v/\mathrm{d}t = -GM(r)/r^2, \tag{9.104}$$

where $M(r)$ is again the mass within a sphere of radius r:

$$M(r) = 4\pi m \int_0^r \nu(s)\, s^2\, \mathrm{d}s.$$

Since

$$\frac{\mathrm{d}v}{\mathrm{d}t} = \frac{\mathrm{d}v}{\mathrm{d}r}\frac{\mathrm{d}r}{\mathrm{d}t} = v\frac{\mathrm{d}v}{\mathrm{d}r} = \frac{1}{2}\frac{\mathrm{d}(v^2)}{\mathrm{d}r},$$

eqn. (9.104) can be written

$$\mathrm{d}(v^2)/\mathrm{d}r = -2GM(r)/r^2.$$

Let r_0 be the greatest distance reached by the star in its oscillatory motion, i.e. the distance from the centre to the turning point of the motion. At this point the instantaneous velocity of the star is zero. Integration of the last equation over r from 0 to r_0 gives

$$v_0^2 = 2G \int_0^{r_0} \frac{M(r)}{r^2}\,\mathrm{d}r, \tag{9.105}$$

where v_0 is the velocity of the star as it passes through the centre of the system.

The number of stars whose turning points lie at distances between r_0 and $r_0 + \mathrm{d}r$ and which are within an infinitesimal solid angle $\mathrm{d}\omega$ is, by (9.80),

$$f(0, 0, r_0)\, r_0^2\,\mathrm{d}r\,\mathrm{d}\omega = K \exp(-2h^2 W_0)\, r_0^2\,\mathrm{d}r\,\mathrm{d}\omega. \tag{9.106}$$

Omitting the suffixes in r_0 and W_0, which are now unnecessary, we find the velocity variance of these stars to be

$$\frac{1}{2h^2} = \sigma_R^2 = \frac{2G \int_0^a \exp(-2h^2 W)\, r^2\,\mathrm{d}r \int_0^r \frac{M(s)}{s^2}\,\mathrm{d}s}{\int_0^a \exp(-2h^2 W)\, r^2\,\mathrm{d}r}.$$

Since by (9.88) $\exp(-2h^2 W) = h^3(1 + \varkappa^2 r^2)\,\nu(r)/2\pi^{3/2} K$, we have

$$\frac{1}{2h^2} = \sigma_R^2 = \frac{2G \int_0^a \nu(r)(1 + \varkappa^2 r^2)\, r^2\,\mathrm{d}r \int_0^r \frac{M(s)}{s^2}\,\mathrm{d}s}{\int_0^a \nu(r)(1 + \varkappa^2 r^2)\, r^2\,\mathrm{d}r}, \tag{9.107}$$

which is a convenient form, since the right-hand side no longer involves h.

The last equation may be written in a clearer form by again using the quantities

$$M(r) = 4\pi m \int_0^r \nu(r)\, r^2\,\mathrm{d}r$$

and $M = M(a)$, which are respectively the mass within a sphere of radius r and the total mass of the system. We may also use the moments of inertia about the centre of the system,

$$J(r) = 4\pi m \int_0^r \nu(r)\, r^4\,\mathrm{d}r$$

and $J = J(a)$. Then formula (9.107) becomes

$$\sigma_R^2 = 2G \frac{\int\limits_0^a \frac{M(r)\,[M - M(r)]}{r^2}\,\mathrm{d}r + \varkappa^2 \int\limits_0^a \frac{M(r)\,[J - J(r)]}{r^2}\,\mathrm{d}r}{M + \varkappa^2 J}. \qquad (9.108)$$

Finally, in terms of the dimensionless mass, moment of inertia and radius: $\mu(r) = M(r)/M$, $j(r) = J(r)/J$, $\xi = r/a$, we obtain

$$\sigma_R^2 = \frac{2GM}{a} \frac{\int\limits_0^1 \frac{\mu(\xi)\,[1 - \mu(\xi)]}{\xi^2}\,\mathrm{d}\xi + \varkappa^2 \varrho^2 \int\limits_0^1 \frac{\mu(\xi)\,[1 - j(\xi)]}{\xi^2}\,\mathrm{d}\xi}{1 + \varkappa^2 \varrho^2}, \qquad (9.109)$$

where ϱ denotes the radius of gyration of the system about its centre: $\varrho^2 = J/M$. Formula (9.109) can be used as a refinement of the corresponding formula $\overline{v^2} = 3GM/5a$ obtained by means of the virial theorem in Chapter V, § 12. It becomes simpler if applied to systems with a high concentration of mass towards the centre ($\varrho \ll a$) or with an almost spherical velocity distribution ($\varkappa\varrho \ll 1$).

§ 7. The Distribution of Star Orbits. Lindblad and Bottlinger Diagrams

It has already been pointed out that the treatment of a stellar system as a continuous medium, i.e. the hydrodynamical approach to the problems of stellar dynamics, does not do away with the necessity of also treating the system as an ensemble of uniform particles; that is, the hydrodynamical approach must be used in combination with the statistical approach. This we have done in the preceding sections, and we have obtained a number of results concerning the nature of the phase distribution.

There is yet a third possible approach to the study of stellar systems. This may be called the celestial-mechanics treatment, and consists in examining the regular orbits described by the individual stars in the system.

The individual regular orbits of stars have been discussed in Chapter VI, §§ 1, 3, 5. Here we shall consider the whole family of regular orbits, whose nature is defined by what is called the *distribution of orbits*.

It has been mentioned several times that the distribution of orbits in a system in a quasi-steady state is maintained regardless of the action of the irregular forces, since in that case the action of these forces amounts to transferring stars from one orbit to another, without affecting the nature of the orbits themselves. In other words, in a quasi-steady stellar system with irregular forces acting, the motion of the stars takes place in the same orbits as in a system where there are no irregular forces. The only difference is that, when irregular forces are present, we cannot be certain that the same individual star is moving all the time in any particular orbit. This uncertainty, however, does not affect our statistical conclusions. That is, the principle of individual indistinguishability of stars is valid in the dynamics of stellar systems. This principle, which has been mentioned in Chapter V, § 1, will be more fully discussed in Chapter X, § 2 in connection with the significance of the corona as a kind of "adiabatic envelope" which replaces the "vessel walls" of the theory of gases. We now see that the same principle holds in the internal parts of stellar systems, within the main body.

The only case which we have met where the principle of indistinguishability does not hold arises in connection with the problem of local dynamics. Here we considered only the stars of types O to B5, which are therefore "marked" stars.

Each regular orbit is entirely determined by specifying the initial data or the arbitrary constants (elements of the orbit), which may be defined in various ways. Hence the distribution of star orbits may be regarded as known if the distribution of some system of arbitrary constants is given.

It has been shown in Chapter VI, § 5, that in a spherical system all the orbits are plane curves. We shall not take account of the orientation of the orbital plane in space or the position of the orbit in that plane, since these are unimportant because of the spherical symmetry. In the case considered, we are therefore interested only in the distribution of orbits as regards size and shape.

In the present section we shall consider three ways of solving the problem.

The first way follows mainly the results obtained as early as 1913 by Eddington [25], in an investigation both profound and elegant. It has been shown in Chapter VI, § 5, that almost all regular orbits are elongated in shape. Let r_0 be the distance from the centre O of the system to the apocentre C and T_0 the transverse velocity at the point C. The corresponding radial velocity R_0 is evidently zero. We shall show that r_0 and T_0 may be taken as elements of the orbit. The integrals of the motion in this case may be written as

$$\begin{aligned} R^2 + T^2 + 2W &= T_0^2 + 2W_0, \\ rT &= r_0 T_0, \end{aligned} \tag{9.110}$$

where for brevity we have put W_0 in place of $W(r_0)$. From (9.110) it follows that, if r_0 and T_0 are given, the values of $\pm R$ and $\pm T$ are completely determined for all values of r between the apocentre r_0 and the pericentre r_1. Conversely, if the velocities R and T are given (and, of course, if they lie within the permissible limits) for any one value of r, then r_0 and T_0 are uniquely determined.

Let us now find the distribution function for the orbital elements r_0 and T_0. Let

$$\varphi(r_0, T_0)\, \mathrm{d}r_0\, \mathrm{d}T_0 \tag{9.111}$$

be the number of stars whose apocentres are at distances between r_0 and $r_0 + \mathrm{d}r_0$ from the centre, and whose transverse velocities at apocentre lie between T_0 and $T_0 + \mathrm{d}T_0$. We describe about the centre O two concentric spheres of radii r and $r + \mathrm{d}r$ and calculate the number of stars lying in the spherical shell between them. This number is the product of (9.111) and the quantity $2\,\mathrm{d}r/|RP|$, which corresponds to the two intervals of time during which each star is within the spherical shell near the points A and B (Fig. 35). P is the period of the orbital motion defined by r_0 and T_0. Finally, in order to obtain the number of stars per unit volume, we must divide this product by the volume $4\pi r^2\, \mathrm{d}r$ of the spherical shell.

Thus the number of stars per unit volume at a distance r from the centre whose apocentric distances and transverse velocities are in the ranges mentioned is

$$d\nu = \varphi(r_0, T_0)\, dr_0\, dT_0 \frac{2\, dr}{|R|\, P} \frac{1}{4\pi r^2\, dr},$$

or

$$d\nu = \frac{1}{2\pi} \frac{\varphi(r_0, T_0)}{r^2\, |R|\, P}\, dr_0\, dT_0. \tag{9.112}$$

On account of the spherical symmetry of the system, the direction of the radius r and the orientation of the plane containing r and r_0 are unimportant.

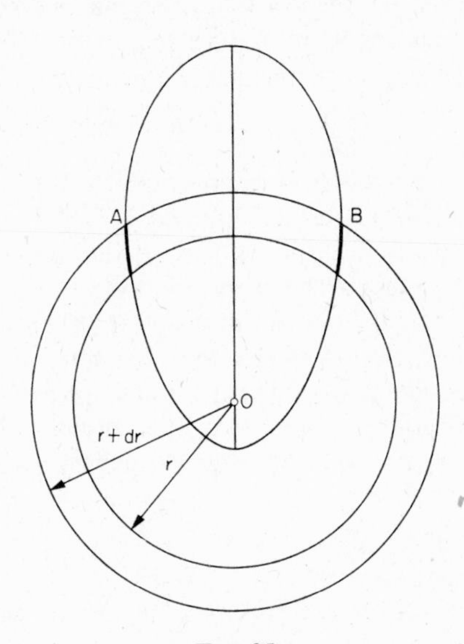

FIG. 35

Formula (9.112) may be written in terms of the variables R and T. To do this, we calculate the Jacobian $\partial(R, T)/\partial(r_0, T_0)$. Writing T in terms of T_0 in the first formula (9.110) by using the second formula, we obtain

$$R^2 = -2(W - W_0) - T_0^2 \left(\frac{r_0^2}{r^2} - 1\right),$$

$$rT = r_0 T_0,$$

whence

$$\begin{aligned} R\frac{\partial R}{\partial r_0} &= \frac{dW_0}{dr_0} - T_0^2 \frac{r_0}{r^2}, \quad r\frac{\partial T}{\partial r_0} = T_0, \\ R\frac{\partial R}{\partial T_0} &= -T_0 \left(\frac{r_0^2}{r^2} - 1\right), \quad r\frac{\partial T}{\partial T_0} = r_0. \end{aligned} \tag{9.113}$$

Thus

$$rR\frac{\partial(R, T)}{\partial(r_0, T_0)} = r_0\frac{\mathrm{d}W_0}{\mathrm{d}r_0} - T_0^2. \tag{9.114}$$

The expression $r_0\,\mathrm{d}W_0/\mathrm{d}r_0$ has a simple physical significance: it is the square of the circular velocity T_{c0} at the distance r_0. This velocity is clearly the greatest possible value of the transverse velocity T_0 at the apocentre.

Hence (9.114) can be written

$$\left|\frac{\partial(R, T)}{\partial(r_0, T_0)}\right| = \frac{T_{c0}^2 - T_0^2}{r\,|R|},$$

and so

$$\mathrm{d}r_0\,\mathrm{d}T_0 = \frac{r\,|R|}{T_{c0}^2 - T_0^2}\,\mathrm{d}R\,\mathrm{d}T. \tag{9.115}$$

From (9.112) and (9.115) we find the distribution law in the form

$$\mathrm{d}\nu = \frac{1}{2\pi rP}\frac{\varphi(r_0, T_0)}{T_{c0}^2 - T_0^2}\,\mathrm{d}R\,\mathrm{d}T. \tag{9.116}$$

In a spherical stellar system the distribution law is of the form $\mathrm{d}\nu = f(R, T, r)\,\mathrm{d}\omega\,\mathrm{d}\Omega$, where $\mathrm{d}\omega$ and $\mathrm{d}\Omega$ are volume elements in coordinate space and velocity space respectively, and $f(R, T, r)$ is the distribution function in R and T, depending on r as a parameter. If $\mathrm{d}\nu$ is referred to unit volume in coordinate space as in (9.116), we must put $\mathrm{d}\omega = 1$. In velocity space (R, Θ, Φ) we take the volume element in the form of a cylindrical shell of radius $T = \sqrt{(\Theta^2 + \Phi^2)}$, thickness $\mathrm{d}T$ and height $\mathrm{d}R$. Then $\mathrm{d}\Omega = 2\pi T\,\mathrm{d}T\,\mathrm{d}R$, and therefore

$$\mathrm{d}\nu = 2\pi T f(R, T, r)\,\mathrm{d}R\,\mathrm{d}T. \tag{9.117}$$

A comparison of (9.116) with (9.117), replacing rT by r_0T_0 (the area integral), gives finally the following formula for the distribution function of apocentric distances r_0 and apocentric transverse velocities T_0:

$$\varphi(r_0, T_0) = 4\pi^2 r_0 T_0 P(T_{c0}^2 - T_0^2) f(R, T, r). \tag{9.118}$$

It can be shown in exactly the same way that the elements of the orbit may also be taken as the pericentric distance r_1 and the pericentric transverse velocity T_1. To do so, we replace r_0 and T_0 in formulae (9.110) by r_1 and T_1, and use the fact that the circular velocity $T_{c1} < T_1$. The distribution function of these elements is then found to be

$$\varphi(r_1, T_1) = 4\pi^2 r_1 T_1 P(T_1^2 - T_{c1}^2) f(R, T, r), \tag{9.119}$$

where T_{c1} is the circular velocity at the pericentre.

The relations (9.118) and (9.119) lead to some important conclusions. Firstly, since the left-hand side of (9.118) depends only on r_0 and T_0, the same must be true of the right-hand side. This can happen only if $f(R, T, r)$ depends on R, T, r through combinations of these three variables which can be expressed in terms of r_0 and T_0. Such combinations of coordinates and velocities are integrals of the motion. Thus we return to Jeans' theorem: the phase density is a function only of integrals of the motion.

Secondly, we see from (9.118) that the frequency (probability density) of circular orbits with $T_0 = T_{c0}$ and linear orbits with $T_0 = 0$ is zero. In other words, such orbits are extremely rare in a spherical stellar system in a steady state. The same conclusion follows from (9.119).

Thirdly, it follows from (9.118) that $\varphi(r_0, T_0) \to 0$ as $r_0 \to 0$. This means that the number of stars which remain permanently near the centre of the system is small.

Fourthly, we conclude from (9.119) that $\varphi(r_1, T_1) \to 0$ for $r_1 \to 0$. This means that very few stars pass near the centre in their motion.

These conclusions are evidently independent of the particular form of the function $f(R, T, r)$, and are therefore applicable to any stellar system which is spherically symmetrical.

Let us now suppose that the system is statistically homogeneous. Then, as has been shown in § 3, the most probable phase distribution is given by formula (9.80). We shall find it more convenient, however, to use formula (9.74), which, with the notation (9.77)–(9.79), can be written

$$f(R, T, r) = K \exp[-h^2(\varkappa^2 l^2 + 2H)]. \tag{9.120}$$

According to (9.110), the energy and area integrals can be put in the form $l^2 = r_0^2 T_0^2$, $2H = T_0^2 + 2W_0$. Hence

$$f(R, T, r) = K \exp(-2h^2 W_0) \exp[-h^2 T_0^2(1 + \varkappa^2 r_0^2)]. \tag{9.121}$$

From (9.118) and (9.121) we derive the distribution function of the orbital elements r_0 and T_0 for a statistically homogeneous spherical stellar system:

$$\varphi(r_0, T_0) = 4\pi^2 r_0 T_0 (T_{c0}^2 - T_0^2) \exp(-2h^2 W_0) \exp[-h^2 T_0^2(1 + \varkappa^2 r_0^2)]. \tag{9.122}$$

Similarly, we can obtain the distribution function of r_1 and T_1:

$$\varphi(r_1, T_1) = 4\pi^2 r_1 T_1 (T_1^2 - T_{c1}^2) \exp(-2h^2 W_1) \exp[-h^2 T_1^2(1 + \varkappa^2 r_1^2)], \tag{9.123}$$

where $W_1 = W(r_1)$ is the value of the potential energy at the pericentre.

Let us now consider the second way of finding the distribution of stellar orbits. The shape and size of each orbit are, as we know, uniquely defined by the values of the energy integral and the area integral. Hence the distribution of orbits may be characterised by the distribution function of these integrals.

In a statistically homogeneous spherical system, the most probable distribution is given by formula (9.74) or (9.120), the latter being the more convenient in the present case. The number of stars in a phase volume element $d\omega\, d\Omega$ is therefore given by

$$dn = K \exp[-h^2(\varkappa^2 l^2 + 2H)]\, d\omega\, d\Omega.$$

The volume element $d\Omega$ in velocity space can, as we have seen, be represented as

$$d\Omega = 2\pi T\, dR\, dT. \tag{9.124}$$

From the last two formulae we find

$$\mathrm{d}n = 2\pi K T \exp[-h^2(\varkappa^2 l^2 + 2H)]\, \mathrm{d}\omega\, \mathrm{d}R\, \mathrm{d}T. \tag{9.125}$$

To find the distribution of the integrals, we must express the volume element $\mathrm{d}\Omega$ in terms of H and l. We have

$$\mathrm{d}R\, \mathrm{d}T = \left| \frac{\partial(R, T)}{\partial(H, l)} \right| \mathrm{d}H\, \mathrm{d}l,$$

where $\partial(R, T)/\partial(H, l)$ is the Jacobian. Expressing T in the energy integral in terms of l, we obtain

$$R^2 = 2H - l^2/r^2 - 2W, \quad T = l/r, \tag{9.126}$$

whence $R\,\partial R/\partial H = 1$, $\partial T/\partial H = 0$, $R\,\partial R/\partial l = -l/r^2$, $\partial T/\partial l = 1/r$, and so $\partial(R, T)/\partial(H, l) = 1/rR$ and

$$T\, \mathrm{d}R\, \mathrm{d}T = \frac{l}{r^2\,|R|}\, \mathrm{d}H\, \mathrm{d}l. \tag{9.127}$$

From (9.125) and (9.127) we obtain the required expression for $\mathrm{d}n$:

$$\mathrm{d}n = \frac{2\pi K}{r^2} \exp[-h^2(2H + \varkappa^2 l^2)] \frac{l\, \mathrm{d}H\, \mathrm{d}l}{\sqrt{[2H - (l^2/r^2) - 2W]}}\, \mathrm{d}\omega. \tag{9.128}$$

The quantity $\mathrm{d}n/\mathrm{d}\omega$ is the distribution function of the constant integrals H and l among the stars in unit volume of coordinate space at a given distance r from the centre of the system:

$$F(H, l) = \frac{2\pi K}{r^2} \frac{l}{\sqrt{[2(H - W) - l^2/r^2]}} \exp[-h^2(2H + \varkappa^2 l^2)]. \tag{9.129}$$

This function may be represented in a convenient manner by the use of a *Lindblad diagram*. We plot l as abscissa and H as ordinate; since the possible values of H and l are limited by the conditions

$$2H - l^2/r^2 - 2W \geqslant 0, \quad H < 0, \tag{9.130}$$

the points in the lH-plane which correspond to possible values of l and H lie in a region bounded above by the axis $H = 0$ and below by Lindblad's parabola [68]

$$l^2 = 2r^2(H - W). \tag{9.131}$$

The shape of this parabola depends on r. Its latus rectum is $2r^2$, and its vertex (C in Fig. 36) is on the H-axis at a distance W from the origin. Its points of intersection with the l-axis have coordinates $l = \pm r\sqrt{(-2W)}$.

Since $R = 0$ at every point on the parabola, the stars corresponding to such points pass through apocentre or pericentre at the distance r which corresponds to the diagram. In particular, the vertex C corresponds to stars moving in straight lines and having turning points at that value of r.

The points on the parabola include two, D and D_1, which correspond to circular orbits. It is evident that all points on the parabola above the points D and D_1 pertain to stars passing through pericentre, which have the greater area integral, and those lying below pertain to stars passing through apocentre.

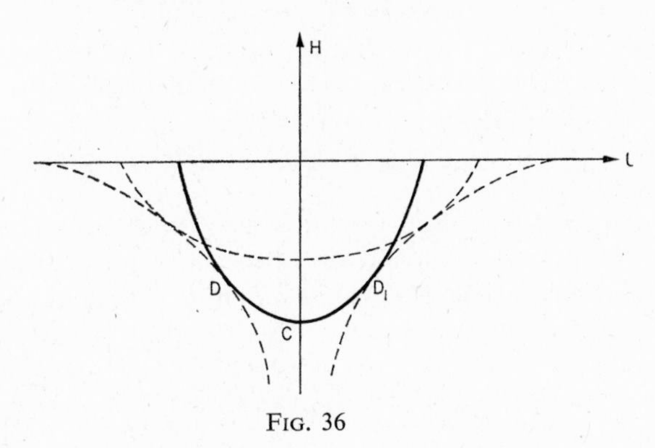

FIG. 36

We shall show that the points D and D_1 are the points where the parabola touches the envelope of all such parabolas for various r. Differentiating eqn. (9.131) with respect to r, we have

$$2r(H - W) - r^2 \, \mathrm{d}W/\mathrm{d}r = 0. \tag{9.132}$$

Eliminating $H - W$ from (9.131) and (9.132), we have the equation of the envelope in parametric form:

$$l^2 = r^3 \, \mathrm{d}W/\mathrm{d}r, \quad H = W + \tfrac{1}{2} r \, \mathrm{d}W/\mathrm{d}r. \tag{9.133}$$

It is easy to see that the angular momentum l which satisfies the first of these relations is that which corresponds to circular motion. In other words, the envelope of the family of parabolas is the locus of the points corresponding to circular motion at various distances from the centre of the system.

In the Lindblad diagram, the distribution function $F(H, l)$ given by formula (9.129) represents the density of points within the region bounded by the parabola and the axis $H = 0$. The radial velocity R, expressed in terms of integrals of the motion,

$$R = \sqrt{[2(H - W) - l^2/r^2]}, \tag{9.134}$$

vanishes on the parabola, and therefore the function $F(H, l)$ becomes infinite at every point on the parabola; it remains integrable, however, and this is one of the fundamental properties which any distribution function must have.

It is also easily shown that $F(H, l)$ is a monotonically decreasing function of H in the region where it is defined. The sign of the logarithmic derivative $\partial \log F/\partial H$ is that of the quantity $[(l^2/r^2) - 2(H - W)]^{-1} - 2h^2$, and this is negative, because from (9.130) and the equation of the parabola (9.131) the expression in square brackets is negative at all points lying within the parabola.

The sign of the product $l \, \partial \log F/\partial l$ is that of the expression $l^4 - 2r^2(H - W)l^2 + r^2(H - W)/h^2\varkappa^2$. For the limiting values $l^2 = 0$ and $l^2_{\max} = 2r^2(H - W)$ this is positive. Its discriminant is, apart from a positive factor, $h^2 \varkappa^2 r^2 (H - W) - 1 = \frac{1}{2} h^2 \varkappa^2 l_{max}^2 - 1$. Hence, if the maximum value of l is less than $\sqrt{2}/h\varkappa$, the discriminant is negative, and $F(H, l)$ increases on each side of $l = 0$. In the lower part of the diagram, near the point C, where the range of variation of l is small, the variation of the function $F(H, l)$ (see (9.129)) depends on that of the denominator, which decreases monotonically, so that the function F increases. In the upper part of the diagram, where l varies over a wide range, the decrease of the exponential factor in the numerator may initially outweigh the decrease of the denominator. As we approach the boundary of the region, however, the denominator tends to zero and F therefore increases rapidly.

Finally, let us consider the third way of finding the distribution of star orbits.

The motions of the stars are most easily observed in the outer part of a stellar system. Here the force field is quasi-Newtonian, and the distribution of orbits which lie entirely in the outer parts of the system is determined by the distribution of the Keplerian elements of elliptical orbits, namely the major semiaxis a and the eccentricity e. To find the distribution function, we must express the constants H and l in (9.128) in terms of a and e.

The relations between H, l and a, e are well known:

$$2H = -GM/a, \quad l^2 = GMa(1 - e^2), \tag{9.135}$$

whence

$$l \, \mathrm{d}H \, \mathrm{d}l = l \left| \frac{\partial(H, l)}{\partial(a, e)} \right| \mathrm{d}a \, \mathrm{d}e = \frac{G^2 M^2}{2a} e \, \mathrm{d}a \, \mathrm{d}e.$$

Then (9.128) gives

$$\begin{aligned} \mathrm{d}n/\mathrm{d}\omega &= F(a, e) \, \mathrm{d}a \, \mathrm{d}e \\ &= \frac{\pi K}{r^2} \frac{(GM)^{3/2}}{a} \frac{\exp\{-h^2 [\varkappa^2 GMa(1 - e^2) - GM/a]\}}{\sqrt{[(2/r) - (1/a) - (a/r^2)(1 - e^2)]}} e \, \mathrm{d}a \, \mathrm{d}e. \end{aligned} \tag{9.136}$$

At first sight it might appear that this formula gives a probability of zero for a circular orbit, since the expression for $F(a, e)$ in the numerator contains e as a factor, and $e = 0$ for a circle. This is not so, however, because, when $e \to 0$, $r \to a$ and the radicand in the denominator also tends to zero. Since

$$\lim_{r \to a} \left[\frac{2}{r} - \frac{1}{a} - \frac{a}{r^2} (1 - e^2) \right] = \frac{e^2}{a},$$

we have from (9.136)

$$F(a, 0) = \frac{\pi K (GM)^{3/2}}{a^{5/2}} \exp \left[- h^2 GM \left(\varkappa^2 a - \frac{1}{a} \right) \right]. \tag{9.137}$$

It should be mentioned, however, that this expression is valid only for large values of a, i.e. the region where there is a quasi-Newtonian force field.

The distribution of the elements of elliptical orbits in a Newtonian field may be represented by means of a *Bottlinger diagram* (Fig. 37), first used by Bottlinger [16] to illustrate stellar motions in the Galaxy in the neighbourhood of the Sun. This diagram can be applied, with slight modifications, to spherical systems also.

Let the line AGB (Fig. 37) be in the direction of the radius vector r of the point in the system for which the diagram is drawn. It is evident that, on account of the spherical symmetry of the system, the same diagram will serve for every point on the sphere of radius r. This, however, does not affect the drawing of the diagram itself.

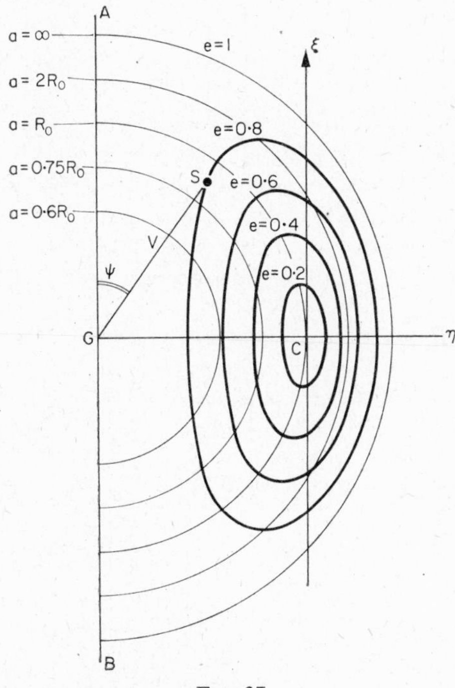

FIG. 37

We take the point G as the origin in measuring the absolute velocities of the stars. By the absolute motion of a star we mean its motion relative to non-rotating coordinate axes moving with the centre of mass of the system. Let ABC be an arbitrary diametral plane passing through the radius vector r. We mark off, on an arbitrary scale, a segment GC equal to the circular velocity V_c at the given distance from the centre. We can say that the point C represents every star which moves in a circular orbit of radius r. There is evidently an infinity of circular orbits of radius r passing through the point chosen. The planes of all such orbits are obtained by rotating the plane ABC about AB.

Let us now consider an arbitrary star having a velocity V at an angle ψ to the radius vector. Let the point S represent this star and all other stars whose velocities and directions of motion are the same. Since we are interested in the number of elliptical orbits with various values of a and e, all these stars are equivalent in the present discussion.

If V and ψ, or $R = V\cos\psi$ and $T = V\sin\psi$, are given, the shape and size of the elliptical orbit, i.e. a and e, are uniquely determined. This follows from the area and energy integrals, which for elliptical motion are

$$r^2T^2 = GMa(1 - e^2), \tag{9.138}$$

$$V^2 = GM\left(\frac{2}{r} - \frac{1}{a}\right). \tag{9.139}$$

These two formulae give the elements a and e if the velocity and direction of motion of the star are known. For convenience in the subsequent calculations, we may substitute for GM in (9.138) and (9.139) from

$$V_c^2 = GM/r. \tag{9.140}$$

Then

$$T = V_c\sqrt{[a(1 - e^2)/r]}, \tag{9.141}$$

$$V = V_c\sqrt{[2 - (r/a)]}. \tag{9.142}$$

According to (9.142), by giving a various values we obtain semicircles in the diagram, these being the loci of the points S representing stars with a given major semi-axis, and the ratio of the radii of these semicircles to the fundamental length $GC = V_c$ in the diagram is easily found. The semicircles are the level lines a = constant.

To draw the second family of level lines, e = constant, it is more convenient to move the origin to C. Denoting the new coordinates by ξ and η, we have

$$\xi = R = \sqrt{(V^2 - T^2)}, \quad \eta = T - V_c. \tag{9.143}$$

Substituting (9.141) and (9.142) for V and T gives

$$\xi = V_c\sqrt{\left[2 - \frac{r}{a} - \frac{a}{r}(1 - e^2)\right]}, \tag{9.144}$$

$$\eta = V_c\left\{\sqrt{\left[\frac{a}{r}(1 - e^2)\right]} - 1\right\}.$$

Elimination of a gives the required equation of the family:

$$\frac{\xi^2}{V_c^2} + \frac{V_c^2}{(\eta + V_c)^2}(1 - e^2) + \frac{(\eta + V_c)^2}{V_c^2} - 2 = 0. \tag{9.145}$$

From this equation the curves round the point C in Fig. 37 are plotted; they have been taken from Trumpler and Weaver's book [132]. If we consider only values of ξ and η which are small compared with V_c, expand the left-hand side of eqn. (9.145) in powers of ξ/V_c, η/V_c and e, and neglect terms above the second order, we have the equation of an ellipse:

$$\frac{\xi^2}{V_c^2} + \frac{4\eta^2}{V_c^2} = e^2. \tag{9.146}$$

Hence it follows that the curves nearest C are almost a family of concentric similar ellipses with the η-axis half the ξ-axis. Both axes are proportional to the corresponding eccentricity.

§ 8. Spherical Systems with an Inhomogeneous Star Population

Until now we have begun from the assumption that the stellar composition of spherical systems is statistically homogeneous. A study of actual stellar systems, however, shows that their composition is not homogeneous. The Galaxy is an instance where the star population can be divided into sub-systems of various types: plane, intermediate and spherical. The stars in these different sub-systems differ as regards the dispersion of their residual velocities, the velocity with which they participate in the rotation of the Galaxy, and their degree of concentration towards the galactic plane. The greater the velocity dispersion, the less the concentration.

Unfortunately, we have as yet no similar data for spherical systems. Apart from the fairly rare spherical galaxies, which are usually regarded as a particular type (E0) of the more general class of elliptical galaxies, there are only the globular clusters. Star counts in even the brightest of these clusters are as yet practically limited to giant stars and some of the brightest dwarfs. The question whether the star population of the spherical systems can be divided into sub-systems like the stars in the Galaxy therefore cannot yet be finally decided, but it is highly probable that such sub-systems exist, at least in the globular clusters. In this connection the results obtained by P. N. Kholopov [48, 49] on the spatial distribution of stars in globular clusters are of great importance. He has discovered that these clusters can be divided into three regions: the nucleus, the intermediate region and the corona, separated by discontinuities in the star density gradient. The mere existence of these regions makes it very probable that their stellar compositions differ in kinematical and morphological properties.

If the star population of spherical systems can be divided into sub-systems, the dynamics of such systems must be accordingly modified. For, in homogeneous systems, the force field is determined by the spatial distribution of the very stars on which it acts, i.e. it is self-consistent; in heterogeneous systems the field is not self-consistent, since it is determined by the distribution of all the stars in the system, including those belonging to other sub-systems.

In considering a stellar system which is not homogeneous, it is natural to assume that the phase distributions of the various constituent sub-systems are independent. This corresponds to Dalton's law in the theory of gases: in a mixture of gases the distribution of each gas, and in particular the partial density of each gas, are not affected by the presence of the other gases.

The applicability of Dalton's law to stellar systems is confirmed not only by analogy, although this is a possible argument in its favour. An example where Dalton's law holds is afforded by the distribution of stars in the Galaxy. In numerous investigations beginning with the pioneering work of Baade, and especially in the work of B. V. Kukarkin, P. P. Parenago and their colleagues, it has been found that the distribution of stars in the z-direction, i.e. in the direction perpendicular to the galactic plane, is fairly well represented by the *barometric formula*

$$D(z) = D(0)\, e^{-z/\beta}, \tag{9.147}$$

where $D(z)$ and $D(0)$ are the star densities at z and in the galactic plane, and β is the half-thickness of an equivalent homogeneous layer. The same relation is observed in the distribution of the various gases in the Earth's atmosphere with respect to height above sea level. Similar results are obtained for the distribution with respect to distance from the centre of the Galaxy. This distribution is usually characterised by the quantity m, the logarithmic gradient of the star density at the position in the galactic plane occupied by the observer (the Sun):

$$m = -\partial \log D / \partial R. \tag{9.148}$$

The mean values of the root-mean-square residual velocity $\sigma = 1/\sqrt{2} \,.\, h$ and of m and β for the three principal classes of the stellar population of the Galaxy are given in Table XII, taken from the data of Parenago [107].

TABLE XII. DYNAMICAL PARAMETERS OF GALACTIC SUB-SYSTEMS

Type	σ (km/s)	m (kps^{-1})	β (ps)
Plane	20	0·11	50
Intermediate	42	0·20	350
Spherical	110	0·26	2000

In the work of Kholopov [49] the idea of dividing the stellar population into sub-systems has been successfully applied to the globular clusters.

Let us now proceed to find the most probable phase distribution, and consider first a spherical stellar system consisting of only two statistically homogeneous sub-systems, S_1 and S_2. Let the stars belonging to S_1 have a smaller velocity dispersion, and a greater concentration towards the centre of the system, than those of S_2, the spatial distribution in both sub-systems being symmetrical. We can say that the stars of the sub-system S_1 form a spherical system of radius a_1 placed within a second such system (S_2) of greater radius a_2. Since the stellar velocities in S_1 are supposed generally smaller than those in S_2, the above two properties of the sub-systems can be concisely stated by saying that the phase volume Γ_1 of the first sub-system lies entirely within the phase volume Γ_2 of the second sub-system.

Applying Dalton's law to the two statistically homogeneous sub-systems S_1 and S_2, we obtain for each a phase distribution like (9.80):

$$\left.\begin{aligned}
\mathrm{d}n_1 &= K_1 \exp(-2h_1^2 W) \exp\{-h_1^2[R^2 + (1+\varkappa_1^2 r^2)T^2]\}\, \mathrm{d}\omega\, \mathrm{d}\Omega \quad (r \leqslant a_1),\\
\mathrm{d}n_1 &= 0 \quad (r > a_1),\\
\mathrm{d}n_2 &= K_2 \exp(-2h_2^2 W) \exp\{-h_2^2[R^2 + (1+\varkappa_2^2 r^2)T^2]\}\, \mathrm{d}\omega\, \mathrm{d}\Omega \quad (r \leqslant a_2),\\
\mathrm{d}n_2 &= 0 \quad (r > a_2).
\end{aligned}\right\} \tag{9.149}$$

In a spherical shell of radii a_1 and a_2 the second of the two distributions (9.149) is found; within the sphere of radius a_1 there is a combined distribution

$$\begin{aligned} dn &= dn_1 + dn_2 \\ &= (K_1 \exp(-2h_1^2 W)\exp\{-h_1^2[R^2 + (1 + \varkappa_1^2 r^2)T^2]\} \\ &\quad + K_2 \exp(-2h_2^2 W)\exp\{-h_2^2[R^2 + (1 + \varkappa_2^2 r^2)T^2]\})\, d\omega\, d\Omega. \end{aligned}$$

If the system consists of m sub-systems, the ith sub-system ($i = 1, 2, \ldots, m$) has the distribution

$$\begin{aligned} dn_i &= K_i \exp(-2h_i^2 W)\exp\{-h_i^2[R^2 + (1 + \varkappa_i^2 r^2)T^2]\}\, d\omega\, d\Omega \quad (r \leqslant a_i), \\ dn_i &= 0 \quad (r > a_i), \end{aligned} \tag{9.150}$$

and the distribution for the whole system is given by the sum of the partial distributions:

$$dn = \sum_{i=1}^{m} dn_i.$$

In general, a stellar system may consist of a continuous sequence of concentric spherical sub-systems, one within the other. Replacing K_i, h_i and $\varkappa_i$ by continuously varying parameters $K(s)$, $h(s)$ and $\varkappa(s)$, we obtain the distribution in the integral form

$$dn = \int_r^a K(s)\exp[-2h^2(s)W]\exp\{-h^2(s)[R^2 + (1 + \varkappa^2(s)r^2)T^2]\}\, ds\, d\omega\, d\Omega. \tag{9.151}$$

Here $K(s)\, ds$ denotes a quantity which, apart from a constant factor, is equal to the total number of stars of all sub-systems whose outer boundaries lie between s and $s + ds$, and $h^2(s)$, $\varkappa^2(s)$ denote the corresponding parameters for these sub-systems.

Integrating over all velocity space and using the fact that $d\Omega = T\, dT\, d\psi\, dR$, where ψ is the polar angle in the transverse plane, we find for the star density the expression

$$\nu(r) = \pi^{3/2} \int_r^a \frac{K(s)\, ds}{h^3(s)\,[1 + \varkappa^2(s) r^2]}, \tag{9.152}$$

which is correct but for the number of stars escaping.

The expressions for the velocity variances in the radial direction $\sigma_R^2(r)$ and in the transverse direction $\sigma_T^2(r)$ are the following functions of r:

$$\sigma_R^2(r) = \frac{\pi^{3/2}}{2\nu(r)} \int_r^a \frac{K(s)\, ds}{h^5(s)\,[1 + \varkappa^2(s) r^2]}, \tag{9.153}$$

$$\sigma_T^2(r) = \frac{\pi^{3/2}}{2\nu(r)} \int_r^a \frac{K(s)\, ds}{h^5(s)\,[1 + \varkappa^2(s) r^2]^2}. \tag{9.154}$$

Since, as r increases towards the value a, an increasing number of "internal" systems (namely all those of radius less than r) must be omitted in calculating the variances from formulae (9.153) and (9.154), these variances must increase as $r \to a$. This result can easily be verified by using l'Hospital's rule to resolve the indeterminacy of the right-hand sides of (9.153) and (9.154) when $r \to a$. This gives $\sigma_R^2(r) \to 1/2h^2(a)$ and $\sigma_T^2(r) \to 1/2h^2(a)\,[1 + \varkappa^2(a)a^2]$ as $r \to a$. Thus we can draw two conclusions.

(1) A constant dispersion of radial velocities in a spherical system corresponds to an entirely uniform star population of the system. In actual systems the radial-velocity dispersion is a monotonically increasing function of r, whose value at the outer boundary of the system is equal to its maximum value for the individual sub-systems.

(2) For composite systems the phase distribution is the most probable one only for each sub-system separately, and is not the most probable for the whole system. This is because, according to the law of multiplication of probabilities, the maximum probability for the simultaneous occurrence of several independent random events (the phase distributions for the individual sub-systems) is equal to the product of the maximum probabilities for the separate events. Thus we should obtain the product of the most probable distributions for the individual sub-systems, and not their sum. The product of exponential functions is itself an exponential function, and the exponent is the sum of the exponents of the factors. We should therefore have the result that the dynamical parameters for the total distribution would be the sums of those for the sub-systems:

$$k_1 = k_1' + k_1'' + \ldots, \quad k_2 = k_2' + k_2'' + \ldots \quad . \tag{9.155}$$

This leads to a physically impossible conclusion: for example, since the radial-velocity variance is given by $-k_2 = 1/2\sigma_R^2$, we should find, since $|k_2| > |k_2'|$ etc., that the dispersion for the total distribution was less than those for the individual sub-systems.

CHAPTER X

DYNAMICS OF ROTATING STELLAR SYSTEMS

§ 1. The Masses and Rotation of Various Types of Galaxy

AMONG stellar systems the great majority are rotating systems. These include almost all galaxies. Direct spectroscopic examination has shown that all the spiral galaxies, and probably also the elliptical galaxies and the barred spirals, are rotating systems. The non-rotating galaxies include only a very few irregular ones (2·5% of the total number of observed galaxies, according to Hubble) and the spherical (E0) galaxies. The latter, however, must certainly be regarded as a limiting case of the elliptical galaxies, for which the rotation may be regarded as established. In the irregular galaxies any tendency towards regularity of form is accompanied by a tendency to rotate; for example, in the Magellanic Clouds structural features have recently been discerned which indicate that they should be classified as barred spirals.

The spiral galaxies are the most massive. According to the estimates of Zwicky [140], Page [94] and others, the masses of the spirals are of the order of 10^{10} to 10^{11} solar masses. The masses of the elliptical galaxies are of the same order of magnitude, but there are also many dwarf elliptical galaxies whose masses are less by two orders of magnitude, and are about the same as the masses of the nuclei of the spiral galaxies, i.e. of the order of 10^8 to 10^9 solar masses. The masses of the barred spirals are not accurately known, but are probably intermediate between those of the elliptical and the spiral galaxies.

Thus the galaxies may be divided according to mass into three groups: (a) the spirals, (b) the giant and dwarf elliptical galaxies and the nuclei of the spiral galaxies, and (c) the barred spirals.

The observed rotation of galaxies is almost always rigid rotation. Let us consider first the nuclei of the normal spirals. In all eight cases where observational data are available (M 31, NGC 4594, 4559, 3556, 3034, 7640, 6015, 4244) the rotation is that of a rigid body within the limits of experimental error.

The law of rotation in elliptical galaxies can be judged from only one member, namely NGC 3115 (type E 7), for which the measurements of Humason [39] at Mount Wilson also indicate rigid rotation.

The rotation of the normal spiral galaxies has been measured in two cases, M 31 and M 33, for which radial velocities at various distances from the centre have been obtained by Babcock [9] and by Mayall and Aller [70]. In both cases there was agreement with the theoretical formula deduced from Schwarzschild's law (cf.

Chapter V, §7): $V_0 = k_1\varrho/(1 + k_2\varrho^2)$, where V_0 is the circular velocity of the centroid and ϱ the distance from the axis of rotation. However, within the apparent boundaries of the galaxies, given by the condition $\varrho < 1/\sqrt{k_2}$, this formula becomes $V_0 \approx k_1\varrho$, corresponding to rigid rotation.

Finally, there are the barred spirals. For these, unfortunately, there are no spectroscopic measurements from which the law of variation of velocity of rotation with distance from the axis could be established. This is mainly because, when a galaxy is seen edgewise, it is difficult to assign it to a particular type; in other cases, the velocity component in the line of sight is too small to be observed. Yet it is obvious without any observation of the barred spirals that, if they are rotating, they cannot be rotating otherwise than as rigid bodies, because in that case the elongated body would be completely deformed after a single revolution.

Thus the law of rotation of the ordinary (S) spirals is the most complex; the remaining galaxies rotate as rigid bodies. Since the latter form (together with the nuclei of the spiral galaxies) the majority of stellar systems we may conclude that rigid rotation is the normal and most frequent type of internal motion in stellar systems.

§ 2. The Boundaries of the Galaxies

The problem of the outer boundaries of stellar systems arose only when it had been shown by the use of astronomical photography and self-recording microphotometers that the photographic images of galaxies extend considerably beyond their apparent boundaries, gradually falling away and vanishing at distances several times the visible dimensions. In theory, the absence of any sharp boundary of a stellar system should mean that its radius is infinite; and in fact, as we have seen in Chapter V, the ellipsoidal dynamics of galaxies, based on the hypothesis that Schwarzschild's law is universally valid, leads to an "isothermal" model† of the stellar system, with an infinite radius. Since the star density in this model decreases inversely as the distance, the mass of the system is infinite.

Thus both observation and theory seem to lead to the conclusion that the dimensions of stellar systems are entirely indeterminate. In terms of models, a good representation would be given by a model with a star density tending asymptotically to zero with increasing distance, i.e. having an infinite radius and a finite mass, but no such model has yet been devised.

We see that the question of the boundaries of stellar systems involves a fundamental contradiction: observation indicates that there is no sharp boundary and that the radius is practically infinite, but then theory requires an infinite mass, which clearly is physically impossible.

In the author's opinion [84] this contradiction does not really arise, because there is evidence that stellar systems have definite objective dimensions. First of all, it may be noted that what we call the apparent boundary of a galaxy certainly has an objective significance. This is shown by Hubble's well-known equation giving a one-

† That is, to a constant velocity dispersion in the direction of the cylindrical radius vector.

FIG 38

to-one relation between the diameter d of a galaxy and its total apparent magnitude m_T: $m_T + 5 \log_{10} d = c$.

When photographs of galaxies are carefully examined (for example, that of the elliptical galaxy NGC 4621 shown in Fig. 38), it is seen that the spherical component forms a *corona*, in which is embedded the denser plane component forming the main body of the galaxy. Only the latter has structural features; the corona has no structure. Hence any reference to structural differences between the different types of galaxy (S, SB, E) can apply only to their main central parts; in the spiral galaxies these are the plane components, and in the other galaxies they are the parts within the apparent boundaries.

We shall show that in stellar dynamics the volume of any stellar system may be naturally divided into two qualitatively different parts: an inner dense region (the main body) and an outer less dense region (the corona). The fundamental arguments of stellar dynamics assume that a phase density can be defined in a stellar system. This requires the definition of a macroscopic volume element at every point of the system. This volume element must be small compared with the structural features of the galaxy in the neighbourhood of the point considered, and at the same time it must be large enough to contain a sufficient number of stars. It is evident that, in any stellar system of finite mass, the region in which the concept of the macroscopic volume element is applicable is of finite diameter. As the star density decreases to zero, a point must be reached beyond which the macroscopic volume element does not exist, and then the phase density also is no longer meaningful. This limit is the boundary of the main part of the system; beyond it lies the corona, which in theory extends to infinity.

In the dynamics of stellar systems, the corona plays a part similar to that of the walls of a vessel in the theory of gases. The effect of the walls is that a molecule striking them rebounds with a velocity determined in magnitude and direction by the condition that a steady phase distribution should be maintained. The corona has exactly the same effect. While one star leaves the main body and enters the corona, another star which has completed the part of its quasi-Keplerian orbit lying in the corona returns to the main body. If the phase distribution is steady, the exchange of stars between the main body and the corona will maintain it exactly as the walls of a vessel do. The only difference between a gas and a stellar system is that in the former, when the boundary is reached, the same molecule returns into the gas, whereas in the latter a star reaching the boundary is replaced by a different star. Within the main body, however, the stars are indistinguishable, in the sense that only the distribution of phase paths, and of stars in those paths, are of significance in the dynamics of a stellar system. Which particular stars are in which paths at any given instant is of no importance in stellar dynamics, especially as the stars are continually changing from one orbit to another under the action of the irregular forces.

Another important fact is that the surfaces of equal star density in the corona resemble a system of concentric ellipsoids; see, for example, Baade's well-known three-colour photographs of the galaxy NGC 4594, or the photograph of the elliptical galaxy NGC 4621 in Fig. 38. The attraction of the corona at points within the main body, which is in any case small because of the small mass of the corona, is therefore entirely negligible because the attraction of ellipsoidal layers at internal points is zero.

Thus galaxies may be regarded as enclosed in rigid shells which can be penetrated only by stars whose velocities exceed the velocity of escape.

We shall see later that statistical mechanics leads, in agreement with the conclusions of the classical ellipsoidal dynamics, to the result that the velocity distribution in galaxies is "isothermal". Hence, by analogy with the theory of gases, we can say that the corona forms an adiabatic envelope surrounding the galaxy. The above discussion shows that in the inner part (main body) of the galaxy, where its structure is defined by specifying the phase density, the system resembles a continuous medium, but in the outer part (corona) it is discrete. Within the main body, the orbits of individual stars are of almost no importance, like the paths of the individual molecules in a gas in the laboratory. In the corona, the orbits of individual stars are of great importance, and the internal parameters of the corona, such as the star density, the velocity ellipsoid, etc., are determined by the distribution function of the elements of the individual orbits. The motions of the stars in the corona are largely determined by the attraction of the main body of the galaxy.

Of course, the above arguments do not imply that there are no paths of individual stars within the main body. This is evident from the fact that, in a steady state, any phase point uniquely defines an orbit. Hence the phase density uniquely defines the distribution of (elements of) orbits. In a steady state (when the phase density is independent of time) the distribution of orbits remains constant, and therefore the irregular forces can only cause stars to change from one orbit to another, and the number of stars in every orbit at every instant is the same. Hence a study of individual orbits within the main body is possible, but it does not suffice to resolve the fundamental problem of stellar dynamics, namely to deduce the theoretical equilibrium figures of stellar systems.

The definition of a specific boundary between the main body and the corona involves, of course, some indeterminacy, but observationally the determination of the outer boundaries of the galaxies offers no difficulty. The task of the theory must be to determine the shape and size of galaxies of various types, given their dynamical parameters (mass, energy, angular momentum, etc.). In a correctly constructed theory, the theoretical boundary should approximately coincide with the observed boundary. We shall see that this may be in fact fairly well achieved.

Thus the development of a theory of the structure of galaxies falls into two related but essentially different parts: the study of the structure of the inner region (the main body), in which the methods of hydrodynamics are pre-eminent, and that of the structure of the corona, in which the methods of the statistics of discrete ensembles are the more important. The purpose of investigating the internal structure is to set up a theory of the equilibrium figures and laws of evolution of the central parts of galaxies, lying within the apparent boundaries. The purpose of investigating the corona is to determine the distribution of individual star orbits. The common boundary between the main body and the corona, which may be called the *effective boundary* of the galaxy, serves to link the solutions to the two problems.

In this respect the dynamics of galaxies exhibits an analogy with the theory of stellar structure, which may be divided into the theory of the internal structure of the stars and the theory of stellar atmospheres. These regions in stars are qualitatively different, and therefore require the use of different methods of investigation. As in stellar physics, so in galactic dynamics it is incorrect to attempt to examine the whole range from the centre to the outermost layers by the same methods.

In this chapter we shall consider only the dynamics of the main body of a galaxy. The potential of the main body at an external point must be known in order to investigate the structure of the corona. Thus the dynamics of the main body must take precedence.

§ 3. The Method of Additive Parameters

It has already been pointed out that the main body of a stellar system may be regarded both as a discrete medium in which particles (the stars) move and as a continuous medium in which centroids move. Thus we have two approaches to the problems of stellar dynamics, the statistical and the hydrodynamical.

In Chapter VIII it was shown that a purely hydrodynamical treatment is insufficient, since the hydrodynamical equations involve quantities which are essentially statistical, namely the variances of the residual stellar velocities. To solve problems of stellar dynamics it is therefore necessary to combine the hydrodynamical and statistical methods.

The same conclusion is reached if we start by considering the nature of the regular and irregular forces acting in a stellar system. It has been shown in Chapter IV that there must exist in stellar systems an efficient action of irregular forces equivalent to that of viscosity in a liquid. Thus a stellar system may be regarded as a continuous medium, and so the equations of stellar hydrodynamics may be applied to it; on the other hand, any considerable irregular forces in a stellar system act as a statistical process which tends to establish the most probable phase distribution.

In finding the most probable distribution in rotating systems, we shall use the method of additive parameters. This has already been applied (though without any detailed theoretical justification) to spherical systems. We shall now give a more complete account of the method.

According to Jeans' theorem, the phase density f must be a function only of the one-valued integrals of the motion of a free particle. For example, in a stellar system where the mass distribution has rotational symmetry, there are two such integrals: the energy integral and the angular-momentum integral. In cylindrical coordinates these are

$$I_1 = \mathrm{P}^2 + \Theta^2 + Z^2 - 2U(\varrho, z), \quad I_2 = \varrho\Theta, \tag{10.1}$$

where $U(\varrho, z)$ is the gravitational potential. Hence

$$f = F(I_1, I_2), \tag{10.2}$$

where F is an arbitrary function.

The method of additive parameters [86] consists in forming functions $\varphi(I_1, I_2)$ of the integrals I_1 and I_2 such that their sum over all stars in the system is an *additive parameter*, i.e. a quantity expressing some characteristic property of the system.

Let

$$\left.\begin{aligned} P_1 &= \sum_{i=1}^{N} m_i \, \varphi_1(I_{1i}, I_{2i}), \\ P_2 &= \sum_{i=1}^{N} m_i \, \varphi_2(I_{1i}, I_{2i}), \\ &\quad \ldots, \\ P_n &= \sum_{i=1}^{N} m_i \, \varphi_n \, (I_{1i}, I_{2i}) \end{aligned}\right\} \tag{10.3}$$

be a set of additive parameters satisfying two conditions: (1) they are independent, (2) they uniquely define the stellar system. In (10.3) I_{1i}, I_{2i} and m_i denote the values of the integrals and the mass for the ith star, and $\varphi_1, \varphi_2, \ldots, \varphi_n$ are *characteristic functions* formed as indicated above.

Any set of additive parameters satisfying the conditions (1) and (2) will be called a *defining set*. Each stellar system has in general an infinity of mutually equivalent defining sets of parameters. This is seen, for example, from the fact that any linear combination of additive parameters of a given system is itself an additive parameter.

If there is, in a stellar system, an efficient relaxation mechanism in the form of irregular forces, the phase density cannot differ greatly from its most probable value corresponding to a system with the same physical characteristics and therefore the same values of the additive parameters.

Having found some defining set of additive parameters such as (10.3), we then seek the most probable phase distribution such that the parameters $P_1, P_2, \ldots, P_n$ have given values. Then, as we shall see, Jeans' theorem is necessarily satisfied.

The simplest type of stellar system is the *dynamically determinate system* or *D system*. This concept may be described as follows. The simplest additive parameters are

$$\left.\begin{aligned} (1)\ P_1 &= \sum_{i=1}^{N} m_i, \\ (2)\ P_2 &= \sum_{i=1}^{N} m_i \, I_{1i}, \\ (3)\ P_3 &= \sum_{i=1}^{N} m_i \, I_{2i}, \end{aligned}\right\} \tag{10.4}$$

where I_{1i} and I_{2i} are the right-hand sides of (10.1) for the individual stars. We shall show that the constancy of these parameters follows from general theorems of mechanics for a closed system of particles.

The physical significance of the parameter P_1 is evident: it is the mass of the system. Thus the constancy of P_1 corresponds to the conservation of mass. We shall call it the *mass parameter* and henceforward denote it by M.

The second parameter P_2 is twice the sum of the energies of the individual stars in the total gravitational field of the system. Explicitly,

$$P_2 = \sum_{i=1}^{N} m_i \, (\mathrm{P}_i^2 + \Theta_i^2 + Z_i^2 - 2U_i). \tag{10.5}$$

Here

$$U_i = U(\varrho_i, z_i) = G \sum_{k=1}^{N}{}' m_k / r_{ik}, \tag{10.6}$$

where G is the gravitational constant and r_{ik} the distance between the ith and kth stars; the prime signifies that the term with $i = k$ is not included in the sum.

In the quasi-steady stellar systems which are considered here, the gravitational force is independent of time. Under these conditions every star has an energy integral just as if it were moving in some external field of force. The sum of these integrals, P_2, is therefore constant. P_2 can easily be expressed in terms of the energy of the whole system. Firstly, we have

$$\sum_{i=1}^{N} m_i U_i = -2W, \tag{10.7}$$

that is, twice the potential energy of the system (with reversed sign), since in the summation each star appears twice (once as attracting and once as being attracted).†
Next, the sum

$$\sum_{i=1}^{N} m_i (\mathrm{P}_i^2 + \Theta_i^2 + Z_i^2) = 2T \tag{10.8}$$

is twice the kinetic energy of the system, and so

$$P_2 = 2(T + 2W). \tag{10.9}$$

Thus, although the parameter P_2 is not twice the total energy of the system, it is nevertheless constant. This is because, for a system of particles in a steady state, we have both law of conservation of energy ($T + W =$ constant) and the virial theorem ($2T + W = 0$). In a steady state, both the kinetic and the potential energy therefore have constant values, but only the combination P_2 is additive with respect to the energy integrals of the individual stars. We shall call P_2 the *energy parameter* and henceforward denote it by E.

Finally, P_3 is the sum of the angular momenta of the individual stars about the axis of symmetry of the system. In what follows we shall consider only stellar systems rotating about the axis of symmetry.†† Then the constancy of P_3 follows from the law of conservation of angular momentum. We shall call P_3 the *angular-momentum parameter* and denote it by K. The quantities M, E and K will be referred to as *dynamical parameters*.

In general the dynamical parameters do not form a defining set, i.e. they are insufficient for a unique definition of a stellar system of the type considered. The general theorems of mechanics give no integrals, except those of energy and angular momentum, for an arbitrary closed system of particles.††† Hence it follows that all additive parameters except those mentioned above must be non-dynamical, i.e. must be derived

† See Chapter V, § 11, for further details regarding the additivity of the potential energy.

†† In other words, we exclude the case where, besides a pure rotation, there are also precessional and nutational motions. These could hardly occur except in a stellar system subject to external agencies.

††† We ignore, of course, the integrals of uniform motion of the centre of mass, which do not play any part in the present discussion.

from statistical physics. Thus stellar systems can be divided into two classes: dynamically determinate and dynamically indeterminate. For dynamically determinate systems all the additive parameters needed for a unique definition of a system are integrals of the motion. It is evident that every stellar system must have the three dynamical parameters among its defining parameters. Thus the dynamically determinate systems are the simplest from the viewpoint of statistical mechanics, and we shall therefore begin by considering only such systems; the discussion of dynamically indeterminate systems is postponed until the end of the chapter.

§ 4. The Most Probable Phase Distribution for the Simplest Types of Galaxy

Let us consider a D system of stars (a galaxy), defined by the three dynamical additive parameters: the mass M, the energy E and the angular momentum K [85]. We shall suppose that the system rotates about its axis of rotational symmetry, and for simplicity we assume that the mass of every star is the same. Then, if N is the total number of stars in the system, the three parameters are

$$N,\ E = \sum_{i=1}^{N} (\mathrm{P}_i^2 + \Theta_i^2 + Z_i^2 - 2U_i), \quad K = \sum_{i=1}^{N} \varrho_i \Theta_i . \tag{10.10}$$

Since the main body of the galaxy (which is here considered) occupies a finite volume and the velocities are limited by the velocity of escape, we can suppose that the phase volume Γ of the system is finite. Furthermore, in accordance with Chapter IV, §§ 8, 9, we shall assume that there is an efficient action of irregular forces which tends to bring the phase distribution into its most probable form.

To find the most probable phase distribution, we use the *method of phase cells*. To do this, we divide the phase volume occupied by the system into a large number m of very small phase cells, which must yet be such that

$$1 \ll m \ll N; \tag{10.11}$$

that is, the mean population N/m of each phase cell must be considerably greater than unity (for example, some tens or hundreds). Let γ_μ be the volume of the μth cell. Clearly

$$\sum_{\mu=1}^{m} \gamma_\mu = \Gamma . \tag{10.12}$$

We now make an assumption usually made in such calculations, namely that the *a priori* probability of a star's being in the μth cell is the same for every star, independent of the position of the cell in the phase volume Γ, and equal to the geometrical probability:

$$p_\mu = \gamma_\mu / \Gamma . \tag{10.13}$$

This hypothesis is evidently equivalent to assuming that, without the restrictions imposed by the conditions $E =$ constant, $K =$ constant, the phase distribution would be uniform (statistically homogeneous) throughout the volume Γ.

The binomial law shows that the probability of the respective cell populations being $n_1, n_2, \ldots, n_m$ when stars are randomly placed in them is

$$P = \frac{N!}{n_1!\, n_2! \ldots n_m!}\, p_1^{n_1}\, p_2^{n_2} \ldots p_m^{n_m}. \tag{10.14}$$

Within each cell the phase coordinates of all points may be regarded as the same. Hence the eqns. (10.10) may be written

$$\left.\begin{aligned} N &= \sum_{\mu=1}^{m} n_\mu, \\ E &= \sum_{\mu=1}^{m} n_\mu\, (\mathrm{P}_\mu^2 + \Theta_\mu^2 + Z_\mu^2 - 2U_\mu) \\ &= \sum_{\mu=1}^{m} n_\mu\, E_\mu, \\ K &= \sum_{\mu=1}^{m} n_\mu\, \varrho_\mu\, \Theta_\mu = \sum_{\mu=1}^{m} n_\mu\, K_\mu, \end{aligned}\right\} \tag{10.15}$$

where the suffix μ denotes the value corresponding to the μth cell.

The problem now reduces to that of finding the maximum of the probability as a function of the variables $n_1, n_2, \ldots, n_m$, subject to the conditions (10.15). Since our purpose is to use the maximum to change to a continuous differentiable phase density (satisfying the Boltzmann equation), the arguments may be regarded as continuous.

The Lagrangian function for the conditional extremum problem can be taken as

$$L = \log P + k_1 \sum_\mu n_\mu + k_2 \sum_\mu n_\mu\, E_\mu + 2k_3 \sum_\mu n_\mu\, K_\mu, \tag{10.16}$$

where k_1, k_2, k_3 are Lagrangian multipliers. In the terminology of physics $\log P$ is the entropy, and the problem is seen to be equivalent to that of determining a conditional maximum of the entropy. Differentiation with respect to n_μ gives

$$\frac{\partial L}{\partial n_\mu} = -\frac{\mathrm{d} \log n_\mu!}{\mathrm{d} n_\mu} + \log p_\mu + k_1 + k_2 E_\mu + 2k_3 K_\mu.$$

To simplify the derivative of $\log n_\mu!$ we assume that in the cell considered

$$n_\mu \gg 1. \tag{10.17}$$

Then we can put with sufficient accuracy

$$\frac{\mathrm{d} \log n_\mu!}{\mathrm{d} n_\mu} = \frac{\log (n_\mu + 1)! - \log n_\mu!}{(n_\mu + 1) - n_\mu} = \log (n_\mu + 1) = \log n_\mu. \tag{10.18}$$

Substituting this in the formula for $\partial L/\partial n_\mu$, we find, after some obvious transformations, and omitting the suffix μ, which is now unnecessary,

$$n = p \exp (k_1 + k_2 E + 2k_3 K). \tag{10.19}$$

Instead of the cell population we may use the phase density f, where

$$n = fp. \tag{10.20}$$

Then the most probable phase density is

$$f = \exp(k_1 + k_2 E + 2k_3 K). \tag{10.21}$$

The condition $n_\mu \gg 1$ (10.17) which has been imposed is of fundamental importance. It arises from the necessity of changing from a discrete to a continuous distribution, and so is due not to the particular mathematical treatment used but to the nature of the physical problem. From this condition, moreover, there follows the very important result that all conclusions concerning the phase distribution law are valid only for values of the phase coordinates such that the population of the phase cells is fairly large. In particular the expression (10.21) which we have found for the phase density is no longer meaningful for large values of the coordinates (beyond the main body of the system) or large values of the velocities (beyond the velocity of escape).

From (10.21) we see that the phase density depends on the velocities and coordinates only through the integrals of the motion. Thus we return to Jeans' theorem.

To ascertain the physical significance of this solution, we substitute in (10.21) $E = \mathrm{P}^2 + \Theta^2 + Z^2 - 2U$, $K = \varrho\Theta$. This gives the most probable phase distribution for a statistically homogeneous D system with rotational symmetry:

$$f(\mathrm{P}, \Theta, Z, \varrho, z) = \exp(k_1) \exp[-2k_2(U + \tfrac{1}{2}\Theta_0^2)] \exp\{k_2[\mathrm{P}^2 + (\Theta - \Theta_0)^2 + Z^2]\}, \tag{10.22}$$

where

$$\Theta_0 = -k_3\varrho/k_2. \tag{10.23}$$

From (10.22) it is seen that in this case the velocity variances in all three directions are equal:

$$\sigma^2 = -1/2k_2 = \text{constant}. \tag{10.24}$$

Hence formula (10.22) gives a Maxwellian distribution of residual velocities, and also the distribution of coordinates if the potential $U = U(\varrho, z)$ is known. The distribution law (10.22) may therefore be called the *Maxwell–Boltzmann* law. This completes the proof of

THEOREM I: *The most probable law of distribution of the residual velocities and coordinates of stars at every point of a dynamically determinate and statistically homogeneous galaxy is the Maxwell–Boltzmann law, with a dispersion the same at every point – an "isothermal" distribution.*

§ 5. The Phase Distribution for Dynamically Determinate Galaxies

Let us now proceed to investigate the phase distribution found in § 4, and first of all note that the quantity Θ_0 which appears in formula (10.22) is just the centroid velocity. We see that this velocity is perpendicular to the galactic meridian through the point considered. In accordance with our initial hypotheses, we thus find a rotation of the galaxy about the axis of symmetry, and Θ_0 gives the circular velocity of the centroid.

Formula (10.23) shows that Θ_0 is proportional to ϱ. Thus the angular velocity of the centroid is

$$\omega_0 = \Theta_0/\varrho = -k_3/k_2 = \text{constant}; \tag{10.25}$$

we therefore have

THEOREM II: *The most probable form of rotation of a dynamically determinate and statistically homogeneous galaxy is rigid rotation.*

It has been shown at the beginning of this chapter that the existing observational data are in agreement with Theorem II.

To obtain the star density, we must integrate the phase density (10.22) over the part of velocity space which corresponds to the phase volume occupied by the system. However, as shown in Chapter IX, § 4, the truncation of the distribution law in velocity space is in general small.† Hence we can obtain an approximate value for the star density by integrating the phase density (10.22) over all velocity space. Then, using (10.23) and (10.24), we have

$$\nu = C_0 \exp\left(\frac{U}{\sigma^2} + \frac{\omega_0^2 \varrho^2}{2\sigma^2}\right), \tag{10.26}$$

where

$$C_0 = (2\pi)^{3/2} \sigma^3 \exp(k_1) \tag{10.27}$$

is a constant determined from the condition that the integral of ν over the main body of the galaxy should equal N, the number of stars in it.

The above treatment makes it possible to ascertain the physical significance of the Lagrangian multipliers. By (10.24) k_2 determines the dispersion, by (10.25) k_3 gives the angular velocity of rotation, and by (10.27) k_1 determines the mass; to find the mass we must know, besides k_1, the potential (and therefore the mass distribution) and the size and shape of the galaxy, i.e. the dynamical problem must have been solved completely.

From (10.26) it follows immediately that the equation for the level surfaces of star density is

$$U + \tfrac{1}{2}\omega_0^2 \varrho^2 = C. \tag{10.28}$$

This is exactly the same as the corresponding equation for the classical problem of the equilibrium figures of rotating liquids, and we shall make extensive use of it.

Formula (10.26) still does not give the star density explicitly, because the potential U appears on the right-hand side. The potential is, however, related to the star density ν by Poisson's equation:

$$\Delta U = -4\pi G m \nu, \tag{10.29}$$

where G is the gravitational constant and m the mean mass of one star.

To eliminate the potential from formula (10.26), we take logarithms and then apply to both sides the Laplacian operator in cylindrical coordinates:

$$\Delta = \frac{1}{\varrho}\frac{\partial}{\partial \varrho}\left(\varrho \frac{\partial}{\partial \varrho}\right) + \frac{1}{\varrho^2}\frac{\partial^2}{\partial \theta^2} + \frac{\partial^2}{\partial z^2}.$$

† The estimates given in Chapter IX, § 4, for spherical systems are *a fortiori* applicable to the case considered here, since they are based on Maxwell's law.

Then, using Poisson's eqn. (10.29), we obtain a differential equation for the star density:

$$\sigma^2 \triangle \log \nu = -4\pi G m \nu + 2\omega_0^2. \tag{10.30}$$

This is the fundamental equation to determine the mass distribution in D galaxies. In explicit form it is

$$\sigma^2 \left[\frac{1}{\varrho}\frac{\partial}{\partial \varrho}\left(\varrho \frac{\partial \log \nu}{\partial \varrho}\right) + \frac{1}{\varrho^2}\frac{\partial^2 \log \nu}{\partial \theta^2} + \frac{\partial^2 \log \nu}{\partial z^2}\right] = -4\pi G m \nu + 2\omega_0^2.$$

In the case considered (a system with rotational symmetry), the middle term in the brackets must be omitted. Then the basic equation becomes

$$\sigma^2 \left[\frac{1}{\varrho}\frac{\partial}{\partial \varrho}\left(\varrho \frac{\partial \log \nu}{\partial \varrho}\right) + \frac{\partial^2 \log \nu}{\partial z^2}\right] = -4\pi G m \nu + 2\omega_0^2. \tag{10.31}$$

This equation may also be derived "hydrodynamically" if we start by assuming that the system is rotating rigidly and has an isothermal velocity distribution. Then we must put $\sigma_\varrho^2 = \sigma_\theta^2 = \sigma^2 = \text{constant}$, $\Theta_0 = \omega_0 \varrho$ in the hydrodynamical eqns. (8.48) for a rotationally symmetrical system, and this gives

$$\sigma^2 \frac{\partial \log \nu}{\partial \varrho} - \omega_0^2 \varrho = \frac{\partial U}{\partial \varrho}, \quad \sigma^2 \frac{\partial \log \nu}{\partial z} = \frac{\partial U}{\partial z}.$$

Differentiating these equations with respect to ϱ and z respectively and using Poisson's eqn. (10.29), we again obtain the fundamental eqn. (10.31).

Equation (10.31) is too complex to warrant an attempt at integrating it in the general case, but one particular solution is obvious. The star density appears on the left-hand side only as a differentiated logarithm, and so, if we consider only sufficiently "smooth" models, the magnitude of the left-hand side is small. Putting it equal to zero, we have the equation $2\pi G m \nu - \omega_0^2 = 0$, which gives the particular (and exact) solution

$$\nu_0 = \omega_0^2 / 2\pi G m. \tag{10.32}$$

This result is very important, and opens the way to further deductions. It may be stated as

THEOREM III: *In any dynamically determinate galaxy, the star density, if sufficiently smooth, is constant in a first approximation and proportional to the square of the angular velocity of rotation.*

Before going on to further deductions, we should make one general remark. The solution $\nu = \text{constant}$ is an exact mathematical solution of the equation, but it cannot be an *exact* solution of any physical problem. Equation (10.26) shows that, when $\nu = \text{constant}$, the potential depends only on ϱ. This is physically impossible, because then the z-component of the gravitational force (along the axis of symmetry of the galaxy) would be zero, and the component along the radius ϱ would be independent of

z. Thus the galaxy would be a homogeneous circular cylinder of infinite length rotating about its axis. This case has no physical significance. To obtain a physically meaningful model of a galaxy, the solution derived above must be regarded as a first approximation, and the second approximation to a solution of (10.31) must be found.

However, if we assume that the solution obtained gives the approximate value of the star density, some conclusions concerning the dynamical stability of D galaxies may be drawn from (10.32). In Chapter V, § 12, we deduced from the virial theorem that, in a stellar system in a steady or linearly non-steady state rotating with angular velocity ω, the quantity

$$\Omega = \omega^2/2\pi G m \nu = \omega^2/2\pi G \delta \tag{10.33}$$

must be less than unity (Poincaré's inequality). If the angular velocity ω increases to the point where the inequality (5.176) ceases to hold, the system becomes dynamically unstable.

The product $m\nu$ is the density of matter δ, and so formula (10.32) can be written

$$\Omega_0 = \omega_0^2/2\pi G \delta_0 = 1. \tag{10.34}$$

Hence we see that the value of the star density which we have obtained as a first approximation does not satisfy Poincaré's condition regarding the maximum angular velocity.† This demonstrates

THEOREM IV: *A dynamically determinate galaxy with a quasi-constant star density (i.e. one which is constant to a first approximation) is dynamically unstable.*

Here and henceforward the term "instability" signifies a state of the system in which it cannot be in equilibrium. If the system has been formed in such a state, it will, under the conditions stated, begin to disintegrate. Thus the first approximation to the solution of eqn. (10.31) leads to a dynamically unstable model of a galaxy.

To obtain a dynamically stable model, a more accurate value for the star density ν must be derived. We put $\nu = \nu_0 + \nu_1$, where ν_0 is the first approximation (10.32) already found, and ν_1 is a quantity small compared with ν_0. The uniform density ν_0 will be called the *basic density* of the galaxy, and ν_1 the *additional density*. If the latter is positive the quantity

$$\Omega = \omega_0^2/2\pi G m \nu \tag{10.35}$$

is reduced, and so the dynamical instability of the galaxy is lessened. Hence we have the following important conclusion: if $\nu_1 \geqslant 0$ in a stellar system, the system may continue to be an equilibrium figure although ν_1 is so small that the system is in general near the limit of stability.

We shall see later that the principal effect of the additional density is that it causes the formation of a central nucleus such as is in fact observed in the majority of galaxies. Thus we should expect that the presence of a central condensation of this type is, at least in the simplest cases, a necessary condition for the equilibrium of galaxies.

† Poincaré's condition as used here was derived in Chapter V, § 12, only for ellipsoidal equilibrium figures, but it is applicable to the present case because, as we shall show in § 6, the equilibrium figures of D galaxies are almost ellipsoidal.

It will also be shown that the additional density leads to the appearance of certain structural features of galaxies, in particular a ring structure and a series of condensations resembling beads which are sometimes observed in galaxies (for example, NGC 4324).

§ 6. Equilibrium Figures of *D* Galaxies

Throughout this chapter we are assuming that the stellar system considered is in a quasi-steady state. This means that at every instant the system is in dynamic equilibrium. In such a state, a rotating system takes up some equilibrium figure analogous to that of a rotating mass of liquid. The analysis given above has shown that dynamically determinate galaxies have the following properties: (1) they rotate as rigid bodies, (2) they are quasi-homogeneous as regards mass distribution (the density being almost the same everywhere), (3) the equation of the level surfaces of the star density has the same form as the corresponding equation in the theory of equilibrium figures of rotating liquids. In Chapter VIII it has been shown that stellar systems behave as continuous media in the sense that the methods of hydrodynamics can be applied to them. These results make possible the application to galaxies of the classical theory of the equilibrium figures of rotating homogeneous liquids.

A detailed account of the extensive literature on the classical theory is beyond the scope of the present book; in this section we shall briefly recount the principal results of the theory, which can then be used to develop a theory of equilibrium figures of stellar systems in rigid rotation. For further details the reader is referred to the specialised literature.†

The condition for a homogeneous ellipsoid

$$\frac{x^2}{a^2} + \frac{y^2}{b^2} + \frac{z^2}{c^2} = 1 \tag{10.36}$$

with constant density

$$\delta = M/\tfrac{4}{3}\pi abc \tag{10.37}$$

to be an equilibrium figure is the same as the condition for the surface (10.36) to be a level surface of the gravity force potential (the gravity force being the resultant of the gravitational force and the centrifugal force):

$$U + \tfrac{1}{2}\omega^2(x^2 + y^2) = C, \tag{10.38}$$

where U is the potential at an internal point of the ellipsoid and C a constant.

It is known from the theory of the potential of a homogeneous ellipsoid that U is of the form

$$U = A - Px^2 - Qy^2 - Rz^2, \tag{10.39}$$

† See, for example, M. F. Subbotin ([127], especially Chapter VI); P. É. Appell, *Equilibrium figures of a rotating homogeneous liquid* (*Figures d'équilibre d'une masse liquide homogène en rotation*), *Traité de Mécanique Rationelle* **4** (1), 2nd ed., Gauthier-Villars, Paris, 1932. These works have been extensively utilised in the discussion given here.

where P, Q, R are constants given by the formulae

$$\left.\begin{aligned} P &= k \int_0^\infty \frac{ds}{(a^2 + s)\,\Delta}, \\ Q &= k \int_0^\infty \frac{ds}{(b^2 + s)\,\Delta}, \\ R &= k \int_0^\infty \frac{ds}{(c^2 + s)\,\Delta}, \end{aligned}\right\} \tag{10.40}$$

with

$$k = \pi G \varrho abc, \quad \Delta = \sqrt{[(a^2 + s)(b^2 + s)(c^2 + s)]}. \tag{10.41}$$

The constant A is unimportant. These formulae are valid for any point within or on the ellipsoid.

In order that eqns. (10.38) and (10.36) should be identical, we must have

$$a^2(P - \tfrac{1}{2}\omega^2) = b^2(Q - \tfrac{1}{2}\omega^2) = c^2 R. \tag{10.42}$$

Putting

$$u = c^2/a^2, \quad v = c^2/b^2, \quad t = s/c^2, \tag{10.43}$$

we obtain

$$\left.\begin{aligned} P &= \pi G \varrho \int_0^\infty \frac{u\,dt}{(1 + ut)\,D}, \\ Q &= \pi G \varrho \int_0^\infty \frac{v\,dt}{(1 + vt)\,D}, \\ R &= \pi G \varrho \int_0^\infty \frac{dt}{(1 + t)\,D}, \\ D &= \sqrt{[(1 + t)(1 + ut)(1 + vt)]}. \end{aligned}\right\} \tag{10.44}$$

Again putting

$$\Omega = \omega^2/2\pi G \delta \tag{10.45}$$

and taking the first and second members of (10.42) with the third, we find two different expressions for Ω:

$$\Omega = \int_0^\infty \frac{u(1 - u)\,t\,dt}{(1 + t)(1 + ut)\,D} = \int_0^\infty \frac{v(1 - v)\,t\,dt}{(1 + t)(1 + vt)\,D}, \tag{10.46}$$

which must be simultaneously valid. From these equations we see immediately that the ellipsoid can be an equilibrium figure only if $u < 1$ and $v < 1$, i.e. if the rotation is about the shortest principal axis (about which the moment of inertia is greatest).

Equating the last two members of (10.46), we obtain the condition for an equilibrium ellipsoid to exist:

$$(u - v) \int_0^\infty (1 - u - v - uvt)\, t\, \mathrm{d}t / D^3 = 0, \tag{10.47}$$

where D is given by the last eqn. (10.44). This equation can hold only if either $u = v$, or $u \neq v$ but the integral is zero. The first condition obviously gives a spheroid; the second, as we shall see, gives an ellipsoid with three unequal axes.

(a) $u = v$. The equilibrium ellipsoid is a spheroid, called a *Maclaurin ellipsoid.* In this case both formulae (10.46) give

$$\Omega = \int_0^\infty \frac{u(1 - u)\, t\, \mathrm{d}t}{(1 + ut)^2 (1 + t)^{3/2}}.$$

The integral is easily evaluated, and gives

$$\Omega = \frac{1 + 2u}{1 - u} \sqrt{\frac{u}{1 - u}} \tan^{-1} \sqrt{\frac{1 - u}{u}} - \frac{3u}{1 - u} = \varphi(u), \text{ say.} \tag{10.48}$$

To find the flattening of the spheroid corresponding to a given angular velocity, we must solve eqn. (10.48) for u. The resulting value of $u = c^2/a^2$ gives the shape of the spheroid. By examining the function $\varphi(u)$ it is easy to see that this function is zero for $u = 0$ and $u = 1$, and within that range has a single maximum at $u_0 = 0{\cdot}1352\ldots$, which gives the maximum value of Ω:

$$\Omega_0 = 0{\cdot}22467\ldots. \tag{10.49}$$

Thus the Maclaurin equilibrium ellipsoids do not exist for $\Omega > \Omega_0$. For $\Omega = \Omega_0$ there is one such ellipsoid, and for any $\Omega < \Omega_0$ there are two such ellipsoids, one with $u < u_0$, corresponding to the ascending part of the curve $\Omega = \varphi(u)$, and one

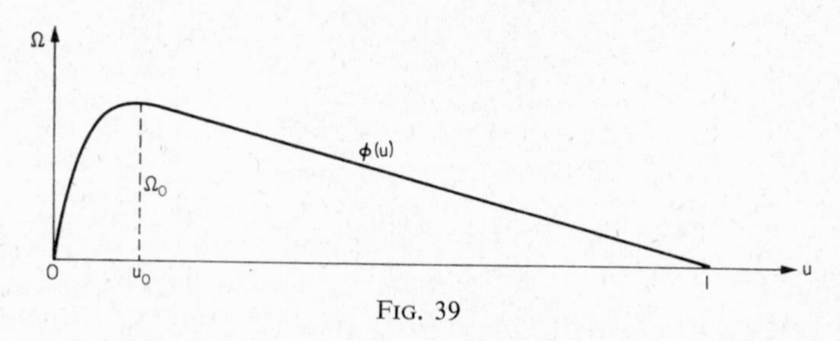

FIG. 39

with $u > u_0$, corresponding to the descending part (Fig. 39). As $\Omega \to 0$, the first ellipsoid tends to a sphere and the second to a thin disc of infinite radius. The first series are called *Maclaurin ellipsoids of the first kind* or *planet-like*, and the second series are called *Maclaurin ellipsoids of the second kind* or *disc-like*.

(b) $u \neq v$. The condition for the existence of an ellipsoid of equilibrium with three unequal axes, corresponding to the condition $u \neq v$, is that the equation

$$\int_0^\infty (1 - u - v - uvt)\, t\, \mathrm{d}t/D^3 = 0 \tag{10.50}$$

should hold. The corresponding equilibrium ellipsoids are called *Jacobi ellipsoids*. The corresponding value of Ω is given either by eqn. (10.46), or by the equivalent symmetrical formula obtained by multiplying the first by v, the second by u, subtracting, and dividing the result by $v - u$:

$$\Omega = uv \int_0^\infty \frac{t\, \mathrm{d}t}{(1 + ut)(1 + vt)\, D}. \tag{10.51}$$

The geometrical significance of this equation may be made clearer by using the functions

$$\begin{aligned} \Phi(u, v) &= uv \int_0^\infty \frac{t\, \mathrm{d}t}{(1 + ut)(1 + vt)\, D}, \\ \Psi(u, v) &= \int_0^\infty (1 - u - v - uvt)\, t\, \mathrm{d}t/D^3. \end{aligned} \tag{10.52}$$

Then the conditions for a Jacobi equilibrium ellipsoid to exist are

$$\begin{aligned} \Phi(u, v) &= \Omega, \\ \Psi(u, v) &= 0. \end{aligned} \tag{10.53}$$

In rectangular coordinates (u, v, Ω), eqns. (10.53) represent a curve in space. It is easily seen that the end-points of the curve are $A(0, 1, 0)$ and $B(1, 0, 0)$. A more detailed investigation, which we omit, shows that the curve (10.53) has a single maximum when

$$u = v = u_1 = 0{\cdot}3396\ldots, \tag{10.54}$$

at which

$$\Omega = \Omega_1 = 0{\cdot}1871\ldots. \tag{10.55}$$

The corresponding Jacobi ellipsoid is also a Maclaurin ellipsoid.

The curve on the surface $\Omega = \Phi(u, v)$ which is given by the equation $u = v$ gives a sequence of Maclaurin ellipsoids, and that given by the equation $\Psi(u, v) = 0$ gives a sequence of Jacobi ellipsoids. The point of intersection of these curves corresponds to the ellipsoid given by (10.55), which is both a Jacobi ellipsoid and a Maclaurin ellipsoid. This point (E in Fig. 40) is a *branch point* or *bifurcation point* with regard to the class of all possible equilibrium ellipsoids. It is easily seen that the curve AEB of Jacobi ellipsoids is symmetrical about E, since both eqns. (10.53) are symmetrical in the parameters u and v. Thus points on the curve which lie symmetrically about E correspond to the same Jacobi ellipsoid, differing only by a rotation through 90° about the c-axis.

Summarising, we may say that to each value of Ω from 0 to $\Omega_1 = 0{\cdot}1871\ldots$ there corresponds one and only one Jacobi ellipsoid rotating about its least axis. As Ω decreases, either u or v tends to zero and the other tends to unity, and so one of the axes a and b tends to c and the other to infinity. Thus the Jacobi ellipsoids tend towards elongated circular cylinders (needle-shaped), rotating about a transverse axis.

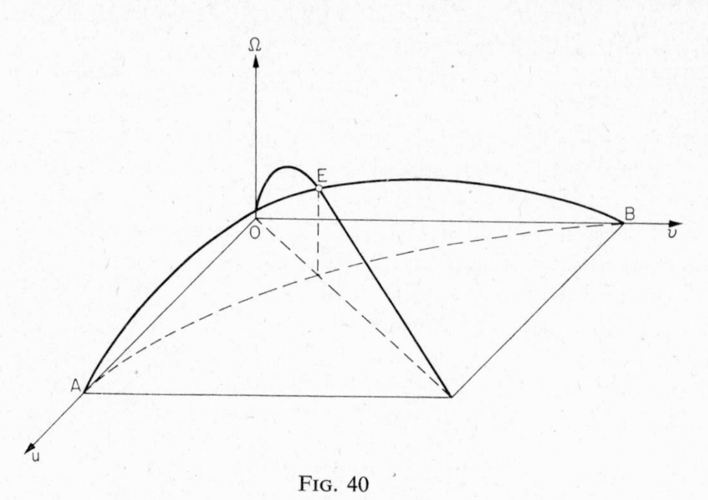

FIG. 40

We may summarise the above analysis as follows. The equilibrium ellipsoids existing for various values of Ω are:

(1) $0 < \Omega < \Omega_1 = 0{\cdot}1871\ldots$	Two Maclaurin, one Jacobi
(2) $\Omega_1 < \Omega < \Omega_0 = 0{\cdot}2247\ldots$	Two Maclaurin
(3) $\Omega = \Omega_0$	One Maclaurin
(4) $\Omega > \Omega_0$	None.

When there are two Maclaurin ellipsoids ($\Omega < \Omega_0$), one planet-like and one disc-like ellipsoid correspond to each value of Ω. The former is not greatly flattened, and tends to a sphere as $\Omega \to 0$; the latter is always considerably flattened, resembling a circular disc and tending to an infinitely thin disc as $\Omega \to 0$. As Ω increases towards $\Omega_0 = 0{\cdot}2247\ldots$, the flattening of the first ellipsoid increases and that of the second decreases, and at $\Omega = \Omega_0$ the axis ratio of both ellipsoids is such that $\varepsilon_0 = (a - c)/a = 0{\cdot}6323\ldots$.

The Jacobi ellipsoid is an elongated ellipsoid with three unequal axes rotating about its least axis. This ellipsoid exists only for $\Omega < \Omega_1 = 0{\cdot}1871\ldots$. If we take $a > b > c$, then as $\Omega \to 0$ the semiaxes b and c become equal and $a \to \infty$. Thus for small Ω the Jacobi ellipsoid is needle-shaped, resembling a long thin needle with pointed ends. As Ω increases towards $\Omega_1 = 0{\cdot}1871\ldots$, the semiaxes a and b become equal, and when $\Omega = \Omega_1$ the Jacobi ellipsoid becomes a Maclaurin ellipsoid (a spheroid) whose flattening is $\varepsilon_1 = (a - c)/a = 0{\cdot}4172\ldots$.

When $\Omega = 0{\cdot}1072$ the flattening is 0·7412. Figure 41 shows (a) planet-like and (b) disc-like Maclaurin ellipsoids and (c) a Jacobi ellipsoid.

Without attempting a mathematically rigorous discussion, we may seek to compare the results of the theory of ellipsoidal equilibrium figures with observation. In what follows, the equilibrium figures of stellar systems will be given names showing their resemblance to the various equilibrium ellipsoids, and will accordingly be called Jacobi or Maclaurin ellipsoidal bodies.

As shown in § 5, for *D* galaxies these configurations are usually almost or completely unstable.

The photographs at the end of this book include some barred galaxies. In accordance with our theoretical conclusions, we assume that the main body of such galaxies is a Jacobi ellipsoidal body with a moderate degree of central condensation. On account of its rotation it is unstable. The region of instability is at the ends of the main body, which are furthest from the axis of rotation. Here part of the mass of the galaxy is detached and forms two spirally coiled arms which often are rapidly dissipated into the surrounding space and lose their form.

Plate IV shows the barred galaxy NGC 7741, which is of a simple type. Two jets of matter are leaving the ends of the central ellipsoidal body, which in the photograph is vertical; it is a Jacobi ellipsoid with $\Omega = 0{\cdot}11$ to $0{\cdot}12$. The rotation of the main

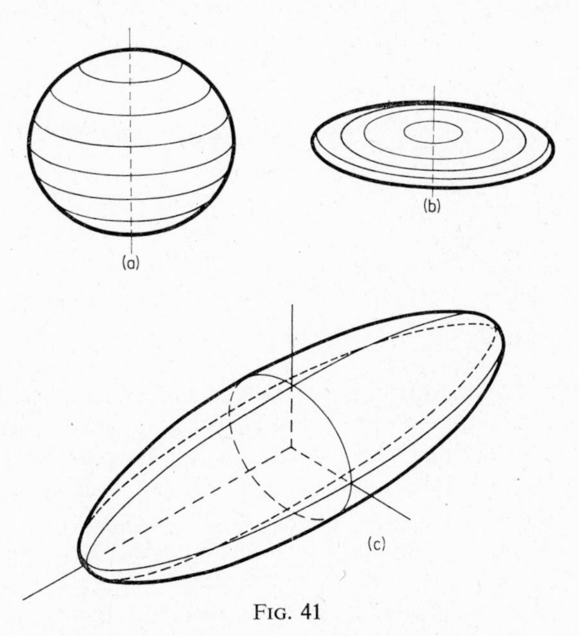

FIG. 41

body is counterclockwise in the photograph, and so the spiral arms are winding up. This galaxy must be very unstable: there is no central condensation and the arms are thick in comparison with the dimensions of the main body. Its mass must be relatively small, since the density of the arms (jets) is small and they are rapidly dissipated into the surrounding space.

Plate V shows a more stable system, NGC 4725. The central body has a considerable condensation, and corresponds to a Jacobi ellipsoid with $\Omega = 0{\cdot}16$ to $0{\cdot}17$. The arms are thinner and also denser, since they can be followed through one and a half turns.

Plate VI shows another rapidly disintegrating barred galaxy, NGC 7479. The unsymmetrical pear-shaped form of the main body is noticeable. This shows that the theory of pear-shaped equilibrium figures of rotating liquids (apioids), developed in the classic work of Lyapunov, Poincaré, Darwin and others, is also probably applicable to galaxies.† When the central body is more markedly unsymmetrical, we may expect an outflow from one end only, i.e. the formation of one-armed spirals. Such spirals are observed in exceptional cases, but they are usually classed as irregular (for instance, NGC 4038).

Plates VII, VIII and IX show elliptical galaxies of various types. Plate VII (a, b, c) shows galaxies corresponding to Maclaurin ellipsoids of the first kind; Plates VII (d), VIII and IX show spindle-shaped galaxies. These should apparently be regarded as elliptical galaxies corresponding to a large flattening (and angular velocity) and seen edgewise.

The absence of fluctuations in the surface brightness of elliptical galaxies is noteworthy. This effect is to be expected when the density is almost constant. As we shall see, this is very probably an indication that at least the dwarf elliptical galaxies represent one of the latest stages in galactic evolution.

§ 7. The Formation of Sharp Edges in Elliptical Galaxies

The presence of sharp edges along the equators of some elliptical galaxies of the Maclaurin ellipsoid type, and the outflow of matter along the equator, is easily accounted for by what is called the *Roche model*. This will now be briefly described.

We have seen that the equation for the level surfaces of star density has the form

$$U + \tfrac{1}{2}\omega^2\varrho^2 = C, \tag{10.56}$$

where U is the gravitational potential, ω the angular velocity of rotation, and ϱ the distance from the axis.

The distribution of mass in a galaxy may be regarded as approximately spherical in two cases: (1) when the velocity of rotation is small, and the mass distribution in the galaxy has approximately spherical symmetry, (2) when the density in a nearly spherical region in the centre of the galaxy is so high that the attraction of the peripheral mass may be neglected. In either case the potential is approximately given by $U = GM/\sqrt{(\varrho^2 + z^2)}$. Then the equation of the level surfaces (10.56) is

$$GM/\sqrt{(\varrho^2 + z^2)} + \tfrac{1}{2}\omega^2\varrho^2 = C. \tag{10.57}$$

Putting

$$k^2 = \omega^2/2GM, \tag{10.58}$$

† The possibility of pear-shaped and needle-shaped D galaxies is discussed in § 9.

we may write, with C in place of C/GM,

$$\frac{1}{\sqrt{(\varrho^2 + z^2)}} + k^2\varrho^2 = C. \tag{10.59}$$

By giving C various values (>0) we obtain various level surfaces. Figure 42 shows diagrammatically the cross-section of these surfaces by a meridian plane. For very large C, the level lines in the meridian planes are almost the small circles†

$$\varrho^2 + z^2 = 1/C^2. \tag{10.60}$$

One such circle is shown as curve 1 in Fig. 42. As C decreases, the central curve increases in size. It intersects the z-axis at the points $z = \pm 1/C$ and the ϱ-axis at the points given by the roots of the cubic equation

$$\varrho^3 - (C/k^2)\,\varrho + 1/k^2 = 0. \tag{10.61}$$

An examination of eqn. (10.61) shows [127] that for $C > 3.2^{-\frac{2}{3}}k^{\frac{1}{3}} = 1{\cdot}890k^{\frac{1}{3}}$ it has three real roots, one negative (having no physical significance) and two positive. Hence it follows that, besides the closed curve round the origin, the level surface has two other branches not shown in Fig. 42.

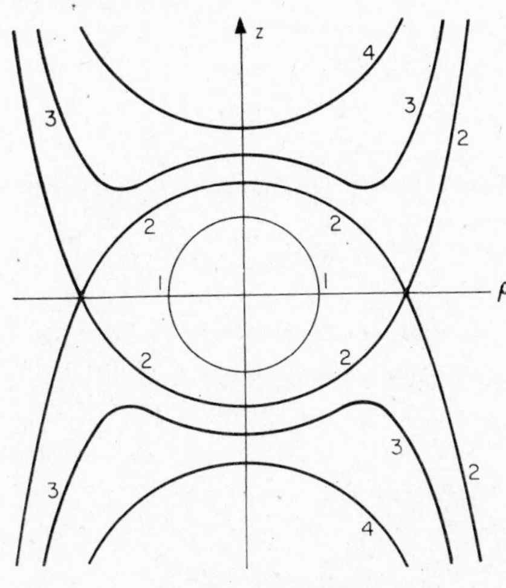

FIG. 42

When $C < 1{\cdot}890k^{\frac{1}{3}}$, the positive roots become imaginary, and so the level surfaces do not intersect themselves and are not closed; they form two hollow surfaces lying symmetrically about the plane $z = 0$. This case obviously does not correspond to an equilibrium figure. The level surfaces of this type are as shown by curves 3 and 4 in Fig. 42.

Finally, when

$$C = C_0 = 1{\cdot}890k^{\frac{1}{3}}, \tag{10.62}$$

† Together with two infinite branches which are almost the straight lines $\varrho = \pm\sqrt{C/k}$; these are of no interest here.

the level surfaces intersect themselves and form an edge. The corresponding curve is 2 in Fig. 42, and the edge is a sharp intersection on the ϱ-axis.

If the angular velocity ω of the galaxy increases for some reason, for example by gravitational contraction with constant angular momentum, then k will increase, by (10.58). The value of C_0 given by (10.62) also increases. If initially $C_0 < C$ for the outer level surface corresponding to a given C, it may happen that C_0 increases and becomes equal to C, and in that case an edge will appear at the equator.

At every point on the edge, since it is a singular point of the surface

$$U_1 = GM/\sqrt{(\varrho^2 + z^2)} + \tfrac{1}{2}\omega^2\varrho^2, \tag{10.63}$$

the partial derivatives $\partial U_1/\partial \varrho$ and $\partial U_1/\partial z$ are zero. These are equal to the corresponding components of the gravity field, and it is therefore clear that, when there is an edge at the equator, the gravity force at every point on the edge is zero, i.e. the centrifugal acceleration exactly balances the gravitational attraction. We call this a *critical surface*.

As the rate of rotation increases further, C_0 becomes greater than C and the surface is no longer critical. It is also no longer a closed surface, and a different level surface lying nearer the centre of the galaxy becomes the critical surface. In the outer layers of the galaxy, beyond this critical surface, an outflow of matter takes place.

It is very probable that the dark bands observed along the equators of some spindle-shaped galaxies indicate an outflow of matter along an equatorial edge. We should expect that the outflow of stellar matter would be accompanied by an outflow of dark diffuse matter. In projection against the bright background of the galaxy itself, only the latter is seen.

§ 8. The Formation of Central Nuclei and Ring Structure in Galaxies

The above discussion has been based on the hypothesis that a D galaxy has an approximately constant density; that is, we have used the first approximation (10.32) to the solution of the fundamental equation (10.30) for the star density. However, it has been shown in §§ 4 and 5 that the first approximation is insufficient to obtain a physically meaningful and dynamically stable model of a galaxy.

To derive the second approximation, we assume that

$$\nu = \nu_0 + \nu_1, \tag{10.64}$$

where ν_0 is the basic density given by formula (10.32) (the first approximation) and ν_1 the additional density, which is small compared with the basic density (the second approximation). We substitute (10.64) in equation (10.30) and use the obvious relation

$$\log \nu = \log \nu_0 + \log\left(1 + \frac{\nu_1}{\nu_0}\right). \tag{10.65}$$

If we expand the right-hand side of (10.65) in series as far as the first power of ν_1/ν_0, substitute (10.64) and (10.65) in (10.30) and use (10.32), we obtain after some simple transformations the following linearised equation for the additional density ν_1:

$$\Delta \nu_1 + 2(\omega_0^2/\sigma^2)\nu_1 = 0. \tag{10.66}$$

The unknown function ν_1 may be either positive or negative in the main body, but its negative values cannot exceed in magnitude the basic density ν_0, since otherwise the star density would be negative. The equation (10.66) has a form familiar in mathematical physics, called *Helmholtz' equation*. It has been used, in particular, by Poincaré and by Jeans to develop a theory of ellipsoidal equilibrium figures.

Putting

$$\nu_1 = u, \quad \sqrt{2}\omega_0/\sigma = k, \tag{10.67}$$

we can write equation (10.66) as

$$\Delta u + k^2 u = 0. \tag{10.68}$$

To solve eqn. (10.68) we must formulate appropriately the boundary conditions, i.e. specify the value of either u or its normal derivative on some closed surface which forms the outer surface of the galaxy. We shall consider, as an example, only the simple particular case of a highly flattened spheroid, regarding it as a circular disc.

Equation (10.68) can be written explicitly in cylindrical coordinates as

$$\frac{1}{\varrho}\frac{\partial}{\partial \varrho}\left(\varrho \frac{\partial u}{\partial \varrho}\right) + \frac{1}{\varrho^2}\frac{\partial^2 u}{\partial \theta^2} + \frac{\partial^2 u}{\partial z^2} + k^2 u = 0. \tag{10.69}$$

The solution of this equation is discussed in textbooks of mathematical physics.† We separate the variables by putting

$$u = \mathrm{P}\Theta Z, \tag{10.70}$$

where P, Θ, Z are respectively functions of ϱ, θ and z only. For brevity we shall call these the radial, azimuthal and vertical factors. Substituting (10.70) in (10.69), we obtain for P, Θ, Z the equations

$$\left.\begin{aligned} \mathrm{P}'' + \frac{1}{\varrho}\mathrm{P}' + \left[(k^2 + l^2) - \frac{s^2}{\varrho^2}\right]\mathrm{P} &= 0, \\ \Theta'' + s^2\Theta &= 0, \\ Z'' - l^2 Z &= 0, \end{aligned}\right\} \tag{10.71}$$

where s and l are further constants and s takes only non-negative integral values.

Since Θ must be a periodic function of θ with period 2π, and the origin of θ may be chosen arbitrarily, we have the azimuthal factor

$$\begin{aligned} \Theta &= \sin s\theta \quad \text{for} \quad s \neq 0, \\ &= 1 \qquad\quad \text{for} \quad s = 0. \end{aligned} \tag{10.72}$$

† See, for example, H. Margenau and G. M. Murphy, *The mathematics of physics and chemistry*, van Nostrand, New York, 1943, p. 216.

For the vertical factor we have the formal solution $Z = Ae^{lz} + Be^{-lz}$. Z must satisfy the obvious condition that $Z \to 0$ for $|z| \to \infty$, and $Z(0) = 1$, since the value of the density at $z = 0$ is determined by the value of the function P. Thus the vertical factor is given by the "barometric" formula†

$$Z = \exp(-l|z|). \tag{10.73}$$

The first eqn. (10.71) is known from the theory of linear differential equations to have only one solution finite at $\varrho = 0$, namely the Bessel function J_s of order s. If we put

$$k^2 + l^2 = q^2, \tag{10.74}$$

the solution is

$$\mathrm{P} = J_s(q\varrho). \tag{10.75}$$

Since the density on the equator at the outer boundary of the galaxy must be zero, q can take a series of values such that $q\varrho_0$ is a root of $J_s(x) = 0$, ϱ_0 being the equatorial radius of the galaxy. If $x_{i,s}$ is the ith zero of the sth Bessel function, we have

$$q_{i,s} = x_{i,s}/\varrho_0. \tag{10.76}$$

The general solution of Helmholtz' equation is the sum of particular solutions such as (10.70), i.e.

$$u = A_0 u_0 + A_1 u_1 + \dots, \tag{10.77}$$

where $A_0, A_1, \dots$ are constant coefficients.

It is known from the theory of Bessel functions that $J_s(x) = 0$ has an infinity of positive roots for every s. The zero-order function $J_0(q\varrho)$, corresponding to $s = 0$, is of particular importance in the present discussion: it is the only Bessel function whose value at the origin is other than zero, and $J_0(0) = 1$. Hence the solution (10.77) must contain a term corresponding to $s = 0$ for any galaxy having a central nucleus. If there are no other terms ($A_s = 0$ for $s \neq 0$), the galaxy will have a ring structure with continuous rings, since for $s = 0$ the azimuthal factor Θ is constant. An example of such a continuous ring structure is, perhaps, the galaxy NGC 488 (Plate X). This has at least three concentric rings round a central nucleus. The ellipticity of the nucleus and rings is probably due to their inclination to the line of sight. Two concentric rings are quite clear. The first ring is divided at two diametrically opposite points. If this division really exists and is not merely due to the absorption of light by dark matter, this must mean that the solution (10.77) for this galaxy has a term with $s = 1$.

The galaxy NGC 4324 (Plate XI) has a still more complex form, in which the ring surrounding the nucleus has bead-like condensations which are equidistant and approximately equal in size and brightness, resembling the ring nebula in the nebular theory of Kant and Laplace. If we assume that the beads are uniformly distributed on the ring, there must be sixteen of them, and the corresponding solution has $s = 16$.

Two difficulties arise here. One is that the first maximum of the Bessel function $J_{16}(x)$ occurs at $x = 18{\cdot}1$, whereas the first few maxima of $J_0(x)$ are 3·83, 7·02, 10·17,

† It is known that such a law of decrease of the star density in the z-direction is in fact observed in the Galaxy; see, for example, [97, 98].

13·32, 16·47. In other words, it would seem that the ring corresponding to $J_{16}(x)$ should contain at least five rings corresponding to the solution $J_0(x)$, which must be included on account of the presence of the nucleus. The difficulty is easily avoided if we recall that the eigenvalue q on which the solution depends is given by (10.74), and this equation involves besides k, which depends only on the properties of the galaxy itself, the coefficient l which appears in the "barometric" formula. The greater l, the more rapid is the decrease in density with increasing z. For large values of l (and therefore of q) the first root of the equation $J_{16}(x) = 0$ may occur for relatively small ϱ. Thus the first difficulty is easily dealt with, at least in principle.

The second difficulty is more awkward. Like the galaxy NGC 4826 (Plate XII), the ring in NGC 4324 is broken at two diametrically opposite points. In the neighbourhood of these discontinuities there are also gaps in the sequence of beads. It is noticeable that, as in NGC 4826, the breaks are exactly at the ends of the greatest apparent diameter of the ring. This casts doubt on the "natural" supposition that the ellipticity of the nucleus and rings is due only to the inclination of the galaxy to the line of sight. Probably non-gravitational (e.g. magnetic) forces are of importance here. We shall see in § 13 that similar features may be found in S0 galaxies also.

This discussion shows that the nature of the ring structure of galaxies is very complex. Quite frequently such a structure occurs together with a spiral structure, forming a superposition of the two.

§ 9. The Existence of Needle-shaped and Pear-shaped Galaxies

In deriving the most probable phase distribution in § 4, we have assumed that the mass distribution in the stellar system is rotationally symmetrical. The angular-momentum integral about the axis of symmetry exists only when this condition is satisfied.

We shall now show that the phase distribution obtained remains valid for a wider class of systems, of which a particular case is an ellipsoid.

THEOREM V: *Any stellar system in rigid rotation, having a density distribution symmetrical about some plane through the axis and a velocity distribution at every point symmetrical about the perpendicular plane through the axis, is a dynamically determinate system.*

Let $f(x, y, z;\ \dot{x}, \dot{y}, \dot{z})$ be the phase density. We take the z-axis along the axis of rotation, and the symmetry planes mentioned in the theorem as the xz and yz planes respectively. Then the symmetry properties stated may be written

$$\begin{aligned} f(x, y, z; \dot{x}, \dot{y}, \dot{z}) &= f(x, -y, z; \dot{x}, \dot{y}, \dot{z}) \\ &= f(x, y, z; -\dot{x}, \dot{y}, \dot{z}) \\ &= f(x, -y, z; -\dot{x}, \dot{y}, \dot{z}). \end{aligned}$$

If in dividing the phase volume Γ we make symmetrically placed cells alike, the above-mentioned symmetry properties will also apply to the phase cell populations $n_\mu = n(x_\mu, y_\mu, z_\mu;\ \dot{x}_\mu, \dot{y}_\mu, \dot{z}_\mu)$.

In a coordinate system rotating with angular velocity ω_0, the differential equations of motion of a free particle are

$$\ddot{x} - 2\omega_0^2 \dot{y} - \omega_0^2 x = \partial U/\partial x,$$
$$\ddot{y} + 2\omega_0^2 \dot{x} - \omega_0^2 y = \partial U/\partial y,$$
$$\ddot{z} = \partial U/\partial z,$$

where U is the potential. Evidently $\partial U/\partial y$ has antisymmetry corresponding to the symmetry in the mass distribution, i.e. $\partial U(x, y, z)/\partial y = -\partial U(x, -y, z)/\partial y$. Multiplying the first differential equation by y, the second by x and subtracting, we obtain after some familiar transformations

$$\mathrm{d}(x\dot{y} - y\dot{x})/\mathrm{d}t + 2\omega_0(x\dot{x} + y\dot{y}) = x\,\partial U/\partial y - y\,\partial U/\partial x.$$

We now write this equation for a star with phase coordinates $(x, -y, z; -\dot{x}, \dot{y}, \dot{z})$:

$$\mathrm{d}(x\dot{y} - y\dot{x})/\mathrm{d}t - 2\omega_0(x\dot{x} + y\dot{y}) = -(x\,\partial U/\partial y - y\,\partial U/\partial x),$$

and add, obtaining for the system consisting of two symmetrically placed stars

$$\mathrm{d}(x_1\dot{y}_1 - y_1\dot{x}_1)/\mathrm{d}t + \mathrm{d}(x_2\dot{y}_2 - y_2\dot{x}_2)/\mathrm{d}t = 0,$$

or

$$(x_1\dot{y}_1 - y_1\dot{x}_1) + (x_2\dot{y}_2 - y_2\dot{x}_2) = C,$$

where the suffixes 1 and 2 denote the coordinates and velocities of the two stars.

Thus, for the symmetry specified in Theorem V, the angular-momentum integral exists for every *pair* of symmetrically placed stars. Since the populations of symmetrically placed phase cells are taken to be equal, the expression for the additive parameter $K = \sum n_\mu \varrho_\mu \Theta_\mu$ remains unchanged. The subsequent conclusions concerning the most probable phase distribution are likewise unaffected. The only difference, which is not important, is that the particles in the stellar system must now be taken as being not individual stars but pairs of symmetrically placed stars.

The range of the theorem just proved is considerably wider than what is necessary for the physical problem under discussion. In the most probable phase distribution, the velocity distribution at every point is Maxwellian, i.e. it is spherically symmetrical. It is therefore symmetrical about any plane through the centre of the velocity distribution (the origin of the rotating coordinate system). The condition of symmetry of the mass distribution about a plane through the axis of rotation is met not only by ellipsoids and similar bodies symmetrical about both principal planes through the least axis, but also by pear-shaped figures of equilibrium.

The existence of galaxies in the form of ellipsoids with three unequal axes has already been mentioned. The case of pear-shaped asymmetry also is by no means rare among galaxies. It may even be said to occur to some extent in almost all galaxies. It is seen, for example, in the needle-shaped galaxies in Plates XIII and XIV, and also in a normal spiral galaxy seen edgewise, NGC 4565 (Plate XV). In the most marked type of asymmetry the spiral galaxies have only a single arm.

In terms of our theory the barred spirals are the result of rotational instability of ellipsoidal or pear-shaped bodies. The various forms depend on the star density, the angular velocity of rotation and the presence or absence of a central dense nucleus. If the nucleus is absent and the mean star density and angular velocity are high, there is a large outflow in the form of broad jets, as in NGC 7741 (Plate IV). If there is a marked central condensation we have thin jets, with low star density at the ends of the main body, as in NGC 1300, NGC 1530, etc.

The theory of this outflow makes possible a relatively simple explanation of some details of barred spirals which are at first sight unimportant. For example, strong jets (as in NGC 7741) are considerably coiled by the progressive shortening of the main body as its mass decreases. When the jets are thin there is little coiling and the shape is almost circular, since the shortening of the main body takes place much more slowly.

Many barred spirals are observed to have jets which do not begin from the ends of the main body. This may be explained as follows, on the basis of the outflow process as described above. In this process the stars detached from the main body are, of course, principally those with the largest absolute velocities (i.e. velocities relative to the fixed coordinate system with origin at the centre of mass of the galaxy), since for a star to be detached it is necessary that its velocity should exceed the local velocity of escape. Thus the jets of stars must consist mainly of those whose absolute velocity exceeds the circular velocity of the local centroid. Hence some stars will continually overtake the main body of the galaxy as it rotates. The circular velocity of the centroid in the remaining part of the main body will decrease because of the loss of these stars which overtake the centroid, and so the jet as a whole will overtake the main body to some extent, a process which continues after the outflow ceases.

We frequently observe not one but two or even more thin jets forming a series of rings or, more precisely, a series of segments of very slightly coiled spirals, as in NGC 488 (Plate X) and NGC 1398 (Plate XVI). This can be explained by the fact that, after the end of a period of outflow, the main body takes up another equilibrium figure under the action of gravitational contraction. Its moment of inertia about the axis of rotation thereby decreases, and the angular velocity correspondingly increases, which may lead to a fresh outflow.

These arguments naturally need to be confirmed by quantitative calculations based on the theory of non-stationary processes.

It follows from the above discussion that in barred spirals the arms must be coiling up, though in some cases only very slightly. In this respect the conclusions are different from those of Lindblad. Lindblad's theory may be applicable to galaxies such as NGC 2841 (Plate XIX), NGC 7217 (Plate XX) and others, which are often classed as S0 galaxies, the type S0 being somewhat heterogeneous. These have a massive nucleus in the form of a flattened ellipsoid, along whose equator thin spiral jets emerge at various points (not necessarily diametrically opposite) on account of the instability of circular orbits *outside* the main body. We propose to call these galaxies *Crab-like*; they certainly form a separate dynamical class.

The problem of the normal spirals is discussed Subsequently.

§ 10. Kapteyn–Lindblad Systems

In the preceding sections we have considered rotating stellar systems of uniform star population. We have found that the most probable state of such systems is a rigid rotation with a Maxwellian distribution of residual velocities. In a first approximation such galaxies have a constant density, and their shapes may be those of the equilibrium figures of a rotating homogeneous liquid: Maclaurin ellipsoids, Jacobi ellipsoids, pear-shaped (apioid), etc. We have seen that the observed forms of some galaxies, namely the elliptical and barred spiral galaxies, are in general qualitative agreement with these theoretical conclusions. The nuclei of the normal spirals may also be included in the same class.

Observation of the spiral galaxies themselves shows that the star population is not uniform and the rotation is not rigid. Our Galaxy itself has a star population which is not homogeneous. It is a superposition of sub-systems differing in their degree of concentration to the galactic plane, in the dispersion of residual velocities and in the velocity of rotation. The greater the residual-velocity dispersion, the smaller the circular velocity of the centroid of the sub-system at any point in the Galaxy. This is called the *asymmetry* of stellar motions; it was discovered empirically in 1923 by Strömberg [125]. Lindblad made the concept of sub-systems the basis of his well-known theory of galactic rotation (1926). Still earlier, in 1922, the idea of stellar sub-systems rotating in opposite directions had been put forward by Kapteyn [47] as part of his rough outline of a theory of galactic structure, based on the two-stream theory, which finds a place in the history of astronomy as the "Kapteyn universe".

The existence of sub-systems of stars in the spiral galaxies was experimentally demonstrated in 1944 by Baade, who discovered by photographic means that the stars in the nearest galaxies belong to two populations, I and II. Subsequently this view was further developed by the work of B. V. Kukarkin, P. P. Parenago and others, who established the presence in our Galaxy of three stellar populations: the plane component, which is practically identical with Baade's Population I; the intermediate component; and the spherical component, which is practically identical with Baade's Population II. At the present time the division of the star population of the spiral galaxies into sub-systems of the Kapteyn–Lindblad type is fundamental to the study of the structure of these systems by the use of statistical astronomy.

We shall use the name *Kapteyn–Lindblad systems*, or simply KL *systems*, to denote stellar systems such as the Galaxy, with a non-uniform star population which can be divided into sub-systems with different residual stellar velocity dispersions.

The structure of KL systems can be illustrated in terms of the simple "Kapteyn universe". Let us imagine a non-rotating stellar system with a star density which is given at every point and a velocity dispersion whose distribution may, for the sake of simplicity, be assumed spherical. We now divide the stars into two sub-systems, I and II, in such a way that the star density in each sub-system is half the original density. Let the two sub-systems begin to rotate in opposite directions with the same angular velocity. Then the system as a whole is still at rest, since the mean velocity of the stars in each macroscopic volume element is zero. The energy of the motion of the stars is increased by the rotation of the sub-systems, but this increase may be annulled by reducing the dispersion of residual velocities. Thus we have another system which is dynamically equivalent to the original one, in the sense that the values of all three dynamical parameters are the same for both. That is, the KL systems are not D systems as defined in § 3.

To define the dynamical state of a homogeneous rotating stellar system it suffices, as we have seen, to specify three dynamical parameters: (1) the mass M (or the number of stars N), (2) the sum of the energies of the stars (the energy parameter E of the system), (3) the angular momentum K.

It is easy to see, however, that these three parameters do not uniquely define every rotating system. The "Kapteyn universe" has already been used to illustrate this. Let us now consider a more general case. We take as basis some rigidly rotating homogeneous stellar system of the type considered in the preceding sections, and suppose it to be rotationally symmetrical about the axis of rotation. We shall show that, by making the star population non-uniform, we can define a second stellar system (and therefore many systems) different from the first but having the same values of all three fundamental parameters. This demonstrates that the values of the three dynamical parameters do not uniquely define the dynamical state of a non-uniform stellar system.

Let $\mathrm{d}N$, $\mathrm{d}E$, $\mathrm{d}K$ denote respectively the number of stars, twice the sum of their energies and the sum of their angular momenta about the axis of symmetry of the system, for a macroscopic volume element $\mathrm{d}\omega$ chosen arbitrarily within the stellar system. Regarding the system as a continuous medium, we have the following hydrodynamic expressions for these quantities:

$$\mathrm{d}N = \nu\,\mathrm{d}\omega, \quad \mathrm{d}E = \nu E_\nu\,\mathrm{d}\omega, \quad \mathrm{d}K = \nu K_\nu\,\mathrm{d}\omega, \tag{10.78}$$

where E_ν and K_ν are the mean values of E and K for all stars in the volume considered. It is easy to see that we can replace the original group of stars belonging to the macroscopic volume element $\mathrm{d}\omega$ by another group in such a way that the quantities (10.78) remain unchanged. To do this, it is sufficient to replace the group of stars in $\mathrm{d}\omega$ by two sub-groups whose partial star densities ν_1 and ν_2, mean energies E_1 and E_2 and angular momenta K_1 and K_2 satisfy the relations

$$\nu = \nu_1 + \nu_2, \quad \nu E_\nu = \nu_1 E_1 + \nu_2 E_2, \quad \nu K_\nu = \nu_1 K_1 + \nu_2 K_2. \tag{10.79}$$

Then the quantities (10.78) remain unchanged.

The same argument can be repeated with respect to each point in the system. As a result, the homogeneous stellar system is replaced by a composite system which is a superposition of two sub-systems formed by the sub-groups substituted in the individual macroscopic volume elements. The inhomogeneous system thus obtained necessarily has the same values of the three dynamical parameters as the original homogeneous system. In this sense the two systems are dynamically equivalent.

Finally, the argument can be generalised to the case of an arbitrary number of discrete sub-systems; (10.79) is then replaced by the following conditions of equivalence between the homogeneous system and the assembly of sub-systems:

$$\nu = \sum_j \nu_j, \quad \nu E_\nu = \sum_j \nu_j E_j, \quad \nu K_\nu = \sum_j \nu_j K_j. \tag{10.80}$$

Let us now ascertain the dynamical significance of the conditions (10.79) and (10.80). To do so, we first derive explicit formulae for the quantities E_ν and K_ν which characterise the dynamical state of a homogeneous system in a given macroscopic volume element. Let $\mathrm{P}'_i, \Theta'_i, Z'_i$ be the residual velocity components of the ith star, and Θ_0 the circular velocity of the centroid. In the present case we have $\mathrm{P}_i = \mathrm{P}'_i, \Theta_i = \Theta_0 + \Theta'_i$, $Z_i = Z'_i$. Hence

$$E_\nu = \overline{\tfrac{1}{2}(\mathrm{P}_i^2 + \Theta_i^2 + Z_i^2) - U_i}$$

$$= \tfrac{1}{2}\Theta_0^2 + \tfrac{1}{2}(\overline{\mathrm{P}_i'^2} + \overline{\Theta_i'^2} + \overline{Z_i'^2}) - U, \tag{10.81}$$

where $\overline{\mathrm{P}'^2}, \overline{\Theta'^2}, \overline{Z'^2}$ are the residual-velocity variances along the three coordinate axes, and U the force potential, which is constant within each macroscopic volume element.

Similarly

$$K_\nu = \overline{\varrho_i\Theta_i} = \varrho\overline{\Theta_i} = \varrho\Theta_0. \tag{10.82}$$

Here we have used the facts that the coordinates of all the stars within a macroscopic volume element may be regarded as equal (so that $\varrho_i = \varrho$), and that the mean value of Θ_i is Θ_0.

With two sub-groups of stars in the volume considered, we have the relations

$$\left.\begin{aligned} E_1 &= \tfrac{1}{2}\Theta_1^2 + \tfrac{1}{2}(\overline{\mathrm{P}_1'^2} + \overline{\Theta_1'^2} + \overline{Z_1'^2}) - U, \\ E_2 &= \tfrac{1}{2}\Theta_2^2 + \tfrac{1}{2}(\overline{\mathrm{P}_2'^2} + \overline{\Theta_2'^2} + \overline{Z_2'^2}) - U, \\ K_1 &= \varrho\Theta_1, \\ K_2 &= \varrho\Theta_2. \end{aligned}\right\} \tag{10.83}$$

Here Θ_1 and Θ_2 denote the velocities of the *partial centroids* or *sub-centroids*, i.e. the centroids of the separate sub-groups.

Substituting (10.81)–(10.83) in the second and third eqns. (10.79), we find

$$\nu\Theta_0 = \nu_1\Theta_1 + \nu_2\Theta_2, \tag{10.84}$$

$$\begin{aligned} \nu(\Theta_0^2 + \overline{\mathrm{P}'^2} + \overline{\Theta'^2} + \overline{Z'^2} - 2U) &= \nu_1(\Theta_1^2 + \overline{\mathrm{P}_1'^2} + \overline{\Theta_1'^2} + \overline{Z_1'^2} - 2U) \\ &\quad + \nu_2(\Theta_2^2 + \overline{\mathrm{P}_2'^2} + \overline{\Theta_2'^2} + \overline{Z_2'^2} - 2U), \end{aligned} \tag{10.85}$$

where $\overline{\mathrm{P}'^2}$, $\overline{\Theta'^2}$ and $\overline{Z'^2}$ denote the values averaged over all stars in the original group, and $\overline{\mathrm{P}_1'^2}, \overline{\Theta_1'^2}, \overline{Z_1'^2}; \overline{\mathrm{P}_2'^2}, \overline{\Theta_2'^2}, \overline{Z_2'^2}$ the values for the first and second sub-group respectively.

In eqns. (10.84) and (10.85) we substitute the deviations $\Delta\Theta_1 = \Theta_1 - \Theta_0$, $\Delta\Theta_2 = \Theta_2 - \Theta_0$ of the partial centroid velocities Θ_1, Θ_2 from the original centroid velocity; using the first eqn. (10.79), we obtain

$$\nu_1\Delta\Theta_1 + \nu_2\Delta\Theta_2 = 0 \tag{10.86}$$

and hence

$$\begin{aligned} \nu(\overline{\mathrm{P}'^2} + \overline{\Theta'^2} + \overline{Z'^2}) &= \nu_1(\Delta\Theta_1)^2 + \nu_2(\Delta\Theta_2)^2 \\ &\quad + \nu_1(\overline{\mathrm{P}_1'^2} + \overline{\Theta_1'^2} + \overline{Z_1'^2}) + \nu_2(\overline{\mathrm{P}_2'^2} + \overline{\Theta_2'^2} + \overline{Z_2'^2}). \end{aligned} \tag{10.87}$$

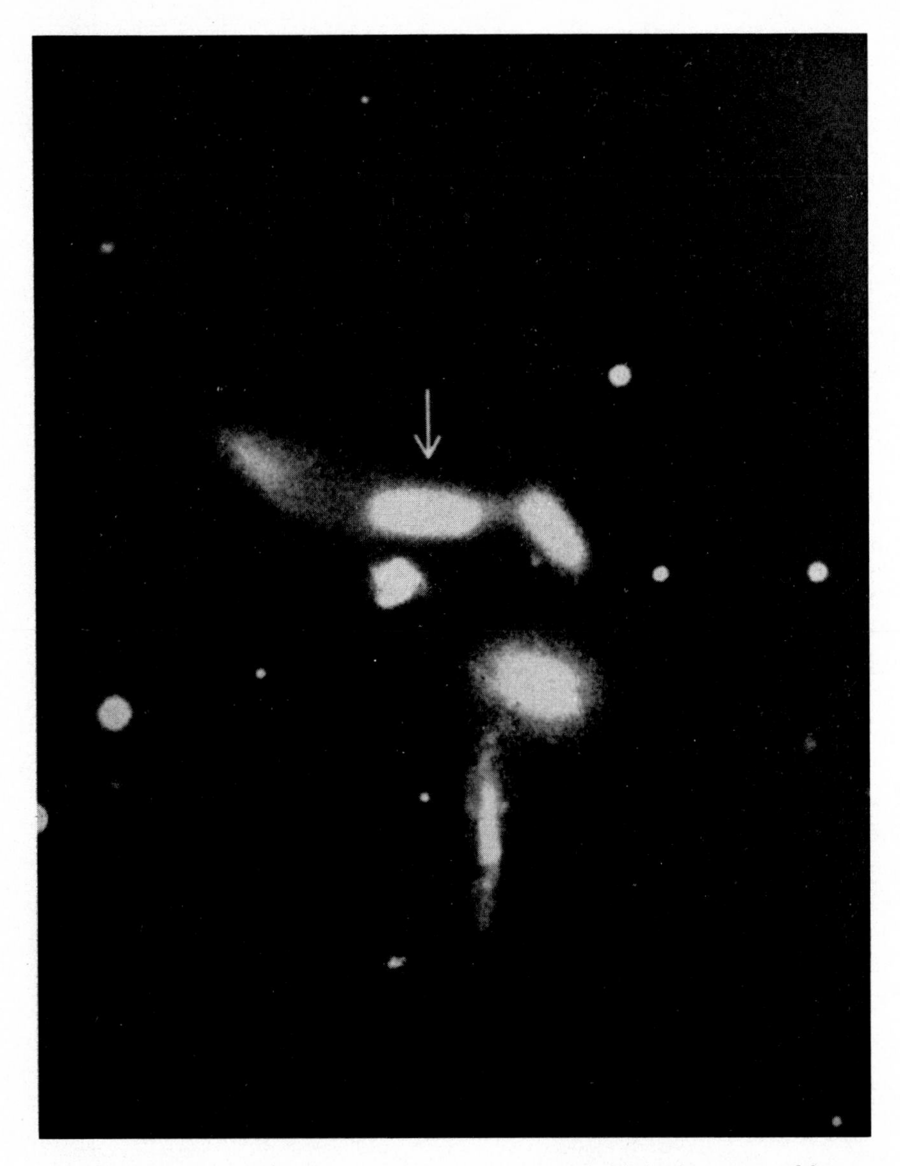

PLATE I. NGC 6027: a group of six galaxies which are probably in the process of formation from a single long filament of diffuse matter. Note that the galaxies at the extremities of the filament are more elongated than the others, and presumably belong to the new type of "needle-shaped" galaxies predicted by the theory.

PLATE II. M 13 = NGC 6205: a typical globular cluster. These are not independent galaxies, but lie within galaxies and in the space around them. Rather more than 100 such clusters are known in the Galaxy.

PLATE III. NGC 2682: a typical open cluster. They appear as local condensations in the Galaxy and in other neighbouring galaxies, and contain from a few dozen to several hundred stars, intermingled with stars belonging to the galaxy but not to the cluster.

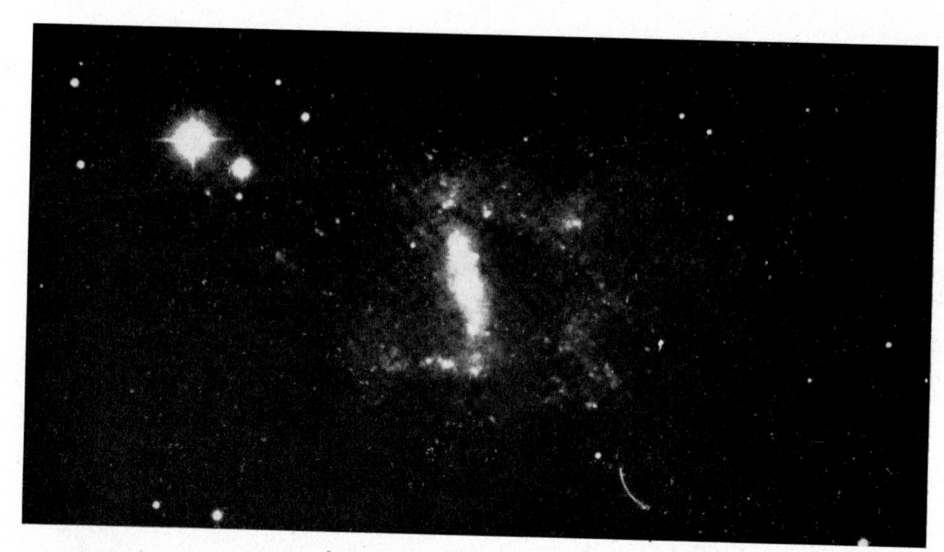

PLATE IV. NGC 7741: a barred spiral galaxy of a simple kind. Note the well-defined straight thick main body with practically no central nucleus. In spite of the apparently rapid rotation, the main body retains the form of an elongated Jacobi ellipsoid. The ejection of matter is proceeding violently and the length of the main body is therefore decreasing rapidly. Although it has turned through only about 130–140°, its length has decreased by a factor of approximately 3, as may be seen by comparing the distance between the ends of the arms with the present length of the main body.

PLATE V. NGC 4725: a barred spiral of quiescent type. Compare with Plate IV. Here the main body has a massive nucleus formed from matter taken from the ends of the main body. The ejection of matter from the ends is slow, and the arms are therefore almost circular.

PLATE VI. NGC 7479. This picture shows the process of formation of an ordinary spiral galaxy from a very long and thin needle-shaped body. The latter is dynamically very unstable, and the stars have therefore easily escaped from it, leaving behind only the diffuse matter. The latter is retained by the magnetic forces remaining from the time when the galaxy formed an elongated condensation in a filament of matter, as shown in Plate I. The main body, except for its innermost part, contains no stars, and is twisted by the rotation of the galaxy. The stars that have escaped form a corona, while those now being formed from the matter remaining in the arms become members of Population I of the galaxy.

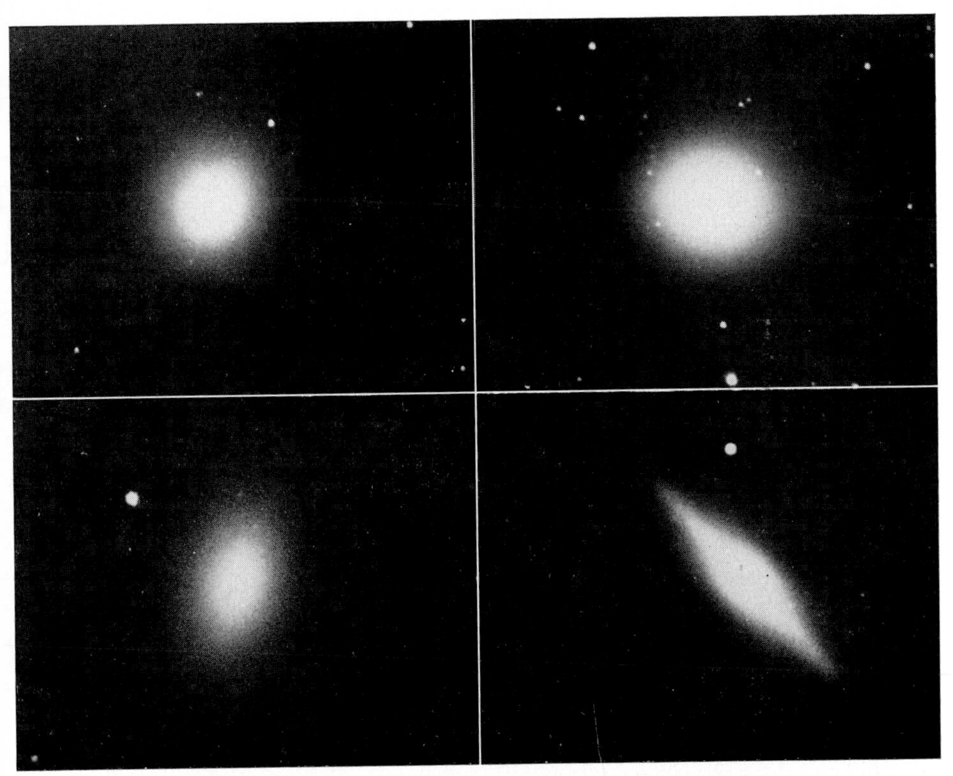

PLATE VII. Elliptical galaxies of varying degrees of oblateness. Note their uniformity and absence of structure, which result in a long lifetime.

PLATE VIII. NGC 3115: a spindle-shaped elliptical galaxy, probably a disc-shaped object seen edgewise and resulting from evolution according to the Roche model.

PLATE IX. NGC 5866: an elliptical galaxy with a sharp edge and a dark equatorial filament of matter. A further illustration of the Roche model.

PLATE X. NGC 488: a galaxy with a well-developed ring structure.

PLATE XI. NGC 4324: a galaxy with a ring structure and a system of equidistant bead-like condensations in the ring.

PLATE XII. NGC 4826: a galaxy with a ring structure.

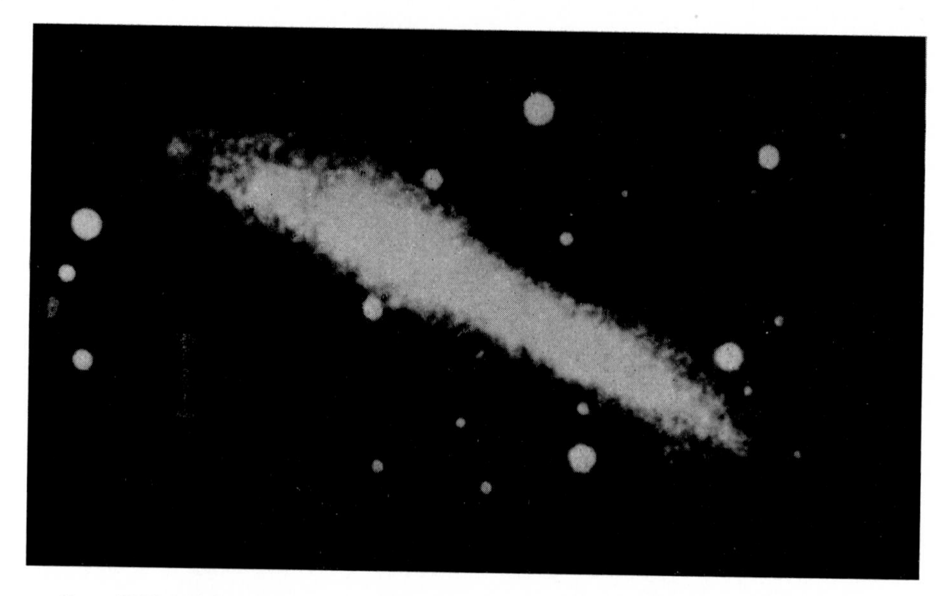

PLATE XIII. NGC 784: a needle-shaped galaxy. Note that it is elongated, but has no trace of the spiral pattern usually observed in actual spiral galaxies seen edgewise; of. the edge-on galaxy in Plate XV. Reproduced from the Palomar Atlas by B. A. Vorontsov-Vel'yaminov.

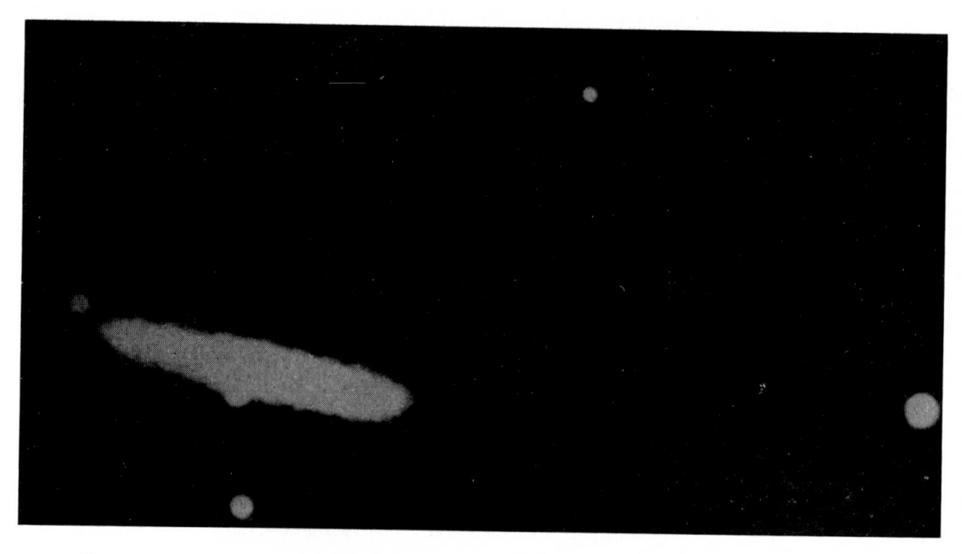

PLATE XIV. A needle-shaped galaxy (unnamed). Enlarged reproduction from the Palomar Atlas by B. A. Vorontsov-Vel'yaminov.

PLATE XV. NGC 4565: a typical regular spiral galaxy seen edgewise. Note the bulge at the centre, which is the projection of the nucleus, and the thin strip of dark matter along the equator. Neither of these features is seen in the needle-shaped galaxies in Plates XIII and XIV.

PLATE XVI. NGC 1398: a barred spiral galaxy with a ring system. The matter in the rings very probably originated by ejection from the ends of the bar, and was prevented from dissipating by gravitational forces: according to the theory, the ring structure is a regular feature of the equilibrium figures of rotating masses. The outer ring was formed when the bar was as long as the diameter of that ring. The matter ejected at intermediate lengths of the bar was dissipated, the theory not allowing the formation of rings with intermediate radii.

PLATE XVII. NGC 1566: a regular spiral galaxy with very clear-cut arms. The original photograph shows that the arms continue far beyond the visible ends and form two indistinct extensions reaching to at least four times the diameter of the central nucleus. These are the remnants of the filament of matter in which this galaxy was first formed as a condensation. The galaxy is a young regular spiral in the next stage of evolution after that shown in Plate VI.

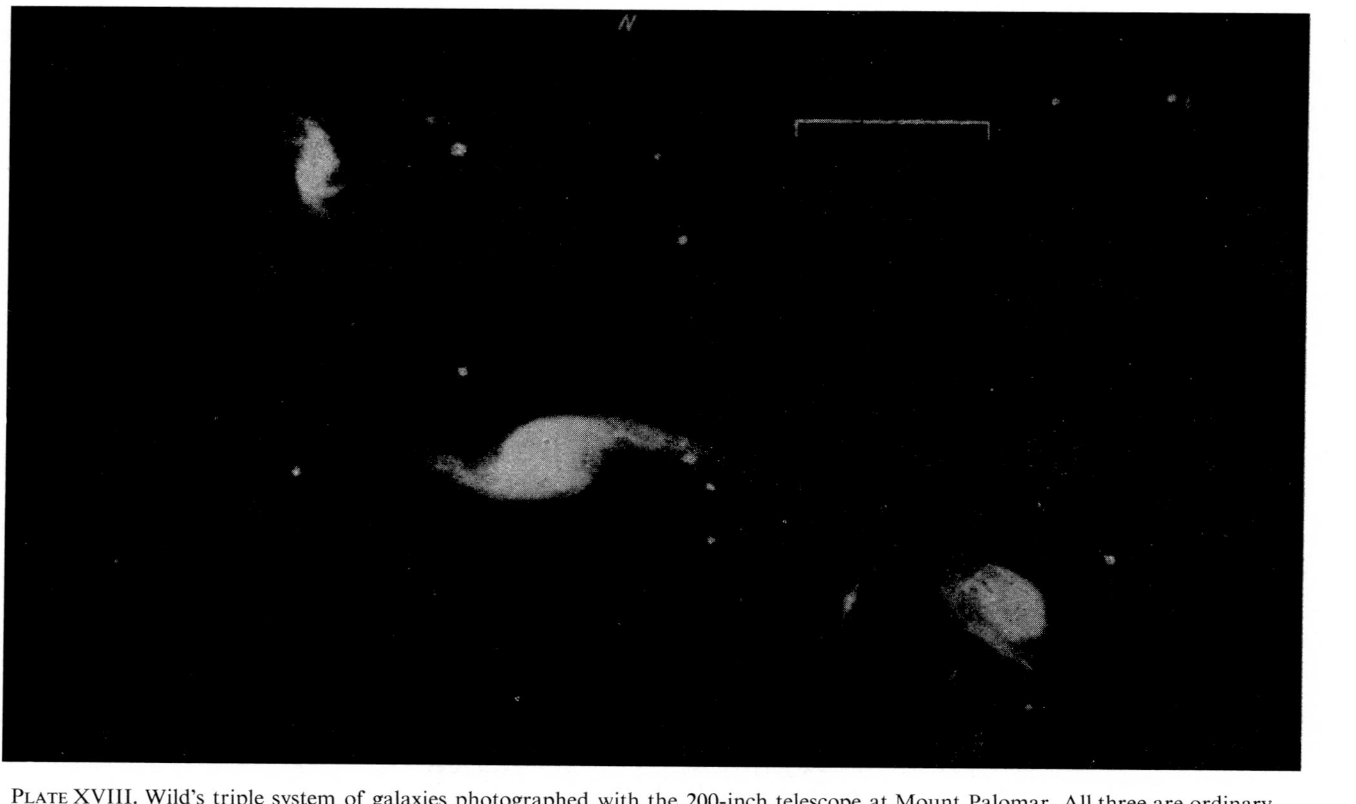

PLATE XVIII. Wild's triple system of galaxies photographed with the 200-inch telescope at Mount Palomar. All three are ordinary spirals, apparently in course of formation, and are connected by a single filament of matter, which is more conspicuous between the two lower ones. The filament or "bridge" between these two galaxies is evidently a direct continuation of their respective arms. An interesting object is seen between the two galaxies, just below the "bridge". In shape it resembles the symbol for a contour integral. A similar but smaller object is also seen along and at the base of the arm of the left-hand galaxy. A third such object, less distinct than the other two, is seen across and at the base of the arm of the right-hand galaxy, making this arm seem broken. It can scarcely be doubted that these three objects are physically connected with the main galaxies, and have probably been ejected by them.

PLATE XIX. NGC 2841: a spiral galaxy of a type dynamically quite different from all those previously considered. See also Plate XX.

Plate XX. This galaxy and NGC 2841 (Plate XIX) are of the Crab-like type, with a very large and massive lens-shaped central nucleus and many very thin spiral jets emerging from equidistant points on the equator of the nucleus. At their outer ends the jets run parallel and give the impression of a ring. Lindblad has developed a theory of such galaxies. In a galaxy in fairly rapid rotation, the orbits of individual stars are almost circular. In general, motion in almost circular orbits is stable except at the extreme circumference of the equator, where an infinitesimal perturbation causes a change in the orbit of a star, which then in turn perturbs other stars, leading to a streaming of stars. In this case the spiral arms are leading in the rotation of the galaxy, whereas in previous examples they are trailing.

Plate XXI.

Plates XXI and XXII. M 51 = NGC 5194 (and 5195): a "classical" spiral galaxy, one of the brightest in the sky. The main features of galaxies of this type are very distinctive: a fairly small central nucleus, and very long and bright spiral arms. These two photographs were taken by Zwicky in blue light (Plate XXI) and yellow light (Plate XXII). It is seen that the brightness of the spiral arms is due primarily to light sources with a high proportion of blue light, i.e. early-type stars. Another remarkable feature is the presence of the satellite elliptical galaxy NGC 5195 at the end of one of the arms. Plate XXII shows that the satellite is actually almost as large as the main galaxy. It could scarcely have been formed as a kind of swelling at the end of the arm, nor can it be an accidental projection of an independent elliptical galaxy, since Vorontsov-Vel'yaminov has found at least thirty such objects in the Palomar Atlas. In his *Atlas of interacting galaxies* he calls these "M 51-type galaxies". Thus we are led to suppose that the two galaxies have always been linked by a long "bridge", which as a result of the rotation of the more massive galaxy was wound round the latter and so formed the spiral arms. This idea is in accordance with the origin of galaxies as condensations in long filaments of matter.

PLATE XXII: for description see Plate XXI.

PLATE XXIII. NGC 3109: an "irregular" galaxy, which, though indeed having no regular form, does exhibit a tendency to form a rod-like elongated structure. In such galaxies, owing to their small mass, gravitation is too weak to bring the galaxy to a regular form, and the dissipation of stars is rapid.

PLATE XXIV. The Large Magellanic Cloud, one of the most conspicuous objects in the southern sky. It is an irregular galaxy, like that in Plate XXIII. There is again a tendency to form an elongated main body, and also a tendency to form arms like those of a barred spiral.

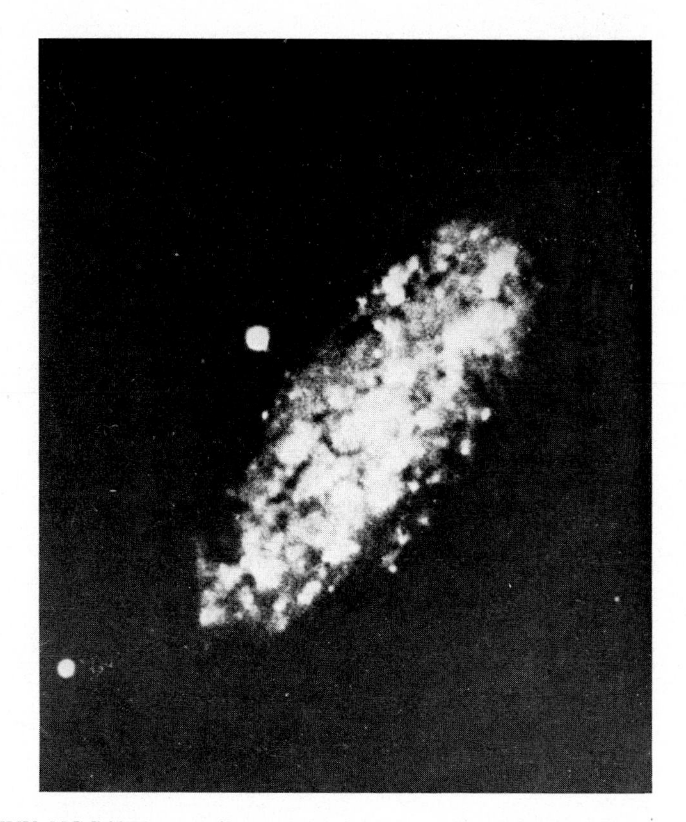

PLATE XXV. NGC 2976: a needle-shaped galaxy in which the distribution of matter has not yet been smoothed out. The appearance is very patchy, with no regular structure. In such systems the relaxation time must be comparatively short. It may be in an early stage of evolution.

PLATE XXVI. M 82 = NGC 3034: a galaxy with a patchy structure. This is interpreted by de Vaucouleurs ("Classification and morphology of external galaxies", *Handbuch der Physik* **53**, 303, 1959) as "the 'broad-side', nearly edge-on view of the bar and lens of an armless SB (s) O/a object". This is quite close to what we have called a needle-shaped galaxy. He also notes that Mayall has observed a rapid rigid rotation about the major axis, in full agreement with the theoretical predictions. This is the only case where rigid rotation of a needle-shaped galaxy has been confirmed by direct observation.

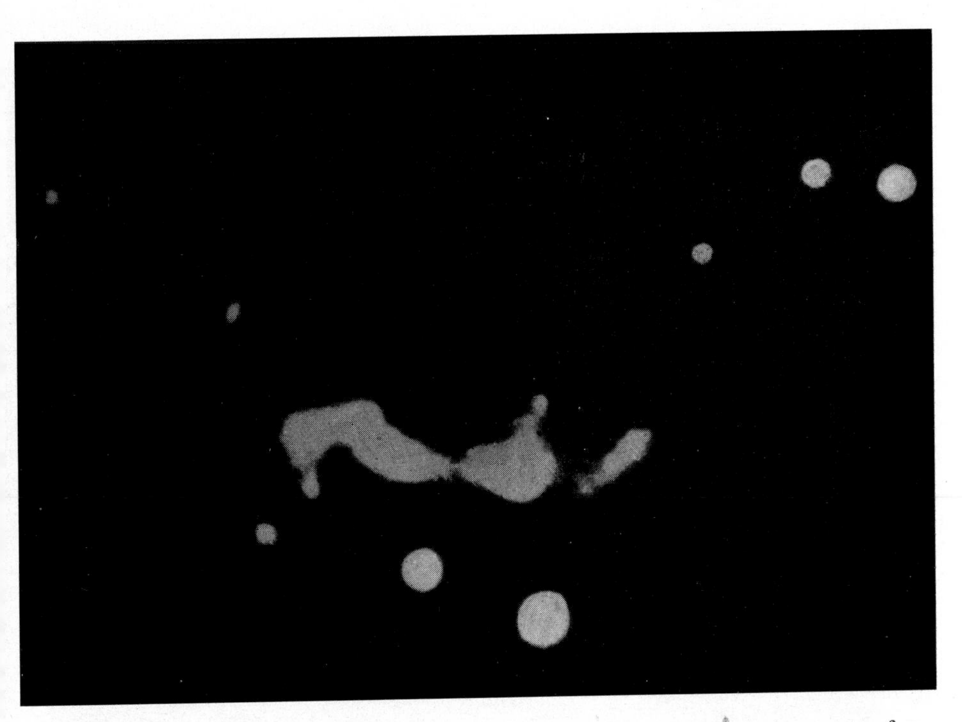

PLATE XXVII. A group of at least five intermingled galaxies, apparently in a state of violent formation from a single long filament of matter. A comparison of an elongated galaxy at one end of the chain with two lateral protuberances strongly indicates that, at least in some cases, galaxies are formed by "budding", as has been suggested by Ambartsumyan.

PLATE XXVIII. The Andromeda nebula. The broken line shows its outer boundary as found by photographic means. The cross marks the point corresponding to the position of the Sun in the Galaxy. Reproduced from the *Larousse Encyclopedia of Astronomy* (Batchworth, London, 1959).

When the first eqn. (10.79), (10.84) and (10.87) are satisfied, the inhomogeneous system consisting of two sub-systems is dynamically equivalent to the original homogeneous system.

Similarly we obtain from (10.80) the conditions for dynamical equivalence between the original homogeneous system and an inhomogeneous system consisting of an arbitrary number of sub-systems:

$$\left.\begin{aligned} \nu &= \sum_j \nu_j, \\ \nu\Theta_0 &= \sum_j \nu_j\Theta_j, \\ \nu(\overline{P'^2} + \overline{\Theta'^2} + \overline{Z'^2}) &= \sum_j \nu_j(\Delta\Theta_j)^2 + \sum_j \nu_j(\overline{P_j'^2} + \overline{\Theta_j'^2} + \overline{Z_j'^2}). \end{aligned}\right\} \quad (10.88)$$

In these equations the left-hand sides may be supposed known. Hence it is clear that the total number of sub-systems, the corresponding values of ν_j and Θ_j, and the partial residual-velocity variances $\overline{P_j'^2}$, $\overline{\Theta_j'^2}$, $\overline{Z_j'^2}$, may be chosen in an infinity of different ways so as to satisfy eqns. (10.88).

The conditions of dynamical equivalence (10.88) have an evident mechanical significance. The first condition is clearly that dynamically equivalent systems must have equal star densities in each macroscopic volume element. The second condition may be written $\sum \nu_j\Delta\Theta_j = 0$, where $\Delta\Theta_j$ is the velocity of the partial centroid with respect to the original centroid. This equation signifies that the mean velocity of the partial centroids relative to the original centroid is zero, i.e. there is no ordered motion of the partial centroids relative to the original one.

The left-hand side of (10.88), $\nu(\overline{P'^2} + \overline{\Theta'^2} + \overline{Z'^2})$, gives twice the kinetic energy density of the residual (disordered) stellar velocities. On the right-hand side are two terms. The second term, $\sum \nu_j(\overline{P_j'^2} + \overline{\Theta_j'^2} + \overline{Z_j'^2})$, is again twice the kinetic energy density of the residual velocities of the stars by sub-groups. The first term, $\sum \nu_j(\Delta\Theta_j)^2$, is twice the kinetic energy of the residual motion of the partial centroids. Thus we have a general criterion which distinguishes a dynamically determinate system from a KL system: whereas, in a dynamically determinate system, all the energy of the disordered motion of the stars is the energy due to their residual velocities, in a KL system part of this energy is due to the disordered motion of the partial centroids.

As has been mentioned above, the division of KL systems into sub-systems is closely related to the phenomenon of asymmetry of stellar velocities, first discovered for the neighbourhood of the Sun in the Galaxy by Strömberg in 1923. In order to ascertain the exact nature of this relation, let us consider the case where a KL system consists of sub-systems corresponding to a continuously varying parameter p. Let $d\nu = n(p)\,dp$ be the partial star density due to sub-systems whose parameters lie between p and $p + dp$. Evidently

$$\nu = \int_{p_1}^{p_2} n(p)\,dp,$$

where p_1 and p_2 are the extreme values of the parameter. The velocity variance in the sub-system is a function of the parameter p:

$$\sigma_p^2 = \overline{P_p'^2} + \overline{\Theta_p'^2} + \overline{Z_p'^2},$$

as is the velocity of the corresponding sub-centroid, which we here denote by Θ_p.

Then the conditions (10.88) become

$$\left.\begin{aligned} \nu &= \int_{p_1}^{p_2} n(p)\,\mathrm{d}p, \\ \nu\Theta_0 &= \int_{p_1}^{p_2} \Theta_p\, n(p)\,\mathrm{d}p, \\ \nu\sigma^2 &= \int_{p_1}^{p_2} (\Delta\Theta_p)^2\, n(p)\,\mathrm{d}p + \int_{p_1}^{p_2} \sigma_p^2 n(p)\,\mathrm{d}p. \end{aligned}\right\} \tag{10.89}$$

An important particular case which occurs in the Galaxy is that where the residual-velocity variance of the sub-system serves as a kinematic parameter. As already mentioned, the differences in the residual-velocity variance correspond to the division of the star population of the Galaxy into plane, intermediate and spherical sub-systems. The variance also determines the asymmetry of stellar velocities, which may be described as a dependence of the circular velocity of the centroid on the dispersion of residual velocities.

In this case formulae (10.89) become

$$\left.\begin{aligned} \nu &= \int_{\sigma_1}^{\sigma_2} n(\sigma)\,\mathrm{d}\sigma, \\ \nu\Theta_0 &= \int_{\sigma_1}^{\sigma_2} \Theta_\sigma\, n(\sigma)\,\mathrm{d}\sigma, \\ \nu\sigma_0^2 &= \int_{\sigma_1}^{\sigma_2} (\Delta\Theta_\sigma)^2\, n(\sigma)\,\mathrm{d}\sigma + \int_{\sigma_1}^{\sigma_2} \sigma^2\, n(\sigma)\,\mathrm{d}\sigma, \end{aligned}\right\} \tag{10.90}$$

where σ_1^2 and σ_2^2 are the minimum and maximum values of the root-mean-square residual velocities for the various sub-systems (σ_1 may, in particular, be zero), and σ_0^2 is the variance for the whole system.

The first term on the right of the last eqn. (10.90),

$$I = \int_{\sigma_1}^{\sigma_2} (\Delta\Theta_\sigma)^2\, n(\sigma)\,\mathrm{d}\sigma, \tag{10.91}$$

is equal to twice the kinetic energy of the motion of the sub-centroids with respect to the total centroid. Hence I characterises the degree of statistical non-uniformity of the star population of the system considered. If $I \neq 0$, the system is a superposition of a number of sub-systems, whose distribution function with respect to the parameter σ (or, in general, p) is $n(\sigma)$ (or $n(p)$).

An example of a system described by the three eqns. (10.90) is our own Galaxy. As has been mentioned more than once in earlier chapters, an asymmetry of stellar motions is observed in the Galaxy, and consists in the fact that the circular velocity Θ_σ of the centroid is a function of the root-mean-square residual velocity σ for the group of stars considered. The greater σ (i.e. the greater the residual-velocity variance), the less is the centroid velocity in absolute magnitude.

According to Strömberg, the asymmetry of the velocities,

$$S = \Theta_c - \Theta_\sigma, \tag{10.92}$$

is given by $S = a\sigma^2 + b$. Here Θ_c is the circular velocity for a particle (i.e. for a motion with $\sigma^2 = 0$) and Θ_σ is the circular velocity of the sub-centroid, i.e. the centroid of the sub-system (sub-group) considered; a and b are functions of the coordinates.

It has been shown in Chapter VIII that the asymmetry S may be approximately represented by the formula

$$S \approx (\Theta_c^2 - \Theta_\sigma^2)/2\Theta_c = a\sigma^2 + b. \tag{10.93}$$

The factor $1/2\Theta_c$ in this expression is unimportant, since it is a function of the coordinates only; the mathematical form of the equation is unaltered when both sides are multiplied by $2\Theta_c$, and only the values of the coefficients a and b are changed. Hence the expression $\Theta_c^2 - \Theta_\sigma^2$ may equally well be taken as a measure of the asymmetry. The quantity

$$\frac{1}{\nu}\int_{\sigma_1}^{\sigma_2} (\Theta_c^2 - \Theta_\sigma^2)\, n(\sigma)\, d\sigma \tag{10.94}$$

is equal to the mean asymmetry of velocities at a given point in the system.

Thus we see that Strömberg's asymmetry, given by formula (10.93), is a particular case of the dependence of the partial centroid velocity on a continuous parameter, since this formula may be written

$$\Theta_\sigma = \sqrt{[\Theta_c^2 - 2\Theta_c(a\sigma^2 + b)]} = f(\sigma),$$

where Θ_c, a and b are constant within each macroscopic volume element.

We shall show that the mean asymmetry may also be characterised by the integral I (10.91). Since $\Delta\Theta_\sigma = \Theta_\sigma - \Theta_c$, we may write

$$I = \int_{\sigma_1}^{\sigma_2} (\Theta_\sigma^2 - \Theta_c^2)\, n(\sigma)\, d\sigma + 2\nu\Theta_c^2 - 2\Theta_c \int_{\sigma_1}^{\sigma_2} \Theta_\sigma\, n(\sigma)\, d\sigma,$$

or, using the second eqn. (10.90),

$$I = -\int_{\sigma_1}^{\sigma_2} (\Theta_c^2 - \Theta_\sigma^2)\, n(\sigma)\, d\sigma + 2\nu\Theta_c(\Theta_c - \Theta_0). \tag{10.95}$$

Here Θ_0 denotes the circular velocity of the mean centroid, given by the second eqn. (10.90). Equation (10.95) shows that I, like (10.94), is a measure of the asymmetry of velocities at a given point in a stellar system: it differs from (10.94) only by a term involving ν, Θ_c and Θ_0, and these are "macroscopic" functions of the coordinates, i.e. they have definite values at every point in the system and are independent of any particular sub-system.

Summarising the results of this section, we may note the following properties of statistically non-uniform systems.

(1) The state of a statistically non-uniform rotating stellar system is not uniquely determined by the values of the "macroscopic parameters" (functions of position only) namely the star density ν, the circular velocity Θ_0 of the centroid, and the kinetic energy of the residual velocities.

(2) A characteristic feature of statistically non-uniform systems is the presence at every point of an "asymmetry of stellar velocities"; that is, the kinetic energy of the disordered motion of the stars (analogous to the thermal motion of gas molecules) can be represented as the sum of two terms different in nature: the kinetic energy of the residual velocities proper and that of the disordered motion of the partial centroids or sub-centroids. The sub-centroids are the various centroids pertaining to the same volume element but to different sub-systems of stars.

(3) A statistically non-uniform system is a superposition of a number of sub-systems rotating about a common axis but with different velocities. The Galaxy is an example of such a system, and so, apparently, are the other spiral galaxies of kindred type.

§ 11. The Fourth Additive Parameter for KL Systems

By analogy with the cases already discussed, we may attempt to derive a set of additive parameters sufficient for a unique definition of a statistically inhomogeneous rotating stellar system.

The first three parameters can be the same as for homogeneous systems.

(1) The first parameter is the mass M of the system or, if all the stars are of equal mass, the total number of stars N in the system.

(2) The second parameter is the angular momentum $K = \sum_i K_i$ or, explicitly,

$$K = \sum_i \varrho_i \Theta_i = \sum_v \nu \varrho \Theta_0 = \int \nu \varrho \Theta_0 \, \mathrm{d}v. \tag{10.96}$$

Here the summation over i includes the values for individual stars whatever their position in the system, while the summation over v is of quantities in which the stellar coordinates and velocities have previously been averaged over the individual macroscopic volume elements. The fact that, on account of the finite dimensions of these volume elements, each individual star is regarded as lying within an infinity of macroscopic volume elements which necessarily overlap and all include that star need cause no concern, since this is simply a way of "smoothing" by averaging over overlapping regions.

(3) Finally, the third parameter may be taken as the energy parameter of the system:

$$E = \sum_i (\mathrm{P}_i^2 + \Theta_i^2 + Z_i^2 - 2U_i) = 2(T + 2W), \tag{10.97}$$

where $W = -\frac{1}{2}\sum U_i$ is the potential energy of the forces of interaction between the stars (that is, the internal energy of the system).

As we have seen in § 10, for a statistically inhomogeneous stellar system the value of the kinetic energy is insufficient to define the system, since, according to the third eqn. (10.88), or the equivalent eqns. (10.80) or (10.90), it is necessary to specify what fraction of the kinetic energy pertains to the dispersion of sub-centroid velocities. This part of the kinetic energy corresponds to the asymmetry of stellar motions – that is, to the non-uniformity of the star population of the system.

However, it is easily seen that a knowledge of the value of I (10.91) does not really solve the problem, since a contradiction of Jeans' theorem arises. We shall omit the calculations for the sake of brevity, and simply mention that in this case the phase density is not an integral of the motion.

At the present time there is no single method of finding a set of additive parameters sufficient to define uniquely an inhomogeneous stellar system. The many-valuedness of the solution corresponds to the variety of types of galaxy, and in particular of normal spiral galaxies. The problem of the extent to which any particular set of additive parameters uniquely and precisely defines some species of galaxy must be resolved by a comparison of theory and observation.

Unfortunately, the relative populations of the various possible sub-systems are unknown, both for the Galaxy and for the neighbouring galaxies which resemble it (such as M 31). We must therefore attempt to define a fourth additive parameter, using some other characteristic properties of galaxies of this type. Such a property is the law of rotation, which is fairly accurately established by empirical means for the normal spirals:

$$\Theta_0 = k_1\varrho/(1 + k_2\varrho^2). \tag{10.98}$$

The validity of this formula has been demonstrated for the Galaxy for $\varrho < 11$ kps by P. P. Parenago [99, 100], and also for the nearest spirals M 31 and M 33 [101, 102].

The law of rotation (10.98) is undoubtedly due to the division of the star population of galaxies into sub-systems, since otherwise the system would rotate as a rigid body.

The maximum value of the circular velocity (10.98) is reached when $\varrho = \varrho_0 = 1/\sqrt{k_2}$. As a rough description of the behaviour of the circular velocity, we can say that Θ_0 increases from the centre to $\varrho = \varrho_0$ approximately in proportion to ϱ, and beyond $\varrho = \varrho_0$ it decreases as $1/\varrho$. The first range ($\varrho = 0$ to ϱ_0) corresponds to rigid rotation, and the second range to a constant value of the product $\varrho\Theta_0$, the angular momentum of the centroid.

Thus formula (10.98) signifies that spiral galaxies consist of two regions: an inner region of rigid rotation, i.e. constant angular velocity of the centroids, and an outer region of constant angular momentum of the centroids.†

To find the fourth additive parameter, we consider the angular-momentum variance

$$S_1 = \sum_i (K_i - \overline{K})^2/N = \sum_i (\varrho_i\Theta_i - \overline{K})^2/N, \tag{10.99}$$

† The real significance of this important statement will be shown below.

where $\varrho_i \Theta_i$ is the angular momentum of the ith star, and $\bar{K}$ denotes for brevity the mean angular momentum, i.e.

$$\bar{K} = K/N = \sum_i \varrho_i \Theta_i / N. \tag{10.100}$$

Here K denotes the total angular momentum of the system. In the numerator of (10.100) the summation is over all N stars, taken in any order.

We shall show in § 12 that a defining set of additive parameters for a KL system with the law of rotation (10.98) is obtained if the angular momentum variance (10.99) is taken as the fourth additive parameter. In the present section the physical significance of the quantity S_1 will be elucidated.

We first transform the sum in (10.99), dividing the system into a number of non-overlapping macroscopic volume elements and, in the summation, grouping the stars in each volume according to sub-systems. The suffix i relates as before to a summation over all stars regardless of their positions in the system and their membership of sub-systems; the suffixes j, σ and ν denote respectively the stars in the various sub-systems, the sub-systems themselves, and the macroscopic volume elements.

In formula (10.99) we first sum over j with σ and ν fixed, i.e. over the stars of a given sub-system in a given macroscopic volume element. The summation over σ is next effected, with ν kept constant, i.e. we sum over the various sub-systems in a given macroscopic volume element. Lastly, the summation over ν is effected. Since

$$K_\sigma = \sum K_j / n(\sigma), \quad K_\nu = \sum n(\sigma) K_\sigma / \nu, \quad \bar{K} = \sum \nu K_\nu / N,$$

we have

$$\begin{aligned}\sum_i (K_i - \bar{K})^2 &= \sum_i [(K_i - K_\sigma) + (K_\sigma - K_\nu) + (K_\nu - \bar{K})]^2 \\ &= \sum_{j,\sigma,\nu} (K_j - K_\sigma)^2 + \sum_{\sigma,\nu} n(\sigma)(K_\sigma - K_\nu)^2 \\ &\quad + \sum_\nu \nu (K_\nu - \bar{K})^2.\end{aligned}$$

Division of both sides by N gives†

$$S_1 = \frac{\sum_{j,\sigma,\nu} (K_j - K_\sigma)^2}{N} + \frac{\sum_{\sigma,\nu} n(\sigma)(K_\sigma - K_\nu)^2}{N} + \frac{\sum_\nu \nu (K_\nu - \bar{K})^2}{N}. \tag{10.101}$$

Each of the three terms on the right-hand side of eqn. (10.101) has a definite meaning. The first term is the variance of the angular momenta of the stars themselves; it relates to the random part of the distribution of angular momenta of the individual stars (the dispersion of their "residual angular momenta"). The second term gives the variance of the angular momenta of the sub-centroids (partial centroids) relative to the corresponding total centroids (the dispersion of the "residual angular momenta" of the sub-centroids). Thus the second term represents the asymmetry of

† In formula (10.101) ν is used both as a suffix indicating the variable of summation and (in the last term) in its usual significance as the star density. All the sums should really be written as Stieltjes integrals.

stellar motions. The third term, as we shall now see, gives the angular velocity of rotation of the galaxy as a function of the coordinate ϱ (see (10.98)).

Since $\sum_\nu \nu K_\nu = N\bar{K}$ and $\sum_\nu \nu = N$ we can write the third term as $(1/N)\sum_\nu \nu K_\nu^2 - \bar{K}^2$, or, changing the sums to integrals,

$$\frac{1}{N}\int_v \nu\varrho^2\Theta_0^2\,\mathrm{d}v - \left[\frac{1}{N}\int_v \nu\varrho\Theta_0\,\mathrm{d}v\right]^2, \tag{10.102}$$

where the integration is over the volume v of the galaxy. It is evident that this expression is zero if

$$\Theta_0 = B/\varrho, \tag{10.103}$$

with B a constant. From the expression (10.98) for the circular velocity we see that Θ_0 in fact satisfies this condition for $\varrho > 1/\sqrt{k_2}$, i.e. outside the region of rigid rotation. Thus the third term defines this region. We should expect that the size of the nucleus and of the central region relative to the whole galaxy would be an important characteristic of the various types of galaxy. Anticipating a little, we may mention that the most probable law of phase distribution, determined by means of the new parameter, gives $\varrho_0 = 1/\sqrt{k_2}$ as the outer boundary of the main body of the galaxy.

Before proceeding to find the most probable phase distribution for KL systems by means of the parameter S_1, we can simplify the latter somewhat. Since $\sum_i K_i = N\bar{K}$ and $\bar{K} = K/N$, we can write

$$S_1 = \frac{1}{N}\sum_i K_i^2 - \frac{K^2}{N^2}.$$

The values of N and K may be supposed given, because they are dynamical parameters, i.e. they must necessarily appear in a defining set for the whole rotating galaxy. Hence the additive parameter S_1 can be replaced by an equivalent parameter

$$S = \sum_i K_i^2 = \sum_i \varrho_i^2\Theta_i^2. \tag{10.104}$$

As we should expect from the arguments given at the end of § 3, the new (fourth) parameter S is statistical and not dynamical.

At first sight it appears that a KL system should become a D system if the variance of angular momenta of the sub-centroids is zero. This is not so, however, as we easily see from formula (10.101), of which only the middle term disappears. For the system to become a D system, the partial velocity dispersions of the stars in the various subsystems must also become equal. Otherwise we have a superposition of distributions with different dispersions, which cannot be equated to a single distribution. For example, the sum of Gaussian distributions with a common vertex is not itself a Gaussian distribution.

We shall see later that KL systems (apart from their nuclei) never become D systems. For the latter, the fourth parameter can be derived from the first three, and is not independent: all the sub-systems are identical, and we then have, since j denotes summation over all the stars in a given macroscopic volume element (suffix k),

$$S = \sum_i \varrho_i^2 \Theta_i^2 = \sum_{j,k} \varrho_k^2 \Theta_j^2 = \sum_{j,k} \varrho_k^2 (\Theta_j - \Theta_0)^2 + \sum_k \nu_k \varrho_k^2 \Theta_0^2 .$$

For a D system $\sum_j (\Theta_j - \Theta_0)^2 = \nu_k \sigma^2$, where σ^2 is the "isothermal" variance and $\Theta_0 = \omega_0 \varrho_k$, ω_0 being the angular velocity. In the first approximation $\nu_k = \nu_0 =$ constant. Hence, changing to integrals, we have for any D system

$$S = \nu_0 \sigma^2 \int_v \varrho^2 \, dv + \nu_0 \omega_0^2 \int_v \varrho^4 \, dv = \sigma^2 J_2 + \omega_0^2 J_4 ,$$

where J_2 and J_4 are the moments of inertia of the second and fourth orders about the axis of rotation. Since σ^2, ω_0 and the shape, size and mass distribution in a D system are uniquely defined by the values of the first three parameters, it follows that the parameter S is also uniquely determined by them.

To conclude this section, we shall return to the problem of determining the most probable phase distribution for spherical systems, already discussed in Chapter IX. There we determined the phase distribution on the basis of general arguments concerning the one-valuedness of the integrals of the motion in Jeans' theorem, and derived three additive parameters: mass, energy and a third parameter equal to the sum of the squared angular momenta of the individual stars. Whereas the position regarding the first two parameters is entirely clear, the third parameter still needs a physical explanation, which we can now give by analogy with the rotating systems just discussed, generalising to them the concept of Kapteyn–Lindblad systems.

As the homogeneous system we take a non-rotating spherical stellar system with a Maxwellian velocity distribution. This is defined by the two dynamical parameters of mass and energy. Now let this system be replaced by another, also non-rotating and having the same values of these two parameters, but fundamentally different in the following way: half the stars are made to rotate as a rigid assembly about an axis through the centre of the system, and the other half to rotate about the same axis at the same rate in the opposite direction. This other system is, as a whole, non-rotating, and has the same mass as the first system. The kinetic energy of the stars, however, consists of, firstly, the energy of their residual (disordered) velocities relative to one or other of two sub-centroids; secondly, the energy of the motion of each of these sub-centroids relative to the total centroid, which is at rest. The sub-centroids are to be regarded as particles, of mass equal to half the total mass of the stars in the macroscopic volume element considered.

Evidently the residual-velocity dispersion can be reduced in such a way that, when the energy of the sub-centroids is added, the result is equal to the original value of the kinetic energy, and it is clear that the rate of rotation of the sub-systems remains arbitrary and so must be specified if the whole system is to be uniquely defined.†

Nevertheless, the system so obtained is fundamentally different from the original system, since, on account of the rotation of the sub-systems, it will be not spherical

† For example, in the "Kapteyn universe" the velocity of the two sub-centroids in the neighbourhood of the Sun was $\pm 19{\cdot}5$ km/s.

but spheroidal, even though as a whole it is not rotating. This difference may be removed by using the fact that the common axis of rotation of the sub-systems may be chosen arbitrarily. Hence, by taking an arbitrary number $n(\gg 1)$ of axes of rotation distributed uniformly as regards direction, allotting to each axis $1/n$ of the stars in the system, and forming for each nth part two sub-systems rotating oppositely about its axis in the manner described above, we again obtain a spherical system.

Finally, instead of two oppositely rotating sub-systems we could take any number of pairs of such sub-systems rotating about each axis.

Hence it is clear that in the most general case, for a unique definition of an inhomogeneous spherical stellar system, it is necessary to specify in addition to the mass and energy the transverse velocity variance $(1/N)\sum T_i^2$, or, since N is known (being proportional to the mass), the sum of the squared transverse velocities. Since the square of the residual velocity is not an integral of the motion, in order to satisfy the requirements of Jeans' theorem we must specify the third additive dynamical parameter as the sum of the squared angular momenta $\sum r_i^2 T_i^2$. This is entirely in agreement with the results of Chapter IX. Here we have deliberately discussed spherical sub-systems in detail, because previously, rotating systems having not yet been considered, it was not possible to discuss the physical aspect of the problem properly, and the definition of the third dynamical parameter for spherical systems was necessarily formal to some extent.

For rotating systems the problem of the fourth dynamical parameter is simpler than for spherical systems, since the circular motion of the sub-centroids (rotation of the sub-systems) can take place only about one particular axis, namely the axis of symmetry of the mass distribution in the system. For non-uniform systems the fourth dynamical parameter may therefore be taken as $S = \sum \varrho_i^2 \Theta_i^2$.

§ 12. The Most Probable Phase Distribution for KL Systems

Let us now go on to determine the most probable phase distribution for the case of a dynamically inhomogeneous rotating system, and first rewrite the expressions for the dynamical parameters in terms of the phase cell populations n_μ. The first three parameters are

$$N = \sum_\mu n_\mu, \quad E = \sum_\mu n_\mu E_\mu, \quad K = \sum_\mu n_\mu K_\mu. \tag{10.105}$$

Expressing the star density in the fourth dynamical parameter (10.104) in terms of the phase cell populations n_μ, we have

$$S = \sum_\mu n_\mu K_\mu{}^2. \tag{10.106}$$

The subsequent calculations are facilitated by the fact that they are largely similar to some carried out previously.

The corresponding Lagrangian function to give the maximum probability of the phase distribution is obtained in exactly the same way as (10.16) and is

$$L = \log P + k_1 \sum_\mu n_\mu + k_2 \sum_\mu n_\mu E_\mu + 2k_3 \sum_\mu n_\mu K_\mu + k_4 \sum_\mu n_\mu K_\mu^2. \tag{10.107}$$

The solution for the phase density is

$$f = \exp(k_1 + k_2 E + 2k_3 K + k_4 K^2). \tag{10.108}$$

For $k_4 = 0$ this phase density becomes that for homogeneous D systems (10.21). We therefore call k_4 the *inhomogeneity coefficient* of the galaxy.

To ascertain the physical significance of this solution, we transform the exponent in (10.108). We have

$$k_2 E + 2k_3 K + k_4 K^2 = k_2(\mathrm{P}^2 + \Theta^2 + Z^2 - 2U) + 2k_3 \varrho\Theta + k_4 \varrho^2 \Theta^2$$
$$= k_2 [\mathrm{P}^2 + \lambda^2 (\Theta - \Theta_0)^2 + Z^2 - 2U_1], \tag{10.109}$$

where

$$\left.\begin{aligned} \lambda^2 &= 1 + k_4 \varrho^2 / k_2, \\ \Theta_0 &= \frac{-k_3 \varrho / k_2}{1 + k_4 \varrho^2 / k_2}, \\ 2U_1 &= 2U + \lambda^2 \Theta_0^2. \end{aligned}\right\} \tag{10.110}$$

Using (10.108) and (10.109), we can represent the phase distribution as

$$F(\varrho, z, \mathrm{P}, \Theta, Z) = \exp\{k_1 + k_2 [\mathrm{P}^2 + \lambda^2 (\Theta - \Theta_0)^2 + Z^2 - 2U_1]\}. \tag{10.111}$$

It is easily seen that (10.111) and (10.110) are the same, apart from the notation, as the Schwarzschild law commonly used in the ellipsoidal dynamics of the Galaxy (see (5.60)–(5.64)). In particular, the second formula (10.110) shows that the most probable phase distribution for KL systems in fact gives the formula (10.98) for the circular velocity as a function of distance.

From formulae (10.111) and (10.110) the physical significance of the coefficients in the phase distribution can be elucidated. The coefficient k_2, as in a D system, gives the "isothermal" residual-velocity variance:

$$k_2 = -1/2\sigma_\varrho^2 = -1/2\sigma_z^2 = -1/2\sigma^2, \tag{10.112}$$

where σ_ϱ^2 and σ_z^2 are the variances in the ϱ and z directions, i.e. the "meridian variances".

For $k_4 = 0$ the second formula (10.110) gives the circular velocity of rigid rotation:

$$\Theta_0 = -k_3 \varrho / k_2. \tag{10.113}$$

Hence the inhomogeneity coefficient k_4 may also be called the *rigid-rotation parameter*. The coefficient k_3, as is seen from (10.113), defines the angular velocity of rigid rotation; for $k_3 = 0$, the centroid velocity $\Theta_0 = 0$, and so k_3 is the *angular-velocity parameter* of the system.

In the case considered the velocity variances along the axes are $\sigma^2, \sigma^2/\lambda^2$ and σ^2, and the centroid velocity components are $0, \Theta_0, 0$. Hence the phase distribution can be put in the form

$$f(\varrho, z, \mathrm{P}, \Theta, Z) = \frac{\nu\lambda}{(2\pi)^{3/2}\sigma^3} \exp\{-[\mathrm{P}^2 + \lambda^2(\Theta - \Theta_0)^2 + Z^2]/2\sigma^2\}. \tag{10.114}$$

A comparison of the two expressions (10.111) and (10.114) for the phase density, using the last eqn. (10.110), gives a relation between the star density ν and the potential U:

$$\sigma^2 \log (\nu \lambda) = U + \tfrac{1}{2}\lambda^2 \Theta_0^2 + C, \tag{10.115}$$

where $C = k_1 \sigma^2 + \sigma^2 \log [(2\pi)^{3/2} \sigma^3] =$ constant.

From (10.115) we can easily derive an equation which involves only the star density ν. The Laplacian in cylindrical coordinates for a function independent of θ is

$$\triangle = \frac{1}{\varrho} \frac{\partial}{\partial \varrho} \left(\varrho \frac{\partial}{\partial \varrho} \right) + \frac{\partial^2}{\partial z^2};$$

applying this to eqn. (10.115) and using Poisson's equation $\triangle U = -4\pi G m \nu$, we obtain

$$\sigma^2 \triangle \log (\nu \lambda) = -4\pi G m \nu + \frac{1}{2} \frac{1}{\varrho} \frac{\mathrm{d}}{\mathrm{d}\varrho} \left[\varrho \frac{\mathrm{d}(\lambda^2 \Theta_0^2)}{\mathrm{d}\varrho}\right]. \tag{10.116}$$

In eqn. (10.116) all the functions except ν are known. Thus this equation is the required equation for ν and serves to determine the equilibrium figures of KL systems.

In order to achieve a general idea of the nature of the solution, we can seek to apply the method of successive approximations as in the case of D galaxies. Neglecting in the first approximation the derivatives of $\log (\nu \lambda)$ and using (10.110), we find as a first approximation

$$\nu_0 = \frac{1}{8\pi G m} \frac{1}{\varrho} \frac{\mathrm{d}}{\mathrm{d}\varrho} \left[\varrho \frac{\mathrm{d}(\lambda^2 \Theta_0^2)}{\mathrm{d}\varrho}\right] = \frac{\varkappa_3^2}{2\pi G m} \frac{1 - \varkappa_4 \varrho^2}{(1 + \varkappa_4 \varrho^2)^3}, \tag{10.117}$$

where

$$\varkappa_3 = k_3/k_2, \quad \varkappa_4 = k_4/k_2. \tag{10.118}$$

The general form of the function $f(\xi) = (1 - \xi^2)/(1 + \xi^2)^3$, which is obtained from (10.117) by a simple change in the coordinate scales, is shown in Fig. 43. The function (10.117) involves two parameters, $\varkappa_3$ and $\varkappa_4$, by choosing which we can vary the height and width of the curve.

From this first approximation some conclusions may be drawn. Figure 43 shows that, in an inhomogeneous system, the density decreases slowly away from the axis of rotation. From (10.117) we see that, for

$$\varrho = \varrho_0 = 1/\sqrt{\varkappa_4}, \tag{10.119}$$

ν_0 is zero, and for $\varrho > \varrho_0$ it becomes negative. Since the solution of the basic eqn. (10.116) cannot differ greatly from ν_0, and ν cannot be negative, the value of ϱ_0 gives approximately the equatorial radius of a KL galaxy. We know that ϱ_0 corresponds also to the maximum circular velocity Θ_0 of the centroid (see (10.98)), and for $\varrho \leqslant \varrho_0$ the rotation is approximately rigid. Thus we conclude that the central region of rigid rotation coincides with the main body of a KL galaxy.

It has been shown in Chapter V that for our Galaxy $\varrho_0 \approx 6$ kps, which is less than the distance of the Sun from the galactic centre. It is known that the corresponding point in M 31 (the galaxy in Andromeda) is at the boundary of the part of the galaxy which can be observed photographically. This is shown in Plate XXVII, taken from the *Larousse Encyclopedia of Astronomy* (Batchworth, London, 1959). The broken curve shows the outer boundary of the Andromeda nebula as found by photographic means, and the cross marks the point corresponding to the position of the Sun in the Galaxy.

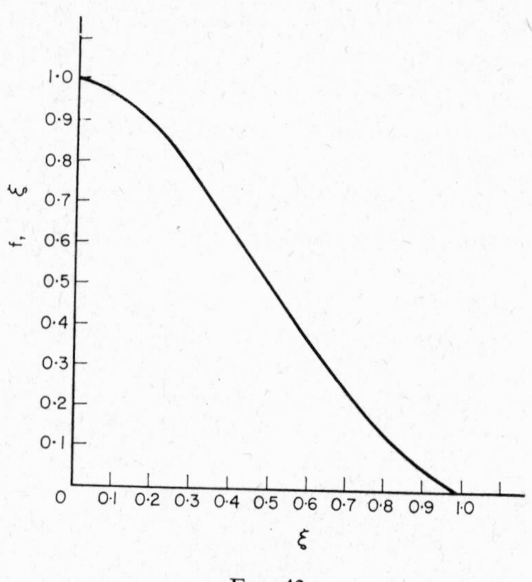

FIG. 43

This shows the range of distances within which our theoretical discussion is valid. Thus ϱ_0 is to be regarded as the apparent outer boundary of a spiral galaxy, since, firstly, the continuous spiral arms of galaxies, which are a distinctive feature of the structure of these systems, extend only up to such distances; secondly, beyond this point the rigid rotation ceases and is replaced by an entirely different mode of rotation which leads to the presence of Oort terms in stellar velocities, namely the "differential" rotation; thirdly, ϱ_0 is the directly observed distance from the centre of the galaxy to its outer boundary.

In solving eqn. (10.116) we encounter the same situation as in solving the similar equation for homogeneous rotating D systems: the first approximation is insufficient to give a physically real model of a galaxy. The first approximation for ν depends only on ϱ. Hence a stellar system corresponding to this approximation would be an infinite cylinder, which is physically impossible. The KL galaxies, like the D galaxies, are dynamically unstable, since they do not satisfy Poincaré's criterion regarding the angular velocity: $\Omega = \omega^2/2\pi G m \nu < 1$. For the second formula (10.110) and (10.118) show that $-\varkappa_3$ is the angular velocity ω_0 at $\varrho = 0$. Hence, putting $\varrho = 0$ in formula (10.117), we have $\Omega = [\omega_0^2/2\pi G m \nu]_{\varrho=0} = 1$. For $\varrho > 0$, the angular velocity ω remains practically constant in the region of rigid rotation, whereas from (10.117) ν_0 decreases, and hence, for $\varrho > 0$, $\Omega = \omega^2/2\pi G m \nu_0 > 1$, and Poincaré's condition is violated. Thus the model of a stellar system based on the first approximation is physically unreal and unstable, and a second approximation must be sought.

The following point is also of interest. It might seem natural to expect that non-rotating systems can be obtained as a particular case of rotating systems when the rate of rotation tends to zero. This is not so, however. If we take first the simple case of D systems and put the angular velocity equal to zero in the expression for the phase density, we obtain a distribution which depends only on the two integrals of mass and energy, and is valid whatever the law of distribution of mass. Thus we obtain not spherical systems as required, but galaxies of "arbitrary", i.e. irregular, form.

For KL systems we likewise do not have a spherical system when the angular velocity becomes zero. The explanation of this apparent inconsistency is as follows. If $\Omega = \omega_0^2/2\pi G m \nu$ remains approximately constant, then as $\omega_0 \to 0$ the star density ν must also tend to zero. This invalidates some of the fundamental hypotheses on which the derivation of the most probable phase distribution is based. One of these is the premise that macroscopic volume elements can be defined; another is that some effective mechanism of the action of irregular forces operates in the system: these are evidently incompatible with a star density which tends to zero.

If ν does not tend to zero, Ω must do so. We know from the previous discussion that the limiting state with $\Omega = 0$ is in general indeterminate and depends wholly on the initial state of the system – for example, on the nature of the ellipsoidal equilibrium figure.

Let us now derive the second approximation. We put

$$\nu = \nu_0 + \nu_1, \tag{10.120}$$

where ν_0 is the first approximation already found (the basic density), and ν_1, whose modulus is less than ν_0, is the additional density. Substituting (10.120) in (10.116) and using the expression (10.117) for ν_0, we find, after calculations like those in § 8, the linearised equation for ν_1:

$$\sigma^2 \triangle (\nu_1/\nu_0) + 4\pi G m \nu_1 = -\sigma^2 \triangle \log (\nu_0 \lambda). \tag{10.121}$$

With a new unknown function

$$u = \nu_1/\nu_0 \tag{10.122}$$

and the above values for ν_0 and λ, we obtain the linear equation

$$\triangle u + 2\frac{\omega_0^2}{\sigma^2}\frac{1 - \varkappa_4 \varrho^2}{(1 + \varkappa_4 \varrho^2)^3}\, u = -\frac{1}{\varrho}\frac{\mathrm{d}}{\mathrm{d}\varrho}\left\{\varrho \frac{\mathrm{d}}{\mathrm{d}\varrho} \log \frac{1 - \varkappa_4 \varrho^2}{(1 + \varkappa_4 \varrho^2)^{5/2}}\right\}, \tag{10.123}$$

where $\omega_0 = -\varkappa_3$. When $\varkappa_4 = 0$ this equation becomes Helmholtz' equation already discussed in connection with D galaxies:

$$\triangle u + 2(\omega_0^2/\sigma^2)\, u = 0. \tag{10.124}$$

Thus the fundamental equation determining the density in a KL galaxy is a generalised Helmholtz equation depending on the small parameter $\varkappa_4$. A detailed solution of this equation will not be given here, but the procedure is as follows.

According to the general theory of linear partial differential equations, the general solution of eqn. (10.123) is $u = u_0 + U$, where u_0 is a particular solution of the ordinary equation

$$\frac{1}{\varrho}\frac{d}{d\varrho}\left(\varrho\frac{du_0}{d\varrho}\right) + 2\frac{\omega_0^2}{\sigma^2}\frac{1-\varkappa_4\varrho^2}{(1+\varkappa_4\varrho^2)^3}u_0 = 2\varkappa_4\frac{7-6\varkappa_4\varrho^2+7\varkappa_4^2\varrho^4}{(1+\varkappa_4^2\varrho^4)^2},$$

and U is the general solution of the homogeneous equation

$$\Delta U + 2\frac{\omega_0^2}{\sigma^2}\frac{1-\varkappa_4\varrho^2}{(1+\varkappa_4\varrho^2)^3}U = 0,$$

which is a generalised Helmholtz' equation. Like Helmholtz' equation itself, it can be solved by separation of the variables. In particular, for the case of a flat disc we put $U = \mathrm{P}\Theta Z$, where P, Θ, Z are respectively functions of the variables ϱ, θ, z only. For Z we obtain the "barometric" formula $Z = e^{-l|z|}$, for Θ the periodic solution $\Theta = \sin p\theta$, and for P a generalised Bessel's equation:

$$\frac{1}{\varrho}\frac{d}{d\varrho}(\varrho \mathrm{P}') + \left[l^2 - \frac{p^2}{\varrho^2} + 2\frac{\omega_0^2}{\sigma^2}\frac{1-\varkappa_4\varrho^2}{(1+\varkappa_4\varrho^2)^3}\right]\mathrm{P} = 0.$$

We may expect that the derivation of the second approximation corresponding to $\varkappa_4 = 0$ will make possible, as in D galaxies, an elucidation of some structural features. By the principle of continuity, for example, we should expect that, at least for small values of $\varkappa_4$, the additional density in KL galaxies should lead to a tendency to the formation of nuclei and rings. We shall not give further details, and conclude this section with some general comments.

We know that for $\varkappa_4 = 0$ the inhomogeneous system becomes a homogeneous rigidly rotating D system. For small $\varkappa_4$ the equilibrium figure of a KL galaxy will differ only slightly from that of the corresponding D galaxy. The equilibrium figures of D galaxies may, as we have seen, be various bodies similar to the three classical ellipsoids (planet-like Maclaurin, disc-like Maclaurin, needle-shaped Jacobi) and also the pear-shaped Lyapunov–Poincaré bodies akin to the Jacobi ellipsoids.

It is easy to see that neither the Jacobi ellipsoids nor the pear-shaped bodies can be even approximately figures of equilibrium for KL galaxies: firstly, they are not rotationally symmetrical, and secondly, the generalisation considered in § 8 is not applicable to the fourth parameter S.

From formula (10.119) we see that for small $\varkappa_4$ the equatorial radius of a KL galaxy becomes very large. Thus the figures of KL galaxies cannot be planet-like, and they must consequently approach the disc-like Maclaurin ellipsoids. These ellipsoids are greatly flattened and resemble thin discs. The smaller the characteristic quantity $\Omega = \omega^2/2\pi G m \nu$, the thinner the disc and the greater its radius.

The photographs of galaxies seen edgewise indicate that they are of precisely this form. The similarity is somewhat lessened by the presence of a central nucleus, but this arises, of course, from the second approximation in the solution of eqn. (10.116). Another dissimilar feature is the spiral arms, which deprive the distribution of matter of the rotational symmetry necessary for the existence of Maclaurin ellipsoids. However, studies of the nearest galaxies show that the mass of the spiral arms is very small compared with the mass of the galaxy itself; the arms are not of great dynamical importance, despite their high luminosity due to the fact that the stars forming them are the bright giants. The arms should be regarded as small perturbations of the rotationally symmetrical gravitational field.

§ 13. The Galaxy and Similar Spirals as Kapteyn–Lindblad Systems

As we have seen, the *D* galaxies include the SB, needle-shaped and E types. In the three preceding sections we have given a theoretical discussion of the dynamics of KL systems, which differ from *D* systems in that the star population is not uniform. Our interest lay chiefly in stellar systems like the Galaxy, i.e. the normal spirals of types Sa, Sb, Sc. Here we shall consider in more detail to what extent the observed galaxies of these types are in accordance with the theoretical description derived above.

The fundamental properties of the normal spirals are as follows.

(1) They have the form of very thin discs, with a mean axis-ratio of about 1:12. This is well shown by the appearance of the galaxies which are seen edgewise (for instance, NGC 4565 (Plate XV)). In this respect these galaxies resemble KL systems.

(2) At the centre of each galaxy is a nucleus which is small in size compared with the galaxy itself. From two diametrically opposite points of the nucleus come two thick spiral arms extending to the boundary of the visible disc of the galaxy. Their appearance led Jeans to propose his famous hypothesis of the tidal origin of the arms. A characteristic feature of the arms is the numerous branchings and condensations, as a result of which the arms of galaxies of type Sc are not even continuous.

The example of our Galaxy shows that the spiral arms are not of great importance as regards the distribution of mass. The concept of the dynamical "insignificance" of the spiral structure of the galaxies, first propounded in 1933 by M. S. Éĭgenson [30], has in recent years been confirmed by the work of G. A. Shaĭn [115], B. A. Vorontsov-Vel'yaminov [137], F. Zwicky [141] and others. From their work it is clear that the size, shape and structure of the spiral arms are determined not by the distribution of stars but by the presence of magnetic fields, along whose lines of force diffuse matter (gas and dust) is distributed. Within the spiral arms, nevertheless, there occurs a continuous formation of stars, leading to the groups of hot giants of early spectral types which constitute the O associations of Ambartsumyan. On account of the radiation from these hot giants, grouped in chains and strings (to use the terminology of Vorontsov-Vel'yaminov), the gas in the arms is rendered luminous, and consequently the arms, despite their relatively small mass, are the most noticeable part of the structure of the galaxy in photographic light.

(3) The spiral arms are *static*. They cannot be the result of an outflow from the nucleus. For their form is that of logarithmic spirals, for which the angle between the tangent and the radius vector is constant; according to Danver [22], its value is between 54° and 86°. If we take as a mean value 73°, then, if the spiral arms were the result of outflow, the radial velocity of matter would be about $\frac{1}{3}$ of the circular

velocity. This means that in the Galaxy, where the circular velocity in the neighbourhood of the Sun is about 250 km/s, the radial velocity would be 83 km/s, and the tangential velocity along the arm would be 260 km/s. Such velocities, whether of stars or of diffuse matter, could not possibly have remained unobserved.

Furthermore, if we assume for a moment that such velocities of outflow could exist, the problem then arises of what forces could cause them. No such forces are known. The repulsion forces which sometimes cause the ejection of matter from the nuclei of galaxies (the classic example being NGC 4486) cannot be responsible for the spiral arms, for the following reasons: first, they appear to act instantaneously, whereas the outflow of matter in the arms seems to be continuous; second, all the ejections hitherto observed are unilateral, while the arms almost always extend bilaterally from the nucleus.

No difficulty arises from a necessity to explain the permanence of the spiral arms in the presence of differential rotation, because the continuous arms lie within the region of rigid rotation. This has been clearly demonstrated by Vorontsov-Vel'yaminov [137].

The forms of the spiral arms are probably determined in the main by the gas and dust, whose dissipation is prevented by the magnetic field. The distribution of early-type bright giants along the arms is to be regarded as a secondary phenomenon arising from the formation of these stars out of diffuse matter.

(4) A very instructive example as regards the explanation of the origin of the spiral structure in the normal spirals is the galaxy NGC 7479 (Plate VI) in Pegasus. This exhibits the process of conversion of a needle-shaped galaxy into a spiral one. What was originally a long thin needle, of length 15 to 20 times its central thickness, has become curved and almost formed two spiral arms (it is of interest to note a perceptible pear-shaped asymmetry). At the same time we see that a lateral outflow of stars is occurring almost throughout the length of a central thicker part which has remained straight.

This gives reason to put forward the following hypothesis concerning the formation of S-type galaxies. Initially there is a very long thin needle-shaped galaxy in slow rotation. Gradually, as a central thicker part forms, its rate of rotation increases, and the needle becomes less and less stable. Unlike the case of the short high-density elongated ellipsoids which become SB-type galaxies, the outflow takes place almost throughout the length except for the central part. The needle loses relatively quickly almost all its stars, which are dispersed and probably form Baade's Population II. These are eventually the corona of the galaxy. At the same time as the older stars are lost, a new process of star formation begins in the diffuse matter which is retained in the needle by the magnetic field, and the young stars have small residual velocities. These form the plane component of the galaxy (Baade's Population I). These stars also would be quite rapidly scattered into a corona, but for the existence of disc-like equilibrium figures such as the Maclaurin ellipsoids of the second kind. Hence, instead of simply being dispersed, these stars mingle with some of the Population II stars and form a thin disc which is the main body of the galaxy.

The picture just outlined is in fair agreement with our information concerning the structural features of the Galaxy and of the neighbouring giant spirals. It also enables us to form some idea of the future evolution of the normal spiral galaxies. The KL spiral system initially formed is dynamically fairly stable, since the two different star populations in it interact hardly at all: near the equatorial plane the Population I stars (plane component) are predominant, while far from this plane we have almost entirely Population II stars (spherical component). Gradually, however, as a result of the action of irregular forces, more and more stars of an intermediate component appear. The star population slowly becomes more uniform, and the inhomogeneity coefficient $\varkappa_4$ decreases. In consequence, as we have seen, the equatorial radius of the disc-like body of the galaxy increases, some of the stars enter the corona, and the thickness of the disc decreases. After a time, this thickness falls almost to zero, and only the nucleus remains, surrounded by a corona of stars. This is a typical dwarf E-type galaxy, for we know that, in their planet-like form, rigid rotation, star population (type II) and mass, the nuclei of the spirals are exactly like the dwarf E galaxies, which are much smaller than the spirals.

The spiral arms evolve at the same time as the galaxy. Outside the region of rigid rotation, the arms are disrupted by the differential rotation. Within that region they tend to undergo repeated ramification along the magnetic lines of force.

A consideration of various galaxies seen "full face" shows that all the stages of evolution of the spiral structure are represented in them. The galaxy NGC 1566 (Plate XVII) is a good and fairly unusual example of a continuous spiral structure which has been retained in its entirety. From it we can estimate the shape of the original needle-shaped body, for which the ratio $c:a$ was (neglecting the nucleus) at least 12:1.

If these arguments are correct, there is a natural relation between the spiral structure and the "bridges" between galaxies (Plate XVIII) discovered by Zwicky [140, 141]. These bridges undoubtedly show that on the metagalactic scale matter is often in the form of filaments extending over tens or hundreds of kiloparsecs. This explains why the original form of galaxies is often needle-shaped, since in their initial stages of development galaxies may well be formed as condensations in a long filament of matter. The group of galaxies in Plate XXVII is an instance of this, having five elongated galaxies on a common filament. The two extreme galaxies are the most elongated. Vorontsov-Vel'yaminov's *Atlas of interacting galaxies* also shows many other examples in a great variety of forms.

The time spent by a galaxy in any given stage of evolution, as well as its degree of stability, is proportional to the relaxation time, which in turn depends on the patchiness of the structure. The relative numbers of galaxies in various stages in unit volume of space at any instant are proportional to the times for which they remain in those stages.

The needle-shaped phase, which is the initial phase in galaxies of the type under consideration, has the shortest lifetime. This is due to the great dynamical instability of the phase. As a rule, such galaxies are observed in the course of fairly rapid disruption, i.e. as SB galaxies. Their lifetime may be 10^7 to 10^8 y.

Spirals such as the Galaxy are more uniform in structure, but within the disc-like main body, whose existence defines this phase of the galaxy's evolution, the mass distribution is by no means uniform, and so the time spent in this phase may be estimated as 10^9 to 10^{10} y.

Finally, in E galaxies the system is highly uniform in structure, and the relaxation time is considerably greater than for spirals. The relative lifetime of E galaxies may be estimated from their relative abundance, estimated in 1948 by Yu. I. Efremov [29]. He found that for every spiral galaxy there are about 100 E galaxies.† Hence it follows that the lifetime of the E galaxies is of the order of 10^{11} to 10^{12} y.

A little should be said about the S0 galaxies. These have been mentioned at the end of § 9 in connection with the solution of Helmholtz' equation for *D* galaxies. We have seen that this solution gives at least a qualitative explanation of the formation of central nuclei and ring structure in *D* galaxies. We now see, however, that such a structure should also be expected in galaxies whose star population is not uniform, i.e. in KL galaxies, especially when the inhomogeneity coefficient $\varkappa_4$ is small. It is probable that S0 galaxies are the most frequent result of the evolution of SB spirals by a slow outflow of thin jets from the ends of the main body. The main body has a marked central nucleus whose presence reduces the mass of the peripheral part. Throughout the evolution of such a galaxy its nucleus remains ellipsoidal, and the fraction of Population I stars remains small.

It is not, however, dynamically impossible that a ring structure should also be formed in *D* galaxies with a rotationally symmetrical mass distribution. In this case the galaxy will be distinguished by a spheroidal central nucleus and a uniform composition of Population II stars. The structure of the ring system will probably be different from that formed by the evolution of barred spirals. In particular, the rings may be very thin. The galaxy NGC 488 (Plate X) may be an example of this type. Of course, further refinement of the theory, and observational studies of such galaxies, are necessary.

The theory given above naturally makes no attempt at completeness; it is merely a first and mainly qualitative approximation. It likewise does not purport to describe every galaxy. In particular, the evolution of the E galaxies is sufficiently well described by the familiar Roche model. Again, no account has been given of those galaxies which have a massive central body in the form of a much flattened spheroid along whose equator relatively slight and narrow spiral jets emerge at numerous points. These galaxies certainly form a class dynamically unrelated to either the normal spirals (S-type) or the E-type galaxies. Typical members of this class are NGC 2841 (Plate XIX) and NGC 7217 (Plate XX). Their dynamics has been satisfactorily explained by the well-known theory of Lindblad, based on a consideration of the instability of circular orbits in the equatorial plane of a flattened spheroid and near its outer surface. In such cases the spirals should be unwinding. These galaxies also show that Maclaurin ellipsoids of the first kind (planet-like) can be equilibrium figures of galaxies as well as of galactic nuclei (which may be the main body of all E galaxies). We propose to call such galaxies *Crab-like*.

† Quoted from Parenago ([107], p. 319).

§ 14. Some Conclusions Regarding the Processes of Evolution of Stellar Systems

It would be as yet premature to draw any further conclusions, since our arguments are only preliminary and are based on the steady-state theory. Nevertheless, we should end by seeking to delineate the ways of future study of the dynamics of stellar systems, and to judge, from the results of observations already carried out, the correctness of the theoretical arguments given in this book.

As has been stated in the Introduction, our treatment has been consciously restricted to problems of the structure and motion of stellar systems in a steady state. Yet the consideration of these problems has led us, and rightly, to see as the next step the analysis of non-steady processes, which must indeed be the foundation of a consistent theory of the evolution of stellar systems. It need only be recalled that, in considering the dynamics of galaxies in rigid rotation, we reached the conclusion that they must always be at the limit of instability, and usually must exhibit an outflow of matter at points along the equator of a Maclaurin ellipsoidal body or from the ends of a needle-shaped Jacobi ellipsoidal body.

The foregoing considerations lead to the following dynamical classification of galaxies.

Type 1: Irregular galaxies. These have masses and star densities which are small in comparison with those of galaxies of other types. Their disintegration half-life is of the same order as the relaxation time. The gravitational force does not suffice to bring the galaxy to a regular equilibrium figure. According to Jeans' theorem, the state of the galaxy is in general determined by only two dynamical parameters: mass and energy. Stellar systems of this type have no regular internal motions. The most probable phase density distribution is that given by the Maxwell–Boltzmann law. This is easily seen by maximising the corresponding Lagrangian function:

$$L = \log P + k_1 \sum_{\mu} n_\mu + 2k_2 \sum n_\mu E_\mu .$$

The irregular galaxies include, among others, galaxies like the Magellanic Clouds, which sometimes show a marked tendency to form barred spirals, e.g. NGC 3109 (Plate XXIII) and the Large Magellanic Cloud itself (Plate XXIV). These galaxies also have a tendency to form a main body which is a Jacobi ellipsoid. The main mass of the galaxy consists of stars and star clusters.†

Thus there is a continuous transition from the irregular galaxies to the barred spiral galaxies, represented by the "unfinished" barred spirals.

Type 2: Homogeneous rotating galaxies. These exist in two essentially different forms: (a) elliptical galaxies and (b) barred galaxies, which may also include needle-shaped galaxies having an elongated main body like the barred galaxies but apparently no outflow of matter from the ends; see, for instance, NGC 4826 (Plate XII).

To define a rotating uniform system, three dynamical additive parameters must be specified: mass, energy and angular momentum. The elliptical galaxies are, to a first approximation, homogeneous and rigidly rotating Maclaurin ellipsoidal bodies, i.e.

† Galaxies with patchy structure will be discussed below.

spheroid-like†. The homogeneous ellipsoidal mass is combined with an additional mass distribution given by the solution of Helmholtz' equation $\triangle u + 2(\omega_0^2/\sigma^2)u = 0$, where ω_0 is the angular velocity. Solutions of this equation exist only for a discrete series of values of the ratio ω_0/σ; in other words, for a given residual-velocity dispersion the angular velocity of the galaxy cannot take arbitrary values. In the simplest case, with the smallest possible angular velocity, the additional mass forms a central condensation; the solution of Helmholtz' equation for this case has no zero within the galaxy. In other cases – for example, in the galaxy NGC 4826 (Plate XII) – the central condensation is accompanied by a number of concentric condensations separated by regions of low density.

The theory shows that both the elliptical and the barred spiral galaxies are at the limit of dynamical stability, given by Poincaré's criterion $\Omega = \omega^2/2\pi G m v < 1$. Hence a specific feature of these galaxies is the formation of a sharp edge along the equator and the outflow of matter from it. The spindle-shaped galaxies are examples of this type, seen sideways.

The barred spirals are Jacobi ellipsoidal bodies, i.e. elongated needle-shaped ellipsoids with three unequal axes. The slower the rotation for a given density (i.e. the smaller Ω), the more elongated are the bodies. The rotation takes place about the least axis. On account of their closeness to the limit of instability there is almost always some outflow of matter in two jets at the ends of the "needle".

When there is a more marked concentration of matter towards the centre of the ellipsoidal body, the outflow takes the form of a fairly thin jet, which is rapidly (during one or two rotations of the galaxy) dispersed and forms a more or less homogeneous cloud in the plane of the equator of rotation.

In other cases there is sufficient mass at the ends of the ellipsoidal body, but the region of instability is restricted to the very points of the "needle". Then the jet is quite strong but is not highly condensed, and so is again rapidly dissipated.

Finally, in a third case the region of instability extends over a considerable part of the length of the body. Then the "needle" itself takes up the characteristic curved form of the arms of an equiangular (logarithmic) spiral. The dissipated matter occupies the space between the turns of the spirals, forming a continuous cloud in the plane of the equator within the volume of the corresponding Maclaurin ellipsoid of the second kind. This is apparently the way in which the normal spiral galaxies are formed.

Type 3: Normal spiral galaxies. These are dynamically the most complex type of galaxy. To define uniquely such a galaxy it does not suffice to specify only the three dynamical parameters mentioned above (mass, energy and angular momentum), since a characteristic property of these galaxies, first noted by Lindblad, is the statistical inhomogeneity of their star populations. Each galaxy is a combination of several subsystems with different residual-velocity dispersions, degrees of concentration to the galactic plane, and rates of rotation. We have called these Kapteyn–Lindblad systems. Whereas in a homogeneous stellar system the kinetic energy of stellar motions relative to the centroid of any macroscopic volume element is due only to the residual velocities of the stars, in Kapteyn–Lindblad systems this kinetic energy is due not only to

† It may be recalled that by "ellipsoidal bodies" we mean those which differ only slightly from the corresponding homogeneous ellipsoids of equilibrium.

the residual velocities but also to the dispersion of the velocities of the sub-centroids (i.e. the centroids for stars belonging to the individual sub-systems).

Let us now consider the origin of the spiral arms in normal spirals. A possible hypothesis is that the normal spirals are formed by the evolution of barred spirals of suitable types. These should be comparatively slowly rotating and have a very long thin Jacobi ellipsoidal body. The galaxy NGC 7479 (Plate VI) gives an idea of this, and may be regarded as a transitional form between a barred and a normal spiral. Evidently this theory necessarily implies that the spiral arms are winding up, i.e. that their ends are trailing in the rotational motion of the galaxy.

We therefore have an evolutionary succession of galaxies whereby barred spirals evolve into normal spirals. The question arises of what precedes the barred spirals, and what follows the normal spirals. In other words, what are the first and last members of the evolutionary sequence in the development of stellar systems? Of course, only certain fairly cautious assertions can be made, but even these may be of some use in the planning of future studies.

In Chapter IV, which dealt with irregular forces in stellar systems, and in particular with the definition of the relaxation time, we have noted the importance of various kinds of condensation of matter in establishing the most probable stellar velocity distribution under given dynamical conditions; that is, the importance of the non-uniformity of galactic structure in the evolution of galaxies, at least in its early stages. Hence we can suppose that examples of the "pre-spiral" stage are furnished by those patchy and often incompletely formed objects such as NGC 2976 (Plate XXV) and NGC 3034 (Plate XXVI). These should all show a tendency to become more or less elongated Jacobi ellipsoidal bodies, as indeed they do. If our arguments are correct, these two objects are galaxies in the process of formation from primeval matter which, apparently, does not emit radiation of optical frequencies.†

But if the normal galaxies are formed from the barred spirals, the pre-spiral galaxies should include some which are very elongated, with lengths at least 8 to 10 times their thicknesses. Such a ratio of dimensions is observed in the galaxy NGC 7741, which we have regarded as marking a transition between the two types.

There remains to be considered the last link in our argument, namely the establishment of the stage in galactic evolution which follows the normal spirals. The answer is perhaps obvious, because in the evolutionary sequence of galaxies outlined above only one type has not been included, namely the elliptical galaxies. With the necessary reservations concerning the provisional nature of our theories, we may assume that the final stage in galactic evolution is represented by the dwarf elliptical galaxies.

At least two arguments may be advanced in favour of this supposition. Firstly, the dwarf elliptical galaxies show no trace whatever of patchiness of structure. This indicates their greater age in comparison with the other types of galaxy. Secondly, the elliptical galaxies are extremely similar to the nuclei of the normal spirals as regards structure, rotation, and stellar population. We may therefore suppose that the dwarf elliptical galaxies are the residual nuclei of normal spirals which have disintegrated. As a spiral galaxy evolves, the stars belonging to the different sub-systems should gradually intermingle, and in consequence the stellar population should become more

† It would be interesting to see whether such galaxies emit at radio frequencies.

and more uniform, the inhomogeneity coefficient $\varkappa_4$ tending to zero. The radius of the galaxy $\varrho_0 = 1/\sqrt{\varkappa_4}$ increases without limit, and the thickness of the Maclaurin ellipsoidal body decreases to zero. As a result of such evolution, the spiral galaxy must lose its peripheral parts and retain only its nucleus, which thereafter exists independently as a dwarf elliptical galaxy.

In the above discussion we have deliberately stressed the difference between the giant and dwarf elliptical galaxies, since they should be, and indeed appear to be, quite different in origin. Whereas the dwarfs are at least in some cases the remnants of large spirals and are comparable in size with the nuclei of the latter, the giant elliptical galaxies are of the same order of size and mass as the great spirals, and often originate simultaneously with them as condensations in long filaments of matter, as seen, for instance, in Plate I.

This dynamical theory of galaxies may with justice be called a *synthetic dynamics of stellar systems*, since it combines the hydrodynamical and statistical approaches to the problem. If our arguments are correct, it is only now, 220 years after Stirling and Maclaurin first found that the figure of equilibrium of a homogeneous rotating liquid may be a spheroid; 180 years after d'Alembert and Laplace discovered the Maclaurin ellipsoids of the second kind; 120 years after Jacobi discovered the equilibrium ellipsoids named after him; and 50 years after the classic work of Darwin, Poincaré and Lyapunov, that we can point out objects which include all these figures, and even the pear-shaped figures envisaged by Lyapunov and by Poincaré. For so many years, for so many centuries, these objects were never out of sight, yet none guessed that in them lay the key to *that* riddle.

REFERENCES

[1] AGEKYAN, T.A.: The dynamics of the passage of stars through clouds of meteoric matter (K dinamike zvezdnykh prokhozhdeniĭ skvoz' oblaka meteornoĭ materii), *Doklady Akademii Nauk SSSR* **69,** 515, 1949; *Uchenye Zapiski Leningradskogo Gosudarstvennogo Universiteta* No. 136 *(Seriya matematicheskikh nauk* **22**) = *Trudy Astronomicheskoĭ Observatorii Leningradskogo Gosudarstvennogo Universiteta* **15,** 33, 1950.

[2] AGEKYAN, T.A.: The dynamics of the passage of stars through dust clouds (Dinamika prokhozhdeniĭ zvezd skvoz' pylevye oblaka), *Doklady Akademii Nauk SSSR* **75,** 361, 1950.

[2a] VAN ALBADA, G. B.: On a generalization of the integral of angular momentum and its significance for stellar dynamics, Contributions from the Bosscha Observatory No. 1, 1952.

[3] AMBARTSUMYAN, V.A. (AMBARZUMIAN, V.): On the derivation of the frequency function of space velocities of the stars from the observed radial velocities, *Monthly Notices of the Royal Astronomical Society* **96,** 172, 1936.

[4] AMBARTSUMYAN, V.A.: The statistics of double stars (K statistike dvoĭnykh zvezd), *Astronomicheskiĭ zhurnal* **14,** 207, 1937.

[5] AMBARTSUMYAN, V.A.: On the dynamics of open star clusters (K voprosu o dinamike otkrytykh zvezdnykh skopleniĭ), *Uchenye zapiski Leningradskogo Gosudarstvennogo Universiteta* No. 22, 19, 1938; On the gravitational potential energy of open clusters, *Doklady Akademii Nauk SSSR* **24,** 874, 1939.

[6] AMBARTSUMYAN, V.A.: Stellar associations (Zvezdnye assotsiatsii), *Astronomicheskiĭ zhurnal* **26,** 3, 1949; Cool supergiants in O associations (Kholodnye sverkhgiganty v O-assotsiatsiyakh), *Doklady Akademii Nauk Armyanskoĭ SSR* **16,** No. 3, 73, 1953; The association Cassiopeia II (Assotsiatsiya Kassiopeya II), *Astronomicheskiĭ zhurnal* **28,** 160, 1951.

[7] AMBARTSUMYAN, V.A., and MARKARYAN, B.E.: A stellar association near P Cygni (Zvezdnaya assotsiatsiya vokrug P Lebedya), *Soobshcheniya Byurakanskoĭ Observatorii* No. 2, 1949.

[8] BAADE, W.: The resolution of Messier 32, NGC 205 and the central region of the Andromeda nebula, *Astrophysical Journal* **100,** 137, 1944.

[9] BABCOCK, H.W.: The rotation of the Andromeda nebula, *Lick Observatory Bulletin* **19,** 41, 1939.

[10] BALANOVSKIĬ, I. (BALANOWSKY, J.) and SAMOĬLOVA, N.: Determination of the solar motion by Bravais' method (Bestimmung der Sonnenbewegung nach der Methode von Bravais), *Astronomische Nachrichten* **222,** 289, 1924; On the motion of the Sun, *Byulleten' Astronomicheskogo Instituta (Bulletin de l'Institut Astronomique)* **1** (6), 33, 1925.

[11] BLAAUW, A.: A determination of the longitude of the vertex and ratio of the axes of the velocity-ellipsoid from the dispersions of the proper motions of faint stars measured at the Radcliffe Observatory, *Bulletin of the Astronomical Institutes of the Netherlands* **8,** 305, 1939.

[12] BLAAUW, A.: Problems of stellar astronomy connected with the proper motions of faint stars, *Transactions of the International Astronomical Union (1952)* **8,** 784, 1954.

[13] BOK, B.J.: Dimensions and masses of dark nebulae, *Centennial Symposia* (Harvard Observatory Monograph No. 7), Harvard, 1948, p. 53.

[14] BOK, B.J.: The stability of moving clusters, *Harvard College Observatory Circulars* No. 384, 1934.

[15] BOK, B.J.: Note on galactic rotation, *Bulletin of the Astronomical Observatory of Harvard College* No. 876, 1930.

[16] BOTTLINGER, K.: The bright stars and the rotation of the Galaxy (Die hellen Sterne und die

Rotation der Milchstraße), *Veröffentlichungen der Universitätssternwarte zu Berlin-Babelsberg* **8,** No. 5, 1931.

[16a] CAMM, G.L.: The ellipsoidal distribution of stellar velocities, *Monthly Notices of the Royal Astronomical Society* **101,** 195, 1941.

[16b] CHANDRASEKHAR, S.: The dynamics of stellar systems. I–VIII, *Astrophysical Journal* **90,** 1, 1939.

[17] CHANDRASEKHAR, S.: *Principles of stellar dynamics*, University of Chicago, 1942.

[18] CHANDRASEKHAR, S.: The dynamics of stellar systems. IX–XIV, *Astrophysical Journal* **92,** 441, 1940.

[19] CHARLIER, C.V.L.: Notes on statistical mechanics, *Meddelanden från Lunds Astronomiska Observatorium, Serie I*, Nos. 69–70, 1915.

[20] CHARLIER, C.V.L.: Statistical mechanics based on the law of Newton, *Meddelanden från Lunds Astronomiska Observatorium, Serie II*, No. 16, 1917.

[21] CHARLIER, C.V.L.: *The motion and the distribution of the stars* (Memoirs of the University of California No. 7), University of California Press, Berkeley, 1926.

[21a] CONTOPOULOS, G.: On the relative motions of stars in a galaxy, *Stockholms Observatoriums Annaler* **19,** No. 10, 1957; On the vertical motions of stars in a galaxy, *Stockholms Observatoriums Annaler* **20,** No. 5, 1958; A third integral of motion in a galaxy, *Zeitschrift für Astrophysik* **49,** 273, 1960.

[22] DANVER, C.-G.: The form of the arms of the anagalactic nebulae (Die Form der Arme der anagalaktischen Nebel), *Meddelanden från Lunds Astronomiska Observatorium, Serie I*, No. 147, 1937.

[23] DZIGVASHVILI, R.M.: A study of galactic orbits and some regularities in the motions of stars (Izuchenie galakticheskikh orbit i nekotorye zakonomernosti v dvizheniyakh zvezd), *Byulleten' Abastumanskoĭ Astrofizicheskoĭ Observatorii* **18,** 115, 1955.

[24] EDDINGTON, A.S.: *Stellar movements and the structure of the universe*, Macmillan, London, 1914; especially Chapters VI and VII, pp. 86–153.

[25] EDDINGTON, A.S.: The dynamics of a globular stellar system, *Monthly Notices of the Royal Astronomical Society* **74,** 5, 1913; **75,** 366, 1915; **76,** 37, 1915.

[26] EDDINGTON, A.S.: The kinetic energy of a star cluster, *Monthly Notices of the Royal Astronomical Society* **76,** 525, 1916.

[27] EDDINGTON, A.S.: The dynamical equilibrium of the stellar system, *Astronomische Nachrichten* Jubiläumsnummer, 9, 1921.

[28] EDDINGTON, A.S.: *The internal constitution of the stars*, Cambridge University Press, 1926.

[29] EFREMOV, YU.I.: The frequency of galaxies of various types (O chastote galaktik razlichnykh tipov), *Astronomicheskiĭ zhurnal* **26,** 286, 1949.

[30] ÉĬGENSON, M.S.: *The universe at large (Bol'shaya Vselennaya)*, USSR Academy of Sciences Press, Moscow, 1936.

[31] EMDEN, R.: *Gas spheres (Gaskugeln)*, Teubner, Leipzig, 1907.

[32] FESENKOV, V.G., and OGORODNIKOV, K.F.: A determination of the speed and direction of the Sun's motion from the radial velocities of B type stars (Opredelenie skorosti i napravleniya dvizheniya Solntsa po radial'nym skorostyam zvezd tipa B), *Astronomicheskiĭ zhurnal* **1,** No. 2, 1, 1924.

[33] FESENKOV, V.G. (FESSENKOFF, B.) and OGORODNIKOV, K.F. (OGRODNIKOFF): A determination of the solar motion from space velocities of the stars, *Astronomicheskiĭ zhurnal* **2,** No. 1, 37, 1925; **3,** No. 1, 36, 1926.

[34] FINLAY-FREUNDLICH, E., and KURTH, R.: The basis, aims and possibilities of a mechanical theory of stellar systems (Über die Grundlagen, Ziele und Möglichkeiten einer mechanischen Theorie der Sternsysteme), *Die Naturwissenschaften* **42,** 167, 1955.

[34a] FRICKE, W.: Kinematics and dynamics of the galactic system (Kinematik und Dynamik des Milchstraßensystems), *Die Naturwissenschaften* **38,** 438, 1951; Dynamical derivation of the velocity distribution in the stellar system (Dynamische Begründung der Geschwindigkeitsverteilung im Sternsystem), *Astronomische Nachrichten* **280,** 193, 1951.

[35] GREENSTEIN, J.L.: The effect of absorbing clouds on the general absorption coefficient, *Annals of the Astronomical Observatory of Harvard College* **105,** 359, 1937.

[36] GUREVICH, L.É.: The evolution of stellar systems (Évolyutsiya zvezdnykh sistem), *Voprosy kosmogonii* **2,** 150, 1954.

[37] HECKMANN, O., and STRASSL, H.: The dynamics of the stellar system (Zur Dynamik des Sternsystems), *Veröffentlichungen der Universitäts-Sternwarte zu Göttingen* No. 41, 1934; No. 43, 1935.

[38] HUBBLE, E.: *The realm of the nebulae*, Oxford University Press, London, 1936.

[39] HUMASON, M.L., MAYALL, N.U., and SANDAGE, A.R.: Red-shifts and magnitudes of extragalactic nebulae, *Astronomical Journal* **61,** 97, 1956.

[40] IDLIS, G.M.: Cosmical force fields and some problems of the structure and evolution of galactic matter (Kosmicheskie silovye polya i nekotorye voprosy struktury i évolyutsii galakticheskoĭ materii), *Izvestiya Astrofizicheskogo Instituta Akademii Nauk Kazakhskoĭ SSR* **4,** 3, 1957.

[41] JEANS, J.H.: On the theory of star-streaming and the structure of the universe, *Monthly Notices of the Royal Astronomical Society* **76,** 70, 1915.

[42] JEANS, J.H.: *Problems of cosmogony and stellar dynamics*, Cambridge University Press, 1919.

[43] JEANS, J.H.: *The dynamical theory of gases*, 4th ed., Cambridge University Press, 1925.

[44] JEANS, J.H.: The motions of the stars in a Kapteyn-universe, *Monthly Notices of the Royal Astronomical Society* **82,** 122, 1922.

[45] JEANS, J.H.: *Astronomy and cosmogony*, Cambridge University Press, 1928.

[46] KAPTEYN, J.C.: Star streaming, *British Association for the Advancement of Science, Report*, 257, 1905.

[47] KAPTEYN, J.C.: First attempt at a theory of the arrangement and motion of the sidereal system, *Astrophysical Journal* **55,** 302, 1922.

[48] KHOLOPOV, P.N.: The spatial distribution of stars of various types in the globular cluster M3 (Prostranstvennoe raspredelenie zvezd razlichnykh tipov v sharovom skoplenii M3), *Astronomicheskiĭ zhurnal* **30,** 517, 1953.

[49] KHOLOPOV, P.N.: The apparent distribution of stars in twenty globular clusters (Vidimoe raspredelenie zvezd v dvadtsati sharovykh skopleniyakh), *Trudy Gosudarstvennogo Astronomicheskogo Instituta im. P.K. Shternberga* **23,** 250, 1953.

[50] KOVAL'SKIĬ, M.A. (KOWALSKI): The laws of proper motion of the stars in Bradley's catalogue (Sur les lois du mouvement propre des étoiles du catalogue de Bradley), *Uchenye Zapiski Imperatorskago Kazanskago Universiteta* **1,** 47, 1860; *Recherches astronomiques de l'Observatoire de Kasan*, No. 1, 1859.

[51] KRYLOV, N.S.: *The foundations of statistical physics (Raboty po obosnovaniyu statisticheskoĭ fiziki)*, USSR Academy of Sciences Press, Moscow, 1950.

[52] KUKARKIN, B.V.: *The investigation of the structure and evolution of stellar systems from studies of variable stars (Issledovanie stroeniya i razvitiya zvezdnykh sistem na osnove izucheniya peremennykh zvezd)*, Gostekhizdat, Moscow, 1949; German translation, Akademie-Verlag, Berlin, 1954.

[53] KUNITSKIĬ, R.V. (KOUNITZKY): A determination of the solar motion relatively to the centre of inertia of stars lying within 50 parsecs from the Sun, *Astronomicheskiĭ zhurnal* **4,** No. 1, 44, 1927.

[54] KUNITSKIĬ, R.V.: The system of near B and M stars (Sistema blizlezhashchikh B i M zvezd), *Astronomicheskiĭ zhurnal* **12,** 229, 1935.

[55] KURTH, R.: Is there a statistical mechanics of stellar systems? (Gibt es eine statistische Mechanik der Sternsysteme?), *Zeitschrift für angewandte Mathematik und Physik* **6,** 115, 1955.

[56] KUZMIN, G.G.: The distribution of mass in the Galaxy (O raspredelenii mass v Galaktike), *Publikatsii Tartuskoĭ Astronomicheskoĭ Observatorii* **32,** 211, 1952.

[57] KUZMIN, G.G.: Proper motions of galactic-equatorial A and K stars perpendicular to the galactic plane, and the dynamical density of the Galaxy (Sobstvennye dvizheniya galaktiko-ékvatorial'nykh A i K zvezd perpendikulyarno galakticheskoĭ ploskosti i dinamicheskaya plotnost' Galaktiki), *Publikatsii Tartuskoĭ Astronomicheskoĭ Observatorii* **32,** 5, 1952.

[58] KUZMIN, G.G.: A third integral of stellar motion and the steady-state dynamics of the Galaxy. I (Tretiĭ integral dvizheniya zvezd i dinamika statsionarnoĭ Galaktiki, Chast' I), *Publikatsii Tartuskoĭ Astronomicheskoĭ Observatorii* **32,** 332, 1953.

[59] KUZMIN, G.G.: The gravitational potential of the Galaxy and the third integral of stellar motion (O gravitatsionnom potentsiale Galaktiki i tret'em integrale dvizheniya zvezd), *Izvestiya Akademii Nauk Éstonskoĭ SSR* **2,** No. 3, 1953.

[60] KUZMIN, G.G.: The value of the dynamical parameter C and the density of matter in the neighbourhood of the Sun (K voprosu o velichine dinamicheskogo parametra C i plotnosti materii v okrestnostyakh Solntsa), *Publikatsii Tartuskoĭ Astronomicheskoĭ Observatorii* **33,** 3, 1955.

[61] KUZMIN, G.G.: A model of a steady-state Galaxy which allows a velocity ellipsoid with three unequal axes (Model' statsionarnoĭ Galaktiki, dopuskayushchaya trekhosnoe raspredelenie skorosteĭ), *Astronomicheskiĭ zhurnal* **33,** 27, 1956.

[62] LANDAU, L.D.: The Boltzmann equation for a Coulomb interaction (Kineticheskoe uravnenie v sluchae kulonovskogo vzaimodeĭstviya), *Zhurnal éksperimental'noĭ i teoreticheskoĭ fiziki* **7,** 203, 1937; *Physikalische Zeitschrift der Sowjetunion* **10,** 154, 1936.

[63] LEBEDINSKIĬ, A.I.: A hypothesis on the formation of stars (Gipoteza ob obrazovanii zvezd), *Voprosy kosmogonii* **2,** 5, 1954.

[64] LYAPUNOV, A.M. (LIAPOUNOFF): *The general problem of the stability of motion (Problème général de la stabilité du mouvement)*, Annals of Mathematics, Study No. 17, Princeton University Press, 1947.

[65] LINDBLAD, B.: Star-streaming and the structure of the stellar system, *Arkiv för matematik, astronomi och fysik* **19**A, No. 21, 1925; **19**B, No. 7, 1926.

[66] LINDBLAD, B.: On the cause of the ellipsoidal distribution of stellar velocities, *Arkiv för matematik, astronomi och fysik* **20**A, No. 17, 1927.

[67] LINDBLAD, B.: On the dynamics of the stellar system, *Arkiv för matematik, astronomi och fysik* **21**A, No. 3, 1928.

[68] LINDBLAD, B.: The dynamics of the Galaxy (Die Dynamik der Milchstraße), *Handbuch der Astrophysik* **V**/2, 1033, Springer, Berlin, 1933.

[69] MARTYNOV, D. YA.: A forgotten paper by M.A.Koval'skiĭ (Ob odnoĭ zabytoĭ rabote M.A.Koval'skogo), *Astronomicheskiĭ zhurnal* **27,** 169, 1950.

[70] MAYALL, N.U., and ALLER, L.H.: The rotation of the spiral nebula Messier 33, *Astrophysical Journal* **95,** 5, 1942.

[71] MILNE, E.A.: *Relativity, gravitation and world-structure*, Clarendon Press, Oxford, 1935.

[72] MINEUR, H.: The rotation of the local cluster (Sur la rotation de l'amas local), *Comptes rendus* **188,** 1086, 1929.

[73] MINEUR, H.: The solar apex and galactic rotation, *Monthly Notices of the Royal Astronomical Society* **90,** 789, 1930.

[74] MINEUR, H.: Additional studies on the motions of B stars (Recherches complémentaires sur les mouvements des étoiles B), *Bulletin astronomique (Mémoires et Variétés)* [2] **9,** 41, 1933.

[75] MININ, I.N.: The application of the equations of stellar hydrodynamics to globular clusters (Primenenie uravneniĭ zvezdnoĭ gidrodinamiki k sharovym skopleniyam), *Uchenye Zapiski Leningradskogo Gosudarstvennogo Universiteta* No. 153 *(Seriya matematicheskikh nauk* **25***) = Trudy Astronomicheskoĭ Observatorii Leningradskogo Gosudarstvennogo Universiteta* **16,** 60, 1952.

[76] NORDSTRÖM, H.: A study of stellar motions based on radial velocities, *Meddelanden från Lunds Astronomiska Observatorium, Serie II*, No. 79, 1936.

[77] OGORODNIKOV, K.F. (OGRODNIKOFF): A method for combining observations by applying the method of least squares and its application to the statistical investigations of stellar motions, *Astronomicheskiĭ zhurnal* **5,** 1, 1928.

[78] OGORODNIKOV, K.F. (OGRODNIKOFF): On the systematic distance-effect in velocity ellipsoid determinations, *Monthly Notices of the Royal Astronomical Society* **96,** 866, 1936.

[79] OGORODNIKOV, K.F.: Fundamentals of the dynamics of rotating stellar systems (Osnovy dinamiki vrashchayushchikhsya zvezdnykh sistem), *Uspekhi astronomicheskikh nauk* **4,** 3, 1948.

[80] OGORODNIKOV, K.F.: A method of determining the specific form of the generalised ellipsoidal velocity law (Method opredeleniya konkretnoĭ formy obobshchennogo éllipsoidal'nogo zakona skorosteĭ), *Uchenye Zapiski Leningradskogo Gosudarstvennogo Universiteta* No. 116 *(Seriya matematicheskikh nauk* **18***) = Trudy Astronomicheskoĭ Observatorii Leningradskogo Gosudarstvennogo Universiteta* **13,** 80, 1949.

[81] OGORODNIKOV, K.F.: The dynamics of the local system (K voprosu o dinamike mestnoĭ sistemy), *Uchenye Zapiski Leningradskogo Gosudarstvennogo Universiteta* No. 136 *(Seriya matematicheskikh nauk* **22***) = Trudy Astronomicheskoĭ Observatorii Leningradskogo Gosudarstvennogo Universiteta* **15,** 1, 1950.

[82] OGORODNIKOV, K.F.: The kinematics of the Metagalaxy (K voprosu o kinematike Metagalaktiki), *Voprosy kosmogonii* **1,** 150, 1952.

[83] OGORODNIKOV, K.F.: The equation of hydrostatic equilibrium for spherical stellar systems

(Uravnenie gidrostaticheskogo ravnovesiya dlya sfericheskikh zvezdnykh sistem), *Doklady Akademii Nauk SSSR* **116,** 200, 1957.

[84] OGORODNIKOV, K.F.: On the principles of statistical mechanics of stellar systems (O printsipial'noĭ vozmozhnosti obosnovaniya statisticheskoĭ mekhaniki zvezdnykh sistem) *Soviet Astronomy AJ* **1,** 877, 1957 (*Astronomicheskiĭ zhurnal* **34,** 809, 1957).

[85] OGORODNIKOV, K.F.: Statistical mechanics of the simplest types of galaxies (Osnovy statisticheskoĭ mekhaniki galaktik prosteĭshikh tipov), *Soviet Astronomy AJ* **1,** 748, 1957 (*Astronomicheskiĭ zhurnal* **34,** 770, 1957).

[86] OGORODNIKOV, K.F.: Statistical mechanics of galaxies with nonhomogeneous stellar populations (Statisticheskaya mekhanika galaktik s neodnorodnym zvezdnym sostavom), *Soviet Astronomy AJ* **2,** 375, 1958 (*Astronomicheskiĭ zhurnal* **35,** 408, 1958).

[87] OHLSSON, J.: A review of stellar dynamics together with research on steady stellar systems in dynamical equilibrium, *Meddelanden från Lunds Astronomiska Observatorium, Serie II,* No. 48, 1927.

[88] OORT, J.H.: Observational evidence confirming Lindblad's hypothesis of a rotation of the galactic system, *Bulletin of the Astronomical Institutes of the Netherlands* **3,** 275, 1927.

[89] OORT, J.H.: Investigations concerning the rotational motion of the galactic system, together with new determinations of secular parallaxes, precession and motion of the equinox, *Bulletin of the Astronomical Institutes of the Netherlands* **4,** 79, 1927.

[90] OORT, J.H.: The stars of high velocity, *Publications of the Kapteyn Astronomical Laboratory at Groningen* No. 40, 1926.

[91] OORT, J.H.: Dynamics of the galactic system in the vicinity of the Sun, *Bulletin of the Astronomical Institutes of the Netherlands* **4,** 269, 1928.

[92] OORT, J.H.: The force exerted by the stellar system in the direction perpendicular to the galactic plane and some related problems, *Bulletin of the Astronomical Institutes of the Netherlands* **6,** 249, 1932.

[93] OORT, J.H., and VAN WOERKOM, A.J.: The attractive force of the galactic system as determined from the distribution of RR Lyrae variables, *Bulletin of the Astronomical Institutes of the Netherlands* **9,** 185, 1941.

[94] PAGE, T.: Radial velocities and masses of double galaxies, *Astrophysical Journal* **116,** 63, 1952.

[95] VON DER PAHLEN, E.: *Introduction to the dynamics of stellar systems (Einführung in die Dynamik von Sternsystemen)*, Verlag Birkhäuser, Basel, 1947.

[96] PARENAGO, P.P.: The mass–luminosity relation, *Astronomicheskiĭ zhurnal* **14,** 33, 1937.

[97] PARENAGO, P.P.: The dark nebulae and the absorption of light in the Galaxy (O temnykh tumannostyakh i o pogloshchenii sveta v Galaktike), *Astronomicheskiĭ zhurnal* **17,** No. 4, 1, 1940.

[98] PARENAGO, P.P.: The interstellar absorption of light (O mezhzvezdnom pogloshchenii sveta), *Astronomicheskiĭ zhurnal* **22,** 129, 1945.

[99] PARENAGO, P.P.: The motion of the long-period Cepheids and galactic dynamics (Dvizhenie dolgoperiodicheskikh tsefeid i galakticheskaya dinamika), *Peremennye zvezdy* **6,** No. 3, 102, 1948.

[100] PARENAGO, P.P.: The motions of the globular star clusters (O dvizheniyakh sharovykh zvezdnykh skopleniĭ), *Astronomicheskiĭ zhurnal* **24,** 167, 1947.

[101] PARENAGO, P.P.: The similarity of the Galaxy to the Andromeda nebula (O skhodstve Galaktiki s tumannost'yu Andromedy), *Astronomicheskiĭ zhurnal* **25,** 306, 1948.

[102] PARENAGO, P.P.: The structure of the Galaxy (Stroenie Galaktiki), *Uspekhi astronomicheskikh nauk* **4,** 69, 1948; The motion of the long-period Cepheids and galactic dynamics (Dvizhenie dolgoperiodicheskikh tsefeid i galakticheskaya dinamika), *Trudy Gosudarstvennogo Astronomicheskogo Instituta im. P.K. Shternberga* **16,** 71, 1949.

[103] PARENAGO, P.P.: The distribution of density and population of various galactic sub-systems (Raspredelenie plotnosteĭ i chislennost' razlichnykh galakticheskikh subsistem), *Astronomicheskiĭ zhurnal* **25,** 123, 1948; An investigation of the spatial velocities of stars (Issledovanie prostranstvennykh skorosteĭ zvezd), *Trudy Gosudarstvennogo Astronomicheskogo Instituta im. P.K. Shternberga* **20,** 26, 1951.

[104] PARENAGO, P.P.: The gravitational potential of the Galaxy (O gravitatsionnom potentsiale Galaktiki), *Astronomicheskiĭ zhurnal* **27,** 329, 1950.

[105] PARENAGO, P.P.: An investigation of the spatial velocities of stars (Issledovanie prostranstvennykh skorosteĭ zvezd), *Astronomicheskiĭ zhurnal* **27,** 150, 1950; *Trudy Gosudarstvennogo Astronomicheskogo Instituta im. P.K.Shternberga* **20,** 26, 1951.

[106] PARENAGO, P.P.: The gravitational potential of the Galaxy. II (O gravitatsionnom potentsiale Galaktiki. II), *Astronomicheskiĭ zhurnal* **29,** 245, 1952.

[107] PARENAGO, P.P.: *Stellar astronomy (Kurs zvezdnoĭ astronomii)*, 3rd ed., Gostekhizdat, Moscow, 1954.

[108] PARENAGO, P.P.: The density of matter in the neighbourhood of the Sun and a dynamical determination of the mean absolute magnitude of short-period Cepheids (Plotnost' materii v okrestnostyakh Solntsa i dinamicheskoe opredelenie sredneĭ absolyutnoĭ velichiny korotkoperiodicheskikh tsefeid), *Astronomicheskiĭ zhurnal* **31,** 425, 1954.

[109] PLASKETT, J.S., and PEARCE, J.A.: The distance and direction to the gravitational centre of the Galaxy from the motion of the O5 to B7 stars, *Monthly Notices of the Royal Astronomical Society* **94,** 679, 1934.

[110] PLASKETT, J.S., and PEARCE, J.A.: The motions of the O and B type stars and the scale of the Galaxy, *Publications of the Dominion Astrophysical Observatory, Victoria,* **5,** No. 4, 1936.

[111] POINCARÉ, H.: *Lectures on cosmogonical theories (Leçons sur les hypothèses cosmogoniques)*, Hermann, Paris, 1911.

[112] SAFRONOV, V.S.: The density of matter in the Galaxy in the neighbourhood of the Sun (Plotnost' materii v Galaktike v okrestnosti Solntsa), *Astronomicheskiĭ zhurnal* **29,** 198, 1952.

[113] SAFRONOV, V.S.: The statistics of the physical characteristics of visual binary stars (K statistike fizicheskikh kharakteristik vizual'no-dvoĭnykh zvezd), *Astronomicheskiĭ zhurnal* **28,** 172, 1951.

[113a] SANDAGE, A.: Current problems in the extragalactic distance scale, *Astrophysical Journal* **127,** 513, 1958.

[114] SCHWARZSCHILD, K.: The proper motions of the stars (Ueber die Eigenbewegungen der Fixsterne), *Nachrichten von der Königlichen Gesellschaft der Wissenschaften zu Göttingen, Mathematisch-physikalische Klasse*, 614, 1907.

[115] SHAĬN, G.A.: Polarisation of starlight, the magnetic field and the spiral structure of the Galaxy in the region of Sagittarius (Polyarizatsiya sveta zvezd, magnitnoe pole i spiral'naya struktura Galaktiki v oblasti Strel'tsa), *Astronomicheskiĭ zhurnal* **33,** 469, 1956.

[116] SHAPLEY, H.: *Star clusters* (Harvard Observatory Monograph No. 2), Harvard, 1930.

[117] SHATSOVA, R.B.: Asymmetry of proper motions in Boss's *General Catalogue* (Asimmetriya sobstvennykh dvizheniĭ GC Bossa), *Uchenye zapiski Leningradskogo Gosudarstvennogo Universiteta* No. 136 *(Seriya matematicheskikh nauk* **22***)* = *Trudy Astronomicheskoĭ Observatorii Leningradskogo Gosudarstvennogo Universiteta* **15,** 113, 1950.

[118] SHATSOVA, R.B.: Asymmetry of radial velocities of stars (Asimmetriya luchevykh skorosteĭ zvezd), *Uchenye Zapiski Leningradskogo Gosudarstvennogo Universiteta* No. 153 *(Seriya matematicheskikh nauk* **25***)* = *Trudy Astronomicheskoĭ Observatorii Leningradskogo Gosudarstvennogo Universiteta* **16,** 67, 1952.

[119] SHIVESHWARKAR, S.W.: On the direction of star-streaming in the Galaxy, *Monthly Notices of the Royal Astronomical Society* **95,** 655, 1935.

[120] SHIVESHWARKAR, S.W.: Remarks on some theorems in the dynamics of a steady stellar system, *Monthly Notices of the Royal Astronomical Society,* **96,** 749, 1936.

[121] SMART, W.M.: On the *K* term of the radial velocities of B-type stars, *Monthly Notices of the Royal Astronomical Society* **96,** 568, 1936.

[122] SMART, W.M.: *Stellar dynamics*, Cambridge University Press, 1938.

[123] SMART, W.M., and GREEN, H.E.: The solar motion and galactic rotation from radial velocities, *Monthly Notices of the Royal Astronomical Society* **96,** 471, 1936.

[124] SPITZER, L., JR., and SCHWARZSCHILD, M.: The possible influence of interstellar clouds on stellar velocities, *Astrophysical Journal* **114,** 385, 1951; **118,** 106, 1953.

[125] STRÖMBERG, G.: The asymmetry of stellar motions and the existence of a velocity restriction in space, *Astrophysical Journal* **59,** 228, 1924 = *Contributions from Mount Wilson Observatory* **12,** No. 275.

[126] STRUVE, F.G.W.: *Studies in stellar astronomy (Études d'astronomie stellaire)*, Imperial Academy of Sciences, St. Petersburg, 1847.

[127] SUBBOTIN, M.F.: *Celestial Mechanics (Kurs nebesnoĭ mekhaniki)*, Vol. III, Gostekhizdat, Moscow, 1949.

[128] SUNDMANN, K.F.: Theory of the planets (Theorie der Planeten), *Encyklopädie der Mathematischen Wissenschaften* VI. 2. A, 742, Teubner, Leipzig 1905–1923.

[129] TAKASE, B.: On the density distribution in the Galaxy. I, *Publications of the Astronomical Society of Japan* **7**, 201, 1955.

[130] TORONDZHADZE, A.F.: Properties of the motions of stars of spectral classes O and B, and the expansion of the stellar associations (Osobennosti dvizheniĭ zvezd spektral'nykh klassov O i B i rasshirenie zvezdnykh assotsiatsiĭ), *Byulleten' Abastumanskoĭ Astrofizicheskoĭ Observatorii* **15**, 115, 1953.

[131] TORONDZHADZE, A.F.: The kinematics of the local cluster (K voprosu o kinematike Mestnoĭ sistemy), *Byulleten' Abastumanskoĭ Astrofizicheskoĭ Observatorii* **20**, 45, 1956.

[132] TRUMPLER, R.J., and WEAVER, H.F.: *Statistical astronomy*, University of California Press, Berkeley and Los Angeles, 1953.

[133] VAN TULDER, J.J.M.: A new determination of the galactic pole and the distance of the sun from the galactic plane, *Bulletin of the Astronomical Institutes of the Netherlands* **9**, 315, 1942.

[134] VELDT, D.: The rotational motion of the galaxy determined from stars of the tenth magnitude, *Bulletin of the Astronomical Institutes of the Netherlands* **4**, 203, 1928.

[135] VORONTSOV-VEL'YAMINOV, B.A.: The spiral structure of the Galaxy (Spiral'naya struktura Galaktiki), *Astronomicheskiĭ zhurnal* **30**, 37, 1953.

[136] VORONTSOV-VEL'YAMINOV, B.A.: Clouds of hot giants and the clouds in the Milky Way (Oblaka goryachikh gigantov i oblaka Mlechnogo Puti), *Astronomicheskiĭ zhurnal* **30**, 394, 1953.

[137] VORONTSOV-VEL'YAMINOV, B.A.: The spiral structure and rotation of galaxies (Spiral'naya struktura i vrashchenie galaktik), *Izvestiya Astrofizicheskogo Instituta Akademii Nauk Kazakhskoĭ SSR* **3**, 46, 1956.

[138] WILLIAMSON, R.E.: A simple demonstration of star streaming, *Astrophysical Journal* **93**, 511, 1941.

[139] VON ZEIPEL, H.: The constitution of the globular clusters (Recherches sur la constitution des amas globulaires), *Kungliga Svenska Vetenskapsakademiens Handlingar* **51**, No. 5, 1913.

[140] ZWICKY, F.: On the masses of nebulae and of clusters of nebulae, *Astrophysical Journal* **86**, 217, 1937.

[141] ZWICKY, F.: Bright and dark forms of intergalactic matter (Leuchtende und dunkle Gebilde intergalaktischer Materie), *Physikalische Blätter* **9**, 406, 1953.

SUBJECT INDEX

INDEX OF GALAXIES

(NGC numbers)

MADE IN GREAT BRITAIN